W9-BNI-380

GUARANTEE

This book belongs to:

_____

_____

Price $15.25

It is **GUARANTEED** right in price, edition, condition and course *through the first two weeks of the quarter. Guarantee void if this slip is altered or removed.*

DuBOIS BOOK STORE

TRY US FIRST FOR ANY BOOK

332 S. Lincoln, Kent, 673-4730

Yours for lower cost of higher education

EUGENE HECHT

Adelphi University

Physics in Perspective

ADDISON-WESLEY PUBLISHING COMPANY

Reading, Massachusetts • Menlo Park, California

London • Amsterdam • Don Mills, Ontario • Sydney

DU BOIS
BOOK STORE   JAN 14 1982

SPONSORING EDITOR: *Laura Rich Finney*
PRODUCTION EDITOR: *Margaret Cassidy*
TEXT AND COVER DESIGNER: *Marshall Henrichs*
ILLUSTRATORS: *Oxford Illustrators*

This book is in the Addison-Wesley Series in Physics

Library of Congress Cataloging in Publication Data

Hecht, Eugene.
   Physics in perspective.

   Includes index.
   1. Physics. I. Title.
QC23.H3918   530   78–73369

   ISBN 0–201–02830–1

Second printing, July 1980

Copyright © 1980 by Addison-Wesley Publishing Company, Inc. Philippines copyright 1980 by Addison-Wesley Publishing Company, Inc.

All rights reserved. No part of this publication may be reproduced, stored in a retrieval system, or transmitted, in any form or by any means, electronic, mechanical, photocopying, recording, or otherwise, without the prior written permission of the publisher. Printed in the United States of America. Published simultaneously in Canada. Library of Congress Catalog Card No. 78–73369.

ISBN 0–201–02830–1

ABCDEFGHIJ–HA–79

To **Jennifer, Jamey**, and **Amy**
with the hope
that when they someday read it
they'll enjoy it.

Preface

This is an almost nonmathematical introduction to physics for those who don't already know very much about the subject. My primary goal was to produce a lively, effective teaching instrument, one that would complement and enrich the college classroom experience by providing more than just a printed version, however insightful, of the usual lecture material. And so this book contains provocative quotes and anecdotes, paintings and poems, biographical sketches and historical insights, stories of sad moments and whimsical ones, of brilliant moments and foolish ones. Students should be exposed to all of this fascinating material, even if there isn't always time to discuss it in class; after all, it colors and gives depth to the vital appreciation of physics as one of the greatest creations of humankind.

This, then, is primarily a book *of* physics, but it's also *about* physics. It's a book that introduces and explains the basic concepts of both classical and modern physics and applies them in a contemporary context. And yet, more than anything else, it is a book about ideas. Many of the major ideas of physics are traced as they evolved from simple beginnings into their not-so-simple contemporary forms. "How did anyone ever arrive at the concept of kinetic energy?" "Where did the notion of mass come from?" Our refined modern ideas grew out of far more obvious ones (although those early concepts were always incomplete and sometimes partially or even totally incorrect). This unfolding progression gives the student a good sense of the dynamic nature of physics, of how the tremendous insights gradually evolved. Just as important, it provides time for the ideas, which naturally build in subtlety, to mellow into understanding.

Some months after sending the last chapter of this book to the publisher, I came across the sentence by the renowned nineteenth-century physicist J. C. Maxwell that hangs here, quoted in the margin. Surprisingly, and delightfully so, it expresses a thought that guided the writing of this text from the very beginning. And so "we take some interest in the great discoverers and their lives." Students will learn no physics when they read that Tycho Brahe wore a silver nose, that Kepler's aunt was a witch, that Newton had a nervous breakdown, that Einstein was thrown out of high school, or that Feynman broke into the secret A-bomb files to plant comic messages, but perhaps these asides will help the reader better understand scientists (and science), and that, too, I see as a responsibility.

The first chapter, "The Philosophy of Physics," is meant to define the subject, to underscore its humanity, to show that it is as much a thing of logic as of imagination, to develop an appreciation of the role of creativity in science, and in the end, to reinforce in the student a pattern of thinking and doubting. Too many come to physics intimidated, as if standing before the altar of immutable truth. I have tried to lighten that tiresome burden—in particular, by telling it as it is, *straight out*. We don't know what energy is; we don't know what matter is; we guessed about this; we can't define that; indeed, many of our great discoveries were made quite accidentally. Galileo's

. . . in Science it is when we take some interest in the great discoverers and their lives that it becomes endurable, and only when we begin to trace the development of ideas that it becomes fascinating.

JAMES CLERK MAXWELL

v

OBSERVER OR PARTICIPANT?

No theory of physics that deals only with physics will ever explain physics. I believe that as we go on trying to understand the universe, we are at the same time trying to understand man.... The physical world is in some deep sense tied to the human being.

JOHN A. WHEELER

Physicist (1973)

law of free fall, Boyle's law, Ohm's law, Kepler's three laws, Newton's law of gravity are all restricted approximations. Not one of them is true, though all are usually superbly true enough. Just within this century we've radically changed our minds about atoms and elementary particles, about light and mass, and space and time, and gravity and energy and momentum—and *yes*, we may certainly change them again! Students should realize that they come to study physics not because it's the final word, true and complete, but because it's powerful and marvelously wide open to discovery. That sense of excitement that comes with standing close to the edge, however unsettling, energizes and sustains the study.

When it's all over, the student will have met some of the great scientists, learned some of the brilliant insights, and perhaps put physics a little more in perspective.

In order to make life a little easier for teachers, I've prepared an instructor's manual that contains answers to all the questions not already given here. In addition, it examines in a bit more depth some of the stickier physical and philosophical problems introduced in the text.

To hold down the size and cost of the book, I've omitted specifically listing the many hundreds of references used over the years. Still, I would earnestly like to acknowledge my debt to all those unnamed scholars whose works have been an inspiration and from whom I've learned so much. For their suggestions and for taking the time to read portions of the manuscript, I thank Professors Roger Creel of the University of Akron, Margaret Feero of El Camino College, and Bernard Long of Foothill College. I am also grateful to my colleagues at Adelphi University, particularly Professors A. Lemos, H. Ahner, and J. Dooher, for always finding the time to talk about physics.

I am especially indebted to Professors John S. Eck and O. Larry Weaver of Kansas State University and Professor Dietrich Schroeer of the University of North Carolina for poring over the entire manuscript. I still have some of their insightful comments tacked to my office walls. Although I took many of the photographs, wherever possible I used pictures provided by my students, and to them I extend my appreciation. Miriam LaRosa typed much of the work, and I thank her. Carolyn Eisen Hecht tolerated five years of "the book," and I thank her. Lastly, I extend my sincere gratitude to Marshall Henrichs, who, as its designer, created the cover, selected the type, took some of the pictures, laid out the pages, and brilliantly brought to fruition my long-held vision of a graphically beautiful book.

Freeport, New York
October 1979

E. H.

Contents

PART II ENERGY

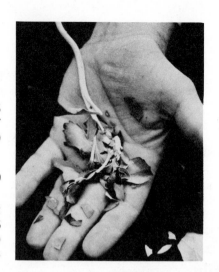

PART III MATTER

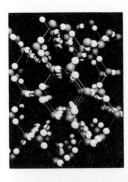

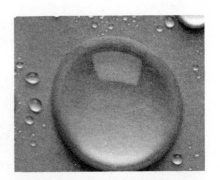

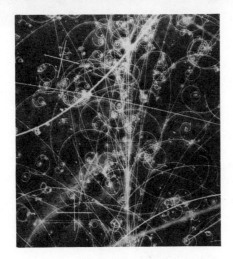

PART IV MATTER-ENERGY

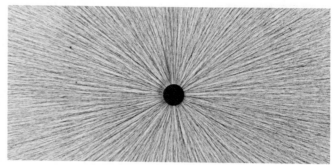

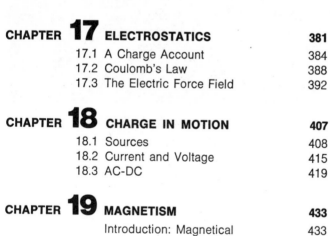

Fifteenth-century tapestry, "The Light" from the *Lady with the Unicorn* series. Here a maiden subdues the mythical unicorn by showing him his own image.

The purpose of this first chapter is primarily to demystify the subject and, at the same time, to underscore its intrinsic humanity. It is an essay in four parts, beginning with a comparison of science and the supernatural. The rise in recent times of the mystical mentality, the irrational as opposed to the scientific, makes this excursion appropriate. The next section is an attempt to define empirical science in part by examining both its **areas of concern** and the **purposes** for doing it. The third bit of business is a summary of the central roles played by **data, law,** and **theory** in the development of scientific understanding.

The last section starts with an examination of the so-called **scientific method** and ends with a discussion of **quantification** and the need for **units** and **standards.**

INTRODUCTION

So you've gone and bought a ticket, or they've gone and sold you one—it doesn't matter which—the trip is the same and the welcome is earnest.

Let's be honest right at the start. Physics is neither particularly easy to comprehend nor easy to love, but then again, *what*—or for that matter, *who*—is? For most it is a new vision, a different way of understanding with its own scales, rhythms, and forms. And yet, as with *Macbeth, Mona Lisa,* or *Lucy in the Sky with Diamonds,* the ride rewards. Surely you have already somehow prejudged this journey. It's all too easy to compartmentalize our human experience: science in one box (brushed chrome, of course, with a black plastic lid), music, art, and literature in another box (gilt wood this time, lined with purple velvet).

The Western mind delights in little boxes—life is easier to analyze when it's disemboweled—small pieces in small compartments (we call it specialization). It is our traditional way of seeing the trees and missing the forest. The label on the box for physics too often reads "Caution: Not for Common Consumption," "Dispassionate." If you can, please tear off your label and chuck the [expletive deleted] box or we will certainly, sooner or later, bore each other to death. There is nothing more tedious than the endless debate between humanist and scientist on whose vision is truer—each of us is less for what we lack of the other.

It is pointless and even worse to separate physics from the body of all creative work, to pluck it out from history, to shear it from philosophy, and then to present it pristine pure, all-knowing, and infallible. *We know nothing of what will be with absolute certainty.*

A Coney Island seer.

There is no scientific tome of unassailable, immutable truth. Yet what little we do know with any surety is heady stuff; it reveals an inspiring grandeur and intricate beauty.

There is no clear beginning to science; it did not spring full-blown out of the head of humankind but rather was birthed, in long labor, from the belly of myth and dark magic, from the toils of astrology and the arcana of alchemy. The scientific world view slowly emerged from mysticism, and we are not even now (nor will we perhaps ever be) wholly "purified." Accordingly, we'll look for a moment at magic—not sleight of hand, but "serious" sorcery. This is the magic performed earnestly enough for centuries by full-time conjurors and witch doctors, the secret powers that so fascinated Shakespeare and blighted Salem, the powers that still fill palmist parlors from Denver to Delhi. The intent is not to lend credence to wizardry but to distinguish it from science and, in so doing, to gain some insight into the business at hand.

Physics, surely the most accomplished of the sciences, is a marvelous creative adventure—powerful and yet elegant and subtle. What does physics know, and what will it never know? How can it be surrounded in mystery and still be held off from magic? What is this science that now surely pervades our lives? Much of what is to follow in this chapter is aimed at confronting these questions and, at the same time, unfolding a sense of what science believes and how the game is played.

In physics, as in all human experience, "law" is usually impermanent, "fact" is generally the result of human interpretation, and "truth," even if attainable, may be unrecognizable. Yet there are far simpler questions than the nature of the cosmos that go unanswered, and we all should be familiar with illusory truth; it pours like sand through our fingers in everyday life. Was Oswald alone? Did Nixon know? We may seek absolute truth, but we settle for less than certainty.

A drawing of a prehistoric cave painting from Les Trois Frères in southern France. The bearded Cro-Magnon sorcerer seems to be wearing a disguise consisting of features from several different animals.

1.1 MAGIC AND MYTH

The south of France is a limestone land channeled by ancient rivers and gouged hollow with innumerable caves. Two hundred centuries ago our ancestor, Cro-Magnon the hunter, worked magic in those great caverns. Secret ritual places abound with magnificent multicolor murals of animals and men. Painted by the light of burning fat in cupped rock lamps, these pictures are Stone Age remnants of magical hunting rites.

James Frazer, in *The Golden Bough*, speaks of the so-called law of similarity—"like produces like . . . an effect resembles its cause." When the Cro-Magnon sorcerer flung his spear at a painted wall image, he knew that if the magic of that moment did not fail him, the spellbound beast would surely fall to the hunt. That's "sympathetic magic"; it assumes "that things act on each other at a distance through a secret sympathy, the impulse being [invisibly] transmitted from one to the other. . . ." There is supposedly yet another kind of sympathetic magic, contingent on the law of contact—"things which have once been in contact with each other continue to act on each other. . . ." A tuft of hair or a scrap of fingernail is all the sorcerer would need to establish continuity and, from there, control.

The desire to influence capricious Nature, whether by toppling a cave bear or by turning a reluctant profit, seems human enough. "Good luck." "Keep your fingers crossed." In any event, magic enveloped the precivilized world, permeated all historic cultures, and still prevails in paperback and powdered aphrodisiac. Check out the *Yellow Pages* under "occult" or "astrology," and don't miss the rerun of *Rosemary's Baby*. Though we can barely suppress a smile, twentieth-century humankind has its dowsers and diviners, purveyors of potions, fortune tellers, magicians, prophets, witches, exorcists, astrologers, healers, mediums, rainmakers, mind readers, and a thousand mystical compulsions.

No subtle logic or clear demonstration could necessarily convince the mystics of this world, now or before, that their vision was gravely in error. Whatever might occur in any test or trial could always be interpreted in a mystical context and only serve to reinforce belief. But this is, of course, a game we all play—we cannot see beyond our own personal sense of reality. When the Zulu magician fails to protect a village from lightning, it is simply because the medicines were bad or a powerful witch hurled the thunderbolt or possibly some taboo was violated by a villager—*magic fails for magical reasons*. When, on a nice fall day in 1940, the 2800-foot center span of the new Tacoma Narrows Bridge in Washington began to wave in the breeze like so much silk ribbon, only to collapse in shreds a few hours later, it was simply a technical oversight—*science fails for scientific reasons*. Perhaps Demosthenes (ca. 348 B.C.) was right after all when he said, "We believe whatever we want to believe."

Our modern-day "ghosts" are usually dispelled with scientific notions because we have more confidence in science than in spooks. But this is not to say that a mystical universe just beyond reach could not exist. A clever ghost hell-bent on evading scientific scrutiny could, no doubt, spend a happy eternity just in the company of believers. Realize that many a scientist would actually like nothing better than some "hard" evidence "confirming" ghosts, ESP, or even little green women.

Nonetheless, let him among us who has slammed a pinball, tossed a horseshoe, or perhaps even flung a spear and *never once* strained to impel it to its target by sheer force of will—let him cast the first stone. You can stand in any bowling alley in the country and watch the ritual dancers try to nudge gutter balls with their personal magic. Still, should you ever actually manage to make such a mid-course correction, it would not be *supernatural;* anything that

The Tacoma Narrows Bridge opened for business July 1, 1940, and collapsed four months later. A wind of from 40 to 45 miles per hour set the central span twisting at about 14 vibrations per minute (p. 482) until the bridge ripped apart about an hour later.

happens is *natural* enough, however unlikely, and *magic that works is reality—the domain of science.*

In some ways the two, science and magic, do seem kindred spirits. Frazer put it most succinctly when he said magic was "the bastard sister of science." After all, the roots of science go back to numerology, astrology, and alchemy, to the dark occult rituals of antiquity. Indeed, there was a long transition period, which ended only a few hundred years ago, when the scientist and the mystic were one and the same.

Vive la Différence

Both science and magic have their invisible fields of force, their "laws" that govern the universe, and yet they are different laws, and they have come out of different visions. Despite the similarities, the dissimilarities are fundamental and overriding. Science sees the universe as *naturalistic*, as unfolding according to internal relationships prevailing among its various parts with no mind to the desires of humanity. We can discover these underlying relationships, these patterns of order known as laws. And though we cannot change Nature's laws, we *can* use them to alter the course of events. The alternative perspective sees the universe operating by purposeful interaction of spirit forces, which may choose to be directly responsive to human desires and thereupon alter events. Science functions as the will of humankind acts on the universe; magic functions as the will of the universe acts on humankind. The sorcerer asks for intervention, the scientist for insight.

There are several less philosophical distinctions. For instance, though science can certainly be done in solitude, in the end it is a *communal adventure.* The modern scientific community has as its eyes and ears the journals and meetings of its various societies. New tests, confirmations, refutations, ideas, theories, techniques, data, discoveries, devices—anything and everything of interest is poured into the literature, into the arena of peer scrutiny and judgment. And since it's almost as amusing to debunk a theory as to develop one, no error is unnoticed, no misconception is spared, no major observation goes unchecked—nothing is overlooked out of compassion for the novice or respect for the master. It is only here, in this uncompromising arena of constant challenge, that *scientific consensus* is welded and the community forges its accepted view of the universe. By contrast, magic, passed from sorcerer to apprentice, is usually secretive and generally individualistic.

Science is predicated on the reproducibility of experimental results—identical systems identically affected will presumably behave identically. All of those "identicals" are really an idealization, one that science strives for though probably never attains. Quite to the contrary, the practitioner of the occult realistically enough sees each event as inherently different; each occurrence is a one-of-a-kind coming together of spiritual influences. If a ritual fails to produce rain today, it may well work tomorrow when the spirits are more willing.

DEAR ANN LANDERS:

For years I have been getting help from your column. Now it is my turn to help someone else. Many people in our community (mostly Spanish) have become involved in "faith movements." They believe saints and spirits come to life through mediums and do many magical things. Two years ago we were told our 16-year-old daughter had leukemia. When we learned of her illness we were willing to listen to anybody who had a "cure."

A friend said she knew the best medium in Miami. I went immediately for a session and was told that our daughter would not die if I joined the group.

I went through the rituals. Animals were killed and their blood sprinkled all over. The ceremony was full of witchcraft. I paid membership dues of $3,000.

Six months later our daughter died. I was told by the medium that it was my fault. "You and your husband didn't have enough faith in the saints," was the explanation.

That very week, I threw out all the witchcraft junk and cleared my head of the nonsense that nearly ruined my life. How I could have been such a crazy fool I do not know. Will you please print this letter and warn others?

—A Former Nut

Since systematic reproducibility is not essential to magic, its "laws" are not susceptible to refutation, and here lies another major distinction between magic and the precepts of science. *Scientific theory must always be formulated so as to be vulnerable to disproof.* In other words, scientific conceptions have to be constructed in a way that will allow them to be "proved" wrong, or at least cast in doubt, if they are wrong. That may seem a strange thing to insist on, but think, for example, of the basic tenets of religion; they are articles of faith, not to be submitted to test and clearly beyond human refutation. It would be unthinkable, indeed presumptuous, to imagine that one might test the fundamental canons of theology. Of course, the crucial difference in approach between science and religion is that one knows it may be wrong and the other knows it cannot be wrong.

As far as being vulnerable to disproof goes, it is an easy matter to construct a physical theory of some phenomenon into which *all* observations *must* fit, one so marvelously open-armed that it will give comfort to any possibility. Not only is such a formulation irrefutable, but it cannot even be considered scientific. If I maintain, along with St. Thomas Aquinas, that invisible angels move the planets in their orbits, we can account for all planetary motion and never run the risk of being proved wrong—this is not science. *A theory in which everything is understandable and which runs no risk of being refuted yields nothing understood.*

Today we no longer beg of nature; we command her because we have discovered certain of her secrets and shall discover others each day. We command her in the name of laws she can not challenge because they are hers; these laws we do not madly ask her to change, we are the first to submit to them. Nature can only be governed by obeying her!
JULES HENRI POINCARÉ (1854–1912)

The Mythopoeic

Another side of our relation with the world is our cherished mythology—part personal, part universal. Virility, youth, and happiness can still be found on the back pages of a dozen magazines, among the ads for "cosmic consciousness," breast-enlarging devices, and machines that make plastic flowers at home. "The president is a consummate politician." "Nature abhors a vacuum." "Employees must wash hands before leaving." "Whatever goes up must come down." And there was something about a kitchen and a woman's place, but I forget that one. Myths help to smooth out the daily chaos of life. They give comfort, and comfort is more appealing than untempered doubt. Some 35,000 years ago, Neanderthal man buried food and tools in mounds along with the dead, presumably so they might journey more comfortably beyond the incomprehensible finality of death. Perhaps that was *Mythos One*.

The very nature of magic suggests spirit powers that exert the wizard's extraordinary control. This idea of cosmic forces, combined with ritualized myth (*a traditional tale explaining some phenomenon*), provides the basis of magic, the substrate of religion, and even mimics the structure of science. The latter two, science and religion, are hardly each other's handmaidens; indeed, some would suggest that for centuries they had hands on each other's throats. But that was "only" a time of defining intellectual turf. ("You take

All things flow, nothing abides.

HERACLITUS (ca. 500 B.C.)

Greek philosopher

human values, piety, morality, and love; we'll take all the physical things, and both of us can draw the line down the middle of cosmology.") I don't mean to imply that this happy harmony came painlessly. Science, too, can claim martyred prophets, like that strange fellow Bruno, burned at the Inquisitor's stake (Section 2.3). Nowadays, although "the progress of science," as Alfred Whitehead put it, "must result in unceasing codification of religious thought," the two function much like a couple who have been married too long—they both understand and politely try to stay out of each other's way.

Science, made by the human hand, is not free of its own myths, since scientists are not beyond constructing them: *The laws of the universe are constant and the same throughout. Nature favors simplicity. Beautiful theories are preferable. The physical world has a fondness for symmetries.* And so on. When we believe these things without question, they cease to be hypothetical, and without doubt we are disarmed.

1.2 SCIENCE—CONCERN AND PURPOSE

By way of definition—though I for one have never seen a satisfactory definition of science—let's first examine its areas of concern and its purposes.

All creative human endeavor is similar in the sense that change is an essential quality. Since to define is to freeze in language, definitions are doomed either through exclusivity or all-embracing vagueness. A contemporary definition of art must encompass Leonardo's *Mona Lisa* as smoothly as it embraces Duchamp's ready-made urinal, *Fountain.* Would any such successful result be even recognizable to a Renaissance mind? Would the grinding electronic wail of acid rock have been music to Mozart? The same is certainly true of science—we redefine it as it grows, as it varies in emphasis and changes in content.

The "science" of Galileo is not the "science" of Einstein, and the path to "now" is scattered with yesterday's discarded definitions. Strangely enough, practitioners rarely stop to describe their own craft. More often than not, the scientist simply does what works, relying by the habit of training on a personal interpretation of the traditional rules of the game. Still, science is not always what scientists do; sometimes—most infrequently, to be sure, but sometimes—that personal interpretation leans more to magic than to science, as we'll see in *The Great N-Ray Fiasco* (Section 1.4).

Error, too, occurs in science, and although it usually runs to disaster, once in a long while oversight may unknowingly cancel misconception and lead the lucky to paradise. One no less prestigious than the great astronomer and mystic Johannes Kepler took this accidental route to profound discovery (Section 2.3). And unhappily, science has always had its outright, outrageous frauds, as well. In recent years we've seen faked tissue transplants on mice (at the Sloan-Kettering Institute) and computerized data-diddling with supposedly psychic rats (at the Institute for Parapsychology in Dur-

All sorts of things have been claimed in the name of science. This frontispiece from *The Cure of Paralysis through Electricity* by Abbé Sans (1772) depicts an early high-voltage hustle. The radiant globe in the sky is part of an electrostatic generator (p. 381).

ham): Old Piltdown man no doubt turned over in his cardboard box.* Still, the basically critical nature of contemporary communal science is a self-correcting mechanism that does not remain deceived for long.

In the late 1950s, when engineering was at a high point in its cyclical swings of public favor, anyone in need of the whole-cloth mantle of prestige would assume, without academic degrees, the name engineer. So it was that the venerable trade of garbage collecting took on the less odoriferous title of sanitation engineering. Science, too, has its uninvited guests: "library science," "military science," "astrological science," and "management science," to name but a few. Like Napoleon, they crown themselves without worrying much about legitimacy.

The *natural sciences*, as distinct from the social sciences, have as their subject *Nature*, i.e., the physical properties of the material universe as revealed, directly or indirectly, through human experience. **Physics is the natural science concerned with matter and energy—** *all matter and all energy and all possible interactions thereof*. It deals with composition, structure, shape, creation, annihilation, interaction, motion, time and light, sound and atoms, screws and levers and scales, glowing reactors, roaring rockets, faint far stars; it deals with the holding together and the tearing apart, with the whole of things physical. Nor can it any longer be strictly limited to the inanimate. With each breath you suck up more than a thousand million million million "lifeless" atoms, and physics might just as well follow them in. The variety of active disciplines involving physics (e.g., biophysics, geophysics, physical chemistry, astrophysics) bespeaks this grand diversity. And it's not inconceivable that someday, when enough is learned, there will be psychophysics; we, too, are matter and energy.

Physics is the study of matter and energy, the study of Nature—but to what end, for what purpose? Almost 4000 years ago, on the heights of the ziggurat temples, Babylonian astrologer-priests scanned the heavens looking for omens of the future. Their intellectual descendants, the astronomers of the Middle Ages and even those of the Renaissance, were usually part-time horoscope casters; mystical astrology, in one way or another, was close in the background. So purpose changes as science changes: The modern astrophysicist, though in the bloodline of the wizard, has a different goal in mind.

---

* The remains of the infamous Piltdown man, a skull fragment and an apelike jaw, were conclusively shown in 1953 to be part of a deliberate hoax carried out about fifty years earlier. It now seems that the scheme, which ultimately backfired, was a trap set by one eminent geologist to embarrass another. When the "missing link" became an instant celebrity, widely accepted as authentic, the perpetrator discreetly decided not to reveal the mischief.

The Subject of Science

An astrologer in Delhi.

To What End?

The aim of science is, on the one hand, a comprehension, as *complete* as possible, of the connection between the sense experiences in their totality, and, on the other hand, the accomplishment of this aim *by the use of a minimum of primary concepts and relations.*

ALBERT EINSTEIN

In general terms, science seeks to understand the universe as we perceive it to be. Realize from the start that our understanding can be profound, it may even be correct, but it is *always tentative* to some extent, regardless of how powerful it makes us (Section 1.3). For example, James Watt turned the primitive steam engine into a machine that changed the very course of history. That work was done out of an "understanding" of heat predicated on the flow of an invisible fluid that in fact never really existed. The theory was wrong, but the engine was right!

It is surely as presumptuous to say *we truly know* as it is to say *we will never truly know*, but science, as far as we can tell, is a marvelously endless game, in which *we do not even know what it is that we do not know*. That lack of foreseen finality happily snatches the pursuit of science from eternal boredom. We have gone a far way from the simpler times of lawyer and courtier Francis Bacon (1561–1626), who once could write, "The particular phenomena of the arts and sciences are really but a handful, the invention of all causes and all sciences would be a labour of but a few years."

Beyond the treasure they give to humankind, the composer and poet work in their art forms for their own sakes, to fulfill individual creative instincts. In the same way, science on some personal yet crucial level serves as a medium for creativity, and the delight in doing is a private purpose in itself.

Scientific truth, though not incontrovertible, nevertheless provides a reasonably reliable basis for human action, and that, too, is a valid purpose. We cannot be absolutely certain that the next object let loose will fall, but we rarely dance on windowsills. Although we do not know with ultimate surety what gravity actually is, we can capture moon dust and bring it home with no fear that physics will deceive us on the voyage. Cro-Magnon lit their caves with burning fat and we with laser beams—surely we know and can do far more. The laws and theories and experiments—the whole structure—give us some sense of what will happen next, some ability to form realistic expectations about the world, and, let us hope, the power to act effectively within it.

The old scientific ideal of *epistēmē*—of absolutely certain, demonstrable knowledge—has proved to be an idol. The demand for scientific objectivity makes it inevitable that every scientific statement must remain *tentative forever*. It may indeed be corroborated, but every corroboration is relative to other statements which, again, are tentative.

KARL RAIMUND POPPER
Contemporary philosopher

Remember, then, that scientific thought is the guide of action; that the truth at which it arrives is not that which we can ideally contemplate without error, but that which we may act upon without fear.

WILLIAM KINGDOM CLIFFORD (1845–1879)
British philosopher

1.3 DATA, LAW, AND THEORY

The aim of science is to understand the universe, but what does that mean? How does science conceptualize the structure of events, the forms of order, the secret rhythms of a seeming chaos—and *understand*? Intellectual creations like art and literature and physics consist of transforming the patterns of events we observe around us into an order of ideas.

The process usually begins with observations, with data—the information perceived. An event is observed, either deliberately or by accident, and things are recorded, *measured:* How much? How long? How many? How big? Physics *quantifies; it associates numbers with its concepts.*

The intellectual life of man consists almost wholly in his substitution of a conceptual order for the perceptual order in which his experience originally comes.

WILLIAM JAMES (1842–1910)
American philosopher

Tycho Brahe in the sixteenth century collected, with incredible precision, the *data* of planetary orbits. *This is exactly where the planets were in the sky yesterday and the day before and before that.* But he did not see the hidden order.

Within the body of observations, patterns are sought out that reveal recurring relationships among the data. An empirical law *is a description of such an observed pattern of occurrences.* "The planets move in elliptical orbits about the sun"—Kepler's first law. It summarizes countless observations and reveals the overall structure into which each bit of numerical data neatly fits. If nothing changes, we can rely on law to tell us something of what will happen tomorrow. For example: Where will Mars be?

Kepler took Tycho's sightings and from them created his three *laws* of planetary motion, which revealed the patterns of recurrence. *This is the way the planets move—in elliptical orbits, etc.* Yet he did not evolve the central concept that ties the laws together, that makes each understandable.

What gives rise to empirical law? What fundamental underlying aspects of the universe guide events so that they manifest themselves in law? Theory *is the rational justification of law, the explanation of phenomena.* Boyle's law relates the pressure and volume of a gas; it is a summary of observations, a statement of a reliable pattern. But it becomes "understandable" only when we theorize that gases are made of atoms and molecules flying around within the chamber.

Newton took the notion of a gravity force as the key idea of planetary motion and with it created a *theory* that allowed us to understand. *This is why the planets follow Kepler's laws.*

That's the basic sequence: data, law, theory. But it doesn't end there; once moderately content, science doesn't close its eyes to the tentative nature of its formalism.

Centuries later, when more refined observations revealed tiny inconsistencies, Einstein's theory of gravity brought an even more refined "understanding": data, law, theory—checked and rechecked—predictions, discrepancies, refinements, new visions, new laws, new theories, new understandings.

Data, the record of our perceptions, are gathered in physics in order to reveal, confirm, and determine the scope of law—rarely, if ever, as an end in itself. And that's done with rulers, clocks, meters, telescopes, scales, and spectroscopes—with all the apparatus of the trade.

Experimentation is the controlled observation of a select aspect of the physical environment. Webster defines experiment in general in terms of its intents: "A trial made to confirm or disprove some-

It is the function of all art to give us some perception of an order in life by imposing an order upon it.

T. S. ELIOT (1888–1965)

American poet

. . . by what words, yea and in what measures, I may avail to spread before your mind a bright light, whereby you may see to the heart of hidden things.

This terror, then, this darkness of the mind, must needs be scattered not by the rays of the sun and the gleaming shafts of day, but by the outer view and inner law of nature.

TITUS LUCRETIUS CARUS (99–55 B.C.)

De Rerum Natura

Empirical Fact

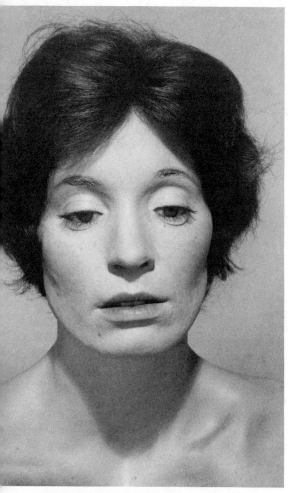

Close-up of "Seated Woman on Stool," a life-sized, incredibly lifelike statue by John DeAndrea. In a situation where you expect to see a woman, you would find it quite shocking to confront one of these frozen figures.

thing doubtful; an operation undertaken to discover some unknown principle or effect, or to test some suggested truth. . . ." On the basis of previous observations and theoretical considerations, an experimenter first isolates a physical area of concern—one, of course, within the reach of the observational talents and resources at hand. Experimentation is generally a directed procedure, which strives not to intrude upon (and thereby to alter as little as possible) the phenomenon being studied. Most often the business is a combination of accident, trial and error, and deliberate methodical probing. Which of the last two tends to dominate depends on how much of the answer is "known" before the experiment starts. The main aspect of experimentation in physics is its quantitative character, its reliance on *measurement*, the association of numerical values with physical concepts.

The human element should not be forgotten in all of this. Modern experimentation has the look of gleaming steel, unerringly automated, tirelessly electronic, silently, coldly objective. But behind the reams of solemn computer scribblings is the researcher, directing, selecting, rejecting, refining, suppressing, and interpreting via prejudice, mood, personal inclination, judgment, blindness, or genius—interpreting and *creating fact*. The investigator must decide what region will be examined and how closely; what will be excluded from view; which apparatus to use; when enough data have been taken so that nothing of consequence is left unseen; which observations are new corroboration of old fact and which are themselves new facts; and of course, which data are to be tossed out as false or put aside as irrelevant.

There is a subtle distinction here that tends often to be overlooked—data and facts are not quite the same. *We make facts out of data by interpreting what we see within the context of what we believe, often within the framework of accepted theory.*

The astronomers of the eighteenth century, fleeing from medieval superstition, stubbornly refused to accept the common knowledge that burning stones were falling from the skies. The French *Académie des Sciences* haughtily denied what ought to have been sufficient evidence and ridiculed the meteoric folklore. Not to be outdone in scientific purity by their Parisian colleagues, museum curators throughout Europe threw away their meteorite collections. And yet the history of meteor activity predates even the Christian era. The ancient Egyptian word for "iron" corresponded to "metal from heaven," and in the treasure inventory of a Hittite king, along with the precious metals silver and gold, is listed "iron from heaven." Meteorites were revered in the temples of Rome and forged into magical weapons in Arabia. But despite this long record of observations, eighteenth-century scientists, who had not themselves witnessed a meteorite fall, were steadfast in their incredulity. "It is impossible for stones to fall from the sky," remarked the famed chemist Lavoisier, "because there are no stones in the sky." On April 26, 1803, the town of L'Aigle in France was battered by thousands of small meteorites, and the testimony of "irreproachable witnesses" ultimately convinced even the Academy of Paris. Still, four years later, the great Virginian intellectual Thomas Jefferson, on hearing of the report of two professors from Yale concerning a meteorite that fell in Weston, Connecticut, quipped that it was easier to believe that the two Yankee professors were liars than that stones had fallen from heaven.

Even seeing is not always believing. Trained observers have seen occurrences and still been blind to the fact of what they saw because it conflicted with their conception of reality. John Couch Adams and Urbain Leverrier quite independently predicted the location of the "eighth planet." On a night in September 1846, the German astronomer Johann Galle, guided by Leverrier's calculations, trained a telescope at a point in the heavens more than 2.5 thousand million miles away. He immediately wrote to Leverrier, "The planet of which you have given the position really exists." Neptune was revealed. By contrast, James Challis, director of the Cambridge observatory, had at his disposal the predictions of J. C. Adams, which were later found to have been within a mere two degrees of the true location of the planet. Yet Challis vehemently distrusted the theoretical arguments, and though his records show that he "saw" Neptune four times that summer and even noticed its apparent disk, he proceeded to disregard his perceptions—for him they were not facts, and the spot of light was just a star.

Someone digging back a half century into the star records of Joseph Lalande in Paris found that he had observed the planet twice on the eighth and tenth of May 1795. Since the positions were different (as indeed they must be for a planet), Lalande discarded one sighting as probably erroneous and designated the other as questionable. He chose not to study the matter further; he *chose* to make the data conform to his preconception, to his own reality. And this sort of human interpretation is a factor in all experimentation.

In the novel *Stranger in a Strange Land*, Robert Heinlein wrote about the ultimate observer, the "fair witness." Where you see a house, a fair witness sees only a shingled wall. Certainly it may be part of a building, but then again it may just be the façade of a Hollywood prop or any one of a number of other unlikely things. We tend not to recognize the unlikely, and what is likely is what is already part of the idiom of our belief; it is what fits the scheme of our accepted theoretical vision. When particular "facts" don't conform to that vision, we tend to be blind to them; we interpret what we see and make fact of it. That flying saucer was only a weather balloon, or perhaps it was a cloud or even someone's hat caught in the wind.

Physics is built on a fundamental reliance on experiment as the ultimate guide to the universe. Even so, this does not mean that experimental results go uncontested or, indeed, that the theoretician

THE UNEXPECTED

What has 18 legs, red spots, and catches flies?

A baseball team with chicken pox.

should necessarily be humbled before empirical "fact." Einstein once remarked, "The facts are wrong," and P. A. M. Dirac, a physicist of great prominence, wrote, "It is more important to have beauty in one's equations than to have them fit experiment." The presumption here is that hidden unknowns can misdirect the interpretation of evidence, and a "correct" theory needs only to wait for "correct" facts.

Measurements are not nearly as straightforward, as unambiguous as one might imagine. When T. W. Richards received the 1914 Nobel Prize in chemistry for his marvelously accurate measurements of atomic weights, how was he to have even guessed that there were such things as isotopes (Section 10.4), that he wasn't really measuring what he thought he was? Unbeknown to science in general and to him in particular, several different forms of each element were present during the nearly three decades when he was making those meticulous determinations—those amazingly careful measurements that soon proved so useless. The possibility of the presence of some hidden unknown must always be recognized, even though there is often little more one can do.

The veracity of data is assumed when empirical results are reproducible and when different independent means and different researchers arrive at the same evidence. Demanding though these criteria may be, they do not preclude misinterpretation. The rigid, immutable clarity of the popular concept of *fact* is, for the most part, an illusion even in the sciences. Yet the game is to pursue truth, being wary of false visions and always bearing "fair witness."

From Law to Theory and Back

Physics is an empirical science; its crucial link with the world it seeks to understand is observation. Yet a catalog of even the most meticulous observations adds little or nothing to our comprehension of Nature. Physics is not a phonebook listing of experiences, not a directory of observations and measurements; rather, it is *a search for the underlying order beneath the surface of those measurements.* An **empirical law** is the end product of such a search; it is *a statement of the form of an observed pattern of occurrences.* Generally these laws apply to a limited range of phenomena, and so we have laws dealing with different aspects of motion, heat, light, sound, electricity, magnetism, etc. One law reveals the rhythms of a pendulum; another traces the long graceful arc of a ballistic missile in free flight; Coulomb's law deals with charge and force; Ohm's law relates current and voltage—all regular occurrences, all patterns of recurrence. To the English poet and critic Samuel Coleridge, *beauty* is "unity in variety." So, too, is physical law.

For practical reasons empirical laws arising directly out of observations tend to be simple, *even if we have to limit their scope to keep them that way.* Hooke's law for springs works beautifully, as

Einstein's space is no closer to reality than Van Gogh's sky. The glory of science is not in a truth "more absolute" than the truth of Bach or Tolstoy, but in the act of creation itself. The scientist's discoveries impose his own order on chaos, as the composer or painter imposes his; an order that always refers to limited aspects of reality, and is biased by the observer's frame of reference, which differs from period to period, as a Rembrandt nude differs from a nude by Manet.

ARTHUR KOESTLER
The Act of Creation

Babylonian religious chronicle dating from the second millennium B.C. The fourteenth line of the cuneiform inscription describes a total solar eclipse: "On the twenty-sixth day of the month Sivan, in the seventh year, day turned to night. . . ."

While you and i have lips and voices which are for kissing and to sing with
who cares if some oneeyed son of a bitch
invents an instrument to measure Spring with?

E. E. CUMMINGS

long as we don't stretch the material *too* far. A pendulum's motion is especially simple, as long as we don't start the swing *too* high. That there are natural laws at all is remarkable enough; that they are all uncomplicated is certainly too much to hope for, despite Kepler's assertion that "Nature loves simplicity." Still, it's one of our scientific myths (certainly one that comes of experience) that a simple formulation is more likely to be true than a complicated one.

Empirical law in physics arises out of *measurement*—not just *counting*, say, the number of beans in a pot, which can be done exactly, but measurement, and that can *never* be done *exactly!* We cannot measure the thickness of this page with ultimate precision or the weight of this book or the time it takes to read a line (assuming each is actually constant, although they are not). There are always unavoidable inaccuracies—*experimental errors* we call them (not to be confused with mistakes). The stop watch in my hand divides each second into five parts, and so I couldn't hope to measure time to within even 1/100 of a second using it. Of course, I could employ a better clock and reduce that error, but it would also have its limits. When we formulate or test a law, the process is always uncertain to within some range of experimental error. In the end, when an empirical law says, "This is exactly the way things are," it must of necessity go beyond its own experimental basis.

The Creative Leap

Astronomers of ancient Babylonia calculated the motions of the visible planets, forecast with precision eclipses of the moon, knew the tidal cycles, and compiled star catalogs dating back to ca. 1800 B.C. They could brilliantly predict celestial events, but they built no theories, and so the pageantry of the heavens was no more intelligible for all their marvelous skill.

The giant and, indeed, audacious step is to attempt to "understand" the basis of empirical law. To that miraculous end, suppose we draw on intuition and imagination and simply guess at what's happening. For example, we might guess at an interrelationship among several fundamental concepts: force, mass, acceleration—whatever. In any event, *guesses*—inspired, maybe, but guesses just the same—**hypotheses** they're called in the trade. In order to comprehend, we leap beyond what we know to what we suspect. Next, we logically deduce some of the physical consequences of the new hypotheses. If we are led to nothing recognizable in the physical world, we have brilliantly constructed one of the infinitude of utterly useless theories. However, if we are brought, by purely logical means, to statements of one or more familiar empirical laws, we have created a viable *physical theory. A construct of hypotheses that explains some observed order in Nature is a* **theory**. These, too, vary in scope; some, like the theory of gravity, are broad and widely appli-

Scientific discovery reveals new knowledge but the new vision which accompanies it is not knowledge. It is *less* than knowledge, for it is a guess; but it is *more* than knowledge, for it is a foreknowledge of things yet unknown and at present perhaps inconceivable.

MICHAEL POLANYI

Contemporary philosopher

Insofar as the propositions of mathematics refer to reality they are not certain, and insofar as they are certain they do not refer to reality.

ALBERT EINSTEIN

cable, but others are very much more specific and limited. If we guess that a spinning Earth and all the other planets revolve around the sun, we have the Copernican theory, which explains a great many observations. Nonetheless, it is only of local interest, revealing little about the fundamental nature of the universe as a whole.

Kepler's planetary laws were known to Newton, and they served as a guide in the formulation of his theory of universal gravitation. Once conceived, that theory was used to arrive at new laws that had not previously been recognized. For example, with it one could predict the motion of objects that swing once about the sun in great open arcs never to return—orbits that do not fall within the limited scope of Kepler's laws. *A powerful theory allows us to deduce statements either that have already been discovered as empirical law or that then stand as predictions to be tested by future observation and, if confirmed, become new law.*

It has been said that "a work of art must narrate something that does not appear within its outline."* And in the same sense, an artful theory must take us far beyond the mere formal statement of its hypotheses. More than 200 years ago, David Hume pointed out that insofar as physical law transcends experience, no amount of empirical evidence can conclusively verify law or theory. What he was talking about and what we are concerned with are fundamental insights, not just the innumerable lists of obvious and inconsequential "laws" that can be formulated. For example, the law of quarts: "Ten cups of water will more than fill a quart bottle." So what!

If by corroborate, verify, or confirm we mean "to make *more* certain" and still not attain certainty, then theory can be corroborated; we can believe in it and use it but never *prove* it true, though we cannot doubt its measure of truth. The theory of relativity, for instance, may well be true, absolutely true and immutable throughout the universe, but we will never know for certain, not without testing *all* the possibilities—and that means *never*. Just as in religion there is no beginning to the testing, in science there can be no ending to it.

The process of refutation of a theoretical system is somewhat more effective than the opposite mode of verification. *Science marches on, not by proving each theory true, but by failing to prove it false, or more accurately, failing to cast it in doubt.* There is no such thing as the single *crucial experiment* that by itself proves or disproves. An experiment usually seems to raise as many questions as it settles. We are far more convinced by several different experiments that yield corroborating results.

It happens on occasion that a clash between prediction and observation can lead to a reformulation of a valid theory and thereby strengthen rather than weaken it. But often it's nothing short of amazing to see how far a theory can be stretched, recast, twisted

---

* Giorgio de Chirico, one of the founders of metaphysical painting.

and reinterpreted to "explain" new observations, despite the fact that the whole thing is totally wrong. Even the resurrection of a dead-and-buried theoretical formulation is not impossible, though it doesn't happen often. When each new theory is accepted, it should encompass the whole range of knowledge of the old, at least in principle, and so each successive formulation should be more effective and far-reaching than its predecessors. But that doesn't mean it will always remain that way (p. 371).

New theories vie for dominance, not by being truer but by "explaining" more or being better at it; by making testable predictions and withstanding meticulous testing; by having the fewest, if any, and least profound conflicts with observation. And yes, if all things are equal, the more mathematically elegant a formulation is, the more it will be favored.

Fantasy or no, theoretical vision is the force that drives research into the mysteries of Nature. Tentative in character, powerful by inspiration, *theory* fills the sails of science. A theory becomes useful through its ability to explain and its power to predict; it remains viable so long as it continues to agree with observations, past and present; it is never uncontestable and rarely uncontested; it is never absolutely true but often truly reliable; it cannot be simply deduced from Nature but is always imaginatively created by the human mind. If it works, we use it, and if it ever fails us, we move to replace it with a new vision.

One might well ask: How is science done? What are the steps in its special process that lead unerringly to happy truth? Well, we sell not truth but understanding; the road is strewn with errors, and there really is no flawless systematic procedure for the doing. Science evolves as much out of great flights of creative imagination as it does out of methodical logic and as much by accident as by purpose. Still, underlying it all is the conviction in physics that every conclusion must ultimately be tested against Nature, and that, if you will, is the *scientific attitude*. The elements are there—data, law, theory—and on to new law, new data, and perhaps new theory. But that no more describes the process than "Find the words and make up a tune" describes how to write a piece comparable to Beethoven's Ninth.

Nonetheless, there is a supposed *scientific method*, which is naïvely taught as if it were the very route to Oz, the definitive, step-by-step cookbook to the joys of science. Sadly, it's only a guide, an outline of well-ordered inquiry. The method is a fourfold ticket to nowhere-in-particular that still lives on, quite overrated and over-peddled. This so-called method of doing science consists of (1) isolating a problem, (2) formulating a hypothesis, (3) deducing predictions from the hypothesis, and (4) testing those predictions in order to verify or refute the hypothesis. All of that is fine, but it does

You don't have to go far even nowadays to find a touch of the mystical. Here's the all-seeing eye of Providence. Incidentally, the Latin inscriptions read "He has looked upon our beginning with favor" and "The new order of the ages."

1.4 A MADNESS TO THE METHOD

The stumbling way in which even the ablest of the scientists in every generation have had to fight through thickets of erroneous observations, misleading generalizations, inadequate formulations, and unconscious prejudice is rarely appreciated by those who obtain their scientific knowledge from textbooks. It is largely neglected by those expounders of the alleged scientific method.

JAMES B. CONANT
Then President of Harvard University

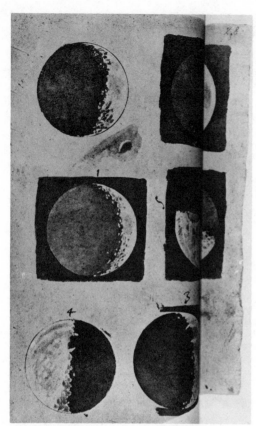

Some of Galileo's early sketches of the moon as he saw it through his telescope. Strangely enough, it still looks as though it could be made of cheese.

not always lead to science, nor is it the way science is always led to knowledge.* The procedure rests on a fallacy known as *affirming the consequent*. If only one hypothesis could lead to a set of predictions, then confirming those predictions would indeed confirm the hypothesis itself. Depressing though it may be, this is not the case; many different conceptions can yield the same prediction. *The verification of a prediction just tells us that our hypothesis may be correct—it is one of many that may be correct.*

Let's try out the scientific method on the problem of the physical structure of the moon. Our hypothesis: *The moon is made of cheese.* An obvious prediction is that it should be moldy and therefore spotted. In addition, it might even have some circular holes if it's the right sort of cheese. A long and studied naked-eye look at the moon reveals its blotchy surface—ergo, the moon is moldy cheese. Well, of course, if we had a telescope or landed an astronaut, we would find out differently, but that's not the point. The point is that even that dumb hypothesis survived the initial test.

On coming across an automobile for the first time, you might wonder about its extraordinary propulsion system. Hypothesizing that there must be a fire-breathing dragon under the hood, you might immediately predict great roaring sounds, a hot breath, belching billows of smoke, and a vile smell. All of those would certainly be observed by anyone standing behind my car. Another prediction— if you are not yet convinced—might call for finding carbon dioxide and water vapor in its exhalant. That, too, would be borne out by very scientific chemical analysis. Even though its pungent vapors are traceable to dinosaurs (via petroleum), our metal monster is true to a different hypothesis.

In much the same way, accurate predictions have been derived from serious theories that were subsequently proved to be totally erroneous. A typical example is the set of equations derived by Augustin Fresnel about 150 years ago that are still in use today. Describing the reflection and transmission of light, these expressions, though in agreement with observation, evolved from the *aether theory*, whose hypothesis we have long since abandoned. The exact same equations, differing only in interpretation, were later derived from a completely changed perspective via the *electromagnetic theory of light*.

If there is a *method* by which science progresses, then it must include *error* and *chance*, for these have always played a significant role. Horace Walpole in 1754 coined the word "serendipity," whose

---

* Take a look at J. Somerville, "Umbrellaology, or, Methodology in Social Science," *Philosophy of Sci.*, **8**, 557 (1941). He satirically uses the method to formulate the "science" of umbrellas, with such far-reaching results as "the Law of Color Variation Relative to Ownership by Sex. (Umbrellas owned by women tend to great variety of color, whereas those owned by men are almost all black.)" Don't overlook the sexist societal implications.

origin stems from the fairy tale *The Three Princes of Serendip* (i.e., ancient Ceylon). "As their highnesses traveled," wrote Walpole, "they were always making discoveries, by *accident* or *sagacity*, of things which they were not in quest of." From the remarkably long list of serendipitous discoveries in science, just a few will make the point: Galvani's accidentally twitching frogs' legs ultimately led to the electric battery; Malus was surprised by his chance observation of the polarization of light via reflection; Oersted just happened to bring a current-carrying wire close to a compass needle, and the result was the marriage of electricity and magnetism; Davisson and Germer didn't even know that they had observed electron diffraction until quite a bit later. Serendipity has played a part in far too many instances of great discovery to be overlooked. Still, the researcher to whom chance presents a pearl must be alert or the moment is lost. Pasteur said as much: *"In the field of observation, chance only favors the prepared mind."*

Science has had its aberrant moments when groups of well-intentioned scholars followed some course or other that seemed reasonable at the time but that in retrospect is surprising and even shocking. Under the cloud of politics some rather strange notions floated around during the Nazi era in Germany. "National Socialist science," as Hitler liked to call it, embraced the *world ice theory*, which envisioned the universe locked in a mystical struggle between ice and fire. Aryan physics rejected Einstein's relativity; indeed, Philipp Lenard, himself a Nobel laureate, referred to the theory as "a Jewish fraud."

During the late 1960s laboratories throughout the world were excitedly studying every aspect of a wonderful new substance known as *polywater*. Polymerized water, a clear Vaseline-like material, was the subject of elaborate measurements and extensive theoretical analyses. And yet it took more than a half dozen years before a consensus was finally reached and researchers realized that all those exotic properties were due to impurities only. Polywater subsequently dribbled into total obscurity, vanishing from the literature as if it had never existed; indeed, it never had. All the *data* were right, all the theorizing was remarkably explanatory, but for much of the time, certainly in the beginning years, all the *facts* were wrong!

There is a great deal to be learned about the doing of science by focusing on its worst moments, when the process went astray. The Great N-Ray Fiasco was one of those peculiar, rare incidents when even the data were erroneous.

René-Prosper Blondlot, Professor of Science at the University of Nancy and *Correspondent* of the *Académie des Sciences*, was 54 years old at the beginning of 1903. Prior to his great discovery, his career in physics, mostly in experimental electromagnetism, had been long, solid, and fairly distinguished. Only eight years earlier

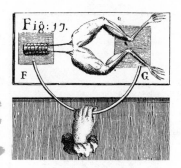

The Great N-Ray Fiasco

...for it is easy in experimentation to be deceived, and to think one has seen and discovered what we desire to see and discover.

LUIGI GALVANI (1737–1798)

PSYCHIC FORCE. 315

PSYCHIC FORCE.

Among the interesting incidents of 1871 was the publication of the report of the investigations of Dr. William Crookes, F.R.S., of London, into the phenomena of modern spiritualism. Dr. Crookes is one of the most distinguished and accurate of scientific investigators, and his experimental deductions upon any of the subjects of his study have heretofore been regarded as almost conclusive.

Dr. Crookes is the editor of the London *Chemist*, the London *Quarterly Journal of Science*, and occupies other prominent positions. His report, entitled "Experimental Investigation of a New Force," was announced in the *Quarterly Journal of Science*, and attracted much attention. In his first he describes a variety of experiments made by him, in the presence of Daniel Douglas Home, the celebrated spiritualistic medium, consisting, among others, of the well-known accordeon experiment, which was held at one end by Home, under a table, and within a wire cage provided by Dr. Crookes. Dr. Crookes affirms that the bellows of the instrument moved up and down, and tunes were played, although no person operated the thing. Subsequently the accordeon floated about in the air, and performed other curious antics.

A spring balance contrivance specially provided by Dr. Crookes was then tried, and the pointer was made to descend by the medium, without the application of visible power.

One of the scientific friends of Dr. Crookes who witnessed these experiments, was the celebrated *savant* and astronomer, Dr. William Huggins, who in a published letter certifies to the correctness of Dr.

A page from *The Science Record For 1872.* We'll be looking at some of Dr. Crookes's more conventional work later on.

Röntgen had accidentally come upon X-rays, and like many of his colleagues, Blondlot was still caught up in the excitement of that moment. Indeed, it was while working with X-rays that he, too, would lunge at fame.

Blondlot had found that the brightness of a small spark would appear to increase ever so slightly when inundated by X-rays, and this effect was then used as a probe to examine the incident radiation itself. "Only after that study," he wrote, "did I become convinced that I was not tackling X-rays but an entirely new type of radiation." As a generous tribute to the city of Nancy, he named his newfound wonder N-rays and plunged zealously ahead discovering new sources and properties in a matter of weeks. From February to the beginning of December 1903, he published twenty papers in the scientific journals of Germany and France. By year's end, another Frenchman, Augustin Charpentier, who proved to be even more prolific a writer than Blondlot himself, began pouring out a deluge of publications.

Not being content simply with qualitative results, Blondlot undertook a precise determination of the wavelength of N-rays. In one particular medium he found it to be 0.0085 millionths of a meter. To verify this result, he once again performed the measurement, this time using yet a different method, and found, to his delight, a value of 0.0081 millionths of a meter—in fine agreement. He was now entirely convinced (his delusions had been corroborated).

As the great work progressed, more and more names were added to the ranks of the active researchers. But within a year, in Britain and Germany, there began to appear occasional reports of failure to reproduce the experiments that were blossoming so magnificently in France. The work continued. Yet another closely related form of radiation (N_i-rays) was found; countless sources (e.g., ice, bent ivory, stretched muscle, the sun, celluloid, anesthetized dogs, tempered steel) were discovered; and the physicians were moving ahead in the development of new diagnostic applications. In response to his critics, Blondlot remarked, "The observation of N-rays and allied phenomena can be made by anyone, almost without exception, and I have only encountered three or four people who were unsuccessful." To the list of believers were added such world-renowned men as D'Arsonval, H. Poincaré, and H. Becquerel. But outside France the number of failures slowly grew larger and the protestations more ominous. Blondlot met the criticisms head-on with more discoveries, and finally he even managed to photograph the faint change in the brightness of a spark. Despite the fact that only about one in forty photos proved useful, Blondlot felt confident that he had swept away all reasonable opposition with this absolute objectivity. It was the height of his success—and the peak of his folly.

Convinced that some lunacy had overtaken French science, a group of physicists, having met at a summer congress in Cambridge, decided to send their own man to Nancy to ferret out the deception.

They chose for the job the brilliant experimentalist Robert W. Wood, then professor of physics at Johns Hopkins (probably best known as the author of *How to Tell the Birds from the Flowers*). Wood was a highly respected professional, meticulous, ingenious, and totally audacious.

Blondlot himself ran the demonstration lecture, showing his American visitor all the wonderful observations of the new physics. By placing a source of N-rays, a hardened steel file, near his eyes, Blondlot was able to read the time from a faintly lit clock, a feat that otherwise would presumably have been impossible. He demonstrated the N-ray spectroscope, which analyzed the radiation, using an aluminum prism. Someone must have been suspicious of Wood, who started milling about the apparatus, but when the lights were abruptly switched on, everything was in order, and Wood left Nancy without incident.

In Wood's letter of September 29, 1904, to *Nature*, he unflinchingly exposed all, and N-rays hit the proverbial fan. Poor Blondlot! He had let Wood hold the file to his head while he, in all honesty, did his famous clock-in-the-dark routine, but that wily Wood palmed the file and used a wooden ruler against Blondlot's forehead instead. Ironically, wood was one of the very few "nonemitters," and though it should not have done a thing, Blondlot was able to read the clock only when the ruler was in place. As if that were not enough, Wood had also surreptitiously removed the prism from the spectroscope prior to the "successful" demonstration. Only after replacing it did he stomp around enough to cause the lights to be put on, thereby verifying his innocence. Hapless Blondlot and his eager colleagues had wasted themselves on a wretched delusion. They had innocently deceived themselves. In the darkness, the exceedingly faint subtle changes in brightness they wanted so much to see they saw—and they believed what they wanted to believe.

Several months later, in December, Blondlot received the Leconte prize of 50,000 francs and the gold medal from the Academy, but in the recital of his scientific accomplishments, scant reference was made to N-rays. Blondlot gave away the prize money to build a park in Nancy and quietly slipped into a deeply troubled obscurity.

No better lesson to guide our science can be bought than the one for which Blondlot so dearly paid the piper.*

A Kirlian photo of fingertips. Many observers claim these mysterious halos are auras revealing nothing less than "fields of life energy." Others see only lovely electrical discharges varying with skin moisture.

Modern physics is a communal structure built on measurement and dependent on reproducibility. And yet it is obvious that before the players can even begin any game but solitaire, there must be agreement on the rules—on how long to make the field and how heavy the ball. The transition from thinking in terms of "a blink of the eye" to

Units, Standards, and Systems

---

* For a complete N-ray bibliography, see the three-part article by G. F. Stradling, *Journal Franklin Inst.*, **164**, pp. 57, 113, 177 (1907).

◀ God taking the measure of the universe in the painting "The Ancient of Days" by the English poet and artist William Blake (1757–1827).

The length of an inch shall be equal to three grains of barley, dry and round, placed end to end lengthwise.

EDWARD II OF ENGLAND (1284–1327)

When you can measure what you are speaking about, and express it in numbers, you know something about it; but when you cannot measure it, when you cannot express it in numbers, your knowledge is of a meagre and unsatisfactory kind: it may be the beginning of knowledge, but you have scarcely, in your thoughts, advanced to the stage of *science.*

LORD KELVIN (1824–1907)

"a tenth of a second" is a profound transition from poetry to physics, but everyone must first agree on what a *second* is.

Standardized measurement is a practical business that developed out of the everyday needs of practical people: builders, merchants, and peddlers. Long before they became the instruments of science, scales and rulers were measuring wheat and cloth. The basic notions of length, volume, weight, and time were quantified in antiquity and simply carried over into physics.

Length

Although long distances could be crudely determined against the extent of a day's journey, the human body itself was the most convenient linear measure in early times. The length of a stride or foot, the breadth of a finger or hand, the span of a forearm—all served as ready reference to the measurers in antiquity. By the time of the great kingdoms of Egypt and Babylon (ca. 2500 B.C.) the *cubit,* which corresponded to the length of a man's forearm from elbow to farthest extended fingertip, had become the most common of linear measures. *This sort of agreed-upon conception by which we quantify anything physical is spoken of as a* **unit.** To ensure some degree of constancy for a widely used measure—since forearms obviously differ—an advanced society *must evolve an unchanging physical embodiment of each unit to serve as a primary reference or* **standard.** The black granite master cubit was such a standard against which all cubit sticks in the land of Egypt were checked and calibrated.

From the Near and Middle East, carried by commerce, the ancient notions of measurement came westward to Greece, then to Rome, and by conquest to most of Europe. The foot, although it varied considerably in length, was in common use by the Greeks and Romans. Its history winds from the length of a Roman sandal to the British boot and on to the familiar contemporary concept.

As the Roman legions marched across the world, they measured their progress in *passus* or paces, each equivalent to five Roman feet. A thousand *passus* or *milia passuum* was the precursor of the British mile (5280 feet). Legend has it that the *yard,* or double-cubit, was fixed in the twelfth century by Henry I of England as the distance from his nose to his extended outermost fingertip.

The Romans made extensive use of a numerical system based on *twelve.* For example, they apportioned the foot into twelve equal parts, or *unciae* (hence the word *inch*), and divided the year into a dozen months. The appeal of this approach derives from the fact that

FATHOM

YARD

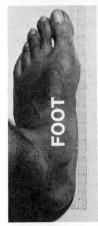

FOOT

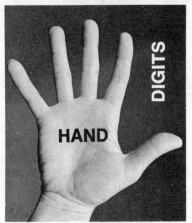

HAND

DIGITS

INCH

THE MEDIEVAL FOOT

To find the length of a rood [sic] in the right and lawful way and according to scientific usage you shall do as follows. Stand at the door of a church on a Sunday and bid sixteen men to stop, tall ones and short ones, as they happen to pass out when the service is finished; then make them put their left feet one behind the other, and the length thus obtained shall be a right and lawful rood [sic] to measure and survey the land with, and the sixteenth part of it shall be a right and lawful foot.

JACOB KÖBEL

Von Ursprung der Teilung (1522)

A pan balance still used to weigh out vegetables on a street in Bombay.

twelve is evenly divisible by quite a few numbers (1, 2, 3, 4, 6, and 12), and it is an easy matter to fold a string into halves, thirds, quarters, etc. With the fall of the Empire, regional variations became commonplace. By the sixteenth century the techniques of standardized measurement had fallen to disuse. Medieval Europe, out of intellectual lethargy and neglect, returned for the most part to primitive body measurements.

The birth of modern science and its rapid development in the next several centuries made the lack of a standardized system of units painfully evident. The climate of radical social and political change that whirled about France during the Revolution was ripe for the bold innovations that science so desperately needed. It was none other than Bishop Talleyrand who in 1790 brought the problem to the *Assemblée Nationale*. The Academy of Sciences, charged with the task of reform, soon adopted a *decimal* approach to weights, where quantities were subdivided into 1000, 100, or 10 equal parts. After several months of study, the Academy decided on a new unit of length, the *metre* (or **meter**), to be one ten-millionth of the distance from the North Pole to the Equator, measured along a meridian line passing through Paris. However patriotic, the choice was fairly uninspired, and since errors made in the difficult measurements were later discovered, the whole thing really turned out to be quite arbitrary. But that didn't matter; any universally accepted standard would do just as well. Spread by the armies of Napoleon Bonaparte, this **metric system** slowly took hold in much of Europe although Napoleon's archenemy, the English, refrained from going along with anything the emperor suggested.

Building from the metric system, the worldwide scientific community in 1960 embraced a new *Système International* (SI). Although every major country still preserves its own carefully made duplicate of the International Prototype Meter (the United States maintains Number 27), the meter is now defined as exactly 1,650,763.73 wavelengths of the orange-red light coming from a special Krypton-86 lamp. The obvious advantage of this sort of natural standard is that it can be equally well maintained in any laboratory that wishes to do so and is, therefore, both convenient and quite invulnerable to ordinary calamity.

Mass

It seems apparent that when humans of prehistory struggled to lift and haul their daily burdens, they could hardly have avoided the concept of weight. Carrying water jars certainly suggests some sort of relationship between volume and weight. The Babylonian unit of capacity, or liquid measure, was the *ka*, the volume of a cube one handbreadth high. When filled with water it became the unit of *weight*, the *great mina*. Sixty *shekels* to a mina, sixty mina to a *talent*—the words are still familiar if only for their biblical ring

(Gen. 23:16). The Roman system of weights and measures pervaded Europe, and its *libra* (or *pound*) survives even now in several English-speaking countries in the abbreviation *lb.*

When the French set out to construct the metric system, they followed the Babylonian example, defining weight in terms of a fixed volume of water. The gram (originally a unit of weight) was thus specified as "the absolute weight of a volume of pure water equal to a cube of one-hundredth part of a meter [i.e., a cube one centimeter on a side]."

But because the Earth is neither uniform nor precisely spherical, the weight of any object varies with its location on the globe, even though the thing itself is otherwise unaltered. Jean Richer in 1671 found that his pendulum clock, when carried to French Guiana, ran slow, losing 2½ minutes each day although neither the bob nor the arm was altered. Newton explained that effect by distinguishing between *weight* and *mass*. *The two are proportional to each other, but whereas weight arises out of gravity and even vanishes where there is no gravity, mass is a property of the object itself and never vanishes!* Weight ceased to be considered fundamental, and in 1889, by international agreement, the **kilogram** (1000 grams) was redefined as the unit of mass. The mass of any object can be determined by simply using a pan balance to compare it with a set of standard masses.

The kilogram in the SI system is the only unit still based on an actual artifact, namely, a platinum-iridium cylinder watched over by the International Bureau of Weights and Measures in Sèvres, outside Paris. The national standard of mass in the United States is a precise copy of that cylinder, designated as Kilogram Number 20. Whenever you buy five pounds of potatoes, that determination *must* be traceable directly back through a long list of calibrations to the primary standard. It's amusing to realize that perhaps the first large organization in the United States to begin to make the transformation to the metric system was the Mafia, which has been dealing in drugs by the *kilo* for years.

COMMON APPROXIMATE EQUIVALENTS

1 millimeter	=	0.04 inch
1 meter	=	39.4 inches
1 meter	=	3.3 feet
1 meter	=	1.1 yards
1 kilometer	=	0.6 mile
1 inch	=	25.4 millimeters*
1 foot	=	0.3 meter
1 yard	=	0.9 meter
1 mile	=	1.6 kilometers
1 gram	=	0.035 ounce†
1 kilogram	=	2.2 pounds†
1 ounce†	=	28.3 grams
1 pound†	=	0.45 kilograms

* Exact
† Weights on Earth of the corresponding masses

As depicted in the Egyptian *Book of the Dead* (ca. 2400 B.C.), the god Thoth serves as scribe while Anubis, the jackal god, balances the heart of each man newly dead against the feather of truth and right.

COMMON UNIT PREFIXES

Multiple	Power of 10	Prefix	Pronunciation	Symbol	Example
1,000,000,000,000	10^{12}	tera-	tĕr′ à	T	terahertz, THz
1,000,000,000	10^{9}	giga-	jĭ′ gà	G	gigavolts, GV
1,000,000	10^{6}	mega-	mĕg′ à	M	megahertz, MHz
1,000	10^{3}	kilo-	kĭl′ ŏ	k	kilowatt, kW
0.1	10^{-1}	deci-	dĕs′ ĭ	d	decibel, db
0.01	10^{-2}	centi-	sĕn′ tĭ	c	centimeter, cm
0.001	10^{-3}	milli-	mĭl′ ĭ	m	millimeter, mm
0.000001	10^{-6}	micro-	mī′ krŏ	μ	microamp, μA
0.000000001	10^{-9}	nano-	nǎn′ ŏ	n	nanometer, nm
0.000000000001	10^{-12}	pico-	pē′ kŏ	p	picosecond, ps

John Harrison's prize-winning fourth marine time keeper (1759). This five-inch-diameter device was the first chronometer accurate enough to "solve" the ancient problem of determining longitude at sea.

The sight of day and night, of months and the revolving years, of equinox and solstice has caused the invention of number and bestowed on us the notion of time.

PLATO

Timaeus

Although an atomic standard for mass is probably inevitable, the platinum-iridium cylinder persists primarily because it's fairly easy to compare objects of that size to within an accuracy of several parts in a thousand million. (Since 1,000,000,000 is a U.S. billion but only a thousandth of a European billion, the common names for numbers beyond a million are rarely used in scientific work.)

Time

"At the tone, Standard Time will be 11:37 and 19 seconds [beep]." They say that Galileo as a young man used the pulse at his wrist to reckon the time of a swinging pendulum. We, of course, have electronic digital wristwatches with little red numbers that light up in the dark. More than 3000 years ago, the Egyptians used the dripping water clock, a cone-shaped vessel with a small hole at the bottom, to determine time into the evening, when their shadow-casting devices were ineffectual. Both their days and their nights were divided into twelve equal hours, which themselves varied in duration with the seasons. The Babylonians also used water clocks, but they divided their day into twelve two-hour intervals, which were subdivided into sixty parts, and each of these, again, was subdivided into sixty parts. The Babylonian delight with the number sixty is still with us in our measures of angles, as well as of time.

Sundials of one sort or another were in common use in ancient Greece and Rome. Aristophanes took a characteristically casual approach to time when he wrote the invitation, "Come and dine when your shadow is ten times as long as your foot." Our *minute* derives from the Roman *partes minutae primae,* and the *second* from their *partes minutae secundae.*

Mechanical clocks driven by falling weights were developed in the Middle Ages, but they were scarcely more accurate than their predecessors. Indeed, when Galileo needed to measure time with any precision in his experiments, he collected and weighed water dripping from a reservoir. Huygens put Galileo's idea of a pendulum clock into practical form (1656), ushering in a new era of improved timekeeping. But really accurate clocks came into being only in the early eighteenth century as a result of the needs of the British fleet. Without such devices a sailor couldn't calculate his ship's longitude with enough precision either to avoid the rocks or to arrive at the battles. Those eighteenth-century clocks were the wind-up ticking kind that survived right up until the integrated circuit made them "obsolete."

Although the *second* is ordinarily thought of as one sixtieth of one sixtieth of one twenty-fourth of the interval from high noon to high noon, that's not really a very precise measure—after all, the Earth's spin rate is gradually slowing, and every day is a trifle longer than the day before. At present the SI unit of time, the *second,* is defined as 9,192,631,770 vibrations of the Cesium-133 atom, as measured via an atomic beam clock. This device offers a precision of one part in 1,000,000,000,000, which is equivalent to an error of about one second in 30,000 years.

"At the tone, Standard Time will be 11:39 and 5 seconds [beep]."

EXPERIMENTS

1. Just for fun, try staring at something for a few minutes in a pitch-black closet to see if you can "see" faint nothings, as Blondlot did.

2. Measure the length from your nose to your farthest outstretched fingertip. How close to a yard is it?

3. Your bathroom scale is probably off by a pound or so. Put a few empty paper cups on it, and slowly fill them with water until the scale begins to register. Is it any better while you're standing on it holding the cups?

REVIEW

Section 1.1

1. What is the place of "supernatural" events in the framework of science?

2. Which ancient disciplines were the forerunners of science?

3. Distinguish between a *naturalistic* and a *spiritualistic* world view.

4. What is meant by the term *scientific consensus*?

5. Science is predicated on the _____ of experimental results.

6. What does it mean for a theory to be vulnerable to disproof?

7. What is a myth? List a few contemporary ones not cited in the text.

Section 1.2

1. Why isn't it reasonable to say simply that science is what scientists do?

2. What was Piltdown man? When I was a kid, he was right there in *The Boy's Golden Book of Anthropology*, but now you would be hard pressed to find a reference to the poor beggar.

3. What is the subject matter of natural science?

4. Physics is the study of _____ and _____. That's not much of a definition since we will have a hard time later on trying to define _____ and _____.

5. What is the primary goal of science?

6. Was Francis Bacon right in his estimate of the rate of progress of science?

Section 1.3

1. You will come across the expression "physics quantifies" in several places in this book. What does it mean?

2. What is a datum?

3. Define the term *empirical law*.

4. Explain the function of *theory*.

5. Discuss what is meant in physics by the process of *experimentation*.

6. Does trial and error play a role in serious research?

7. Why did the museum curators in eighteenth-century Europe toss out their meteorite collections?

8. What did Einstein mean when he said, "The facts are wrong"?

9. List criteria that establish the veracity of experimental results.

10. The theoretical musings of physics must always be checked against _____ by _____.

11. Explain the meaning of the term *experimental error*.

12. In what sense does empirical law transcend experience?

13. Why would we have to say that the motion of the planets was not understood by the Babylonians?

14. Define the term *hypothesis*.

15. What two qualities must a useful theory possess?

Section 1.4

1. What is the *scientific method*?

2. What is the fallacy of *affirming the consequent*?

3. When a prediction arising from a theory is confirmed by observation, we have confirmed the _____.

4. Is serendipity a factor in science?

5. Why might the Romans have used twelve as the basis for their measuring system?

6. What is the definition of a *meter* in the *Système International*?

7. The mass of a cube of water one centimeter on a side is a _____.

8. The mass of a cube of water ten centimeters on a side is a _____.

9. If you put a one-kilogram mass on a scale, what would be its approximate weight in pounds on the Earth?

10. How tall are you in inches? In centimeters?

11. What is your weight in pounds? How massive are you in kilograms?

12. Why isn't the day or any fraction thereof a particularly good unit of time?

QUESTIONS

Answers to starred questions are given at the end of the book.

Section 1.1

1. Make a list of several of the distinctions between science and magic.
* 2. Why do you think the backers of ESP are having such a tough time establishing the validity of the phenomenon?
3. List all the mystical beliefs you have or had at some time—from Santa Claus to rabbits' feet.
* 4. What did Poincaré mean when he said, "Nature can only be governed by obeying her"?
5. What is the main point of the Ann Landers article on p. 4?
6. The popular (TV-movie-comic book) image of the scientist is threefold: the well-intentioned wizard, the mad professor, and the egghead bookworm. The first two are retreads of the white and black magicians. List as many examples of each as you can (strangely enough, the evil madmen will probably be the most numerous).

Section 1.2

1. The nice man sitting on a village street in India (p. 7) thought it was quite reasonable that his sign should read "scientific astrology." Why is that such a witless, though popular, combination of terms?
2. In spelling out the aim of science, Einstein (p. 7) spoke about a comprehension "as complete as possible." Why was he hedging? In other words, don't we seek total comprehension?
3. With reference to the previous question, why does Einstein talk about *the use of a minimum of primary concepts*? *Hint*: There is a myth a-lurking.
4. When W.K. Clifford, more than a hundred years ago, said that "scientific thought is the guide to action," what could he have had in mind?
5. Explain the meaning of the William James quotation on p. 8.
* 6. The planet Neptune was discovered by following the guide of theoretical predictions. As it turned out, the planet happened to be in the appropriate part of the sky but was actually in a much different orbit from the one predicted; the theoretical analysis was way off, but the planet was found by using it. Name a scientist who had similar good luck. Distinguish this from serendipity.

Section 1.3

1. Describe what is meant by the basic quantitative character of experimentation in physics.
2. Experimental discovery is accomplished by a combination of accident, trial and error, and deliberate probing. How does the relative importance of these three elements in each case usually reflect the depth to which the research problem is understood at the outset?
3. When radioactivity was discovered experimentally, there was no prior theoretical understanding of what was occurring or why. How, then, was it probably discovered?
4. Discuss the distinction between *data* and *fact*. The ancient Babylonians tracked the sun god, Lord Shamash, across the sky, and we plot the path of a star called Sol, a thermonuclear furnace. The data are the same; how about the facts?
5. Anaxagoras (ca. 500–428 B.C.) was sentenced to death in Athens in absentia for maintaining that the sun was only a blazing ball of fire. What does that say about accepted fact?
6. In what respect is "unity in variety," as Coleridge put it, applicable to physical law?
7. What could Immanuel Kant (1724–1804) have meant by the assertion, "Our intellect does not draw its laws from Nature but imposes its laws upon Nature"?
* 8. Is it possible for a theory once "proved" false to be resurrected later? What does this imply about the "facts" used in the original rejection?
9. After people had been watching Mercury, Venus, Mars, Jupiter, and Saturn for 4000 years, William Herschel in 1781 discovered the seventh planet, Uranus. Even though he saw its unmistakable disk, the sure sign of a planet, and, moreover, observed its motion across the star-field, he announced that he had found a new *comet*. Why do you think he came to that conclusion? By the way, Flamsteed (1690) and Mayer (1756) both had it listed as a fixed star in their catalogs.
10. On hearing of the marvelous properties of X-rays, a renowned scientist of the last century, Lord Kelvin, denounced the newly discovered phenomenon as a deliberate hoax. How does that response fit into the discussion of this chapter?

11. Having spoken for an hour or so about his recent theoretical work on elementary particles, Wolfgang Pauli settled down for the usual general discussion involving the audience. It was New York, 1958, and the great Niels Bohr was there. After some heated remarks, Bohr rose and said, "We are all agreed that your theory is crazy. The question which divides us is whether it is crazy enough to have a chance of being correct." How would they ultimately find out if it was "crazy enough"?

Section 1.4

1. Let's apply the scientific method to the question of the Earth's configuration. Hypothesizing it to be disk-shaped, we can predict that ships will fall off the edges and that its shadow cast on the moon will be circular. Ships do disappear over the horizon, and circular shadows are easily confirmed. Does that mean the Earth is flat? Explain.

2. Can one systematically follow the tenets of the scientific method and be led to nonsense? Explain.

3. What did Pasteur mean when he said, "Chance only favors the prepared mind"?

4. Lord Kelvin (p. 20) wrote, "When you can measure what you are speaking about . . . you know something about it." Did Blondlot know anything about N-rays? Of course, he only *thought* he was measuring their properties, but they sure fooled him.

5. Did Blondlot follow the scientific method?

6. We have already noted (p. 11) how Challis could not "see" Neptune and how Blondlot could "see" N-rays. What did they have in common?

7. Robert Millikan, the precise experimentalist and Nobel laureate, wrote in 1936, "In science, truth once discovered always remains truth." How can this be understood in light of the tentative nature of science?

8. The first standard weights were almost surely natural seeds used to counterpoise a simple balance in weighing out gold dust. The *grain* is a unit that survives from such measurements. Find something around the house that is specified in grains. *Hint*: Look in your medicine cabinet. The *carat* from the Arabic word *carob*, or bean, is still a common measure used in weighing ——————.

PART I Motion

Traffic hustling along concrete ribbons; a tiny meandering red spider, almost too small to notice; whirling galactic islands of a hundred billion stars; countless molecules in every cell and grain, in every raindrop and tear, all moving: a universe of flow and change, a universe of all-pervading *motion*.

The concept of motion, so familiar, so commonsensical, so obvious, is far more subtle and fundamental than is generally imagined. Being deeply bound to the ideas of matter, space, and time, and so to the very nature of the universe, the seemingly simple concept of motion becomes a central theme in the whole development of physics.

Our study of dynamics, matter in motion, can be effectively divided into three broad segments: all the early work up to the mid-seventeenth century, culminating in the hands of Galileo Galilei; the grand synthesis of Isaac Newton; and the new understanding provided by Einstein in his theory of relativity. Galileo, Newton, and Einstein are the pivotal characters that stand at the great turning points in the history of physics. Although they surely were not alone and others had similar visions, their genius dominates the subject as it unfolds in the next three chapters.

Prior to the twentieth century we believed, along with Galileo and Newton, that matter moved *through* the unaffected void. Space merely served as an environment, a nonparticipating vessel to contain the motion. A ball hurled from one place to another sailed unchangingly through space and time, quite independent of either. The empty void was just the nothingness that indifferently held the ruler marks, the labels *here* and *there*. And time was the detached, immutable *tick-tock* that we superimposed on each event to reckon the succession of its unfolding.

That 200-year-old imagery we're so happy with, even now, was swept aside decades ago by Einstein (Section 6.4). The spatial and temporal aspects of events, long viewed as separate and independent, were welded into a single *space-time* setting. Matter became inseparable from that backdrop; in fact, a material object was then understood to be a manifestation of the "void," something like a local knot, a discontinuity, in the smooth sweep of space-time. But even if we put aside the subtleties of relativity theory, things are not always what they seem. Sitting there "at rest," reading, surely you're not moving—or are you? Now there's a notion that was hammered back and forth for well over 2000 years.

Spaceship Earth spins like a top, and a point at its equator sails around at near 1000 miles per hour. You, now, wherever you are unless you're straddling the icy poles, *you* are whirling about with the planet at hundreds of miles per hour. The book, that old lamp, the building, trees, clouds, the sky, all of it sailing together—imperceptibly. We are on a merry-go-round where no one of us can feel the ride or hear the music, and you can only catch the ring when you go beyond your senses—a merry-go-round hurtling about the sun at roughly 66,600 miles per hour.

If there is a single issue that led to the scientific revolution and beyond to that "one giant leap for mankind," it was the study of motion. In what follows we will examine the development of these insights from the Classical Age of Greece to the French Fried Era of today. And we do so, not out of affection for Aristotle, who is at once the hero and villain of much of the piece, but because, like art or music, physics builds on its past.

Although Aristotle was a far better philosopher than a physicist and his physical theories were more often wrong than right, his influence on human thought, even now, is so pervasive that we cannot overlook it. And so it is with him and "the foolishness of the crystal spheres" that we begin this saga.

This chapter is divided into three main sections, which deal with Aristotle's theories, the work of the Middle Ages, and the coming of the scientific revolution. Some of the material is meant to serve as intellectual scaffolding from which we can see how the "correct" physical notions subtly evolved and so better come to understand them. In that spirit we will briefly look at impetus theory, the not-quite-right medieval precursor of two of our most profound insights—**inertia** and **momentum.** The very practical and important concepts of **speed, velocity, and acceleration** will be developed in detail, and we'll begin to explore the idea of **mass.** The last section follows the development of the new imagery of the solar system from its beginnings in the hands of Copernicus to its culmination in the **three laws of Kepler.**

Alexander the Great was Aristotle's (384–322 B.C.) most famous pupil, and while the older man was writing about the world, the blue-eyed boy took it by force. Greek culture and science, dappled with Babylonian astronomy and astrology, spread on the heels of the conquering armies. The Greeks were the first theoreticians, and

CHAPTER **2**
The First 2000 Years

On facing page, a Greek bust of Aristotle, now in the Louvre.

2.1 FROM ARISTOTLE TO AQUINAS— THE DIVINE COMEDY

Alexander's Empire 336–323 B.C.

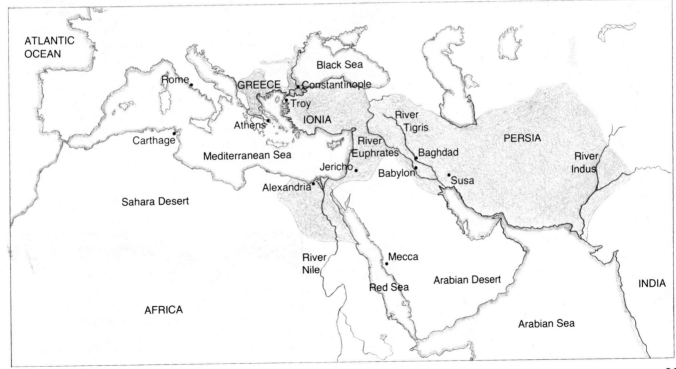

31

that unique perspective continued to flourish long after the young god-king died and his empire crumbled. Great cosmopolitan cities, particularly Alexandria in Egypt, were the centers of learning and commerce. There the monumental accomplishments of Aristotle, his studies of poetry and politics, ethics, physics, and metaphysics were preserved and refined. There the high point of Greek science was later reached at the hands of the geometer Euclid, the physicist Archimedes, the astronomer Hipparchus, and others.

While dust settled on the face of Greece, a new force boiled out of southern Italy—Rome became mistress of the Mediterranean (146 B.C.). Scholars at Alexandria continued to work creatively, but for the most part, Rome became content to take its science in the form of popularizations. The treasure of Aristotle and of those who went before was redigested and often jumbled in the process.

Greco-Egyptian astronomer and astrologer Claudius Ptolemaeus (Ptolemy) was perhaps the last outstanding scientific figure of ancient times. Whereas Aristotle's theorizing represented a search for the essential nature of the universe, Ptolemy's much later masterwork, *The Almagest* (ca. 140 A.D.), was the culmination of practical mathematical astronomy. Charts, tables, diagrams—it was a deluge of observations and numerical relationships. The doctrines of Aristotle and the calculations of Ptolemy would still hold unyielding sway over Western thought almost 2000 years later.

Asleep in Byzantium

Rome was at the height of its power in the second century A.D., and from there it spiraled downward, sweeping away with it the few remaining vestiges of active Greek science. The fourth century triumph of Christianity in Rome angrily wrung the last gasp from "pagan" philosophy. The scholars of Alexandria were driven from the city, and a portion of its magnificent library, the greatest in the world, was put to the torch under the sanction of Bishop Theophilus. In 415 A.D. the mathematician Hypatia was literally torn to pieces in the streets of Alexandria by an enraged mob when she refused to give up her heathen religion. By the time (476 A.D.) the Western Roman Empire fell under the great weight of itself, the Church, and the horrible Hun, the remaining flicker of science had been blown out—and it was dark in Europe.

The Eastern Roman Empire at the mouth of the Black Sea continued to sustain itself, Emperor and all. Constantinople, the great capital built on the old Greek colony of Byzantium, was the "new Rome." And although science did not really bloom there, it at least lay well preserved. Even the official language of this Byzantine Empire shifted during the sixth century from Latin back to Greek.

To be sure, the ancient philosophers themselves often disagreed, yet Aristotle's views were generally embraced in Byzantium with unchallenged reverence. Still, this was a happier state of affairs than existed in Europe, where the Church was busy converting its conquerors and dogma devoured reason. The theologians had inherited the Earth, but they were far more interested in Heaven. Few people in the West were literate, far fewer read Greek, and there was precious little left of it there to read.

Perhaps one reason Aristotle's work had such tremendous appeal then and later was its marvelously broad scope. *The Philosopher*, as he came to be called, formulated both a conception of the structure of the universe and a theory of motion. The two were inseparably and naturally intertwined and interdependent. Together they constituted a whole vision that for a long time seemed, if not too glorious to question, certainly impossible to cast aside.

Things fall—the Greeks knew Earth was spherical, and they even knew that objects always wondrously dropped downward toward its center. Clearly there was something very special about this globe and even more so about its central point. To Aristotle, *the center of the Earth was the very center of the universe*. But of course, one could easily see the "seven planets" of antiquity that had already been recognized for thousands of years (Moon, Sun, Mercury, Venus, Mars, Jupiter, and Saturn) arcing across the sky—"obviously" revolving about the center of the cosmos. *Circular motion was perfection*, and so, reasonably enough, Aristotle envisioned the planets, each hung on one of seven concentric transparent spherical shells. The eighth crystalline shell carried the dome of stars, and they all revolved about a motionless Earth, the universal pivot.

Unhappily for Aristotle, not only is our humble "dirt ball" not the center of the universe; it's not even the center of the solar system. There are no invisible crystal spheres, no perfect circular orbits, and only the moon (as well as an assortment of spy satellites) revolves around this *moving* globe. The seven planets aren't all planets, nor are they all *the* planets. But Aristotle is not yet done with his lovely house of cards. There's more.

Clearly things decay—leaves rot, iron rusts—this is a world of change and imperfection. The Philosopher maintained that all terrestrial matter was composed of different amounts of the four ordinary classical elements—earth, water, air, and fire (Section 9.1). By contrast, the region extending from the moon and beyond was eternal; it had always been and hence must be changeless and incorruptible. The celestial domain must therefore be formed of a fifth perfect element—*aether*.

Falling objects, which race along all by themselves, represented *natural* motion. Each of the ordinary elements had its rightful place in the universe toward which it strained to move. A rock was predominantly earthlike and so would naturally rush at increasing speed downward along a straight line toward the center of the universe, the center of the Earth, like an animal running to its den. Fire resided at the periphery of the lunar sphere, and so, quite reasonably, smoke and tongues of flame licked upward toward their proper place.

Aristotle maintained that the weight (or gravity) of a body and the resistance of the medium through which it moved worked to-

According to Aristotle

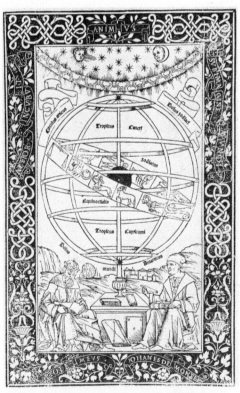

The first "modern" Latin translation of the *Almagest* (1496). That's Ptolemy on the left opposite the man who finished the work, Johannes Regiomontanus.

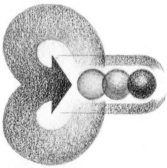

Aristotle's conception of air rushing in behind and thereby propelling a hurled object.

gether to cause it to fall at a rate proportional to its weight. Accordingly, an object ten times heavier than another should fall ten times faster. Certainly a grain of chalk dust drops far slower than does a large chunk of chalk—that's easily observed. He even argued that *all bodies would fall at the same rate, independent of weight, in a vacuum*—and that, as we will see, *is* true!—but then he concluded that this seemingly incredible result must be "impossible" and so insisted that vacuum simply could not exist.

Actually there are more than 100 elements, not five, no one of which is earth, water, fire, air, or aether. All the planets and stars—everything is in constant change. Not one of them is pristine pure, and that will become an important point in the ultimate refutation of the whole Aristotelian scheme. We no longer suppose that rocks have a natural place or quiet desires about where they are going. As for the motion of falling things, that really is rather complicated when some substantive medium is present (Section 3.2). The thicker, more dense the medium, the more closely it approaches Aristotle's description. Pebbles descending through honey are an Aristotelian delight but a rather uncommon one. Generally, though, sizable everyday objects moving through some tenuous substance like air do not begin to approach dropping at rates proportional to their weights until they have reached rather high speeds or, equivalently, have been falling for unusually long times. And last, vacuums of varying degree, instead of being impossible, are now quite commonplace. Although all his conclusions were more or less wrong, none was obviously wrong. For that matter, the erroneous notion that a 400-pound piano will fall much faster than a four-pound rock is widely held even today, more than 2000 years later!

To Aristotle any motion of a stone other than straight down was *unnatural* and required the exertion of some external force. One need only study a horse straining on its harness against a load "to see the truth" as Aristotle saw it: *Without an impelling force there is no motion; end that force and motion in progress ceases.*

A theory that works for horses ought to work just as well for projectiles, but one gets the sense that even Aristotle knew he was on soft ground there. Once a hurled rock leaves the hand, *what force propels it on?* Here the great philosopher saved his theory with yet another ingenious fiction. It seemed that air itself, rushing behind the stone in order to fill the void created by the motion, pushed the projectile forward.

The observation of horse and load is accurate enough; when the horse stops pulling, the load stops moving. When you stop pushing on the table, it stops sliding across the floor. But the conclusion drawn regarding the essential nature of motion—that is, *sustained motion requires a sustained applied force*—that conclusion, however logical, however in seeming agreement with experience, goes beyond experience; it is a hypothesis, and it is wrong! Today, too many of us are still innate Aristotelians, who believe that a body in motion must constantly be pushed. The Apollo astronauts shut down their engines and coasted "pushless" in unpowered flight for much of the 200,000 miles to the moon; at this very instant Pioneer 10 is hurtling toward the stars at tens of thousands of miles per hour, its rockets long ago burned out.

That's a remarkably subtle point, and it took eighteen centuries before it crystallized, before anyone realized that *constant motion was the natural state of things;* that *a body in motion will move forever all by itself* unless something else slows or stops it (Section 3.2). This is the concept of *inertia,* a gift of many of the great minds from Aristotle to Galileo and Descartes. Incredibly, Aristotle correctly concluded that motion would indeed go on *ad infinitum* in a void, but unfortunately he turned that argument around, insisting that such an outlandish result only proved that vacuum could surely not exist.

Hipparchus was an early (second century B.C.) critic of Aristotle's views, and it was he who seems to have inspired, almost seven hundred years later, the brilliant sixth-century Byzantine John Philoponus, "John the Grammarian." Philoponus challenged the notion that objects fall at a rate in proportion to their weight. He confirmed by *experiment* that despite a very large weight difference, two objects take just about the same time to fall in air from the same height. Still, Aristotle's followers saw this not as a powerful refutation, but only as one more of several minor exceptions to the rule. They remained totally committed to their belief.

As for the idea that air rushed behind and propelled things like stones and arrows in flight, Philoponus remarked, "Such a view is quite incredible and borders on the fantastic." Instead, he argued that just as a poker carries heat away from the fire, so a moving body acquires some power or *motive force* when it's initially set into motion. As long as a projectile retains this power, it continues to be propelled. However simplistic, this rudimentary idea carries the earliest trace of our modern view. According to Philoponus, an arrow continues to sail through the air not because of some external force constantly pushing, but because of an internal drive. There it is, a first faint suggestion of inertia way back in the sixth century. Yet Philoponus was in no position to replace the entire structure of Aristotelian theory, nor for that matter did he wish to; the Philosopher's system prevailed.

In the seventh century the Greco-Roman-Christian cultural dominance of the southern Mediterranean gave way to a new host—the rule of Allah was upon the land. There were the inevitable disasters that come with imposed liberation, the worst of which was the total destruction of the library at Alexandria when the city fell to the Muslims in 640. Yet for the most part, Mohammed's messengers were not marauders, and the religion of Arabia spread as much by proselytism as by knife point. Islam soon ranged from India over Persia, Syria, Egypt, all along North Africa, and across the straits into Spain. In time, the hub of substantive scholarship swung to the

Lost and Found

Thirteenth-century scholars. The man in the middle is sighting the stars with an astrolabe while the one on the right reads from an Arabic astronomical table and the one on the left takes notes.

great centers of the Muslim world, particularly Baghdad. There the scientific treasures of antiquity, largely obtained through Byzantium, were eagerly translated and widely read. The multinational Islamic empire was catholic in the strong support given to its science, and its science flourished.

Despite the natural proximity that comes of war, Europe learned little or nothing of its own lost heritage from the enlightened enemy.

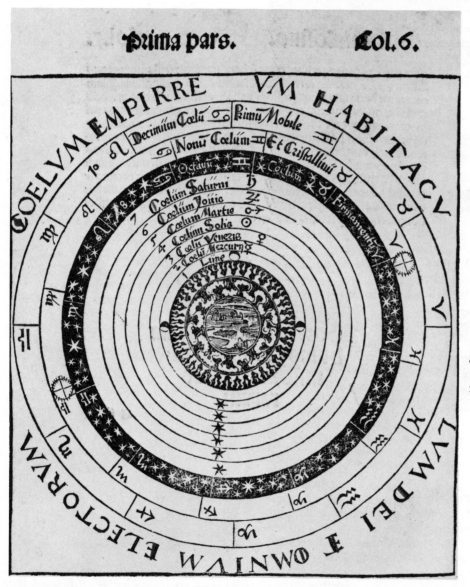

A woodcut from Peter Apian's *Cosmographia* of 1524 depicting the Aristotelian Earth-centered universe. The elements of fire, air, earth, and water form the central terrestrial region.

For centuries it pondered the few pitiful scraps of Greek scholarship still remaining and floundered under the confused, conflicting, and incomplete versions produced by the Latin encyclopedists. Not until the tenth century did complete translations from the Arabic begin to trickle out of Muslim Spain to tantalize Christendom. As Islam was forced to retreat from northern Spain and Sicily toward the end of the eleventh century, the new Christian rulers found themselves inheritors of a surprisingly sophisticated Muslim scholarship. The twelfth century brought a European intellectual revival marked by the Latin translation of a vast array of Greek and Arab treatises. The West glutted itself on Galen, Euclid, Ptolemy, and Aristotle; the fabled masterworks were theirs at last. All across Europe, students gathered to study the new translations, and out of these informal meetings came the new focal points of creative scholarship, the universities.

The monumental task of rendering pagan Aristotelianism palatable to a distrusting Church was the life's work of an Italian Dominican friar, Thomas Aquinas (1225–1274). When Aquinas was done cutting and interpreting, when he was finished balancing the truth of faith on the blade of reason, he had all but reconciled Aristotle's science with the Christian world view. An Earth-centered universe driven by a Divine Mover became the new orthodoxy and the dogged dogma of the next four centuries. The poet Dante (1265–1321) put on the final touch in his *Divine Comedy*, setting the Aristotelian scheme to verse and locating the seven rings of Hell at the center of the Earth. Henceforth everything was at last and forever in its proper place: Devil, man, and God—circles within circles.

2.2 OUT OF THE MIDDLE AGES

After Aquinas's synthesis, Aristotle's system, which had once been banned from the schools, became sanctified dogma—misconceptions and all. Yet it wasn't long before a new creative spirit of critical analysis was in the air, particularly at Merton College, Oxford, and at the University of Paris.

During the fourteenth century scholars thought of motion as a variable quality of matter, like the redness of an apple or the sweetness of a plum. Their concerns were theological and metaphysical, certainly not practical. No one seemed to have been particularly interested in measuring these qualities. Logic (drawing on experience, to be sure, but still logic) was the primary instrument of knowledge.

Our present-day intuitive conception of motion is certainly different than it was in the Middle Ages. We have an easy sense of the unfolding of events, frame after frame. The overlapping of multiple images brings a unique perspective—motion via the camera's eye. And of course, we have speedometers, speed traps, speed limits, and speeding tickets.

Speed and Velocity

The arts work to recreate a sense of motion and thereby to understand it. Instead, physics strives *to quantify*, to describe in terms of measurables, to associate *numbers* with quantities and in so doing establish its own form of understanding.

The more swiftly an object moves, the less time it will take to traverse a given distance. We have clocks for measuring intervals of time and rulers for distance; we have only to somehow relate these primitive intuitive ideas in order to quantify "swiftness" and so pass from poetry to physics. Accordingly, let's pick up on the ancient concept of speed, which Aristotle, among others, specified as distance traveled in a given time. Nowadays we say almost the same thing, defining speed as *the ratio of the distance traveled to the corresponding elapsed time:*

$$\text{speed} = \frac{\text{distance traversed}}{\text{time elapsed}}.$$

If a salmon-colored Edsel traversed *30 miles in 2 hours,* its speed, or more accurately, its *average speed,* was 30 miles divided by 2 hours, or *15 miles per hour.** To be sure, it could have stopped for a red light and then sped along and still covered 30 miles in a total of 2 hours—the reason this is referred to as the **average speed.** In other words, if it traveled at a constant speed equal to this average rate, it would traverse the given distance (30 miles) in the same time (2 hours). Realize, too, that a baseball can be thrown at 15 mi/hr, but it's sure to land in just a couple of seconds, obviously without traveling anywhere near 15 miles.

We certainly should have a feeling for what is meant by *constant or uniform speed.* Just keep the speedometer needle locked at, say, 50 mi/hr on the highway. But how to specify it more precisely? Despite Galileo's boastful claims to priority, a major contribution of medieval physics was just such a definition: *Uniform speed corresponds to the traversal of equal distances in equal intervals of time, of any duration.* You can move at a uniform speed in a car or on a carousel, so long as you cover equal distances in equal times (here there is no question about direction).

The word **velocity** is a common synonym for speed, although in physics there is a significant distinction between the two. *Whereas speed conveys only the rate of motion, velocity specifies both rate (speed) and direction.* There is a big difference between the statements "Move at 1 mi/hr" and "Move at 1 mi/hr south, toward the

A multiple exposure photo of a pole vaulter in his birthday suit. It was taken in 1884 by the American painter Thomas Eakins as a motion study.

---

* To convert from miles per hour to feet per second or meters per second, keep in mind the following relationships.

60 mi/hr = 88 ft/s = 26.8m/s
30 mi/hr = 44 ft/s = 13.4m/s
15 mi/hr = 22 ft/s = 6.7m/s

cliff." Velocity is one of several physical notions that have direction as an inherent aspect.*

A *uniform velocity* prevails when *both the speed and the direction of motion are constant*—in other words, *unchanging speed along a straight line*. Aside from the fact that uniform motion of an object is an idealization (which in the strictest sense occurs only for a very limited time, if at all), the concept is approached in actuality and so remains highly useful. If the speed of an object is constant, we expect the object to cover twice the total distance if we double the time of flight, three times the distance if we triple the time, etc. Had speed been defined in any other way (e.g., as the product or difference of distance and time), this would not be so. Alternatively, you might want to define *slowness* as time over distance—the greater the slowness, the lower the speed. Since that yields nothing new other than a different perspective, no one seems ever to have bothered introducing the notion, and "slowometer" sounds dreadful, anyway. Still, it's amusing to consider that we usually don't think in terms of how slow a given motion is, but rather, how unfast.

Someone on a carousel can be moving at a constant *speed*, round and around, and can of course be changing direction and therefore *velocity* all the while. Actually, if you think about it, living on a spherical planet demands that most ordinary movement not be in a *straight* line. A trip from here to the supermarket follows the curvature of the Earth and simply cannot be taken at a constant *velocity* (without burrowing). Note, too, that traveling at a constant *speed* doesn't guarantee that you will always end up someplace other than where you started.

Not until several hundred years beyond the Middle Ages did it become common practice to represent equations in symbolic form. Despite the dread that often wells up in the heart of the newcomer at the mere sight of such "chicken scratches" (as my wife likes to call them), equations tell a concise shorthand story and are at best marvelous, at least harmless. Do try not to look at them as the foreign language of a hostile tribe.

It's almost traditional to represent speed by v and velocity by **v**. The Greek letter delta, Δ, preceding a quantity usually means *the change in that quantity*. Thus if d stands for distance, Δd is the *change in distance* or length traversed. Similarly with t representing time, Δt stands for the *change in time*, or duration elapsed. The average speed, v_{av}, can therefore be expressed as

$$v_{\mathrm{av}} = \frac{\Delta d}{\Delta t}.$$

---

* Force is another. We talk about pushing, but when we come down to doing it, it always has to be in some specific direction.

TYPICAL SPEEDS

Speed, m/s	Motion	Speed, mi/hr
300,000,000	Light, radio waves, X-rays, microwaves (in vacuum)	669,600,000
210,000	Earth-sun travel around the galaxy	481,000
29,600	Earth around the sun	66,600
1,000	Moon around the Earth	2,300
980	SR-71 reconnaissance jet	2,200
333	Sound (in air)	750
267	Commercial jet airliner	600
62	Commercial automobile (max.)	140
44	Bird in flight	100 (approx.)
29	Running cheetah	65
10	100-yard dash (max.)	22
5	Flying bee	12
4	Human running	10 (approx.)
0.01	Walking ant	0.03

The distance traversed divided by the interval of time elapsed is the average speed.

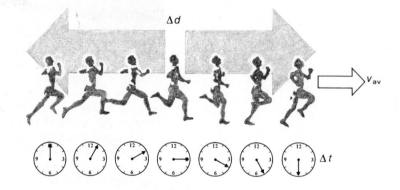

How Fast—Now? Having successfully dealt with uniform speed, the Mertonians attempted, though vainly, to define the concept of **instantaneous speed.** This is supposedly the value we would read off a speedometer at any instant—but how long is an instant? If you actually tried to determine instantaneous speed, you would have to settle for average speed during the shortest possible measuring interval. This is one of those ideas that is a lot easier to talk about than to measure, and it's not very easy to talk about, either. In any event, the instantaneous speed of that cruising Edsel will generally change from one moment to the next over the whole trip. And when you ask, "How fast are we going now?" it's the instantaneous speed you want.

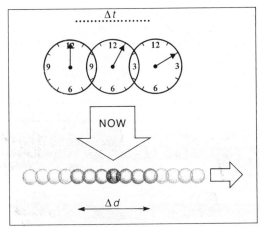

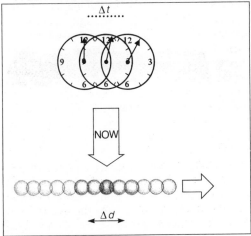

We want the speed of a baseball at the instant someone yells NOW. As the time interval (Δt) decreases, the distance traveled by the ball (Δd) decreases, and the average speed during the interval ($\Delta d/\Delta t$) approaches the desired instantaneous speed—NOW.

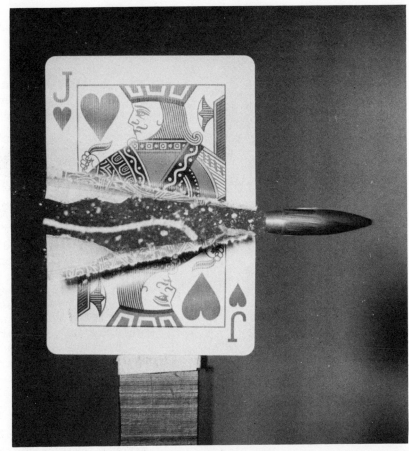

Medieval scholars failed to properly define *instantaneous speed* because they had no conception of the *limiting process*, of *the idea that Δd would become infinitesimally small as Δt became infinitesimally small, and yet the ratio, $\Delta d/\Delta t$, would remain finite.* In other words, the value approached by the average speed during time intervals that become progressively smaller and smaller will be the instantaneous speed at that moment. No matter how fast an object is moving, if the time interval over which we watch it is vanishingly small, the corresponding distance it travels will likewise be vanishingly small. Take as a tiny interval a centillionth of a second (i.e., in the United States $1/10^{303}$ or 1 divided by a 1 followed by 303 zeros). If in that brief time a body moves, say, 10 centillionths of a foot, it's sailing along at $\Delta d/\Delta t = 10$ ft/s, which, for all intents and purposes, is its instantaneous speed at that moment.

Impetus Theory   Volume and weight were familiar ideas in the Middle Ages. They had been important concepts in commerce ever since the time of the ancients. So it was rather remarkable when Aegidius Romanus introduced yet another fundamental measure of matter. This thirteenth-century theologian, a disciple of Aquinas and teacher at the University of Paris, was involved in the contemplation of a tremendously complex religious question. How was it that during the Eucharist the sacramental bread and wine were transformed into the body and blood of Christ despite the obvious discrepancies in weight and volume? Two different visions come together here: the naturalistic and the spiritualistic, Aristotle and the Church, the doctrine and the dogma again in need of reconciliation. Aegidius found the needed reconciliation by maintaining that, independent of weight and volume, the measure of substance, the *how much* of a material entity, was its **quantity-of-matter.** This, then, was the unlikely origin of one of the most profound (and confusing) notions in physics. Almost exactly 400 years later, Newton used the words *mass* and quantity-of-matter interchangeably, just as his contemporaries did.

Although Aegidius's innovation had little or no effect on the course of theology, it did emerge as a useful, though specifically undefined, concept in medieval science. Quantity-of-matter played a rather important role in the teachings of the Parisian Jean Buridan working around 1330 or so. Buridan resurrected and revitalized Philoponus's views on motive force, or *impetus*, as he preferred to call it then. A body maintained its motion because some of the original "push" somehow managed to remain with it as impetus.

Even though impetus theory was wrong, a great many phenomena could be "understood" using it. Such a situation really isn't unusual in science—or anywhere else, as far as that goes. For example, Buridan suggested that God set the heavenly spheres in motion at the time of creation, and they whirl on still because they simply do not dissipate their impetus. His student Nicole Oresme, one of the foremost intellectuals of the period, saw the universe as a grand celestial clockwork, once started, grinding on all by itself untouched forever—a stark, powerful vision that countless others would call up across the centuries from the time of Newton to the French Revolution to yesterday.

Perhaps the major contribution of all this came out of Buridan's attempt to understand the simple observation that one could hurl a stone much farther than a feather. Or equivalently, an iron ball of the same size as a wooden one would travel much farther when launched at the same speed. He reasoned that the more dense material contained a greater amount of matter and so received more impetus at the start. Buridan concluded that the amount of impetus imparted to a body was proportional to both its quantity-of-matter (m) and its speed (v). Medieval scholars liked to argue in terms of proportions and ratios (and of course never used symbols instead of sentences). Nowadays we would say that Buridan's *measure of impetus was the product of quantity-of-matter and speed,* $(m \times v)$. That product, which hundreds of years later would be reinterpreted and renamed *momentum* (Section 4.2), is a fundamental measure of motion and one of the most important of all notions in physics.

ON CELESTIAL MOTION

The situation is much like that of a man making a clock and letting it run and continue its own motion by itself. In this manner did God allow the heavens to be moved continually . . . according to the established order.

NICOLE ORESME

Bishop of Lisieux

Buridan had taken a superb step in the right direction. He knew that motion somehow depended on *both* speed and quantity-of-matter, and he combined the two. A fast-moving cannon ball will roll farther than a slow one, and it will be a whole lot harder to stop than a Styrofoam beach ball at the same speed. Buridan had touched on an essential relationship and brought the theory of motion, however immature, to a turning point.

The velocity of an object in motion can obviously change; variations in speed and/or direction are more the rule than the exception. What, then, is the concept that characterizes *changing* motion? Well, of course, such a notion must certainly incorporate the *change in velocity,* but by itself that's not nearly enough. The speed of something can be altered by a given amount, say 10 mi/hr, in an infinite number of different ways (from very slowly to very rapidly). A speed change of 10 mi/hr can occur over a *time interval* of a second, a year, or even ten years; alternatively, it can be thought of as happening while the body moves over a *space interval* of a foot, a mile, or even ten miles. In short, the actual change in velocity is more often of less immediate interest than is the *rate* at which it changes.*

What we really need, then, is the rate at which velocity varies with, say, distance or perhaps with time. The former concept has generally been of limited concern, curiously, not even warranting a commonly accepted name of its own, though it can be quite informative. For example, my battered '69 Toyota will go from 0 to 80 mi/hr in about 2 miles. Let's (with tongue in cheek) snatch at fame and define *zap* as Δv divided by Δd. My humble vehicle with a Δv of (80–0) mi/hr and a Δd of 2 miles can zap at an average of 40 miles per hour per mile. When a driver is in the process of passing

---

* Certainly the amount of air you breathe during, say, a month is usually of less concern than the rate at which you need to breathe it. Similarly, a ton of soot falling on your head one grain at a time over fifty years is quite different from the disaster that would result if it all came down at once.

Acceleration and Zap

The motion of a uniformly accelerating object.

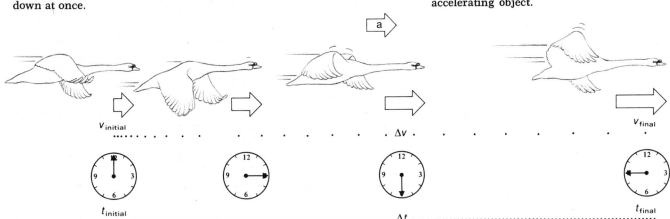

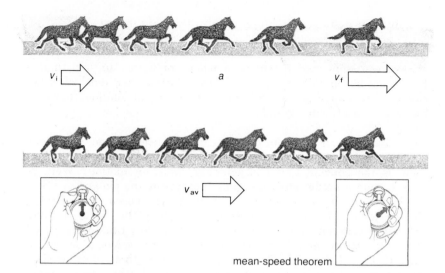

An illustration of the mean-speed theorem. Here the same distance is covered in a given time by two horses, one uniformly accelerating, the other moving at a constant speed.

another car and there's a giant truck bearing down in the oncoming lane, the overriding concern is whether or not the needed Δv can be attained in the distance still remaining, Δd.

The alternative concept, i.e., *the rate of change of speed with time*, is of considerably more importance. Indeed, the basic idea dates back to Greek antiquity. Meriting a legitimate name of its own, it is referred to as **acceleration** and defined by the ratio

$$\text{acceleration} = \frac{\text{change in velocity}}{\text{time elapsed}}.$$

This concept includes changes in speed, direction, or both, but it would be simpler for the moment to limit the discussion to variations only in speed. Accordingly, let's write the average acceleration symbolically as

$$a_{\text{av}} = \frac{\Delta v}{\Delta t}.$$

If in an interval of 10 seconds a runner lopes from rest to a final speed of 22 ft/s (15 mi/hr) the *average* acceleration attained is 2.2 feet per second per second (or 2.2 ft/s²). This is equivalent to uniformly increasing the speed 2.2 ft/s each second. After traveling 1, 2, 3, and 4 seconds, the runner will reach 2.2, 4.4, 6.6, and 8.8 ft/s, respectively, and so on up to 22 ft/s at the end of 10 seconds. By comparison, the acceleration record held by a rocket-propelled dragster blasting across the Bonneville Salt Flats is about 120 ft/s², which is rather formidable. Note that you could be accelerating rapidly and still, after a short time, be moving slowly. Similarly, even with a tiny acceleration, if the travel time is long enough, your speed could become tremendous.

Fourteenth-century scholars at Oxford and Paris successfully dealt with the simplest case of *uniform* or constant acceleration, which they recognized as obtaining when *equal variations in speed occurred during equal intervals of time, of any duration.* Though they pursued the subject of uniform acceleration with considerable facility, they never suspected that this was the all-important link to the motion of freely falling bodies (Section 3.2). Yet it was commonly believed that objects accelerated in some way as they fell. Buridan even went so far as to explain why a thing should speed up, arguing that the motion was caused by "impetus together with the gravity."

Once defined, uniform acceleration had somehow to be related to the more fundamental idea of distance, that is, Δd. In answering that need, the Mertonian scholars made perhaps their greatest contribution to the development of the theory of motion. They knew that the distance traveled at some average speed was expressible as $\Delta d = v_{av}\,\Delta t$. Might it not be possible to find *a uniform speed that would produce the same traversed length in the same time interval as would the actual constant acceleration?* In other words, could they find an average value for the range of speeds that occurred during uniform acceleration? The first derivations were based on a kind of verbal algebra, and soon thereafter, around the year 1350, Oresme produced a geometrical treatment. This mathematization of physical concepts was a major advance toward mature science.

Imagine an object moving at some initial speed, v_i, to be uniformly accelerated up to some final speed, v_f. In each successive unit of time, the speed must increase by the same amount, and so a plot of speed versus time would simply be a straight line (see figure). The average speed is then just the value midway between v_i and v_f, or

$$v_{av} = \frac{v_i + v_f}{2}.$$

If you floor the gas pedal (cleverly known as an *accelerator*) in that Edsel and uniformly accelerate from 40 mi/hr to 60 mi/hr, your average speed during that time interval will be 50 mi/hr. This relationship is by no means special. It applies to many processes, even to the output of a fortune cookie factory, provided the production rate increases by the same amount each interval of time, i.e., provided it is linear. The average will then be half the sum of the two extremes. Indeed, when Oresme did his analysis, it was in terms of any changing quality—speed, whiteness, charity, innocence, etc. He arrived at this result via a general treatment that to a degree anticipated one of the basic conclusions of the calculus. In any event, Galileo would later prove the theorem, using a diagram identical to Oresme's and, building on it, create his "new science of motion."

The Mean-Speed Theorem

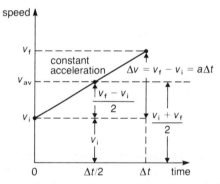

A graphical treatment of the mean-speed theorem.

Nicolaus Copernicus (1473–1543) while still a young man in Torun, Poland.

Substituting the value of v_{av} into the expression for the distance traveled, namely, $\Delta d = v_{av}\Delta t$, we get

$$\Delta d = \frac{v_i + v_f}{2} \Delta t.$$

This is the distance an object would move in the process of uniformly accelerating from v_i to v_f in a time period Δt. A red 1974 Jaguar E-type V-12 sports coupe can roar from a dead stop to 60 mi/hr in 8 seconds flat. Assuming the acceleration to be constant (though it really isn't), we can now calculate the required runway for this sort of gloriously mechanized macho-mayhem. Putting everything in units of feet and seconds (60 mi/hr = 88 ft/s), we get

$$\Delta d = \frac{0 + 88}{2}\, 8 = 352 \text{ ft.}$$

Traveling at an average speed of 44 ft/s (or 30 mi/hr) for 8 seconds, the Jag covers a total of 352 ft before it finally reaches 60 mi/hr.

Various proofs of the mean-speed theorem were widely read throughout fourteenth- and fifteenth-century Europe, particularly in Italy. It seems, though, that the medieval scholar was perhaps as much delighted with such proofs in themselves as with their applicability to the physical environment. The practice of carefully testing Nature had not yet come upon protoscience. And the authority of the Aristotelian world view persisted, unscathed by the challenges of Merton and Paris.

2.3   COMES DE REVOLUTIONIBUS

The philosophical pursuit of *physics* and the practical study of *astronomy* were never totally separate. After all, Aristotle had built a universe in accordance with his laws of motion, and Buridan and Oresme, no less boldly, extended impetus theory to the heavenly spheres.

The interplay of physics and astronomy has always been of the greatest significance, and it will continue to be so as long as there is an "out there" that extends "beyond here." Yet the star-crossed marriage of these two disciplines, held outside the Church, had to wait for Isaac Newton to write the vows.

In what follows we continue to trace two interwoven strands, the development of the theory of motion and the gradual demise of the Earth-centered universe. The motion of the stars and the motion of a tennis ball—one story, one cast of players.

Mikolaj Kopernik

Greek astronomy prevailed into the fifteenth century and beyond, but not without partisanship and conflict. Followers of Aristotle believed, as he did, in an Earth-centered universe of rotating concentric spheres. But although this scheme naturally meshed with Aristotelian physics, it was quite unsatisfactory as a basis for astronomical calculations. The rival scheme of Ptolemy was capable of remark-

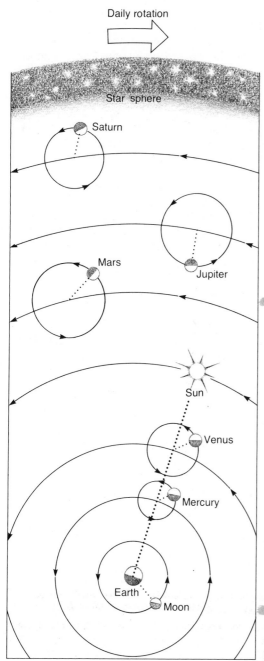

Daily rotation

Star sphere

Saturn

Jupiter

Mars

Sun

Venus

Mercury

Earth

Moon

Ptolemaic system, simplified version.

ably accurate predictions—as indeed it should have been since it was concocted just for that purpose—but it was a horror of intricately compounded motions. Each of the classical "planets" moved on a circle (*epicycle*) centered on yet another circle (*deferent*), which, in turn, was centered at the Earth. Fine-tuning the scheme to bring it into agreement with more precise observations made it a patchwork of further contrivances. And here, faced with the impotent Aristotelian Beauty on one hand and the improvisational Ptolemaic Beast on the other, stood Renaissance scholarship.

The early 1500s were a grand time to be a wealthy young student in Europe; to interact with the flourishing intelligentsia; to joy in the new secular spirit, beguiled by the flesh and free of medieval demons. Mikolaj Kopernik of Torun, Poland, was something of a professional student for about fifteen years, attending the great universities of Krakow, Bologna, Padua, and Ferrara.

Under the guardianship of his uncle, Bishop Lucas, a career in the Roman Catholic Church was an inevitability, which only his prolonged studies kept him from assuming until he was in his thirties. In accordance with custom, Kopernik, rebaptized in the glory of humanism, latinized his name to the form we know him by, Nicolaus Copernicus.

It was at Krakow that the young Copernicus became interested in astronomy, and it was there, at age nineteen, that he heard the news of Columbus's successful voyage to "the Indies." Everyone of any education knew the Earth was round. The ancient Greeks had even measured its diameter, but Columbus had actually braved the dark ocean and the sea monsters and had triumphed. Although the misguided Genoese captain shrank the globe, the young man at Krakow was destined to set it whirling with an even more fundamental insignificance. Indeed, he was destined to set all of Europe and Christendom whirling into the *scientific revolution,* and yet he himself surely was no revolutionary.

His lifework, *De Revolutionibus Orbium Coelestium Libri Sex,* was published in 1543, and story has it that Copernicus held the printed volume for the first time while on his deathbed. This was not the obscure parting gesture of a little known canon of the Church, for by then Nicolaus Copernicus was renowned as a preeminent mathematician-astronomer. *Six Books Concerning the Revolutions of the Heavenly Spheres* was widely anticipated with eager curiosity, if not impending acceptance; reports of his shocking hypothesis had already been stimulating a torrent of criticism for almost thirty years.

Copernicus did not simply suggest that the Earth orbited the sun; *De Revolutionibus* is a complex mathematical treatise fully analyzing all the implications of that hypothesis. As the first European text on a level and scope rivaling Ptolemy's *Almagest,* it demanded respect from those few professionals who could read and understand it.

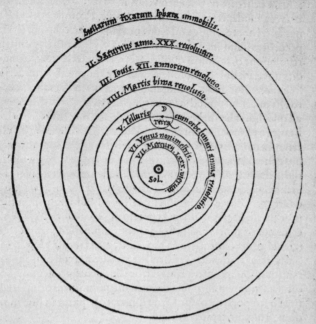

NICOLAI COPERNICI

net, in quo terram cum orbe lunari tanquam epicyclo contineri diximus. Quinto loco Venus nono mense reducitur. Sextum deniq; locum Mercurius tenet, octuaginta dierum spacio circū currens, In medio uero omnium residet Sol. Quis enim in hoc

pulcherimo templo lampadem hanc in alio uel meliori loco po neret, quàm unde totum simul possit illuminare? Siquidem non inepte quidam lucernam mundi, alij mentem, alij rectorem uo cant. Trimegistus uisibilem Deum, Sophoclis Electra intuentē omnia. Ita profecto tanquam in solio re gali Sol residens circum agentem gubernat Astrorum familiam. Tellus quoq; minime fraudatur lunari ministerio, sed ut Aristoteles de animalibus ait, maximā Luna cū terra cognatione habet. Concipit interea à Sole terra, & impregnatur annuo partu. Inuenimus igitur sub hac

Why then do we still hesitate to allow the Earth the mobility naturally appropriate to its spherical shape, instead of proposing that the whole Universe, whose boundaries are unknown and unknowable, is in rotation? And why do we not grant that the daily rotation of the heavens is only apparent, while that of Earth is real? It is like what Aeneas said in Virgil's Aeneid (III.72): "We sail out of the harbour, and the land and cities retire." When a ship floats along on a calm sea, all external things appear to the sailors to be affected by a motion which is really the motion of the ship, while they themselves seem to be at rest along with everything which is with them on the ship.

NICOLAUS COPERNICUS

The sun-centered Copernican universe (in simplified form) as it appeared in *De Revolutionibus Orbium Coelestium* (1543). Note that the outer sphere of stars is immobile. Earth (Terra) is of course the third planet.

It is clear that Copernicus had as his prime intent the desire to bring astronomy into harmony with physics (Aristotelian physics), where any form of celestial movement other than uniform circular motion was unthinkable. Versed in Greek, he studied the classic astronomical writings and returned to the sun-centered scheme of Aristarchus of Samos. To account for the observed flow of celestial events, Copernicus impressed on his mathematical Earth three separate uniform motions. *He set it sailing in a great circular orbit around the sun, spun it about a tilted axis, and then put that axis itself into a conical motion.* Since the planet was hung on a great sphere, the last mechanism was introduced in order to keep the spin axis parallel to itself at all locations in orbit as, in truth, it should be.

Earth revolving, twirling like a flying merry-go-round, was no more intuitively appealing to the Renaissance mind than it is to ours. Regarding Copernicus, Protestant leader Martin Luther remarked in 1539:

> People gave ear to an upstart astrologer who strove to show that the earth revolves, not the heavens or the firmament, the sun and the moon. . . . This fool wishes to reverse the entire science of astronomy; but sacred Scripture tells us that Joshua commanded the sun to stand still, and not the earth.

And Calvin rhetorically demanded to know, "Who will venture to place the authority of Copernicus above that of the Holy Spirit?"

One incredible irony plagued the "upstart astrologer" as he began to flesh out the mathematical details of *planet* Earth. His scheme had a single flaw: It was wrong in its fundamental premise; it was wrong even in the reason for which it was conceived. *The planets actually circulate at varying speeds along elliptical orbits; they do not abide in uniform circular motion.* Copernicus the conservative had unquestioningly followed Aristotle to Wonderland and, unknowingly, set the place ablaze. By the time he was through diddling his uniform circles so that they would match an elliptical reality, Copernicus was back to epicycles and eccentrics. In the full-blown treatment, Earth's orbit was no longer centered precisely on the sun, and this almost heliocentric system was as cluttered, complex, and cumbersome as Ptolemy's had ever been.* Yet it did have an aesthetic appeal in that several qualitative aspects of planetary motion seemed far simpler when viewed from a moving Earth. Copernicus passed on to posterity a planetary system whose virtues were so subtle that to most people they did not warrant the emotional toll of moving humankind from the hub of God's universe and, in so doing, moving God as well. Still, waiting unborn in a time to come were those for whom Copernicus's "bond of harmony" would outweigh all impracticalities—men like Kepler and Galileo, who would at last put the upstart astrologer's torch to Aristotle's house of cards.

---

* Several Muslim scholars already long before had studied a number of non-Ptolemaic alternatives. Among them was Ibn al-Shatir (ca. 1305 – 1375), whose planetary models, though geocentric (i.e., Earth-centered), were otherwise mathematically identical to those of Copernicus. Indeed, it's even possible that some of these details were ultimately transmitted from Damascus to Rome to Poland.

Then spoke Joshua to the Lord in the day when the Lord delivered up the Amorites before the children of Israel, and he said in the sight of Israel, Sun, stand thou still upon Gibeon; and thou, Moon, in the valley of Aijalon. And the sun stood still, and the moon stayed, until the people had avenged themselves upon their enemies.

JOSHUA

10:12–13

Bruno and Brahe

For no reasonable mind can assume that heavenly bodies which may be far more magnificent than ours would not bear upon them creatures similar or even superior to those upon our human Earth.

GIORDANO BRUNO

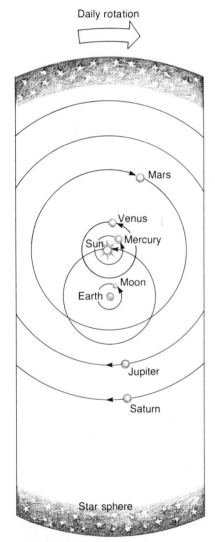

Daily rotation

Mars

Venus

Mercury

Sun

Moon

Earth

Jupiter

Saturn

Star sphere

The Tychonic representation of the universe.

Astronomy abounded with circles upon circles. They were computational devices for the practical astronomer, the tabulator, calendar maker, astrologer, and navigator. And many a Renaissance worker held the sun-centered system to be a useful fiction. As time passed, however, the Copernican world view very slowly proved itself ever more appealing, particularly to the newer generation of less biased practitioners. Yet within about fifty years a far more potent opposition was in foment. The Catholic Church had begun to respond inflexibly to the changing world and to the growing theological implications of the changing universe.

Giordano Bruno, *l'enfant terrible* of the sixteenth century, Italian monk and mystic, was a brilliant, radical, arrogant, penniless prophet of the coming age of science, though by no means a scientist himself. Fleeing Italy when his heretical views on transubstantiation and immaculate conception drew the attention of the wrong people, Bruno wandered about Europe teaching. He was convinced of the reality of the Copernican conception. For him it was no mathematical fiction, and he was quick to call someone who held that view an "ignorant and presumptuous ass."

There had been men in the past, like Nicholas of Cusa (ca. 1401–1464), who believed that an all-powerful Divinity must range over an infinite boundless universe, one with no central body, neither sun nor Earth. But Bruno went even further to suggest that the countless stars had their own planets, and those planets their own living beings. The incomparable revolutionary had transformed the sun into just another star and the entire solar system into an inconsequential string of beads, now only sharing in divine concern.

In 1585, Bruno impudently and imprudently left Elizabethan England and, after again wandering about, returned to the horror house of Italy and the Inquisition. And they did seize him, and though imprisoned for seven years, he did not finally recant. They condemned him, in the spirit of the era, to be "punished with all possible clemency, and without shedding of blood." And on the seventeenth of February 1600, as if to welcome in the new century, they piled high the dried faggots on the Campo dei Fiori and burned the prophet alive, irrevocably spreading his ashes and his ideas.

It was clearly not a good time for unrestrained dissent. The Church, reeling under rising pressure from the new Protestant movements, not surprisingly was becoming blindly defensive of all its dogma. As it always does, war had hardened attitudes into convictions, turned belief to truth and doubt to heresy. By the mid-1500s the Christian West became a bloody battleground of Catholics, Calvinists, Lutherans, Anabaptists, and others, all engaged in ferocious slaughter. Torture was dealt out freely in the search for truth, and the stake served as merciful retribution for error.

Bruno's heresy accomplished two significant things: It provoked the Church into a new hostility toward Copernican doctrine, a stance the Church had not taken before; and it focused attention on the incredible idea of a vast, boundless universe indifferent to the Earth.

A man who mirrored in opposites the free flights of fancy of philosopher Bruno was precisely his contemporary—cautious, meticulous Tycho Brahe of Denmark. Wearing a tyrannical disposition and a false nose (having lost the original in a duel) Tycho stands as the most accomplished naked-eye* astronomer of all time.

In late 1572, a marvelous blazing object appeared in the constellation Cassiopeia. At its most brilliant the "new star," or *nova,* rivaled Venus, and some could see it even in broad daylight. When Tycho first saw it on the night of November 11, while walking home from his uncle's alchemy lab, he could barely believe his eyes. To Brahe it called in doubt the fundamental immutability of the Aristotelian heavens. Using a giant new sextant capable of unprecedented accuracy, he unequivocally found the newcomer to be stationary among the stars of the eighth sphere; the divine unchanging realm had changed!

Unable to resolve his observations of the nova and of the comet of 1577 within the Ptolemaic framework, and being an avowed anti-Copernican, Brahe devised yet another cosmological scheme. The Tychonic system held Earth once again motionless at the center of the revolving stellar sphere. Around it circulated the moon and sun, and around the sun orbited the remaining planets.† As a compromise it was an instant success—far more palatable than the Copernican picture, to which it owed so much and to which it was computationally so similar. Yet a price had to be paid, even for this happy compromise. The spheres carrying Mercury, Venus, and Mars intersected the sphere on which the sun orbited, and so they could no longer be envisioned as the solid crystalline entities of antiquity.

Tycho's greatest contribution to the Copernican revolution, the revolution to which he would not knowingly have contributed at all, was his extensive, precise planetary data, which he passed on to his assistant, Johannes Kepler, and without which no decisive conclusions could possibly have been reached.

Brahe and Kepler met for the first time, face-to-face, on the fourth of February 1600, two weeks before Bruno was executed. The powerful Dane, self-consciously rubbing ointment on his silver nose, was 53, world-renowned, overbearing, and generally despised. Kepler, wiry and dark, was 29, a pauper, weak, provincial, manic, and brilliant.

Tycho's father had been a landed nobleman, his uncle and guardian a country squire, and he himself was then newly ensconced as *Imperial Mathematicus* to Emperor Rudolph II. Kepler, by con-

Tycho Brahe (1546–1601), Danish astronomer.

The Three Laws of Johannes Kepler

---

* The telescope was invented about seven years after his death.

† A contemporary of Aristotle, Heraclides of Pontus, had also suggested a kind of mixed geo-heliocentric system. He proposed that the Earth, *rotating on its axis,* was at the center of the universe. The sun circled it, as did most of the other planets, but Mercury and Venus both orbited the sun. These two bodies are never seen to be far from the sun, and so that model was reasonable, even if unpopular.

Johannes Kepler

If a phenomenon is susceptible of one mechanical explanation, it is susceptible of an infinitude of others which would account equally well for all the features revealed by experience.

JULES HENRI POINCARÉ (1854–1912)

Mathematician

trast, came from a line of impoverished misfits. He was conceived* at 4:37 A.M. on May 16, 1571, by a rogue (who later disappeared, having only narrowly evaded the gallows) and sweet Katherine, the innkeeper's daughter. The great scientist and horoscope caster would someday have to struggle to save his outspoken aged mother from the stake, when she was accused of consorting with the Devil. The Keplers already had one witch in the family, an aunt who was burned alive years before, and one was quite enough.

They came together, these two giants of astronomy, out of dire need, each with the purpose of using the other to his own conflicting ends. Brahe needed Kepler's skill to prove the truth of the Tychonic system, and Kepler needed Brahe's incomparable observations to verify the Copernican scheme. Their tumultuous relationship, punctuated by quarrels and reluctant reconciliations, lasted less than two years. Within weeks of Tycho's death (1601), *Ioannis Kepleri* became *Imperial Mathematicus* (at a much lower salary) and heir to the priceless treasure of data. Ironically, at this point Kepler was quite untrained in mathematics, and that would cause him no little difficulty.

His first book *Mysterium Cosmographicum* had appeared in print in the spring of 1597, and as if to introduce himself to the leading scholars of the day, the brash young man sent them all copies. Galileo in Padua got one, as did Tycho, who was probably the only person to recognize that this was the juvenile work of a budding genius.

The first portion of the *Mysterium* is a mystical treatment of the heavens in the metaphor of the harmonies that were so important to Copernicus. The spheres of the six planets were envisioned as neatly nested on the five perfect geometric solids. Unscientific, to be sure, but still, Kepler was attempting to answer a question that had seemingly never been asked before—that was his gift. Why do the planets have those particular orbits? The answer was wrong, but the question was brilliant. The rest of the book spawned the seeds of his lifework. Accepting Copernicanism as few others dared, he did not challenge it but attempted to improve on it.

The planets, he asserted early, revolve around the sun, not in circles, as had been thought by almost everyone for almost 2000 years, but in oval-shaped paths.† It was well known that the more distant the planet from the sun, the slower it moved and, of course, the larger its orbit. With a marvelous confidence in Nature's order, Kepler concluded that there must be some mathematical relationship

---

* This was the sort of information an astrologer might come by, and in fact, it appears in the family horoscope. Incidentally, his parents were married on May 15.

† In a book written in 1081 by al-Zarqali, "the blue-eyed one," of Toledo, Mercury is pictured moving about in a noncircular orbit. The Arabic text speaks of the path as *baydi*, or oval. Incidentally, Mercury's orbit is far more elongated than is that of any of the other planets known at the time.

between a planet's distance and the time it took to go once around the sun. Again, he probed for the cause of that interdependence, even though he had not yet found the relationship itself. He suggested that some force emanating from the sun drove the planets around like a revolving brush, and that this "force is quasi-exhausted when acting on the outer planets because of the long distance and the weakening of the force which it entails." Once more he was wrong, but close, and yet without a correct theory of motion he could get no closer. Still, the driving power of the solar system moved from the outer sphere of Aristotle to the central sun of Kepler—from spirit mover to physical force.

After six years of incredibly arduous calculation, Kepler published his masterwork, the *Astronomia Nova* (*A New Astronomy Based on Causation/or a Physics of the Sky Derived from Investigations of the Motions of the Star Mars/Founded on Observations of the Noble Tycho Brahe*). It is throughout a rambling though exceedingly frank personal account of struggle, inspiration, failure, and triumph. He quickly proved that *the planes in which the planets orbit all pass through the sun,* and none oscillate as Copernicus had suggested. The sun clearly must command the system. He correctly and radically abandoned the age-old albatross of *uniform planetary motion,* showing that the Earth itself moved in orbit at a nonuniform rate, speeding up and slowing down, in accordance with its distance from the sun. The dark genius wrung from Tycho's data a flood of such immense discovery that it is hard to realize how revolutionary each bold stroke was on its own.

The challenge Kepler took up next was the formulation of a relationship between the varying speed of a planet and its position in orbit. Stumbling along a route strewn with errors in technique and errors in fact, he miraculously arrived at a correct formulation, which is now known as **Kepler's second law:** *A planet moves in such a way that a line drawn from the sun to its center sweeps out equal areas in equal time intervals*. In effect this means that the planets move increasingly quickly as they swoop in near the sun, where the distances are small, and then gradually slow down as they again range far out in orbit. Like a roller coaster rushing downward, each planet hurtles toward its star, reaching a maximum speed at its point of closest approach. Sailing past and away, it slows until, at its farthest distance and lowest speed, it begins to fall back toward the sun once more.

The law maintains that the area swept out by any planet—for example, Earth—in, say, a week will be the same wherever the planet is in orbit, whether it is a week in June or a week in January. Near the sun the speed is great, the arc is long, and the triangular slice of orbital pie is short though broad. Far from the sun the speed is low, the slice is narrow but tall, and still the area remains the same as before. He put aside *uniform* motion, replacing it with a new vision framed in the language of geometry (in the manner of Archimedes).

I am much occupied with the investigation of the physical causes. My aim in this is to show that the celestial machine is to be likened not to a divine organism but rather to a clockwork.

JOHANNES KEPLER

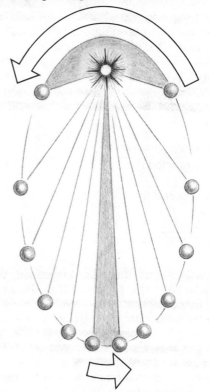

The areas swept out by the Earth moving around the sun in one-month intervals according to Kepler's second law.

Kepler's first law, the fact that *the planetary orbits are ellipses with the sun at one focus,* came to him only later after he had painfully retraced the same route several times without realizing what he had done. He abandoned his early guess that the orbits were oval, calling it a "cartful of dung." Yet he seemed strangely blind to the ellipse, speaking about it and yet avoiding it. By marvelous intuition and brilliant communion with the numerical data, Kepler constructed an equation for the curve that corresponded to the path of Mars. But it was only after abandoning that analysis for yet another new hypothesis (namely, the elliptical orbit) that he finally realized that the equation he had formulated for Mars was actually the equation of an ellipse.

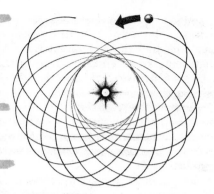

A highly exaggerated view of the unfolding Earth orbit.

The German mystic had recast the Copernican solar system into its modern form, set the sun in command with tentacles of force, and drawn law from the flood of data that otherwise looked like so many unrelated mathematical specks.

In 1618 he completed the *Harmonice Mundi,* the culmination of his lifelong obsession with the mathematical rhythms of the universe, with the harmonies of number and geometric form, of music, astrology, and astronomy. *The Harmony of the World* was finished to a background of cannon as Europe plunged into the Thirty Years' War, yet another religious struggle. All but hidden in this phantasmal garden of mystic delights was the **third law**—the harmonic law. At last, by trial and error and inspiration, he had "found" the relationship between the time it takes a planet to sail once around the sun, the orbital period (T), and average distance from the sun (D)—the relationship he had sought after as a young man so many years before in the *Mysterium. The ratio of the average distance cubed to the period squared is a constant for all planets:*

$$\frac{(\text{average distance})^3}{(\text{period})^2} = \text{constant},$$

or more succinctly,

$$\frac{D^3}{T^2} = K,$$

Planet	Average Distance to Sun (astronomical units) D	Period (Earth-years) T	Period Squared T^2	Distance Cubed D^3
Mercury	0.39	0.24	0.058	0.059
Venus	0.72	0.62	0.38	0.37
Earth	1.00	1.00	1.00	1.00
Mars	1.53	1.88	3.53	3.58
Jupiter	5.21	11.9	142	141
Saturn	9.55	29.5	870	871

where K is a constant to be determined. If we use as the unit of time one Earth-year and as the unit of distance the mean Earth-sun distance (about 93 million miles), known as an *astronomical unit* (AU), then the ratio for the third planet becomes $(1 \text{ AU})^3/(1 \text{ year})^2$, and so $K = 1$. As long as we express time in Earth-years and distance in astronomical units, D^3/T^2 will equal 1 for any planet or artificial satellite in orbit about the sun. The relationship works equally well for moons and machines revolving around planets, but of course, each such system will have its own value of K.

THE ELLIPSE. Point *A* moves along an ellipse when the sum of the distances F_1A and F_2A is constant. Imagine a slack string tacked down at its ends, F_1 and F_2. A pencil holding the string taut will sweep out an ellipse since $(F_1A + F_2A)$ always equals the constant length of the string. Note that as the two foci, F_1 and F_2, are set closer together, the figure more and more resembles a circle.

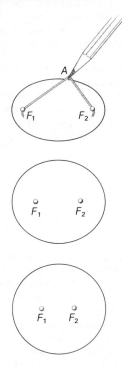

Earth and moon in orbit about the sun. The wiggling motion of the planet is highly exaggerated in this drawing, but it's still very real.

There's the harmony: one ratio, identical for each and every planet sailing around the sun. Notice that the tabulated values of *T* and *D* are fairly crude; they could certainly be made more precise, and the ratio D^3/T^2 would move closer to 1. Yet Tycho could never have determined either *D* or *T* exactly, and neither can we, for that matter. Remember, we can *count* the number of planets exactly, but there is no such thing as an exact *measurement!* There are actually a great many very small but real complications in the planetary motions. The orbits do not even close on themselves precisely. The Earth's orbit ever so slowly rotates counterclockwise about the sun, gradually sweeping out a celestial rosette. Indeed, the moon is so large, it causes Earth to gently weave in and out across the smooth ellipse, generating a relatively tiny serpentine wiggle on the orbit. And as if that weren't enough, tidal effects are also very slowly changing the situation. Clearly it is meaningless to talk about measuring *absolutely precise* values of *D* and *T*, even if we were to assume that they are truly constant. So then, how did Kepler know that *D* was to be raised exactly to the power of 3 and not 3.001 or, perhaps, 2.999? Powers need not be whole numbers, and if my little pocket calculator could try to compute $D^{2.999}$, so could Nature! Kepler guessed. He had faith in his harmonies, he guessed at simplicity, and the third law turns out to be remarkably close to "reality."

Nowadays we might almost feel the great gears churning in his world-machine, in the mechanical cosmos of Buridan and Oresme, but Kepler's contemporaries—including Galileo—were quite unmoved. Even at that period of intellectual revolution, Kepler's meticulous, methodical scientific radicalism was beyond the times. Still, the Imperial Mathematicus never finished the clockwork; that would have taken more physics than he knew. And to be fair, we should not be too hard on Kepler's peers for not appreciating him. It's no mean feat in reading his work to go beyond the mysticism and find the pearls scattered among the pebbles. He himself forgot or changed his mind and often seemed not to remember which his real accomplishments were (or at least which we would come to treasure).

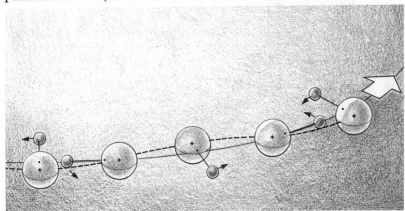

EXPERIMENTS

1. Aristotle maintained that things fall in proportion to their weights. To test this out for yourself, simultaneously drop two objects very different in weight—a big hunk of clay and a tiny ball of it. A nickel weighs ten times as much as a paper clip; does it fall ten times faster? Now try it with an open tissue and a coin; ball up the tissue and repeat the test. Was Aristotle obviously wrong on all counts?

2. Actually Aristotle was referring to falling bodies in resistive media, and so his conceptions get more accurate as the medium gets denser. Drop a few steel balls or clean pebbles in a jar of honey. If you fill the jar to the very top and cap it, you can keep repeating the experiment by just upending the bottle. (Don't allow the honey to warm up appreciably while the jar is capped, or it will expand and shatter the glass.) Now Aristotle is right! The heavier ball falls fastest (provided it doesn't drag along the side of the bottle). Do the balls seem to level off at a constant speed? What happens if you use water instead of honey?

3. Pour a thin continuous stream from a water pitcher, letting the liquid drop through five feet or so. Notice how the smooth stream first crinkles and then breaks into a pattern of dashes after it has fallen for a little while. Try it with your bathroom shower. Up close to the head the stream will be continuous, even to the touch, but near the floor you can feel the staccato drumming of the droplets. Some of the ancient Greeks argued that this effect proved that falling bodies accelerate.

4. If you have a stopwatch and can find a baseball game on TV, time a few pitches. The pitcher-catcher distance is just about 18 meters. Flight times are roughly around 0.6 seconds, and speeds range from 50 to almost 100 miles per hour.

5. Next time there's a thunderstorm, you can calculate how far away the lightning is blasting. Sound travels in air at about 1100 ft/s (roughly 1/5 mi/s), and light moves at 186,000 mi/s. The time elapsed between seeing the flash and hearing the thunder is the sound's travel time (neglecting the minute fraction of a second the light takes). That multiplied by 1100 ft/s is the distance.

REVIEW

Section 2.1

1. Where was the Earth in Aristotle's universe? Why?

2. Did Earth move at all, according to Aristotle?

3. What sort of motion was considered perfection by the Aristotelians?

4. What is "unnatural" motion?

5. According to Aristotle, what was necessary in order to sustain motion?

6. The Aristotelians believed that projectiles were propelled by some mechanism. What was it?

7. What did Philoponus mean by "motive force"?

8. Describe the role played by Islam in the history of science.

9. What was the basis of the European intellectual revival of the twelfth century?

10. How did Thomas Aquinas contribute to the history of science?

Section 2.2

1. What is *impetus,* as conceived in the Middle Ages?

2. The measure of impetus is the product of _____ and _____.

3. Define speed, i.e., *average speed.*

4. List three common units of speed.

5. Define *uniform* speed.

6. What's the difference between *velocity* and *speed*?

7. What's meant by the term *instantaneous speed*?

8. Define uniform *acceleration.*

9. List three possible sets of units for acceleration.

10. What's the average speed of an object that uniformly accelerates from an initial to some final speed?

11. If a body accelerates constantly from rest to v_t in a time Δt, how far will it travel in the process?

12. Driving up and down a hilly but otherwise straight road at a constant speed, is your velocity constant as well?

Section 2.3

1. Did Columbus's voyage surprise everyone with the fact that the Earth was round?

2. Was Copernicus's sun-centered system an original idea of his?

3. Was Copernicus a rebel?

4. What did Aeneas have to say about relative motion? (*Aeneid* III.72 by Virgil)

5. What did Bruno say about extraterrestrial life?

6. Describe the Tychonic system.

7. State *Kepler's three laws* of planetary motion.

8. Which of Kepler's laws are applicable to weather satellites?

9. How did Kepler know that it was exactly the square of the period needed in his third law?

10. What's an *astronomical unit*?

QUESTIONS

Answers to starred questions are given at the end of the book.

Section 2.1

1. Why do you think Aristotle divided the universe into two realms? What sort of observational evidence might he have had for this hypothesis?

* 2. What led Aristotle, quite reasonably, to believe that the center of the entire universe was in the middle of our humble little planet?

3. What was so unnatural about nonvertical motion?

* 4. Imagine that you have two spears that are identical except for their back ends. On one the end narrows down to a point, and on the other there is a large flat disk positioned perpendicularly like the head on a nail. Which do you expect would travel farther? What would Aristotle have said? This example was a serious early problem for the Aristotelians.

5. Aristotle maintained that all space was filled with an invisible material; interplanetary vacuum was impossible. How did his theory of motion force him to that conclusion?

6. For Aristotle and company, fire resided at the periphery of the lunar sphere. How would they have explained the operation of a hot-air balloon?

Section 2.2

1. It's a lot more difficult to push a Cadillac than a Volkswagen. How would Buridan have explained that?

2. Which supposedly has more impetus, a horse or a dog, when they're running neck and neck?

3. What's meant by the phrase "a clockwork universe"? This is an idea we'll come across many times again in considering the centuries after Oresme.

4. How might you measure the average speed of a bowling ball at your local alley?

5. What would I have to specify for you to be able to throw a ball so that it initially moves exactly the way I want it to?

* 6. Why is it unlikely that anything moving on Earth would actually have a uniform velocity for any great distance?

7. After reaching operating altitude, a jet can cruise at an unchanging 500 mi/hr along the shortest path from New York to Tokyo. What dynamical quantity would you say was constant over that part of the flight?

8. An ant sitting motionless on a spinning phonograph record is moving with a uniform _____ but not a uniform _____ .

9. Is it possible to move at the same time with a constant velocity and not a constant speed?

10. If an object travels three times as far while taking three times as long as another object, how do their average speeds compare?

* 11. Suppose that one person takes twice as long to go half as far as another. Compare their average speeds.

12. Can you move from one place to another with some average speed without ever having had an instantaneous speed equal to it anywhere during the journey?

13. Is it possible to have a car accelerate and yet have the speedometer read a constant value?

* 14. Discuss a situation in which a body may have zero speed and yet a nonzero acceleration.

15. A plane goes into a dive, thereafter increasing its speed every second by one tenth its original value. What can you say about its acceleration?

16. Is it possible that at some instant two cars have the same acceleration and different speeds? The same speeds and different accelerations?

17. What's the average speed of a body that uniformly accelerates from rest to a value v_t?

18. A rocket getting lighter and lighter as the fuel is consumed can accelerate for a while at a rate that *increases*. If the rocket starts from rest, what can you say about the *average speed* as compared with the *final speed*?

19. If we took a series of photographs, one each second, of a car uniformly accelerating, what would happen to the distances covered between photos as time went on?

* 20. Two identical cars are at the starting line in a drag race. The first starts and continues to uniformly accelerate up to the finish line some time later. One second after it roars off, the second car follows in identical fashion. Will the cars be the same distance apart throughout the race? If not, when will they be closest to each other? How much time will there be between them crossing the finish line?

21. Galileo's law of inertia maintains that a body in motion will continue moving without change *all by itself*. Why can we consider impetus theory to be a midway step between the Aristotelian and Galilean formulations?

Section 2.3

1. What was Copernicus's goal in creating a new picture of the solar system?

2. Aristotle's Earth-centered system had one major practical failing. What was it?

3. What serious error made by Copernicus ended up complicating his theory, to the point where it was as ugly as Ptolemy's?

4. The Tychonic system did away with the crystalline spheres that supposedly supported the planets. Why?

5. What role did Brahe play in the Copernican revolution?

6. The Earth-sun distance ranges in its extremes from

1.49 × 10⁸ km in January to 1.52 × 10⁸ km in July. To get a feeling for how nearly circular the orbit is, try to draw two concentric circles with radii proportional to these distances.

* 7. Does the Earth move faster around the sun in summer or winter? *Hint:* When is it nearer the sun?

8. An artificial Earth satellite is boosted from a low circular orbit into a higher one. What will happen to its orbital period (the time it takes to fly once around)?

9. Aristotelians of the Middle Ages saw the universe in the organic tradition—as a thing with purposes and desires. The contrasting approach explained phenomena in mechanical terms. Which view do you think Kepler held?

10. *De Revolutionibus* by Copernicus was dedicated to the Pope and written by a churchman in a most complex and mathematical style, and yet, that doesn't account for the fact that it wasn't banned by the Church until 60 years later.

Why did his work take so long to be condemned? For that matter, why was it condemned at all?

11. Brahe, who was an alchemist as well as an astrologer, was drawn back to astronomy by the nova of 1572. The comet of 1577 (Kepler's mother took him to see it when he was five) was also a major event for Tycho. Why were they so important?

12. Which is farther from the sun, Neptune or Uranus? Which has a longer year?

* 13. During the second week in the month of July, the line from the sun to the Earth sweeps out a certain orbital area. How long will it take during *December* to sweep out the same area?

14. Do Kepler's laws apply to artificial satellites orbiting the moon as well as the Earth? Incidentally, in 1978 there were about 4600 pieces of hardware floating about our planet—including roughly 950 satellites.

MATHEMATICAL PROBLEMS

Answers to starred problems are given at the end of the book.

Section 2.2

* 1. A vindictive elephant can run at 25 mi/hr. How fast is that in meters per second?

2. If you travel 1 mile in 10 minutes, what's your average speed per hour?

* 3. A golf ball slammed off a tee by someone who does that sort of thing for a living can move at 170 mi/hr or more. How much is that in feet per second?

4. Suppose you travel 1 mile in 2 minutes, stop for 5 minutes at a drawbridge, and then move on another 9 miles in 13 minutes. What's your average speed?

5. An elevator in the World Trade Center in Manhattan rises to the observatory $\frac{1}{4}$ mile up in just about 1 minute. What's its average speed in mi/hr?

* 6. The duration record for an automobile was set by a Citroën in 1933. A team of drivers kept it going 185,353 miles at an average speed of 58 mi/hr. How long in time did this brilliant display go on?

7. When Orville Wright became the first human to fly in a powered plane, he remained aloft for 12 *seconds* at an average speed of about 8 mi/hr. How many feet long was the historic flight?

8. The fastest bullet ever fired supposedly traveled at 7100 ft/s. How much is that in miles per hour? Shot horizontally from shoulder height, it would strike level ground in about 0.6 seconds. How far will it sail down range?

9. The U.S. Air Force SR-71 reconnaissance aircraft can attain a speed of 2200 mi/hr (and probably even more than that). How far can it travel in 10 minutes?

10. How long does it take light, moving at 186,000 mi/s, to travel 1 foot? Radio signals sailing at the speed of light took 19 minutes to reach the Viking lander on the surface of Mars. How far away was it?

11. A car goes from 0 to 10 mi/hr in 10 seconds; at the end of 20 seconds, it's moving at 20 mi/hr; at the end of 30 seconds it has reached 30 mi/hr. What can you say about its acceleration?

* 12. A VW Rabbit can travel from 0 to 50 mi/hr in 8.2 seconds. Assume that its acceleration is constant. How long will it take to go from 30 mi/hr to 40 mi/hr?

13. A piston-engine dragster (a car designed specifically to accelerate rapidly) set a record in 1972 by hitting a top speed of 244 mi/hr over a measured track of 440 yards in a time of 6.2 seconds, having started from rest. Calculate both the average *zap* and *acceleration* in mi/hr/mi and ft/s², respectively. Compare this with the acceleration of a rocket, which is typically around 60 m/s².

* 14. A ball fired straight upward at a speed of 320 ft/s comes to rest for an instant at its maximum altitude 10 seconds later. Assuming a constant deceleration, calculate its value in ft/s².

15. Superman slams head-on into a train speeding along at 60 mi/hr, bringing it to rest in an amazing 1/100 second and saving Lois Lane, who was tied to the tracks. What's the deceleration of the train? Who always says, "Great Caesar's Ghost"?

16. The ball in Problem 14 decelerates uniformly from 320 ft/s to zero in 10 seconds. What's its average speed? How far up did it travel?

17. Regarding Problem 15, dear Lois was a mere $\frac{1}{2}$ foot down the tracks from the point where Superman struck the train. How far away from her did the "Man of Steel" finally stop the engine?

* 18. Wonder Woman can hurl a Wonder Rock in a straight line at 200 mi/hr. Assuming that the acceleration is uniform and that the rock stays in her hand for 5 feet, calculate how much time the throw takes.

Section 2.3

* 1. What fraction of the total area enclosed by the Earth's elliptical orbit is swept out by a line drawn to the planet from the sun over an interval of one week?

2. Show that Kepler's second law requires that planets in circular orbits move at constant speeds.

3. If an artificial satellite were placed in a circular orbit at a distance from the sun of 2 AU, what would its period be (in Earth years)?

* 4. Calculate the period of an artificial satellite in a 186,000,000-mile orbit about the sun.

* 5. The moon orbits Earth at an average distance of roughly 240,000 miles with a period of 27-1/3 days. What would the period of an artificial satellite be if it were placed in Earth orbit at an average distance of 120,000 miles?

6. With Problem 5 above in mind, calculate the orbital radius of an Earth satellite that will remain at a fixed location, one always directly above the equator. This sort of orbit is called Earth-synchronous and it's particularly useful for communications purposes. *Hint:* What is its period?

* 7. Suppose that we have a communications satellite in orbit and we wish to increase its distance from the center of Earth so that it's four times greater. By how many times will its period increase?

The ill-fated Skylab orbiting the Earth before its fiery downward plunge in July of 1979.

Galileo at the age of about 60, drawn by his contemporary Ottavio Leoni.

This is Galileo's chapter in three parts. It recounts his astronomical discoveries and their relationship to the Copernican revolution, then treats his conclusions concerning motion, and ends with a brief mood piece that recounts the old man's run-in with the Inquisition. Among the Master's major contributions to the theory of motion were his experimental confirmation of the fact that, **neglecting friction, all bodies fall with the same constant acceleration,** his almost right statement of the **law of inertia,** and his analysis of **the flight of projectiles.** We'll look at all these marvelous accomplishments.

CHAPTER 3
Galileo and the New Science

Galileo was a unique figure in the history of science; indeed, only a few others have stood as tall. He was certainly a major influence on the development of the empirical approach to physics, a phenomenon that rapidly evolved with a renewed vigor toward the end of the sixteenth century. At one and the same time he was the vortex of the Copernican whirlwind and the master of "the new science" of motion. His was "the kiss of death" passionately planted on the Aristotelian universe in a flare of brilliantly conceived experiments and superb observations.

Galileo was a robust, vigorous character, who fathered three more children than is generally considered proper for a lifelong bachelor. He was arrogant, charismatic, and self-assured, a shrewd hustler and entrepreneur, a marvelous writer of prose, whose cutting wit led him ultimately to the angry arms of the Inquisition.

There are two main currents that we must follow as they run through Galileo's life: his defense of Copernicanism with all the notoriety and clamor that swirled around it, and his far more profound and less controversial work on the theory of motion.

Galileo Galilei was born in Pisa in 1564, just three days before Michelangelo died in Rome. That was the same year Shakespeare was born in England; nonetheless, the flower of the Renaissance had already begun to wither. Following Tuscan custom for the firstborn, his rather distinctive Christian name was a variant of his surname. When Galileo was 10, his family moved back to Florence, the city they had left years before for financial reasons. Father Vincenzo, a composer, writer, and one of the founders of opera, was an outspoken man of liberal persuasion and little money. At first Galileo wanted to be a painter, but practical Vincenzo sent his young son off to the University of Pisa to study something more lucrative—medicine. Even so, soon after coming in contact with the mathematics of Euclid and Archimedes, Galileo left Pisa and returned home to Florence to devote himself completely to that new passion.

3.1 TWO CHIEF WORLD SYSTEMS

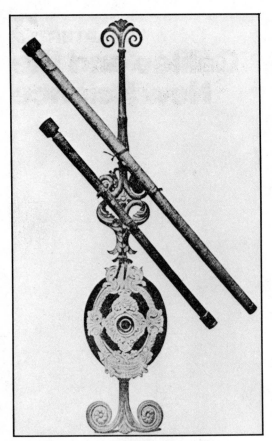

Two of Galileo's telescopes mounted on a rather ornate display stand. One of his hand-ground lenses is at the center of the medallion. Galileo personally made more than one hundred such telescopes.

Not long afterward, he wrote two technical treatises that established his reputation as a competent mathematician and brought him to the notice of the Duke of Tuscany, Ferdinand de Medici. Now, properly befriended at court, the young philosopher promptly received an appointment to the chair of mathematics at Pisa. With considerable zeal, he attacked the mechanical doctrines of Aristotle as they had never been pursued before. Soon there were few of the faculty at Pisa who had any great love left for the pushy Tuscan, and before long he departed in a cloud of rancor, packing off to Padua. It was the fall of 1592, and he was 28. He had already displayed a considerable talent for practical invention, working with the skilled hands that might have made him an artist if the fates had been different. It was probably the latter accomplishment that endeared Galileo to the practical Venetians, who offered him the prestigious chair of mathematics at the University of Padua. Besides, they got the young rascal for a meager 180 florins a year.

At Padua he proved a captivating and highly successful lecturer with a growing international reputation and a considerable following. To augment his modest income, he took on private pupils and began commercially manufacturing some of his inventions. A little shop of his own proved invaluable, since the university maintained no laboratory, and he had to construct his own instruments.

In 1597 he received a copy of an obscure little book written by a young upstart professor in Germany named Kepler, Johannes Kepler. The beginning of that work was the most outspoken praise of the Copernican world view published up to that time. Having just browsed the preface, Galileo wrote in the first of his only two letters to Kepler:

I will only add that I promise to read your book in tranquility . . . and this I shall do the more gladly as I adopted the teachings of Copernicus many years ago. . . . I have written many arguments in support of him and in refutation of the opposite view—which, however, so far I have not dared to bring into the public light.

He was a closet Copernican, afraid not of the Church but of the derision of his learned colleagues. It is doubtful that Galileo ever read Kepler's book, and except for their mutual defense of Copernicanism, the two seem to have been sadly ignorant of each other's major triumphs. Galileo would never have had the patience to wade through Kepler's dreadful Latin prose, and he himself published his masterwork on motion only long after Kepler died. Each held a piece to the puzzle, but neither knew it. Kepler struggled on without the law of inertia, and Galileo held to the anachronism of circular orbits. Had they only met and talked!

Galileo was working on his theory of motion in mid-1609 when word reached Venice that "a certain Fleming had constructed an eye-

glass by means of which visible objects, though very distant from
the eye, were distinctly seen as if nearby." He put everything aside
and very soon produced an incomparable series of increasingly powerful telescopes of his own design. Not one to miss an opportunity,
he marched to the top of the Campanile, the bell tower of San Marco,
accompanied by Venetian senators and noblemen, and there demonstrated his device to everyone's delight and amazement. In appreciation they gave him life tenure at the university and an annual salary
of 1000 florins. They had overpaid for the showmanship, but Galileo
would resign altogether in about a year, moving back to Florence and
a nonteaching position at the court of the Grand Duke.

He was almost 46 when he transformed what had been something of an optical toy into an instrument of science and turned it
skyward. He saw "stars in myriads, which have never been seen
before"—a universe far grander than Ptolemy or Aristotle could ever
have known. The jagged surface of the moon, mountainous and
craggy "just like the face of the Earth itself," made a mockery of
Aristotelian celestial perfection. And then he discovered four moons,
looking like a tiny solar system, "rolling about the sphere of Jupiter"
—more objects unknown to the all-knowing ancients, and these
circling Jupiter and not Earth, the supposed center of the universe.

The heliocentric system naturally led to the conclusion that
Venus, as seen from Earth, should appear to change its size as it circled the sun and, moreover, should display phases resembling those
of the moon. Copernicus himself was sorely disappointed at not being able to observe either of these crucial variations—but of course,
he had no telescope. Toward the close of 1610, Galileo triumphantly
wrote of Venus:

> It went on growing daily in size ... until eventually, arriving at
> a very great distance from the Sun, it started to lose its round-
> ness ... and in a few days diminished to a half circle.

Quite incompatible with the Ptolemaic scheme, these observations
were the most damaging of all.

The *Sidereus Nuncius,* Galileo's compilation of his early startling
findings, was a short, concise, easily read book, quite unlike any of the
scientific tomes that had come before it. And it was an international
sensation that gave "the Man from Florence" world renown. Within
a year he even challenged the purity of the sun, announcing "dark
spots seen in the solar disk." Though sunspots had been observed
long before in the Orient, Galileo used their motion to determine that
the sun actually rotated. It is likely that the strain of this work,
coupled with an ailment suffered in his youth, contributed to his ultimate blindness.

Galileo's telescope, though probably the most powerful in existence at the time, was still a crude, clumsy device, difficult to handle

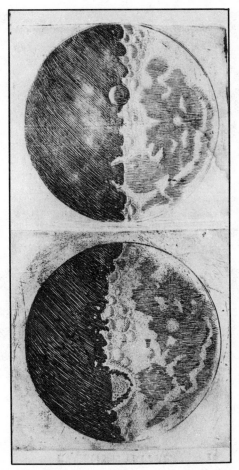

Two of Galileo's drawings of the moon as
they appeared in his *Sidereus Nuncius*
of 1610.

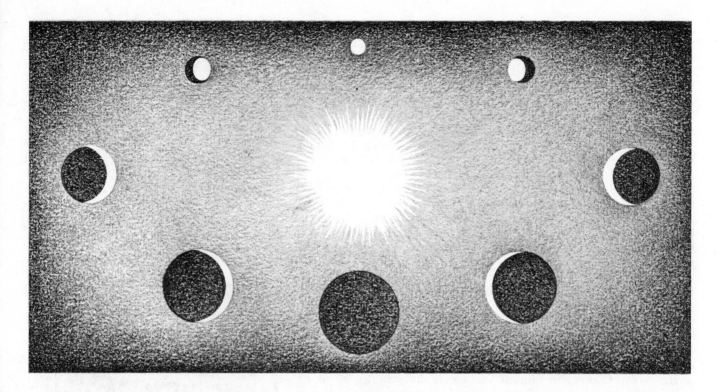

The phases of Venus seen from Earth as
Venus orbits the sun.

And new Philosophy calls all in doubt,
The Element of Fire is quite put out;
The Sun is lost, and th' Earth and no man's
    wit
Can well direct him where to looke for it.
And freely men confesse that this world's
    spent,
When in the Planets, and the Firmament
They seeke so many new; then see that this
Is crumbled out again to his Atomies.
'Tis all in peeces, all coherence gone;
All just supply, and all Relation . . .

JOHN DONNE

The First Anniversary (1611)

and easily shifted out of adjustment. It's not especially surprising that at first the few Italian scholars who condescendingly chose to peer through it—and many did not bother—remained unmoved. Nor was Galileo prepared even to explain how the thing worked. Though he held demonstrations, the blurry little specks of light convinced almost no one of their reality. The scholars of Italy either remained quietly skeptical or turned derisively hostile.

Although Galileo had not responded to Kepler's several previous requests, he now asked the influential Imperial Mathematicus for his opinion of the *Sidereus Nuncius*. Kepler, the leading astronomer of Europe, unbegrudgingly wrote a long open letter to his harried colleague. In it he enthusiastically defended Galileo's discoveries, even though he had no telescope with which to actually confirm them. Kepler had stuck his neck way out in a gesture that helped to turn the tide for the Tuscan dabbler in astronomy. Characteristically, Galileo never took the time even to thank him.

When Galileo later observed Saturn through his rudimentary telescope, he saw a sight so strange that he dared not publish it. To establish immediate priority and yet not risk the humiliation of having been too hasty, he circulated a cryptic message that he intended to translate only when the moment suited him. As if deliberately to tantalize Kepler, who was more than eager to learn of the latest discoveries, Galileo had the Tuscan ambassador to Prague take him this anagram:

smaismrmilmepoetalevmibunenugttaviras.

Patient, generous Kepler struggled with the damned anagram, finally decoding it into a "barbaric Latin verse" describing some nonsense about Mars. Three months later, Galileo sent the solution to the Emperor Rudolph. In translation it read: "I have observed the highest planet in triplet form." Although he did not know it, he was seeing Saturn's rings extending beyond the globe on either side, and he interpreted the vision as if the planet were flanked by two lesser bodies. "The mathematical professor at Padua" was totally dismayed when he again viewed the planet in 1612. Saturn had wheeled about, presenting its thin rings edge-on, and they had disappeared from view. "What is to be said about so strange a metamorphosis?" wrote Galileo. "Are the two lesser stars consumed after the manner of solar spots? Have they vanished or suddenly fled? Has Saturn perhaps devoured his own children?" Although he continued to observe the planet and to sketch it so accurately that we immediately recognize the smooth sweep of the rings in his drawings, he himself never really understood what he was seeing (Section 1.3).

By early 1611, when Galileo journeyed to Rome to demonstrate his discoveries, the initial shock had already worn thin enough to allow for objectivity. His sightings were confirmed, and he was honored by the Jesuits of the Roman College.

For Galileo the torrent of evidence was at last incontestable. He finally came out in print squarely on the side of Copernicanism and, by 1613, was again embroiled in public controversy with the diehard Aristotelians. The Church, not ready for either the tumult or all its implications, summoned Galileo in 1616 to stand before the Inquisition. Lord Cardinal Bellarmine notified him "that the doctrine attributed to Copernicus . . . is contrary to the Holy Scriptures and therefore cannot be defended or held." They had gagged the Tuscan and buried *De Revolutionibus*.

When an old friend and defender, Maffeo Barberini, became Urban VIII, Galileo met with him in Rome in 1624 to pay his respects. Feeling secure in the warm reception he received from the new Pope, he returned to the writing of the *Dialogue Concerning the Two Chief World Systems, Ptolemaic and Copernican*, a labor that would occupy the next six years. Carefully playing by the Inquisition's rules, he wrote the *Dialogue* in such a way that he neither held nor defended the heretical doctrines of Copernicus. Instead, the main characters debated the pros and cons of both systems, with the stronger arguments invariably on the side of heliocentricity. Galileo, flouting the intent while abiding by the very letter of the injunction, fooled no one and infuriated almost everyone—everyone, that is, but the literate public, who gleefully bought up every copy as it came off the presses. This time he had gone too far, and the crafty old man was summarily brought to his knees (Section 3.3).

From Man or Angel the great Architect
Did wisely to conceal, and not divulge
His secrets to be scann'd by them who ought
Rather admire; or if they list to try
Conjecture, he his Fabric of the Heav'ns
Hath left to their disputes, perhaps to move
His laughter at their quaint Opinions wide
Hereafter, when they come to model Heav'n
And calculate the Stars, how they will wield
The mighty frame, how build, unbuild,
  contrive
To save appearances, how gird the Sphere
With Centric and Eccentric scribbl'd o'er,
Cycle and Epicycle, Orb in Orb:

J. MILTON

Paradise Lost (1667)

The frontispiece from Galileo's *Dialogo sopra i due massimi Sistemi del Mondo.* That's Aristotle on the left, Ptolemy at center, and Copernicus on the right.

3.2   ON MOTION

ON FREE FALL

For all things that fall through the water and thin air, these things must needs quicken their fall in proportion to their weights, just because the body of water and the thin nature of air cannot check each thing equally, but give place more quickly when overcome by heavier bodies. But, on the other hand, the empty void cannot on any side, at any time, support anything, but rather, as its own nature desires, it continues to give place; wherefore all things must needs be borne on through the calm void, moving at equal rate with unequal weights.

LUCRETIUS (ca. 94–55 B.C.)
On the Nature of Things

Legend has it that Galileo, while still an undergraduate at Pisa (in 1581), sat musing one day in the cathedral during Mass. An attendant had pulled a hanging candelabrum to one side in order to light it, and it was then gently swinging to and fro. Perhaps guided by his musical background, Galileo began to sense a rhythm in the smooth sweeping motion. At first it moved quickly along large arcs. Then, as the swing gradually diminished, the speed slowed so that the rhythm remained strangely unchanged. The young medical student reached for his wrist to count the steady beats of his pulse, to count out moments as the lamp slowly rose and fell. To his astonishment he found that the time taken to go from one end of a swing, down and back up to the other end, was identical for each sweep, no matter how short the path became.

Galileo went back to his relative's house, where he was staying, and carried out a series of experiments using balls of different weight hung on various lengths of string. He soon discovered that *the time for each oscillation depended on the length of the pendulum, increasing as that length increased, but amazingly, it was quite independent of the weight of the bob!* The string, of course, deflected the ball along a circular arc, but other than that restraint, it seemed that the ball should have behaved as if in free fall. And yet, regardless of the weights of the bobs, two pendulums swung in step, provided they were of the same length. The clear implication was that two such balls, though vastly unequal in weight, would fall freely at the same rate.

By designing experiments that revealed the essential aspects of the phenomenon in a reproducible fashion and by using those results to verify hypotheses, young Galileo made the transition to modern science. He would spend his entire life relying on observation as the ultimate measure of physical truth. In his last years, growing blind and living under house arrest, he finally set down his mature thoughts on motion. The resulting manuscript was smuggled out to Protestant Holland, where in 1638 *Two New Sciences* first appeared in print. Feigning surprise at the publication, the old rascal once again played the innocent, but this time the Church wisely paid him no heed, ignoring the challenge entirely.

All Fall Down   Here again Galileo confronted the sacred ignorance of orthodoxy, of Aristotle's genius, gone rigidly dogmatic in the medieval air. It was foolishness; even Philoponus in the sixth century (Section 2.1) had known it and way back then had challenged the Aristotelian view that *the speeds of falling bodies were proportional to their weights.* In fact, Simon Stevinus, a Flemish dry-goods clerk turned splendid physicist, performed much the same experiment (in 1586) and reaffirmed Philoponus's conclusion of a thousand years before.

So Galileo really wasn't saying anything new when he got around to asserting that, *neglecting air resistance, all bodies fall at the same rate, regardless of weight.* Yet most scholars at the time still held fast to the Aristotelian view. To justify his conjecture, Galileo used an argument that went back to the schoolmen of the Middle Ages, who speculated about what would happen in vacuum, even though they rejected it. The argument goes like this: Imagine three identical objects, say, three balls of clay; they would certainly each fall independently at the same speed. Now simply squeeze two of them against each other so they stick together. You then have just a pair of objects, one weighing twice the other. Yet (neglecting any differences in air friction) these two bodies would obviously still fall together, traversing the same distance in the same time. Clearly, a thousand balls of clay would fall the same way, and it became reasonable to generalize the conclusion to every object, regardless of weight.

If Galileo ever did perform the legendary free-fall experiment, it most probably was not from the Leaning Tower of Pisa, and he surely already knew the answer.

When an object falls, moving faster and faster, plowing through the surrounding air, the resistance to its motion also increases. (You can easily feel the same effect by rapidly moving your hand through water or extending it out the window of a speeding car.) The greater the rate of descent, the greater the resistance exerted by the rushing air as it moves out of the way until a balance is reached where no further increase in falling speed can occur. This is the **terminal velocity,** which depends on the shape and weight of the body. A sky diver, after dropping several hundred feet with arms and legs extended spread-eagle, will reach and maintain a top speed of roughly 130 mi/hr.* With a parachute pluming out, wide open, the cross-sectional area and therefore the drag increase considerably, and the terminal velocity drops to about 20 mi/hr. By contrast, a compact heavy object like a smooth stone might drop five or six hundred feet before deviating appreciably from free fall. Still, air friction spares us from being pelted by raindrops, which ordinarily strike the Earth at only about 15 or 20 mi/hr but might otherwise come slamming down at several hundreds of miles per hour.

Fairly light objects dropping through dense media (e.g., liquids) will have low terminal velocities, and therefore they more nearly match Aristotle's conception of weight-dependent fall. In air, however, it was clearly pure nonsense, and Galileo knew it—just as he knew that air resistance would "render the motion uniform" if the fall was long enough. An actual demonstration that gravity fall was independent of weight did not come until thirty years later, when Robert Boyle pumped the air out of a cylinder and could then drop things in a decent vacuum. Nothing is quite as convincing as *seeing* a little tuft of tissue paper come crashing down alongside a lead ball.

---

* The record 310-foot drop onto an air bag lasted five seconds. The stunt man, who reached 80 mi/hr and burst the bag on impact, walked away smiling.

A cannon ball weighing one or two hundred pounds, or even more, will not reach the ground by as much as a span ahead of a musket ball weighing only half a pound, provided both are dropped from a height of 200 cubits.

GALILEO GALILEI

A multiflash photo showing two objects that differ tremendously in weight falling together.

In the end, one could argue that Galileo's thesis is an idealization, that the perfect vacuum in which to finally *prove* it has not yet been created and, for that matter, is not likely to be. Even so, the extensive confirmations already in hand verify the thesis to within so minute an experimental error that we stand convinced. If there is any deviation at all, it must be minuscule. And so we respond to the logic and the experiments and accept as true the notion that all things fall in vacuum at the same rate—*within the present bounds of experimental error*.

As the Time Squared All right, then, except for the effect of air friction (which tends to markedly retard light objects with lots of surface area like leaves, feathers, and sheet paper), all things fall at the same rate, but what rate is that? Do they accelerate, and if so, does the acceleration vary as well? Galileo correctly came to the conclusion that falling bodies increase in speed *in the simplest possible way*, namely, with *uniform acceleration*. Had he stopped there, his contribution would have been as speculative as those of his Renaissance predecessors. Leonardo da Vinci, for one, held that a free-falling body moves with a speed that is proportional to its time of flight. And this is equivalent to asserting that the object accelerates uniformly (since $\Delta v = a\Delta t$). But Galileo did not stop; instead, the Master designed a series of experiments to actually verify his preconception.

Assuming that gravity fall occurs at constant acceleration, we can apply the general expression for that type of motion, namely, $a = \Delta v/\Delta t$, recalling that $\Delta v = v_f - v_i$. If we simply begin by dropping an object—i.e., its initial speed, v_i, is zero—then at the end of any time interval, Δt, the acceleration is given by $a = v_f/\Delta t$. If, after each of several different durations of fall, we measure v_f and find

FREE FALL ON EARTH IN VACUUM

Duration of Fall (seconds) Δt	Distance (feet) Δd	Final Speed (ft/s) v_f	Final Speed (mi/hr) v_f
0	0	0	0
1	16	32	22
2	64	64	44
3	144	96	65
4	256	128	87
5	400	160	109
6	576	192	131
7	784	224	153
8	1024	256	175
9	1296	288	196
10	1600	320	218

that $v_f/\Delta t$ is always the same value, then we have proved that a is indeed constant. Though conceptually simple, this straightforward measurement has two practical problems and actually could not have been carried out in Galileo's time. First, the speed acquired in free fall becomes very large very quickly, too quickly for the means available to him. Second, he had no way of determining instantaneous speed, v_f, directly—something that is difficult to do even today. The second problem rates restating: There he is with the ball whooshing along, and he has to be able to measure the speed at any instant, *now*, as it flashes by.

The first problem Galileo surmounted by rolling balls down an inclined plane, rather than just dropping them. He argued that the effect was only to "dilute" the motion by reducing the acceleration and thereby slowing the speeds to something manageable. His treatment was not complete enough to establish this point irrefutably, but his assertions were convincing. If he could prove that acceleration was constant for each incline, increasing only as the slope increased, he could argue that in the limit a vertical incline would correspond to free fall.

To handle the second difficulty, he ingeniously reformulated the problem in terms of quantities that are directly measurable, namely, distance and time rather than speed. Years before, in 1545, Domingo Soto had written a widely read book in which he maintained that falling bodies traveled with a speed proportional to time and that the distance so traversed obeyed the already well-known mean-speed theorem (Section 2.2). That theorem (which Galileo either used

If Galileo Galilei were alive he probably would be tugging at his beard with pleasure about the elementary physics experiment Apollo 15 astronaut David R. Scott performed today on the moon.

Scott dropped a hammer and a feather from waist high to illustrate that both objects accelerated equally by the moon's gravity and that both would hit the surface at the same time despite their differences in mass or weight.

The experiment, similar to those Galileo did 300 years ago from the top of the Leaning Tower of Pisa in Italy, was performed as Scott and James B. Irwin were finishing up their final lunar excursion.

"In my left hand I have a feather. In the right hand a hammer," Scott said, standing in front of the camera mounted on the lunar Rover.

"The reason I have these here today is because of Galileo's discourse on falling bodies in gravity fields. Where better to confirm his findings than on the moon."

Then he dropped both objects and, sure enough, they struck the lunar surface simultaneously.

(Associated Press News Release, July 1971)

without acknowledgment or reinvented) states that the distance traveled at constant acceleration is given by

$$\Delta d = \frac{v_i + v_f}{2} \Delta t.$$

Galileo put the separate pieces together in such a way that he could actually measure what was happening. In this particular instance, starting at rest, $v_i = 0$ and $v_f = a\Delta t$; we have

$$\Delta d = \frac{v_f}{2} \Delta t = \frac{a\Delta t}{2} \Delta t = \frac{1}{2} a(\Delta t)^2.$$

The distance varies as the time squared! Since an object in free fall drops at greater and greater speed, we expect the *distance traveled in each successive second to increase* as the fall progresses, and that manifests itself in the time entering as a squared term. In any event, by arguments similar to these but geometrical rather than algebraic, Galileo concluded that distance varied with the square of the time for uniform acceleration. (Oresme knew some of the consequences of this quite a while before but never imagined that the phenomenon actually applied to anything in Nature and certainly not to gravity fall.)

A long piece of wooden molding served as the channel, the inclined plane along which a polished bronze ball could be rolled with a minimum of friction. Timing was accomplished by collecting water from "a thin jet" during each trial and weighing it later, a process far more accurate than using any clock available then.

> Next we tried other distances comparing the time for the whole length with that for the half, or with that for two-thirds, or three-fourths, or indeed for any fraction; in such experiments, repeated a full hundred times, we always found that the spaces traversed were to each other as the squares of the times, and this was true for all inclinations of the plane.

Suppose the ball rolled 1 foot in the first second; after 2 seconds, instead of traveling a total of 2 feet, it traversed 2^2 or 4 feet; after 3 seconds it rolled 3^2 or 9 feet; and so on. The distance covered varied not as the time but as the *time squared*, and that meant that the ball was accelerating and moreover that the acceleration was constant. When the incline was tilted up a bit, the ball rolled faster and covered more ground, but again the distance was proportional to the time squared. The acceleration increased as the tilt increased although it was always constant for each setting.

Neglecting air friction, all bodies, regardless of their weight, fall with the same uniform acceleration. Well, not quite! Despite the fact that this is the customary statement, it is simply not true, and most everyone now knows it. It was Newton who pointed out that acceleration in any earthbound experiment, even in vacuum, actually is not quite constant. As a rule it increases with decreasing height above the Earth. This increase is a relatively gradual one, which is ordinarily quite insignificant in the laboratory. Still, as an object falls, it accelerates at a very slightly increasing rate, *not* uniformly. All this underscores the way physicists (like all scientists) tend to impress simple law on a not wholly compliant Nature.

At sea level on Earth, the acceleration of gravity, usually represented by its own special symbol, *g*, on the average equals about 32 ft/s² or 9.8 m/s². This numerical value was first determined by Christian Huygens, who showed that it could be adduced from the swing of a pendulum with just a ruler and a clock. Galileo himself never bothered to actually measure it. He got what he wanted by using ratios from which *g* cancelled out.

An object at sea level freely falling from rest moves ever more swiftly, covering a total of 16 ft in its first second of flight and attaining a final speed [$v_f = g\Delta t$] of 32 ft/s. That gives it an average speed [$(v_i + v_f)/2$] of 16 ft/s during that first second. By the time the second second elapses, the final speed has increased to 64 ft/s, and the distance fallen [$\Delta d = \frac{1}{2}g(\Delta t)^2$] is 64 ft, and so on.

The Earth's gravitational acceleration is really rather substantial. Compare it to that of the 12-cylinder Jaguar coupe we talked about earlier (p. 49), which, full out and roaring, hits 60 mi/hr in 8 seconds. That's 88 ft/s in 8 seconds, or an acceleration of a mere 11 ft/s², or roughly *g*/3. You personally can do better than that by just jumping off a chair (although you'll land at only about 7 mi/hr).

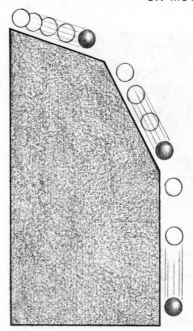

The rate of acceleration of a rolling ball varies with the tilt of the inclined plane. The greater the tilt, the greater the acceleration, which is nonetheless constant for each incline.

The Tuscan Master was by no means done with his inclined planes. He set two together end-to-end, one tilted down, the other up. A ball descending the first continued to roll faster and faster until it swept past the bottom; ascending the second plane, it slowed until it came to rest at very nearly the level from which it was released. Almost regardless of the tilt of the second surface, the ball rolled to a stop at much the same level, the same vertical height.

But why didn't it climb back to exactly its original height? That was a superb question, involving the kind of minute deviation a lesser mind would have overlooked. Galileo studied the matter and concluded that the ball was slowed by *friction*—by rubbing against the plane. To prove the point, he devised a pendulum experiment (see figure), in which friction was greatly reduced. This time the swinging bob rose even closer to the height at which it had been released than had the rolling ball.

Going back to the double incline, he did something next that was simply beautiful. Galileo gradually lowered the second plane so that the ball traveled farther and farther before coming to rest at its original height. And then he asked the crucial question: What would happen if the second plane was made *perfectly frictionless* and *horizontal*? Why, of course, the ball would roll on forever, neither

Inertia

The ancients derived their principles from inviolate basic premises. My theories find confirmation only in their results. In this way, however, it is possible to attain to a degree of probability every whit as valid as strictly logical proof. That becomes so when the results harmonize with reality, and especially when numerous phenomena may be explained by a single theory and when previously unknown phenomena may be predicted and afterwards corroborated.

CHRISTIAN HUYGENS

A portrait of René Descartes, painted by the contemporary master Frans Hals.

An illustration of the logic that led Galileo to the law of inertia.

speeding up nor slowing down! It would sail away with "a motion which was uniform and perpetual." He had gone beyond the bounds of the experiment because, however small the friction might be, he could not get rid of it altogether, and yet he could see past that limitation to the very heart of what was happening! At last, that mystery of motion was revealed—the *law of inertia* was his.

Despite this brilliant insight, he got lost (and with very good reason) in worrying about how an infinite horizontal plane could exist on a spherical Earth. In the end he wrongly concluded that this perpetual motion applied only to circular systems. In one respect, though, he was right. If his frictionless track had run all the way around the planet, the ball, once started, would actually keep revolving about the Earth endlessly, but only because gravity would keep it from sailing off. He rather liked that idea since it seemed to explain the circular orbital motion of the planets, a vision he was totally committed to. Nonetheless, the essence of the concept was sound, and it persisted: *A body in motion, left to its own, uninfluenced in any way, tends to remain in uniform motion.*

The correct formulation of the law evaded several of Galileo's contemporaries until it was finally set right—finally *linearized*—by a professional soldier turned philosopher and mathematician, René Descartes. And this he did, apparently quite independently, not by experiment but by conjecture predicated on an untenable metaphysical conception of God and the universe. In his *Principles of Philosophy* (1644), he maintained "that every body which moves tends to continue its movement in a straight line."

The sanctity of circular motion had bound the cosmos into a finite ball for 2000 years, but now matter could sail out in straight lines forever, out into the infinite universe of Cusa and Bruno.

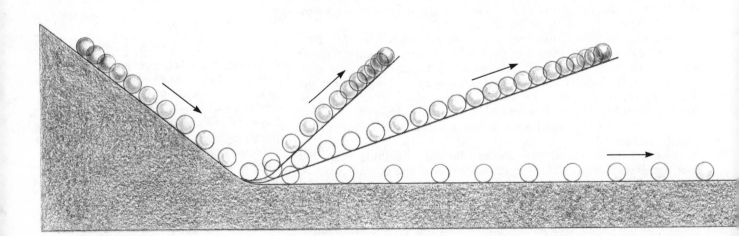

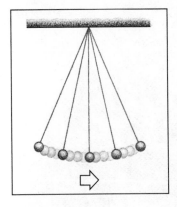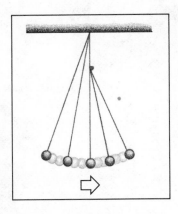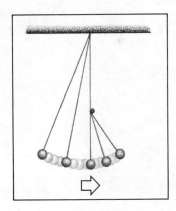

Galileo's pendulum demonstration, showing the bob rising to almost its original level at the end of each swing, whether interrupted or not.

Descartes had intended to publish his first work in 1633, but when he heard of Galileo's fate in Rome, he withdrew the text, not wanting to risk its suppression by the Church. Still, some thirteen years after his death, he joined Galileo and Copernicus on the *Index Expurgatorius*—the Church's list of banned books.

The principle of inertia was profoundly revolutionary; it put aside the millennia-old commonsense conception that motion needs pushing. But then again, common sense exists within an established world view and changes as that perception changes. There is nothing more commonsensical now than the realization that a bowling ball, hurtling down a polished lane, sails on unpushed.

As well you might suspect, this "law," grandly stated, is certainly an idealization. There are no places in the universe where an object can actually move totally free of all external influences, and surely the idea of infinite straight-line motion is quite unrealistic. Despite these practicalities, we believe that matter, once set in motion and thereafter devoid of external influences, will continue to move uniformly and in a straight line. From this distillation of an essential truth, which cannot be directly proved true in this universe, we can "understand" a tremendous range of observed phenomena, and that is its ultimate justification.

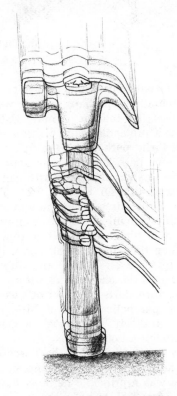

A carpenter will periodically tighten the head on a hammer by slamming the tool on the ground, handle first. The massive head continues to move down somewhat after the handle is jolted to rest, thereby wedging itself firmly in place.

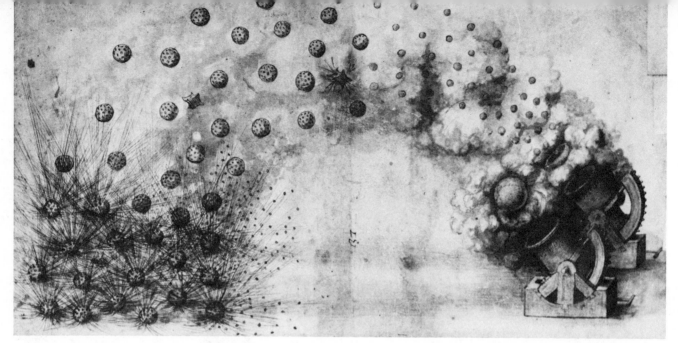

The flight of exploding mortar shells, drawn by Leonardo da Vinci about 500 years ago.

Projectiles

The study of ballistics seems always to have held a particularly compelling fascination—perhaps because we spend so much effort bombing ourselves. Be that as it may, Galileo Galilei was the first to realize that a projectile, sailing through a nonresistive medium, follows a parabolic path under the influence of gravity. (Well, again, not quite—the path is really only approximately parabolic, provided the flight is so short that the Earth can be taken as flat, but we will come back to that in Section 5.2). This remarkable discovery is actually of little direct concern to the modern cannoneer, and it will probably remain so until the military inhabits the moon. The trajectory of a high-speed shell cutting through the air is far more complex than this simple idealization. It takes a powerful electronic computer to track an ICBM; still, an outfielder can effortlessly accomplish much the same thing with a slow-moving baseball. We've all done it with darts or rocks or gushing garden hoses; we've all worked the parabolic magic of little kids and flying pink rubber balls—all intuitively, with never a calculation, just like porpoises leaping through a hoop.

It now seems likely that the Tuscan Master made this discovery quite serendipitously while testing out his principle of inertia in 1608. This was a principle he must have had, along with an understanding of free fall, to grasp the significance of what he saw. Imagine a rock hurled horizontally or even a bullet leaving a pistol in level flight. How would it travel, once free of the initial impelling force? The scholars of the Renaissance saw the trajectory of a projectile as the result of a conflict between gravity and impetus, but because they had no understanding of inertia, their ideas were rendered quite ineffectual.

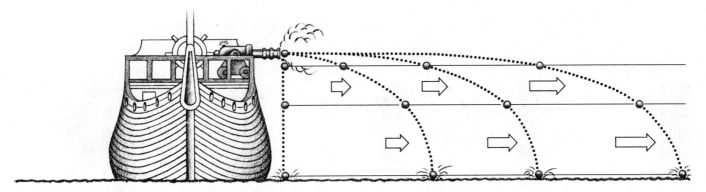

The faster each cannonball is fired, the farther it will go, but all
fall at the same rate and hit the water after the same flight times.

A multiflash photo reveals the parabolic
path of a golf ball.

Galileo astutely perceived the situation as compounded of two
independent motions: a uniform horizontal flight and a vertical gravity fall. The principle of inertia maintains that, left to its own, the
rock should sail off in the direction thrown (horizontally) at constant
speed forever—at least ideally. Moreover, we know that all objects
drop with a constant acceleration. Thus the projectile, free of any
other influences, should fall as it progresses laterally, sweeping out
a graceful curve. The horizontal distance traversed is proportional
to the time of flight $[\Delta d_\mathrm{H} = v\Delta t]$ while the vertical distance varies
as that time squared $[\Delta d_\mathrm{V} = \frac{1}{2}g(\Delta t)^2]$, and that simply yields a parabolic path.

Neglecting air resistance, if you fire a bullet horizontally with
a muzzle velocity of, say, 2000 ft/s and at that instant drop a rock
from the same height as the gun, *both rock and bullet will hit the
ground at the same time!* Despite the fact that the bullet will land
nearly a quarter of a mile away, the only thing causing it to fall is
gravity, and all objects fall at the same rate. During one quarter
of a second, the rock will drop through one foot, as will the bullet,
even though the latter will be some 500 ft downrange by then. After
half a second, both will have fallen four feet; the bullet, still sailing
along horizontally at a constant 2000 ft/s, will be 1000 ft away and
dropping at 16 ft/s.

You might be concerned that the effect of gravity would somehow change because the projectile was moving laterally, but it
wouldn't. For that matter, if you performed the free-fall experiment
inside a uniformly speeding train, the results would be quite unaltered. Another way of appreciating the superposition of these two
independent motions is to envision someone dropping a ball while

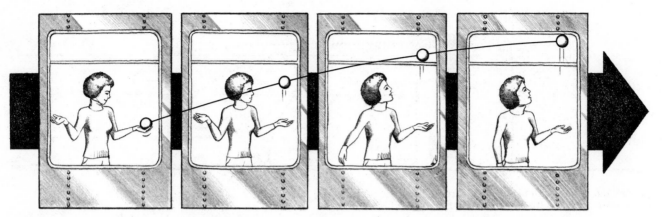

A passenger on a speeding train throws a ball straight up, and naturally enough, she see it move straight up. An outside "stationary" observer, however, sees the ball constantly advancing horizontally *with the train*. To him the ball appears to be following a parabolic path.

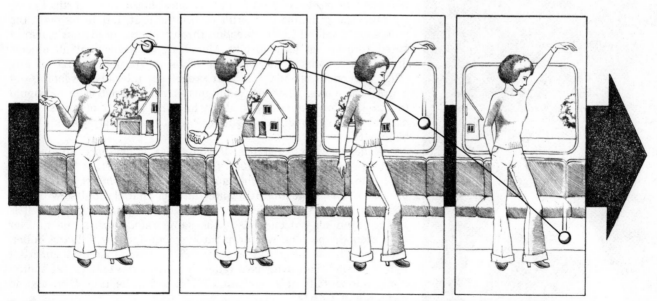

To the passenger on a moving train, the ball falls straight downward, just as you would expect. To an outside "stationary" observer, the ball also travels forward with the train and so appears to follow a parabolic path. The ball is moving forward while she holds it and so continues in that motion after being released.

standing in such a train. To the person dropping it, the object would simply accelerate straight down toward the floor. But to an observer outside looking into the window as the train rushed by, the ball would have the constant horizontal motion of the train in addition to its fall and so would seem to sail along a parabolic course. Clearly, the trajectory of an object is not an absolute; rather, it depends on the relative motion of itself and the observer.

An object thrown straight up slows under the tug of gravity, decelerating constantly as it climbs until the ascent ceases, and for an instant the ball hangs motionless at the point of maximum rise. There is no way for it to move along a straight line and also reverse direction without coming to a stop in the process! Thereafter it falls back downward as if it were simply dropped from rest. Notice that the ball loses speed on the way up at the same rate (32 ft/s²) that it gains speed on the way down. Thus the upward and downward journeys take the same amount of time, correspond to the same distance, and have the same average speed and the same maximum speed. Whatever speed it's fired up with, it comes back down with. If you took a movie of the ball going upward and ran it backward, it would look exactly as though the ball were falling.

When a horizontal component of motion is added to this vertical flight, as when a cannon is fired up at some angle between 0 and 90 degrees to the ground, the ball rises along a parabolic arc. The vertical motion gradually slows until at maximum altitude it terminates for an instant, and the cannonball, still moving forward, begins its parabolic descent. Galileo proved that maximum horizontal distance, or range, is attained for launch at 45° and that altering the angle by the same amount, either increasing or decreasing it from 45°, produces the same range. In other words, for each pair of firing angles that add up to 90°, the projected distance down the course will be equal.

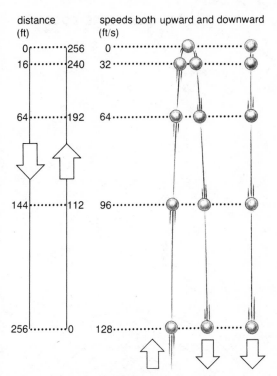

A ball thrown straight up at some speed—in this case 128 ft/s—will (neglecting air friction) return to its launch height at that same speed. In fact, the speeds moving up and down are equal at equal heights.

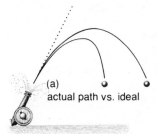

Actual (including air friction) and ideal paths of cannon balls in free flight.

(a) actual path vs. ideal

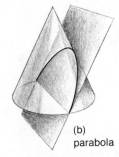

(b) parabola

A parabola is the open curve formed by the intersection of a plane and a cone.

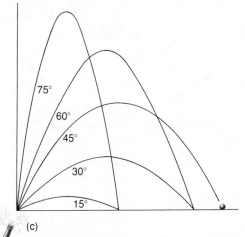

(c)

The maximum range of an artillery shell corresponds to a firing angle of 45°.

3.3 SAID FALSE DOCTRINE

White-shirted in penitence, the old man dutifully dropped to his knees on the smooth floor of the large hall of the convent of Santa Maria Sopra Minerva. Flushed with the heat of the warm summer morning, he began to read aloud softly:

> I, Galileo, son of the late Vincenzo Galilei, Florentine, aged seventy years, arraigned personally before this tribunal and kneeling before you, Most Eminent and Reverend Lord Cardinals Inquisitors-General against heretical pravity throughout the entire Christian commonwealth, having before my eyes and touching with my hands the Holy Gospels, swear that I have always believed, do believe, and by God's help will in the future believe all that is held, preached and taught by the Holy Catholic and Apostolic Church.

Weary and ill, the last grand Renaissance man, humbled before his lessers, recited the litany of his woeful sins:

> But whereas—after an injunction had been judicially intimated to me by this Holy Office to the effect that I must altogether abandon the false opinion that the Sun is the center of the world and immovable and that the Earth is not the center of the world and moves and that I must not hold, defend, or teach in any way whatsoever, verbally or in writing, the said false doctrine, and after it had been notified to me that said doctrine was contrary to Holy Scripture—I wrote and printed a book in which I discuss this new doctrine already condemned and adduce arguments of great cogency in its favor without presenting any solution of these, I have been pronounced by the Holy Office to be vehemently suspected of heresy, that is to say, of having held and believed that the Sun is the center of the world and immovable and that the Earth is not the center and moves.

He had grabbed the golden ring on the cosmic carousel and the Lord Cardinals would have it consumed in the fires of their piety. His old friend Maffeo Cardinal Barberini, now invested with all pontifical power as Urban VIII, was his most furious opponent.*

> Therefore, desiring to remove from the minds of your Eminences, and all faithful Christians, this vehement suspicion justly conceived against me, with sincere heart and unfeigned faith I abjure, curse, and detest the aforesaid errors and heresies and generally every other error, heresy, and sect whatsoever contrary to the Holy Church, and I swear that in future I will never again say or assert, verbally or in writing, anything that might furnish occasion for a similar suspicion regarding me;

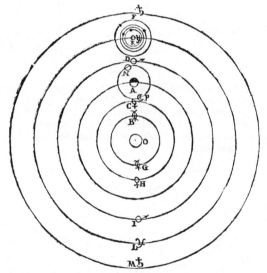

Galileo's version of the Copernican sun-centered system as it appeared in the *Dialogue* (1632). Notice that Jupiter now has four moons and all the orbits are still circular.

---

* Ironically, the general theory of relativity tells us that it's meaningless to ask which *really* rotates, the Earth or the universe. We can say only that there is a motion of one with respect to the other.

They had finally humiliated him, the plotting Jesuits, the compliant Cardinals and the affronted Pope. The Holy Office of the Inquisition had already begun accumulating evidence against Signor Galileo more than twenty years before. And it was Cardinal Bellarmine (i.e., Saint Bellarmine), the same man who had sent Bruno to the stake, who started the secret, irreversible machine of suppression.

The old man haltingly droned on:

> but, should I know any heretic or person suspected of heresy, I will denounce him to this Holy Office or to the Inquisitor or Ordinary of the place where I may be. Further, I swear and promise to fulfil and observe in their integrity all penances that have been, or that shall be, imposed upon me by this Holy Office.

They had threatened him with torture, as was the custom, even though it's unlikely that they would have used it on an aged, sickly man, especially one with his considerable reputation. But the threat suffices—he is not broken, but he recants; he is not purified, but he rescinds; his faith is unshaken, but he is shamed.

> And, in the event of my contravening (which God forbid!) any of these my promises and oaths, I submit myself to all the pains and penalties imposed and promulgated in the sacred canons and other constitutions, general and particular against such delinquents. So help me God and these His Holy Gospels which I touch with my hands.

His ecclesiastic enemies were bitter, relentless, and not beyond falsifying the record in order to silence him. And silence him they would, for there was no reconciliation between Galileo's truth, which could only try to persuade, and the Inquisition's faith, which could also demand. Yet in the end it was more a pitiful clash of personalities than of principles.

The simpleton in Galileo's latest book (*Dialogue on the Great World Systems*) had spoken too many of the Pope's own arguments, and Urban VIII would never forgive him the outrage. As if to make it worse, Galileo had written the book not in Latin, the language of proper science, which few could read, but in Italian, the vernacular of the marketplace. He wrote it for the people, and that was unforgivable.

Rome was the last battleground of the hundred-year-old Copernican revolution, and Galileo was its last political prisoner.

Finishing the recitation of his guilt, he put his signature to the attestation:

> I, the said Galileo, Galilei, have abjured, sworn, promised and bound myself as above; and in witness of the truth thereof I

Pope (exhausted): It is clearly understood: he is not to be tortured. (Pause) At the very most, he may be shown the instruments.
Inquisitor: That will be adequate, Your Holiness. Mr. Galilei understands machinery.

BERTOLT BRECHT
Galileo

have with my own hand subscribed the present documents of my abjuration and recited it word for word at Rome, in the convent of the Minerva, this twenty-second day of June, 1633.

I, Galileo Galilei, have abjured as above with my own hand.

He was confined to his villa, at Arcetri near Florence, under strict house arrest for the rest of his life. There he remained heedless of the injunction not to teach, publish, or even speak to Protestants. Suffering great pain, "totally and incurably blind," he died a prisoner in his own house on January 8, 1642. His body was placed in the basement of the bell tower of Santa Croce, where, at the Pope's insistence, it lay in obscurity for nearly a century.

On Christmas Day of that same year, 1642, in Lincolnshire, England, Isaac Newton was born. The focal point of creative science, which does not thrive in a hostile climate, moved northwestward to the Dutch Republic and Great Britain.

EXPERIMENTS

1. If you stand on a table and extend your arm way up above your head, you can drop something from around 10 ft. An empty soda can conveniently makes a loud thud when it hits ground. If you yell when you drop it, someone else can time the flight. The experiment is rather crude, but it will give you a good feel for what Galileo was up against. Make a list of the sources of experimental error.

2. Try the bullet-and-rock business, discussed on p. 75, using two nickels. Flick one off the edge of a table as hard as you can, just as you drop the other one from the same height. Despite the fact that one will sail four or five feet horizontally, they will both strike the ground with one thud.

REVIEW

Section 3.1

1. Why had Galileo not come out in favor of Copernicanism even before Bruno's death?

2. Who was one of the most outspoken of the early Copernicans?

3. How was Galileo's first book on astronomy, the *Sidereus Nuncius*, received by the public at large? By the way, its title is variously translated as the *Starry Messenger*, the *Star Messenger*, the *Star Message*, the *Messenger from the Stars*, etc.

4. Among the scholars of Europe, who was generally Galileo's staunchest defender?

5. Why did Galileo's scientific colleagues in Italy remain unconvinced of his discoveries, even after looking through his telescope?

6. What did the Tuscan genius discover about Saturn?

7. Kepler's great prestige came not from his laws, which were generally ignored during his own lifetime, but from the fact that he was Tycho's successor as _____ _____ and one of the most successful astrologers in Europe.

Section 3.2

1. Does the time of swing of a pendulum depend on the weight of the bob? What does it depend on?

2. Did Galileo drop weights off the Tower of Pisa?

3. Name at least one person who performed the free-fall experiment before Galileo did.

4. Neglecting friction, what can you say about free fall?

5. What is meant by the term *terminal velocity*?

6. Did Galileo know about the existence of terminal velocities?

7. Why don't hailstones hit the ground at very much higher speeds than they do, considering the great distance through which they fall?

8. Was Galileo the first to realize that heavy things fall in air at a constant acceleration?

9. How did Galileo manage to reduce the acceleration of a falling object to a point where he could measure it?

10. If the initial speed of a uniformly accelerating object is zero, write a symbolic statement of a in terms of the final speed v_f.

11. What did Galileo use as a clock during his experiments with falling objects?

12. What is the numerical value of the average acceleration of gravity at the Earth's surface?

13. How far will an object in free fall travel in one second? In ten seconds?

14. Is g really constant? Explain.

15. Who developed the *law of inertia*?

16. What was Descartes's contribution to the notion of inertia?

17. At what angle should you throw a ball to cause it to land as far away as possible?

Section 3.3

1. Why was Galileo "vehemently suspected of heresy"?

2. Why was Urban VIII particularly annoyed with Galileo?

3. What was the significance of the fact that the *Dialogue* was written in Italian?

4. Galileo spent his last days under house arrest in Arcetri, writing his great work, the _____.

5. The *Dialogue* remained on the Index of prohibited books until 1835. What other indication was there that the Church remained more than a little upset with the Tuscan genius even after he died?

QUESTIONS

Answers to starred questions are given at the end of the book.

Section 3.1

1. What was the significance of the fact that Galileo saw countless new stars with his telescope?

* 2. During his own telescopic observations (1609), the Englishman Thomas Harriot (who at one time was Sir Walter Raleigh's tutor) made maps of the moon that predate Galileo's, but he did not bother to publish them. What was the implication of all these new lunar findings?

3. Although he saw Saturn's rings, Galileo did not recognize what he was seeing. Discuss this as it relates to the considerations of Section 1.3 on making fact. It was C. Huygens much later who first realized he was looking at rings.

4. In what respect was Galileo like a child who, told by his teacher not to chew gum, merely sucks it—and gets in trouble anyway? Despite his worldly wit and cleverness, he seems to have been strangely naïve at times.

Section 3.2

1. What was Galileo's insight regarding pendulum motion, and how does it bear on free fall? By the way, in 1659 C. Huygens showed theoretically that the time of swing was not quite independent of the length of arc swung.

2. Imagine a 100 lb lead weight placed in a sealed box. Will the box fall at the same rate as it would empty? Explain.

* 3. For a ball rolling down an inclined plane, suppose the traveling time is doubled. What will happen to the length of the trip?

4. A polished brass ball rolls a certain distance down an inclined plane. How long will it take to roll twice as far?

5. Explain why the distance covered by an object during its tenth second of fall is greater than the distance traversed during its first second—if, in fact, it is.

* 6. A bus driver slams on the brakes while moving at high speed. Will the passengers be able to detect the resulting change in motion? What law is involved? Explain.

7. A hunter wearing a pith helmet points his rifle directly at a distant monkey sitting on a limb. As soon as the monkey sees the gun flash, it drops from the tree straight down. Would it have been smarter just to sit there? Explain.

* 8. What are the speed and acceleration of a ball at its maximum altitude if you assume it was thrown straight up?

9. Under what circumstances could you actually catch, bare-handed, a bullet fired from a gun?

10. The World War II B-17 shown in the photo has just released a "stick" of bombs. Why do they appear to form a vertical column in the photo? What can you say about the motion of the photographer?

* 11. At what angle must a ball be hurled (neglecting air friction) if it is to land at the same point as when thrown at the same speed but at 20° to the horizontal?

12. Is it possible, while riding a bike at some leisurely constant speed, to bounce a ball (*thrown straight downward*) off the ground and catch it coming straight up, on the move? Explain. Draw a diagram of the ball's path as seen by a kid and his dog watching your crazy behavior.

Section 3.3

1. Kepler, a Protestant, survived the Thirty Years' War, which began in 1618 under the protection of the Catholic emperor. Kepler stayed out of harm's way by casting remarkably prophetic horoscopes for his patrons, not because of his scholarship. Galileo, on the other hand, was brought to trial in 1633 while the war still raged. Discuss the whole political-religious temper of the times as it applied to Galileo's life and work.

2. Legend has it that as Galileo rose from his knees after recanting, he muttered to himself: "Eppur si muove!" (And yet it moves!) What was that supposed to mean?

MATHEMATICAL PROBLEMS

Answers to starred problems are given at the end of the book.

Section 3.2

* 1. A ball rolling down an inclined plane tilted at 30° to the horizontal accelerates at a rate of 16 ft/s² (that is, *g*/2). How far will it travel along the incline in 3 seconds?

2. A toy car accelerated from rest at ½ ft/s² and covered 9 feet before it crashed into a wall. How long did the trip take?

3. Neglecting air friction, how fast will an object be moving and how far will it have fallen after dropping for 15 seconds?

4. The observation deck of the World Trade Center is 1377 feet up. Neglecting air friction, how long will it take to free-fall that far. Can friction be neglected in this case? Explain.

* 5. A car moving at 40 mi/hr drove straight off a sheer cliff, dropping in 3 seconds to the ground below. How high was the cliff? How far from its base did the car land? By the way, the driver, who was found inexplicably smiling, leaped to safety just in time and suffered only minor bruises.

6. A small plaster bust of a politician, who shall go nameless, is thrown straight down from the sixth floor (about 75 ft) of the Watergate Hotel with a speed of 58 ft/s. At what speed will it hit a one-foot-high stack of old issues of the *Washington Post* sitting on the sidewalk (assuming negligible air friction and no interference by the CIA)?

7. A lit firecracker is thrown straight up at a speed of 128 ft/s. How high is it above the ground 5 seconds later when it explodes? How fast is it moving just as it blows up?

* 8. A can of imitation orange juice is flung into the air such that at the end of 2 seconds it's at its maximum altitude, moving horizontally at 64 ft/s. How far from the irate consumer who threw it will it land?

9. Marc Antony is stretched out on the couch, mouth open and waiting. Cleopatra is dashing across the palace straight toward him at 2.213594363 m/s, carrying a bunch of succulent, ripe purple grapes. If she holds one grape exactly 1 m above the level of his head, how far from him will she drop it so that it arcs into his mouth?

* 10. Two 10-meter diving boards are at either end of a 30-meter swimming pool. At what speeds must two clowns run off the boards if they're to collide at the surface mid-pool?

* 11. John Wayne in an old western is standing on top of a train that is moving along a straight run of track at 30 mi/hr while the usual pack of bad guys is in hot pursuit. Cleverly he tosses a lit bomb straight up at a speed of 32 ft/s. When and where will it land (neglecting air friction)? Draw a sketch of the trajectory.

12. Neglecting air friction, a baseball hurled at 45° up from the ground at a speed of 30.7 m/s will attain a maximum altitude of 24 m at a point 48 m down range in an elapsed time of 2.2 s. How far away from the point of release will it strike the ground? Incidentally, this is fine in vacuum, but in air it will be way off. The ball will rise to only about 20 m and hit ground less than 70 m down range, descending at 10° more steeply than it was launched.

The first section of this chapter is a brief introduction to Sir Isaac, the genius and the man. The second segment is the main business. There **mass** is reexamined and **momentum** set out in modern terms; **Newton's three laws** are introduced, and the principle of **conservation of momentum** is examined. The last section is, for the most part, a series of applications of these notions to familiar situations. How do we walk? How do we jump?—that kind of thing. The purpose is to reinforce the new understanding and in so doing get used to thinking of the world around us in terms of these powerful insights— to assimilate them, to make them ours.

CHAPTER 4

Newtonian Mechanics

4.1 LIKE A BOY, PLAYING

Small and frail, his father already dead, Isaac Newton came into the world in 1642, premature and given little chance of survival. It was almost exactly a century after the publication of *De Revolutionibus* and at the very start of the English Civil War. Blood poured over the broad rolling fields not far from Woolsthorpe, where the baby Newton was lord of a modest manor. When he was four, his widowed mother, Hanna Ayscough Newton, a gentlewoman, remarried and went off to live with her new, rich, old husband, leaving Isaac forlorn in the care of his grandmother. Three years later almost to the day, Charles I publicly parted with his royal head, and England became a Commonwealth, kingless and Puritan. Oliver Cromwell, Lord Protector in perpetuity, ruled the British Isles with armed and absolute power. "Merrie England" gave way to the austerity of his dictatorship, to the somber rhythm of his Calvinist devotion.

On the day of Cromwell's death in 1658, a great storm raged across the whole kingdom. And Newton, who was then sixteen, used to tell in his old age how he jumped with the gusting wind that day in order to measure its force. Long afterward, he would quip that it was his first experiment in physics.

Though not particularly precocious he was still better at schooling than farming, and Hanna, widowed again, sent him off to Trinity College, Cambridge. When he left Lincolnshire for the university in 1661, England had a new monarch in Charles II, and Newton a new fiancée in one Miss Storey.

The pious, quiet, suspicious young man from the puritanical countryside arrived at the great university town as it was exuberantly shaking off the repressions of the Commonwealth. He was somewhat taken aback by the raucous, lecherous spirit that prevailed at Cambridge and seemed, for the most part, to have withdrawn into himself—a few visits to the tavern, a few games of cards, but that was it. Newton began at Trinity as a *subsizar*, someone who earned

On facing page, an engraving of Isaac Newton made from a painting by Godfrey Kneller. It gives us a glimpse of a wigless and casual Newton at Cambridge in 1689.

	£	s	d
Drills, gravers, a hone, a hammer and a mandrill	0	5	0
A magnet	0	16	0
Compasses	0	3	6
Glass bubbles	0	4	0
My Bachelor's account	0	17	6
At the tavern several other times	1	0	0
Spent on my cousin Ayscough	0	12	6
On other acquaintance	0	10	0
Cloth, 2 yards, and buckles for a vest	2	0	0
Philosophical Intelligences	0	9	6
The Hist. of the Royal Society	0	7	0
Gunter's Book and Sector to Dr. Fox	0	5	0
Lost at cards twice	0	15	0
At the tavern twice	0	3	6
I went into the country, Dec. 4, 1667.			
I returned to Cambridge, Feb. 12, 1668.			
Received of my mother	30	0	0
For my degree to the College	5	10	0
To the proctor	2	0	0
To three prisms	3	0	0
Four ounces of putty	0	1	4
Lent to Dr. Wickins	1	7	6
Bacon's Miscellanies	0	1	6
Expenses caused by my degree	0	15	0
A Bible binding	0	3	0
For oranges for my sister	0	4	2
Spent on my journey to London, and 4s or 5s more which my mother gave me in the country	5	10	0
Lent Dr. Wickins	0	11	0

April 1669

	£	s	d
For glasses in Cambridge			
For glasses in London			
For aquafortis, sublimate, oyle pink, fine silver, antimony, vinegar, spirit of wine, white lead, salt of tartar	2	0	0
A furnace	9	8	0
Air furnace	0	7	0
Theatrum chemicum	1	8	0
Lent Wardwell 3s, and his wife 2s	0	5	0

his tuition and board by running menial errands and, when called on, fetching a requested quantity or *size* of food from the kitchen. That, too, helped to isolate him, as if his own lack of sociability hadn't been enough.

Three and a half years after his matriculation, Newton took his B.A. degree and a few months later fled Cambridge to escape the Great Plague of 1664–1665. The Black Death had hideously descended over Europe during the fourteenth, fifteenth, and sixteenth centuries, each time growing less severe, less dreadful until the summer of 1665, when it again returned virulently in full terror, raging with such fury that 31,000 souls perished in London alone.

The two years of seclusion at Woolsthorpe were his golden years of incredible creativity, "for in those days," he wrote, "I was in the prime of my age for invention, and minded Mathematicks and Philosophy more than at any time since." In the quiet countryside of his boyhood, the as yet undistinguished young man applied, with marvelous intensity, his remarkable genius. John Wallis at Oxford, the foremost English mathematician of that time, had furthered the analysis of Descartes using infinite series, and in 1665 Newton, extending the work, formulated his *binomial theorem*. By November he had gone on to devise the method of *fluxions*, i.e., the differential calculus. In two months he had "the Theory of colours" with which he would later revolutionize the study of optics. By May of 1666, returning to mathematics, he had constructed the "inverse method of fluxions," now known as the integral calculus. And then back to physics, his true devotion, for the crowning achievement—the *universal law of gravitation*. In one short solitary burst of miraculous creativity he had devised the calculus, uncovered the composite nature of white light, and discovered the law of gravity—and he told no one. Strangely, much of it he simply put aside, not to return to it again for thirteen years. He was twenty-three and a half then. Outside the plague was raging, the Great Fire consumed London, and England and the Dutch Republic were locked in mortal war over a faraway place the British had captured and renamed New York. And in the quiet garden, in the distant world of his own mind, alone, that incomparable intellect flowered beyond the tumult.

Newton returned to Cambridge early in 1667 and was soon elected a Fellow of Trinity College. By then his engagement to Miss Storey had faded into what would be no more than a close lifelong friendship. Newton's master at Cambridge, Isaac Barrow, was the first man to hold the prestigious Lucasian Chair of Mathematics. In 1668, Mercator published a determination of the area under a hyperbolic curve, and Barrow, knowing of his student's interest in infinite series, showed it to young Isaac. It was then that Newton first made known that he had already solved that problem four years before. Barrow, to whom Newton only then revealed his notes on fluxions, stood in amazement at the prodigious accomplishment. Recognizing in Newton "an unparalleled genius," Barrow (who was himself only

39) in 1669 resigned his chair in favor of the young savant. At 27 Newton was appointed to the Lucasian professorship. Now free to follow his own inclinations, he quickly became immersed in the study of light.

In 1672 he sent an account of his work to the Royal Society and awaited, rather confidently, the recognition he felt it deserved. Though the paper was a simple, straightforward descriptive tract setting out the results of his elegant optical experiments, it was widely misunderstood. Newton insisted that he *feigned no hypotheses*—or as he put it in Latin, more picturesque to the modern ear, *hypotheses non fingo*—his conclusions were not speculative, he expounded no theory but merely related observations. The young professor was not prepared for the clamor that results when observation is incompatible with accepted theory, when the long habit of a given perception seems challenged. Even the renowned Huygens, unable to resolve these results within his wave picture of light, preferred to blindly argue against what Newton considered the stubborn facts of unbiased experience. His most energetic detractor was a countryman, the brilliant Robert Hooke: Hooke, whose facile mind touched on so broad a range of phenomena that he viewed all of natural philosophy as his own personal domain; Hooke, who would lamely claim priority over much of Newton's work. Both men were too much alike, inordinately sensitive to criticism, jealous, secretive, and suspicious. They were constantly colliding.

After four years of controversy, four years of reiterating, explaining, and reexplaining, Newton became bitter, disheartened, and hostile. "I see," he wrote in 1676, "a man must either resolve to put out nothing new, or to become a slave to defend it." So enraged was he by Hooke's objections that he refrained from publishing his major treatise on light, the *Opticks*, until after Hooke had died almost thirty years later.

Newton was short and well set, silver-gray even in his thirties, nearsighted, inattentive to his appearance, incredibly forgetful, forever virginal, and understandably, a bit of a nervous hypochondriac. He turned now to the bubbling crucible for, like Hooke and Boyle and Locke, Newton was an alchemist in the old Hermetic tradition.

The next several years were a time of isolation. Barrow and another friend had died, and still another was too ill to correspond with. Newton was almost alone. He stepped away from natural philosophy and busied himself with theology and biblical prophecy. But in 1679 he was drawn back to physics by a conciliatory letter from Hooke, who by then had become Secretary of the Royal Society. The reconciliation was short-lived—Hooke with his marvelous imagination and Newton with his tremendous mathematical power were too covetous of their discoveries, too jealous to be anything but bitter rivals.

Coaxed and cajoled by his friend the young astronomer Edmund Halley, Newton began to write his masterpiece, the *Mathematical*

Louis XIV reigned in France during much of Newton's life. Here the Sun King and his finance minister, Colbert, are visiting the Académie Royale des Sciences. The new observatory founded by the king is going up in the distance.

Principles of Natural Philosophy. The *Principia,* as it's called, set out Newton's theory of motion in what is considered by many to be the greatest single creative accomplishment of all time. He took the law of inertia from Descartes, sensed from Galileo's work the essential role of acceleration, independently formulated the notion of centripetal force, discovered the universal law of gravity, and brilliantly combined all of it in light of Kepler's laws.

When it was finally published in 1687, the *Principia* created a wave of excitement that spread out from the Royal Society and swept across the world. With Newton, we had grasped the grand scheme of the universe; we, humankind, had stretched our power to the stars and beyond. So dazzling and so difficult was the work that few could comprehend it. Yet it was there; it existed, and that was enough for most. It would take a century for the full significance of the theory to permeate the scientific community. Newtonian mechanics would reign unchallenged and supreme until 1905, until Einstein picked up the torch and looked still deeper.

The accession of James II to the throne of England in 1685 sent a tremor of apprehension through the realm. He was an avowed Roman Catholic with an injudicious zeal for "popery" that swiftly led to the Revolution of 1688. Abandoned by his army, James fled to France and with that gesture brought the end of absolute monarchy to the island kingdom. A Convention Parliament was elected to establish the proper succession to the throne and run the affairs of government. Cambridge University sent two Whigs to London to represent it in the House of Commons; one of them was Mr. Isaac Newton. In short order, William and Mary were presented the crown as joint sovereigns. Newton's days as a parliamentarian were brief— not particularly memorable but effective.

These several years were very difficult for him. It was in this period that his mother died of a malignant fever, and a fire in his rooms destroyed many of his manuscripts. And although he sought an official position as some sort of recognition for his loyalty to the new monarch, that, too, was denied him. He became morose; convinced somehow of failure, humiliated and brooding, he again withdrew from the world. Passing his fiftieth birthday in profound depression, he wrote: "I am extremely troubled at the embroilment I am in, and have neither ate nor slept well this twelve month, nor have my former consistency of mind." He suffered a grave psychic disturbance; sinking into delusion, he believed his friends had turned to treachery against him. He wrote insane letters. One accused Locke of trying to "embroil" him with women. The breakdown reached its lowest depths in 1693, but then it seemed to subside.

Several years later, his old friend Lord Halifax secured a position for him as Warden of the Mint, and he moved to London to oversee the recoinage of all monies. In time, Newton became Master of the Mint, a position that allowed him to vent his rage against the counterfeiters and clippers, to pursue, interrogate, and prosecute them and even to bring them to the gallows.

Newton's calculations of the orbit of what would come to be known as Halley's comet.

His archenemy, Hooke, died in 1703, and that year Newton became President of the Royal Society. This was the new London Newton; the aggressive administrator; the wealthy gentleman; the ill-tempered middle-aged godhead of English philosophy; the archetypal, combative, unyielding dictator of science; the knight from Woolsthorpe on a personal quest.

No sooner had death terminated his quarrel with Hooke than he turned to fresh antagonisms. One in particular was with the mathematician Gottfried Leibniz over priority to the calculus. That contest, adjudicated by the Royal Society, was "won" handily by its president in one of the sorriest episodes in science.

In his last years, the truculent prophet showed a sad, sentimental side. As if regaining the sweet naivety of those youthful days running with the wind, he wistfully remarked:

> I do not know what I may appear to the world; but to myself I seem to have been only like a boy, playing on the seashore, and diverting myself, in now and then finding a smoother pebble or a prettier shell than ordinary, whilst the great ocean of truth lay all undiscovered before me.

Sir Isaac, consummate genius, complex and troubled man, died Monday, March 20, 1727, vigorous to the end, in the eighty-fifth year of his age.

4.2 THE CREDO-OF-THE-THREE-LAWS

The *Principia* is a rigorous step-by-step development of the theory of motion in terms of definitions, laws, and propositions. It unfolds with a kind of classical precision, a mathematical precision attained, if at all, by very few previous works in physics.

The very first statement in the book is a definition of *quantity-of-matter*, or **mass.** Of course, the concept by then had been around for centuries (Section 2.2), and as with the word "love," everyone knew what it meant intuitively, but no one could quite put it into words. The Roman poet Lucretius wrote 2000 years ago with remarkable insight:

> Why do we find some things outweigh others of equal volume? If there is as much matter in a ball of wool as in one of lead, it is natural that it should weigh as heavily, since it is the function of matter to press everything downwards.

Fourteenth-century theoreticians of the School of Paris used the concept extensively in their treatment of motion in terms of impetus.

The importance of mass as a fundamental physical quantity further emerged with the determination that the *force of gravity on an object*—i.e., its **weight**—varies with the geographical location of the object in question (Section 1.4). In other words, the weight of a body as measured, for example, by a *spring* balance changes with its distance from the Earth's center, even though the object is otherwise apparently unaltered.

The modern physicist may rightfully be proud of his spectacular achievements in science and technology. However, he should always be aware that the foundations of his imposing edifice, the basic notions of his discipline, such as the concept of mass, are entangled with serious uncertainties and perplexing difficulties that have as yet not been resolved.

MAX JAMMER (1961)

Newton certainly recognized the crucial role that would have to be played by *mass* in a dynamical theory, although his own definition is less than totally satisfying:

> The quantity-of-matter is the measure of the same, arising from its density and bulk conjointly.

Nor is there anything paradoxical about recognizing the importance of something and not being able to define it precisely. It seems to be a general human dilemma that the more significant the notion, the less satisfying the definition. Nonetheless, without an explicit statement of the meaning of *density*—and Newton did not provide one—we are left adrift with all the scholars who have been debating that point now for almost 300 years.

Density was a familiar concept in Newton's time. Boyle had already done his famous work on the compression of gases (Section 14.4), and Archimedes' experiments on floating bodies (Section 13.3) also used the idea. But if we define density in the customary way as the *mass of an object divided by its volume*, then Newton's definition becomes a hopeless circle.

Still, whatever mass is conceptually, it can be measured *relative* to some standard—namely, the standard *kilogram*. Of several ways to do this, the most familiar involves gravity and is therefore less attractive at this point, though simple. A more logically appealing scheme, which has nothing to do with gravity, is more complicated. We will use the former and come back later to the latter—simplicity before subtlety.

Armed with a set of standard masses of various sizes, one needs only to counterpoise the object in question against the appropriate number of standard units on a simple beam balance, a kind of delicate seesaw. Gravity acts equally on both, and so, once the objects are balanced, their masses are equal as well. This is in effect a *how-to-measure-it* or *operational definition* of mass, which rather scrupulously avoids having to evolve a conceptual description. In the end, our working definitions will be operational ones that depend on standards and avoid what cannot be measured. They are dreary but effective.

Rather interestingly, Newton himself did not explicitly make use of the concept of mass in the statement of his three laws of motion.

As the next definition he turned to the specification of some measure of motion. A while before, Descartes had written of the Creator:

> He set in motion in many different ways the parts of matter when He created them, and since He maintained them with the same behavior and with the same laws as He laid upon them in their creation, He conserves continually in this matter an equal quantity-of-motion.

AN UNAPPRECIATED ANTICIPATION
No body begins to move or comes to rest of itself.

ABU ALI IBN SINA, *known as* **AVICENNA**
(980–1037)

For Descartes, *quantity-of-motion* related to the product of matter and speed, but his idea of the essence of matter was volume, not mass. Newton embraced and refined that insight, defining *quantity-of-motion,* or **momentum** (as it came to be known), as *the product of mass and velocity.* This is Buridan's *impetus* physically reinterpreted and very nearly Galileo's *momento* (weight times velocity).

Like his predecessors, Newton realized that the motion of a body must be characterized by more than just its speed; mass must enter the prescription for the *how much* of motion. A firefly and a fire engine, both traveling with exactly the same velocity, respond very differently when we start to change their motion. Furthermore, if each collided with, say, a cannonball, the motion imparted to the ball would again be quite different. Newton framed his dynamics, rather naturally, in terms of this *quantity-of-motion*—its persistence and its change.

The First Law

Having defined the several basic quantities, Newton then set out the "axioms or laws of motion." The first of these was the **law of inertia:** *Every body continues in its state of rest, or of uniform motion in a straight line, except insofar as it is compelled to change that state by forces impressed upon it.*

The Aristotelian understanding was that only the state of rest is enduring; in order for a body to move, a force must be constantly applied. *Now force is recast as an agent of change.* And there is a new equivalence drawn between rest and uniform motion; *altering* either one requires an impressed force, but both, once established, persist interminably in the absence of force.

In fact, rest and uniform motion are only "relatively distinguished," as Newton put it. The peanut is at rest in the palm of my open hand, even though I am sitting in a car speeding along at a constant 50 mi/hr. To be sure, someone "standing still" outside would see the car, me, and the peanut streak past. Even so, the law of inertia clearly holds for the peanut as viewed from both perspectives, both frames of reference. *Rest can be envisioned as simply that particular constant speed equal to zero.* Hence, with respect to my hand, the peanut's speed will remain zero, just as it will remain 50 mi/hr with respect to the outside observer, so long as no applied force alters its motion.

Suppose that you hold a sheet of paper upright by placing your hands flat on either side of it. Now press your hands together, applying an equal force on each surface. Regardless of how great the force exerted by each hand, as long as they are equal, the paper remains at rest.* Forces that are oppositely directed work against

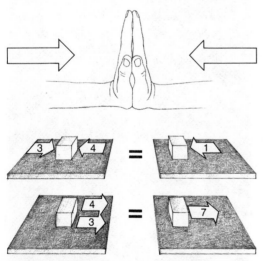

Forces acting in the same or opposite directions add or subtract, respectively.

---

* This is the muscle-building technique of isometrics peddled by Charles Atlas on the back pages of comic books.

each other, partially or (if they are equal) totally canceling each other; forces acting in the same direction combine into one force. With your eyes closed there is no way to tell how many people are pulling against you in a tug-of-war; you experience one resultant force. Forces are *directional quantities* (or as they are called in the trade, *vector quantities*) and cannot simply be added up like ordinary numbers with no regard for direction. It is the *net resulting force* on a body that produces a change in its state of motion.

As long as both teams in a tug-of-war pull equally hard in opposite directions, the net force on the rope is zero, and it remains at rest. If the rope (together with the attached players) starts to move, i.e., accelerates, we can conclude that there is a net force in the direction of the superior team. In the same way, a stone held from falling, at rest, experiences no net force. Galileo knew as much, for he wrote:

> When you support a rock in your hand, what else are you doing but impressing on it just as much of that upward impelling force as equals the power of its heaviness to draw it downward?

The law of inertia applies in all environments at rest or moving at constant speed, which are therefore known as *inertial reference frames*. In other words, if none of the observers are themselves accelerating, they will see persisting in uniform motion all objects on which there are no net forces.

The famous pull-the-tablecloth-off-the-table-without-dumping-the-dishes routine is predicated on the law of inertia. And the purpose of seat belts becomes abundantly clear when "the body in motion tending to stay in motion" after the car screeches to a stop is yours. We don't dawdle on railroad crossings, because freight trains in motion have a nasty tendency to keep on coming long after the brakes are slammed on; nor, for that matter, do we try to catch cannonballs or falling pianos. We are so committed to that first law, the law of inertia, on some unspoken level that if things at rest don't stay at rest when there are no apparent forces at work, most of us are more likely to scream "poltergeist" than to suspect a momentary repeal of inertia. "My God, it moved all by itself!"

Some maintain that the first law arises from the gravitational interaction of all objects in the universe, all drawing on one another. Perhaps it does. In the end we know the way it works, but not what makes it work; we cannot reach in and shut it off, but we believe in it nonetheless.

The state of motion of an object changes with the application of a net force; that much we have already seen. But how? How are the pertinent physical concepts interrelated? How can we quantify that relationship?

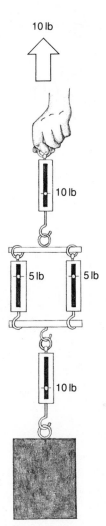

A ten-pound force acting upward is transmitted down the line to the mass.

Eph Equals Em Ay

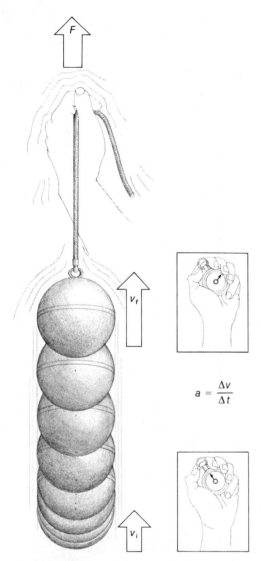

$$a = \frac{\Delta v}{\Delta t}$$

The application of a force F for a time Δt results in a change in momentum. The mass m increases in speed from its initial value, v_i (which may or may not be zero), to some final value, v_f.

Galileo's experiments had shown that the constant force of gravity, i.e., weight, acting on a body (at a given location) produced a uniform acceleration, a constant change of velocity in each successive interval of time. In addition, Lucretius, for one, had long ago suggested a proportionality between mass and weight. Perhaps these were Newton's starting points. It is unfortunate that, unlike Kepler, he wrote nothing of the process by which his insights developed (though he curiously credited Galileo with knowing the first *two* of his laws).

Newton's **second law,** modernized somewhat in its language, reads:

The rate of change of the quantity-of-motion (i.e., the momentum) of a body is equal to and occurs in the same direction as the net applied force.

The momentum of an object (mv) may initially be either zero, if it's at rest, or some finite value; in any case, it will change by an amount $\Delta(mv)$ when a net force F is applied for a time interval Δt. The second law can be recast as an equation in the form

$$\text{force} = \frac{\text{change in momentum}}{\text{time interval}},$$

or symbolically,

$$F = \frac{\Delta(mv)}{\Delta t}.$$

If a body is originally at rest, it will take off in the direction of the net applied force, acquiring a momentum, $\Delta(mv) = F \times \Delta t$. And that is precisely what happens when you pull a wagon, fire a bullet, or kick a soccer ball. As long (Δt) as your foot is on the ball applying force, it gains momentum in the direction of F. Once the ball leaves your foot, there is no more force and no additional change in momentum. The farther back a bow is stretched or the longer the barrel of a gun is, the greater will be the time over which the propelling force acts and the greater the momentum of the projectile.

A force applied to a body already in motion will either increase or decrease the momentum, depending on whether it acts along or opposite to the direction of motion. To slow down the Lunar Excursion Module as it plunged down toward the moon's surface, a retrorocket was fired exerting an upward force, thereby decreasing the downward momentum and speed.

Your hand provides the stopping force on an incoming ball when you catch it. If the ball is soft and deforms, thereby taking a relatively long time to come to rest, Δt is large and F can be small as compared with the force needed to stop an unyielding hard ball sailing along with the same momentum. This is why it's nicer to fall on carpet than on concrete, and why boxers wear gloves. "Roll with the punch, Champ," just reminds "the kid" to move backward, allowing for a longer contact time over which the opponent's fist is brought to rest by his obliging face with less net force.

Jumping with sneakers on gives you a bit longer landing time during which to come to rest while the rubber is compressing. The impact force exerted on your feet will be reduced accordingly. High diving into a pool of water is the same process played out over an even longer time. Similarly, a car slamming into a brick wall will lose all its momentum exceedingly fast, and the impact force exerted on it will be correspondingly large and destructive. Bumpers that compress on impact draw out that stopping time and thus decrease the force of collision.

You may never before have been actually aware that a falling object accelerates, speeding up as it drops. The process occurs quickly, and it's not easy to perceive over short distances. However, while you would surely be willing to have a stone drop into your hand from several inches above, "common sense" would certainly tell you to shy away from trying to catch that same stone after it has fallen a few hundred feet. The impact force would then be considerable, because the momentum would be large, because the velocity would have increased, because the stone would have accelerated —and perhaps you already knew all that on some experiential level.

A rocket blasting off its launching pad is propelled upward via the force exerted on it by the exhaust gases. The longer the engines fire, the greater the total increase in momentum and the greater the maximum velocity attained. As the fuel burns and is depleted, the mass of the rocket decreases, and consequently its speed increases even faster.

Generally, though, we can assume that the mass of an object will be constant. The change in momentum in going from some initial speed, v_i, to a final speed, v_f, can then be expressed as $(mv_f - mv_i)$ or $m(v_f - v_i)$. But this is simply equivalent to $m\Delta v$, and so the second law then becomes

$$F = m \frac{\Delta v}{\Delta t} = ma,$$

force = mass × acceleration.

This is the famous F equals ma, perhaps the single equation most associated with Newton—his hallmark, his signature, if you will—recognized by any physicist anywhere. And yet it appears nowhere in the *Principia*. Strangely enough, that restricted formulation of the second law was not even explicitly given by Newton. It came several decades later in the work of the Swiss mathematician Leonhard Euler. Newton kept mass comfortably buried in the notion of momentum and never succeeded in actually clarifying its meaning, probably because it really wasn't crucial then to go beyond an intu-

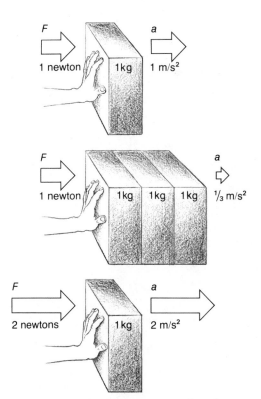

Inasmuch as $F = ma$, varying either the mass or the force will change the resulting acceleration.

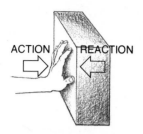

The hand acts on the block; the block reacts on the hand.

Several situations that look different but are equivalent as far as the scale is concerned.

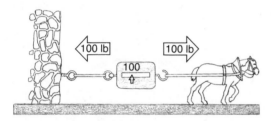

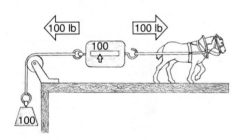

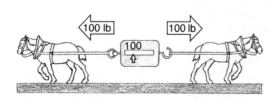

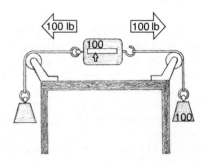

itive sense of the concept. It is a real tribute to Newton's insight that his formulation in terms of the quantity-of-motion is perfectly in tune with modern relativity theory, whereas F equals ma, assuming as it does that mass is constant, is not!

Envision the idealized situation of a standard 1-kg mass (Section 1.4) being pulled along a frictionless plane. This sort of thought experiment is easier said than done, but we could approach it on an airtable or floating around out in space. In any event, if the object is made to accelerate at 1 m/s², the propelling force, from the second law, is

$$F = 1 \text{ kg} \times 1 \text{ m/s}^2 = 1 \text{ kg-m/s}^2.$$

The complicated unit kg-m/s², the unit of force in the *Système International*, is called, for simplicity's sake, a *newton* (N). Thus a force of 1 newton will cause a 1-kg mass to accelerate at 1 m/s²—about the equivalent of a quarter of a pound (that is, 0.225 lb). Without further defining mass, by using the standard kilogram—pulling it about with a spring balance and measuring its acceleration with clocks and rulers—we could calibrate scales and thereafter measure force directly.

Any body of mass m in free fall on Earth accelerates uniformly ($a = g = 32$ ft/s² = 9.8 m/s²) and consequently must be experiencing a constant force, which through long familiarity has become known as **weight,** F_W. The second law gives us the sought-after relationship between weight and mass:

$$F_W = mg.$$

Quite generally, *weight is the gravitational force acting on a body by virtue of its mass*. Geographical changes in the weight of an object just result from the Earth's inhomogeneity and lack of sphericity—g varies slightly from one location to the next. An object far out in space, nowhere near any other material body, will be essentially weightless. Yet if you try to alter its state of motion, make it accelerate, you must apply a force such that $F/a = m$; it may be weightless, but it's never massless. This is sometimes referred to as **inertial mass** because it is manifested by a resistance to the change of motion.

Try to picture yourself floating in the void, effortlessly "holding" a large weightless rock. It will hover in your hand, just as if, back on Earth, it were supported by some invisible cable. But if you try to move it, you will have to strain against its inertia with the same force you would have to exert if you were earthbound.

Let's summarize all this with a practical example. Suppose you come upon a 12,000-lb bull elephant wearing ice skates, stranded on a frozen pond, and charitably enough, you decide to shove the poor beast to shore. Assuming that you need not worry about overcoming friction, how hard must you push to accelerate the pachyderm from rest to 15 mi/hr in, say, 10 seconds? That's a change in speed of 22 ft/s in 10 s, or an acceleration of 2.2 ft/s^2, which isn't very much. The elephant's mass is equal to its weight divided by g, and so the propelling force must be

$$F = ma = \frac{F_w}{g}\,a = \frac{12000}{32}\,2.2 = 825 \text{ lbs.}$$

Get some friends or a tractor and push with 825 lbs for 10 seconds, and the elephant will skate away at 15 mi/hr; use more force and he will just reach that speed sooner. Of course, if you should come across said beastie floating weightlessly in free space, it would still take 825 lbs of force to make it accelerate (or decelerate) at 2.2 ft/s^2, just as it would take 12,000 lbs of force to make it accelerate at 32 ft/s^2. The space-age moral of this tale is that you *can* be squashed between two weightless elephants.

Action-Reaction

Sir Isaac's third principle, his third law of Nature in motion, completes the logical picture of the concept of force. The acceleration spoken of qualitatively in the first law and quantitatively in the second law arises out of impressed force, force exerted *on* the body. Newton defined force, essentially, as *any action that causes a body to accelerate*. But there is an obvious circle here. Whenever a body accelerates, the cause is a force acting on it—by definition. Well, then, how can such a logical system be proved false, if it is false? The answer is, it cannot! Something is missing (Section 1.1), something that ties the notion back to the observable universe.

A body left to its own devices follows the law of inertia. It cannot alter its own uniform straight-line motion; that takes some outside intervention, and that mechanism must now be brought within the discussion. *A body deviates from the first law by the influence of one or more entities external to itself.* Although not necessarily so, it certainly seems reasonable to expect that two interacting objects will *both* be affected by their encounter, that the motions of *both* will be altered. We observe this all the time when things collide. It follows that such objects exert forces on *each other*. Certainly, two identical balls moving toward each other at the same speed must suffer the same alteration of motion on collision—must

exert the same force on each other. But what interaction forces exist in general?

All the plausible arguments don't make a theory; that requires going beyond what is known to what is suspected. Here Newton provides his next creation, the **third law:**

> *For every action force there is an equal and oppositely directed reaction force.*

The interaction of two bodies, however disproportionate their masses, always occurs via an equal *action-reaction pair.* "Whatever draws or presses another is as much drawn or pressed by that other." We tend wrongly to think of force as if it were a singular agent exerted by active entities on passive ones; force is a thing of *pairs.* Whenever we see an object deviate from the law of inertia, we must be able to find (at the other end of the conceptual string) some other body interacting with it via an action-reaction pair.

Many people have very personal (and at times peculiar) ideas about force. It's a common, though erroneous, belief that forces can be exerted only by living creatures. Still others have some unspoken sense that only moving things, living or not, are the force makers. Surely an inanimate automobile, shut off, stopped dead with one of its front tires full on your foot, exerts a crushing downward force. Press on a wall, first lightly, then harder, and watch the flesh on your hand move out and back as the wall's outwardly directed reaction force flattens your hand. Just so, the chair you are sitting on reacts to your downward weight with an upward force that flattens your bottom and holds you suspended two feet above the floor. Raise your feet. If that inanimate chair is not pushing up on you, keeping you from dropping to the ground, what is?

Slap your hands together, moving either or both, and the two will sting equally. Perhaps you've seen a ball hit so hard that its reaction force shatters the bat, or a pistol flung back by the reaction of the bullet, itself thrust forward by the gun.

None of these illustrations *prove* the third law, but they all show how it serves to provide an understandable picture of what is happening. You could argue that a balloon pressed up against a wall is squashed by the pusher, not the wall. Yet it is flattened just as it would be if the wall had *moved* to where the balloon was being

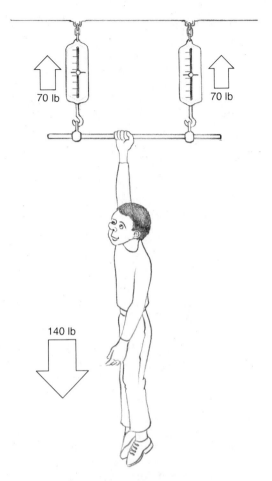

Forces acting in the same direction add in the ordinary way that numbers do.

All the forces involved as a man pulls on a boulder with a rope.

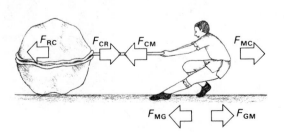

F_{MC} = force exerted by man on cord

F_{CM} = force exerted by cord on man

F_{CR} = force exerted by cord on rock

F_{RC} = force exerted by rock on cord

F_{MG} = force exerted by man on ground

F_{GM} = force exerted by ground on man

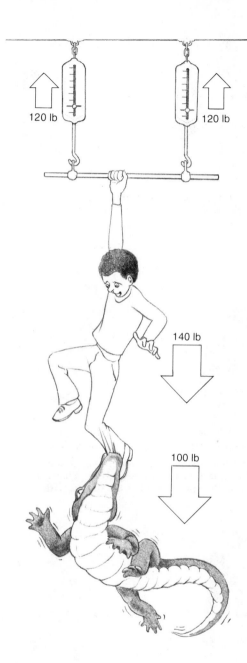

120 lb   120 lb

140 lb

100 lb

held fixed and proceeded to squash it. Force is too often attributed only to the thing that obviously does the moving. If you saw a photo of the wall, balloon, and hand, you would be unable to tell which had moved against the other.

Interaction between two entities means acting and reacting: *Two forces, one on each participant, form the pair.* This cannot be overemphasized; the two interaction forces are always in opposite directions, and they never act on the same body. Your weight (as we will see presently) is due to the Earth's pulling *down on you;* you, in turn, must be pulling *up on the Earth,* however peculiar that may seem.

Tie a heavy cord to a large rock, and tug. You pull on the rope, and it pulls back on you. That force is transmitted along the rope, which then pulls on the rock. The boulder, tugged forward, reacts by pulling back equally on the rope. There are two action-reaction pairs here, four equal-magnitude forces in all: one force exerted backward on you, one forward on the rock, and two opposing forces acting on the cord, tending to stretch it out taut. If all this was occurring in space or on an ice pond, so that there was no friction underfoot, the one force acting on you would propel you backward toward the stone. Under the influence of the single force acting on it, the stone would accelerate forward toward you. By contrast, the rope with no *net* applied force would simply dangle limply.

If this strikes you as strange, remember that in order to make a boat move forward, you can use one of those long poles and push backward on the river bottom, or you can just push backward on the water with a paddle. In either case, the reaction force on the boat is forward. That is, after all, the way we swim—pushing back and being pushed forward.

In the end, the three laws are the inseparable principles of a single theoretical vision. The whole is Newton's triadic poem, which describes the motion of the universe in terms of a thing called force. The theory is as true as the understanding it provides is true to our perceptions, and it continues to be just that.

Conservation of Momentum

The third law leads directly to the fundamental principle of **conservation of momentum** and may, in a restricted sense, be thought of as equivalent to it. Recall that Descartes had framed a wondrous spiritual universe in which the Deity "conserves continually" the quantity-of-motion (p. 91). In other words, the total momentum persists unchanged and will continue to be preserved forever. Charming, isn't it, to see one of the pillars of modern physics bubbling out of the brew pot of metaphysics? In any event, for a very long time, humankind has considered matter as indestructible and therefore conserved. Francis Bacon reiterated the idea in the sixteenth century:

The sum total of matter remains always
the same without addition or diminution.

ZERO TOTAL MOMENTUM

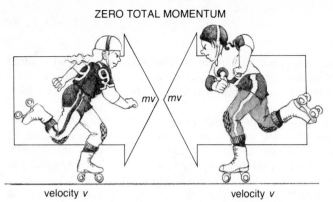

velocity *v*   velocity *v*

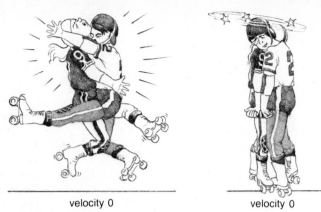

velocity 0   velocity 0

Here the skaters flying toward each other have equal masses and speeds, their momenta are equal and oppositely directed. The total momentum

is therefore initially zero, and it remains zero after the collision, as they simply come to rest and stay that way.

TOTAL MOMENTUM *mv*

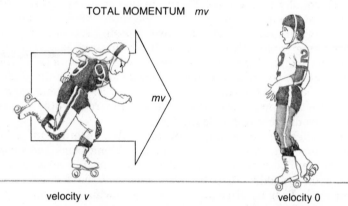

velocity *v*   velocity 0

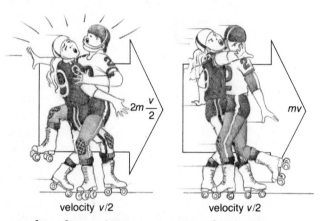

velocity *v*/2   velocity *v*/2

Here number 9, "the Brooklyn Smasher," initially has all of the system's momentum. After impact, since the total momentum is constant and they stay

together, the moving mass is twice what it was, and therefore the speed must be halved. The two sail off at half the speed number 9 came in with.

TOTAL MOMENTUM $2\,mv + mv = 3\,mv$

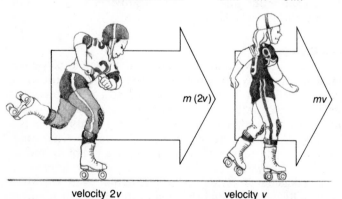

velocity 2*v*   velocity *v*

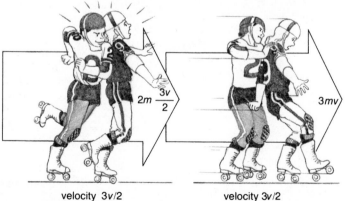

velocity 3*v*/2   velocity 3*v*/2

Taking revenge, "the California Cannonball," number 2, comes flying at "the Smasher," number 9, at twice 9's speed. The total momentum is

$m(2v) + mv$. After the collision the two skaters sail off together ($2m$) with a speed ($3v/2$) that conserves the initial momentum ($3mv$).

Why shouldn't other quantities display this same sort of permanence? Newton, who was educated in a time when Descartes's philosophy reigned supreme, brought the almost spiritual axiom of momentum conservation within the domain of his own mechanics.

Recall that velocity is a directional quantity. Spencer Tracy in an old movie running down the aisle toward the back of a moving train at the same speed with which it's chugging forward can manage to be motionless in front of Katharine Hepburn, who is tearfully waving good-bye from the platform. His velocity is equal and opposite to the train's velocity, and the sum is *zero*. Momentum (which is actually mass times *velocity*) is also a directional quantity—something Descartes never realized although Newton certainly did.

Since interacting objects exert action-reaction forces on each other, by the second law the resulting changes in momentum must be equal and opposite. Just as with forces, opposed momenta tend to cancel. It should therefore be apparent that

the total momentum of a system of interacting masses must remain unaltered, provided no external force is applied.

The momenta of individual members of a system certainly may change, but each change is accompanied by an equal and opposite change in the momentum of the interaction partner. If the whole universe is taken as the system, there are no external forces, and the total momentum must be conserved.

Imagine a closed box containing a half dozen billiard balls all at rest, all floating out somewhere in space. Now if you shake up the system (i.e., the box and its contents) by applying an *outside* force, the momentum will surely change. Once all external force is removed, the newly established momentum of the system will presumably remain constant forever, despite the fact that the balls are now flying about inside the box, smashing into the walls, and rebounding off one another. Unhappily no idealized experiment of this sort could actually *prove* the law of conservation of momentum, but observations confirm it all the time. You already know that there are no places in the universe free of outside influences, and even if there were, waiting "forever" might be tedious. Besides, the box would soon get warm from the banging around and would radiate, but we won't worry about that here (see Question 8.4-3). Colliding real macroscopic bodies do not *exactly* conserve momentum anyway, although the disparity (which can be "accounted for") is exceedingly small and practically negligible—witness Minnesota Fats with a pool stick.

On an atomic scale, particles colliding and rebounding always seem to do so in a manner in which momentum is conserved. We believe in that tenet as a general principle that universally applies, primarily because *it has not yet failed us*. That is surely marching pragmatism, but it's the way the game must be played.

To see how the logic of this new principle works in practice, consider the simple system composed of a bullet in the chamber of a

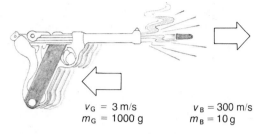

$v_G = 3 \text{ m/s}$ $v_B = 300 \text{ m/s}$
$m_G = 1000 \text{ g}$ $m_B = 10 \text{ g}$

Since the total momentum was zero before the gun was fired, it must be zero afterward as well ($m_G v_G$ directed to the left must equal $m_B v_B$ to the right).

pistol. With both components at rest, the total initial momentum is clearly zero. When fired, the bullet flies out with a momentum ($m_B v_B$). If we assume that the momentum of the escaping gas is negligible, the bullet's momentum must be *equal and opposite* to the momentum ($m_G v_G$) acquired in recoil by the gun if the total is still to be zero. Thus the gun slams backward, and if the experiment were performed while hovering in space, both it and the bullet would sail off in opposite directions. Since the mass of the bullet is small, its velocity will be large, whereas the comparatively massive gun will recoil with a correspondingly small velocity.

A rocket works in just that way. A very high-speed stream of exhaust gas rushing downward thrusts the vehicle up as if it were recoiling from a torrent of emerging bullets. It is completely erroneous to believe that a rocket is propelled by pushing on the ground or the surrounding air. If that were so, it certainly could not accelerate in vacuum—something rocket-driven space vehicles do quite commonly, quite well.

Since billiards was already popular in England in Shakespeare's time, it is not entirely unlikely that our seventeenth-century masters may have idled away a moment or two with those marvelous ivory spheres. Although contemporary pool-hall virtuosos rarely know it, the game is a physicist's delight—a green momentum playground.

There's the deep-red seven ball, the target, at rest. Carefully you strike the white cue ball dead center, and it sails down the table with a speed v_Q. Of course, the harder you hit it, the greater the force and (via the second law) the larger the momentum change (from 0 to mv_Q). Prior to the impending head-on impact, the total initial momentum is mv_Q, directed, say, due north. Since the masses of the balls are equal, the cue ball in a head-on collision slams to a complete stop, transferring all its momentum to the seven ball, which then hurtles straight on, due north, with a speed v_7. Since momentum is conserved, mv_Q must equal mv_7 and $v_Q = v_7$, provided there are no external forces, such as somebody reaching in and grabbing the ball. This is the simplest kind of shot. Off-center collisions result in the balls flying out in different directions, but momentum is still conserved. Hitting the cue ball off center will cause it to spin and do all sorts of marvelous things, but alas, this is not a pool-hall primer, so we must stop here.

Soon after its founding in 1662, the Royal Society began actively encouraging research to determine the nature of the collision process. Hooke performed demonstration experiments at their meetings. So did the architect Sir Christopher Wren, who had returned to his earlier studies on percussion. John Wallis, the mathematician, looked into theoretical aspects of the problem, and Huygens, too, was contacted about his findings. Their independent efforts were all in agreement, although it was Dr. Wallis who first published a modern rendition of the law of conservation of momentum in the *Philosophical Transactions* of the Society (1668). Newton recounted these efforts of his contemporaries in the *Principia*. Whether or not this was the root source of the third law we will probably never know.

The credo-of-the-three-laws is a guide to the physical universe. Ideally, we should be able to apply that theoretical system to all observable interactions, but it has its limitations, as one should expect. *Classical mechanics,* as the system is called, is most applicable to the everyday encounters of ordinary life, where it works superbly well. Its shortcomings are appreciable only at the extremes: in the minute domain of the atom, on the vast scale of the stars, and for motions at tremendous speeds.

Let's now consider some examples of how the three laws work together to make commonplace occurrences understandable.

Stand up for a moment and get a sense of the tugs and strains, the dull pressing feeling on the flattened pads of your feet. The weight of a standing body acts down on the floor, and the floor, in turn, pushes up with an equal and opposite reaction. The net external force acting *on* the body is the difference between the downward draw of gravity and the upward reaction force of the floor. If that net force is zero, the second law demands that the object in question continue to remain motionless in the vertical direction—and there you are, standing still.

Suppose that for structural reasons the floor just cannot exert as large a lift as your weight. Well, then, it simply cannot support you. The nonzero net force acting on you is then straight down, and down you will accelerate ($F = ma$) through the floor, just as if you were trying to stand on a paper box or walk on water.

The floor does not match your weight with an exactly equal opposite force out of inherent cleverness or a deep desire to conform to the third law. It simply distorts underfoot, much as if you were standing on a mattress or a taut horizontal rubber sheet or a trampoline. The greater your weight, the more it stretches, building up a counterforce via springlike atomic interactions within the rubber (or the floor). The deeper you sink, the greater the reaction until that lift cancels the applied force, the body comes to rest, and the material ceases to distort any further—or failing that, rips apart. A strong, rigid floor needs to sag only very slightly, almost imperceptibly, in order to generate an appreciable reaction. Wooden floors often betray the process by squeaking underfoot, whereas concrete hardly distorts at all. The soft, damp sand on the seashore compresses several centimeters before it builds up the appropriate reaction, and quicksand is generally avoided because of its inability to generate a sizable counterforce.

On the other hand, if you wish to accelerate vertically, straight up off the floor, the net *external force* on your body must obviously be nonzero and directed upward. You cannot just reach down, grab your belt, jerk skyward, and expect to accelerate. You yank up, your

4.3 ALTOGETHER NOW

It is the same for me with mechanics as it is with languages. I understand the mathematical laws, but the simplest technical reality demanding perception is harder to me than to the biggest blockheads.
KARL MARX (1818–1883)

Who Only Stand

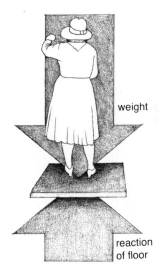

weight

reaction of floor

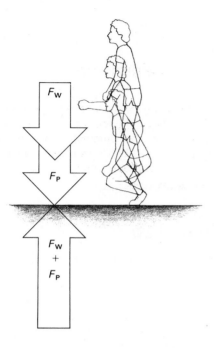

F_W

F_P

F_W
+
F_P

belt yanks down, and nothing much happens. Your body, belt and all, is the system, and these are equal and opposite *internal forces* acting *within* and not *on* that system. Similarly, a person cannot sit inside a stalled car and push it forward by pressing on the dashboard, nor can one stop a dropping elevator by pressing upon its ceiling from inside.

OK, then, if you still want to accelerate upward, simply shove down on the floor with a force greater than your weight; i.e., add some leg-muscle push (F_P) to the ordinary body load (F_W). The floor's upward reaction will then equal this combined thrust ($F_P + F_W$). The floor makes no distinctions between the origins of the forces applied to it—it simply matches them in reaction. This is an external force acting upward on you. Since the only other external influence is your weight, F_W, acting down, the net force (equal in size to F_P) acting *on* you is skyward, and up you'll go. Bending the legs prior to lift-off will extend the thrusting time, (Δt), thereby increasing the change in momentum, $\Delta(mv)$, and consequently the ascent speed. The process is known as *jumping*, and skeptics can try leaping up from a bathroom scale to see whether it reads ($F_P + F_W$) at lift-off. This sounds like a rather involved business, and in fact it is. If you ever see children about a year and a half old trying to jump before they have *learned* how, it will underscore the complexity. Mimicking adult body motions is not enough to get them off the ground despite all their lurching.

Walkin' Our usual means of slow horizontal propulsion over short distances is known as *walking*. The technique depends on somehow causing one's body mass to accelerate in a desired horizontal direction. Evidently from the second law, a properly directed external force that will drive the walker from rest is called for. But what is that? What external force exerted *on* you propels you forward when you begin to walk?

Ordinarily, two solid objects in contact will experience some degree of drag when they are moved past each other. The molecules of the two surfaces tend to cling to one another at high spots, like tiny microscopic welds that need to be broken if sliding is to occur. The resistive force that must be overcome is the so-called frictional drag, *which is always directed in such a way as to oppose an impending or actual motion.*

To start to walk north, one needs only to push back in a southerly direction on the floor, that is, to draw either foot backward, exerting a horizontal force such that there is a forward frictional thrust that will impel you to acceleration. Friction opposing the backward motion of your foot propels you forward. In other words, you push on the floor, and the floor pushes on you. If you don't think you are actually pushing backward as you walk, just think about the cloud of dust a runner or a horse kicks up behind. The backward thrusting process is more obvious when you are crawling on all fours

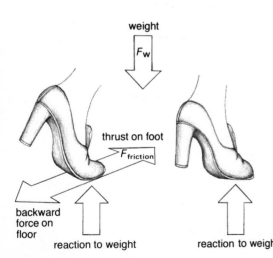

weight

F_W

thrust on foot

$F_{friction}$

backward
force on
floor

reaction to weight   reaction to weight

and your arms do the pulling. Keep in mind that this first operation, *pushing* back as opposed to unimpeded sliding, is impossible without friction.

The maximum amount of friction that can be developed in any given situation is determined by what the individual surface properties are and how hard the two surfaces are pressed together. Slide a chair a few feet across the floor. Now have someone sit on it and push it back. The increased friction will be obvious.

A runner at the starting line must acquire the greatest possible acceleration. That requires a large frictional reactive force, which the runner attains by initially pressing downward, digging in, while thrusting each foot back across the ground. Shy of exceeding that maximum (whereupon your feet would just slip), you can shove backward on the ground with any desired force, and there will be an equal and opposite frictional reaction. The propulsive thrust that can be generated by a smooth, newly waxed surface is so small that one can begin to walk across it only very slowly without slipping. By contrast, wearing cleated shoes that dig into soil will allow much greater action-reaction pairs to be developed and therefore larger accelerations.

Of course, if the surface you are trying to walk on is well greased, no appreciable friction will develop; you can't push back on the floor at all, and except for a lot of fast shuffling in place, you won't do much traveling. Nor could you walk while hovering above the floor of a space ship; that's why science-fiction writers like magnetic boots. In the same way, the old step-on-the-banana-peel number is a classic zero-friction action-reaction ballet.

Note that if momentum is to be conserved, whenever you walk forward, the Earth must move backward. But its mass is so large that any such motion is imperceptible. That's obviously not so for a rowboat, which does indeed accelerate backward when a passenger begins walking forward.

Once you get started walking or running, things become a bit more complicated. The problem arises when you have attained the desired speed and would like to level off at that constant rate. Even though your body as a whole is moving uniformly, your legs must accelerate and decelerate all the while if they are to end up beneath you to support your weight. If either foot hits the ground while moving backward, there will be a backward force on the floor and a forward reaction, which will cause an unwanted acceleration. To avoid that, each foot is ideally planted straight down. That's why you make nice neat footprints trotting (at a constant speed) on damp sand.

This concept may become a bit clearer if you imagine a wheel with a dot of yellow paint on its rim rolling along a perfectly flat surface at a constant speed. As the dot comes around, it will reside for an instant on the forward-most portion of the wheel out in front of the central axis. After that it gradually moves down and *back*, de-

The motion of a point on the circumference of a uniformly rolling wheel.

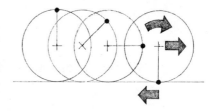

A comparison of the motion of a point on a wheel and the foot of a runner.

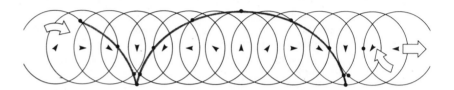

creasing the distance it is in front of the axis as the wheel turns. At the instant the dot hits the ground, it will be moving backward with respect to the center at the same speed as the entire wheel, axis and all, is moving forward. In other words, with respect to the ground, just as the dot touches, it will be *motionless!* A car traveling at a constant speed leaves clear, unsmeared tire marks in damp sand. Ideally the region in contact with the road pushes neither forward nor backward; it does so only when braking or accelerating, respectively.

While traveling at a constant speed, a runner's foot actually moves through space with a motion that very closely matches the path of our yellow paint dot on the uniformly rolling wheel. The law of inertia makes it clear that the runner should continue at a constant speed in a frictionless environment without having to exert any horizontal force at all. In reality, air drag must be overcome, but at ordinary walking speeds that requires only a few ounces of force. Compared with the vertical forces, which are comparable to the person's weight, the backward thrust needed to maintain a constant speed is remarkably small.

Incidentally, tire treads are designed to maintain friction even on a wet road. When the thrust comes not from the ground but, say, from a jet engine, as on an airplane or dragster, the tires only support the weight, and they can therefore be made bald or specially curved to reduce friction as much as possible.

The Old Gray Mare Another famous example of action-reaction is the problem of "the old gray mare." This one presumes confusion to begin with and must be stated with an air of bold arrogance in order to further disarm

The old horse and wagon "paradox."

the victim. It is also usually introduced as a paradox, but that's a ploy to imply legitimacy. Actually it's a simple fake-out that arises from muddled conceptions, not a true paradox. Here it is: If a horse pulling a load exerts a force (F_{HC}) on the cart and the cart exerts an equal and opposite reaction force (F_{CH}) on the horse, don't these cancel, and how, then, does either horse or cart ever manage to move?

The confusion arises only if we overlook the fact that the forces in a reaction pair (F_{HC} and F_{CH}) never act on the same body. Note that F_{HC} and F_{CH} are internal to the horse-cart system and therefore do not directly affect its motion as a whole. Neglecting air friction, there are only two *external* influences acting horizontally: the motive force, namely, the ground thrusting forward (F_{GH}) in reaction to the horse pushing backward, and an impeding friction force (F_{FC}) on the wheels of the cart. So long as the first of these is larger than the second, horse and cart accelerate as one. In other words, since $F = ma$,

$$(F_{GH} - F_{FC}) = (m_H + m_C)a.$$

Alternatively, if you choose to worry about the cart alone, that is, to take *it* as the system, then it will accelerate provided that the net force on it ($F_{HC} - F_{FC}$) is nonzero. Similarly, the *horse* will accelerate forward provided that ($F_{GH} - F_{CH}$) is nonzero. All three approaches to the analysis lead to equivalent results.

As soon as the desired speed is reached, the horse is asked to slacken off a bit, so that $F_{GH} = F_{FC}$. No net force then exists, and the old gray mare trots on, pulling the cart at a constant rate. Ideally, F_{FC} should be zero, but unfortunately the wheels sink into the road a bit, and some friction is unavoidable. By contrast, if the load is

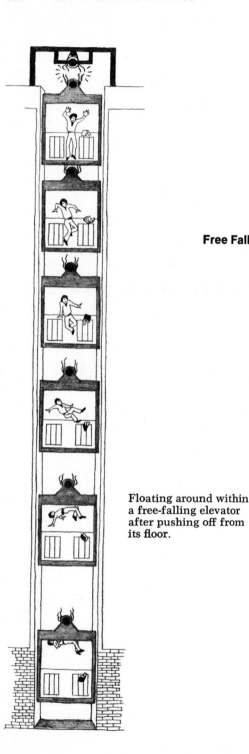

Floating around within a free-falling elevator after pushing off from its floor.

great and the road is soft, it's possible for F_{FC} to equal F_{GH} initially, and then the poor horse, however long it struggles, will not budge the cart. That's when you get off and push.

Incidentally, those wooden rods (thills) that link the horse to the cart serve at least one interesting purpose. They keep the wagon from running over the horse whenever the latter decelerates. Somewhere, very early, people intuitively sensed the significance of the first law.

Note that here, as in the example of walking, friction is the motive force. Similarly, an automobile is moved by the reaction force of the road on its drive wheels, just as a locomotive is thrust on by pushing backward on the rails—all frictional.

Free Fall   Stick out one hand, flat out, palm up. Place a small but appreciable weight on it, such as a paperback book or a wad of keys. Now drop your hand downward, accelerating it at exactly 32 ft/s². Yes, you can, at least for a moment! Think about it. That is the rate at which the book would fall if it were unimpeded; so let it fall freely and have your hand simply precede it down. Under the influence of just its weight, the book accelerates at g. If your hand supports it at all, i.e., pushes up, the net force on the book will be less than F_W, and it will accelerate downward at less than 32 ft/s². Evidently, as you drop your hand faster and faster, just at the point where it accelerates at g, you will no longer feel the book pressing down; it will have become effectively weightless in free fall. If the book were on a scale that in turn rested on your hand and you repeated the g-drop, both book and scale would hover, the latter reading zero. One can accomplish the same thing by walking off a chair while holding the book, and I (having just now done that) can attest to the book's subsequent momentary loss of "weight."

Suppose you were in an elevator that, because its cable had broken, was falling freely; i.e., its acceleration, a_E, equaled g. Not only would the book again hover weightlessly in your hand, but you would be effectively weightless as well. The floor, dropping out from beneath your feet, would exert no upward response unless you pushed down on it with a muscle-generated force; if you did, you would float up within and with respect to the elevator. You could even open the ceiling hatch and thrust yourself up out of the elevator completely. Despite that, you would of course be plunging downward faster and faster toward the Earth all the while, not so fast as the elevator but fast enough. You might try to leap up with the same speed as the car was falling just at the instant before it hit bottom, but unhappily that's not so easy to do.

Another way to appreciate what is happening is to imagine yourself standing on a chair in the free-fall elevator. If you stepped off that perch, you would certainly not approach the floor since it would be dropping away as you dropped. In this instance, with no relative acceleration between passenger and elevator, there can be no measured weight, regardless of whether one stands on or hangs from any sort of scale.

Ordinarily a descending elevator accelerates downward at less than g and for only a second or two before reaching its operating speed. For those few seconds your weight is effectively reduced, and you certainly feel it. If you stepped off the chair during this phase of the ride, you would drop to the floor with a relative acceleration equal to the difference $(g - a_E)$. Your mass times that total acceleration would correspond to your momentary reduced "weight," as measured by a scale bolted to the elevator. That's also the force the floor would have to exert on you as you walked around.

Now imagine that our elevator is fired upward, accelerating for several seconds and then coasting on in free flight. If you stepped off that ever present chair during the first few seconds of powered ascent, gravity would cause you to accelerate downward at g, as always, while the elevator floor was accelerating upward to meet you at a_E. The result is a combined relative acceleration of $(g + a_E)$. Your effective weight (once you alight) would increase, just as it would increase in an ordinary ascending elevator while it was accelerating. This is the source of the space-age g-force jargon. People inside a rocket accelerating upward at, say, $3g$ experience an additional downward inertial force equal to three times their Earth weight.

During the next phase of unpowered ascent, the elevator-turned-rocketship continues to climb, slowing down as it rises because of the tug of gravity. If you get back up on the chair and step off, again you will accelerate earthward at g, but so will the elevator, now itself accelerating downward at g (or equivalently, decelerating in the upward direction of motion). Remember, assuming g to be constant, that the elevator is slowing down at 32 ft/s^2, even though it is ascending. It is sailing upward but going slower and slower, decelerating all the while. There are no constraints, both passenger and craft are in free fall, and you again hover as if weightless. The effect can be reproduced by merely jumping up with a book in your hand—a trampoline might help. As soon as you leave the floor, the book becomes effectively weightless.

We have all seen the astronauts floating about on the way to and returning from the moon. They were not gravityless but they were effectively weightless—freely falling.

Mach and מצה

The word *mass* derives from the Latin *massa* meaning a blob of dough, which in turn comes from the Greek $\mu\alpha\zeta\alpha$ (maza) or barley cake, which may have sprung originally from the Hebrew מצה (matzoh) or unleavened bread. That much is easy, but in a physical context, what is mass?

Two masses being shoved apart, accelerated by a spring.

Mass manifests itself both inertially and gravitationally.

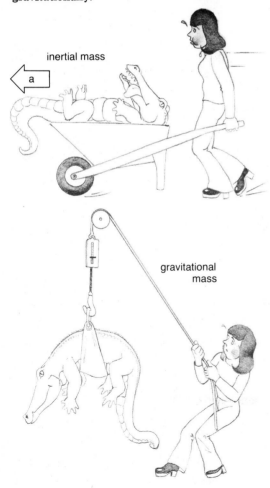

inertial mass

gravitational mass

In the nineteenth century, the Austrian physicist-philosopher Ernst Mach proposed an ingenious thought experiment for determining relative mass, free of any considerations of force, including gravity. The idea was to use an operational definition to recast the Newtonian formulation in a more logical development. The interdependence of the concepts of force and mass and the lack of a practical definition of the latter left both notions unsatisfyingly vague. Newton's own quantity-of-matter, defined in terms of density and volume, seemed to Mach to be framed in the logic of circles, and he rejected it as useless. Instead he turned to the second law.

Two masses, M and m, on a frictionless surface are held together against a spring compressed between them. The bodies are subsequently released, and they fly apart under the influence of the equal and opposite forces exerted by the spring. The second law then tells us that the product of mass and acceleration for each object should be identical; that is, mA equals Ma. And so we may write

$$m = M\,\frac{a}{A}.$$

If we replace the object of mass M by the standard unit kilogram (i.e., set $M = 1$), this statement becomes

$$m = \frac{a}{A}.$$

Now, forgetting the logic that led us here, we can simply take this ratio (along with our apparatus) as an operational definition of the mass of any object—one based on a measuring procedure that, in this instance, is totally independent of the reckoning of force. The object in question and a standard kilogram are pressed against both ends of any spring. With the most fundamental of instruments, clocks and rulers, we need only to determine the two resulting accelerations when the object and the standard fly apart in order to arrive at the ratio of a/A, which equals the sought-after relative mass, m.

We will of course have to establish that at any location (in any inertial reference frame), regardless of which spring is used (its material structure, thickness, length, etc.) and independent of how much it is compressed, the ratio of the two accelerations is constant for a particular body. This has always been found to be so, and therefore m equals a/A becomes more than just an *arbitrary* definition. It is predicated on an intrinsic property of matter and has been verified empirically under a diversity of conditions.

The second law could then be thought of as providing the needed definition of *force*, namely, the product of the mass of an object and the resulting acceleration. If a body is observed to accelerate, we maintain that a force acts on it equal to ma, and since we have m, all is well.

Although the Machian scheme provides a prescription for quantitatively determining relative mass, what is mass conceptually? Is it no more than this single dynamical manifestation—the relative amount of resistance to acceleration? This is very different from the intuitive notion of quantity-of-matter as the amount of stuff. That interpretation is no longer considered to be of any practical value at all, though some dictionaries still cling to it. Nowadays, most physicists will take Mach's approach in defining the mass of an object. Note that this sort of definition involves more than just words. You cannot meaningfully apply it without a standard kilogram in your pocket.

Unhappily, Newton could not have begun the *Principia* any better with Mach's definition of mass, even though it seems to be a fairly good one. After all, hidden in the operational process is the assumption that the spring will exert an equal action-reaction pair on the two masses. Mach's definition is actually equivalent to the third law; it is part of the theoretical structure that it presumably should have preceded. In short, we are not out of the woods yet. Newtonian mechanics seems to defy logical housekeeping, and perhaps that's a clue. Perhaps the definitions of fundamental concepts have to come out of the theory, not before it.

EXPERIMENTS

1. This one is an oldie, but a goodie. With it you can get some idea of your reaction time, that is, the time it takes to transform a visual stimulus into a motor reaction. Have someone hold a ruler vertically between your index finger and thumb, separated by about $1\frac{1}{2}$ inches. Watch the hand holding the ruler. As soon as the ruler drops, try to catch it as it falls between your fingers. Since $\Delta d = \frac{1}{2}g(\Delta t)^2$, it will fall just about two inches in the first 1/10 second, and no one is likely to be able to match that without jumping the gun. This can be fun with youngsters if you use a dollar bill. The drop from Washington's face to the edge of the bill is a sure bet.

2. The first law is easy to illustrate. Place something smooth having a substantial mass, like a metal spoon, on a sheet of paper. Hold the paper by a corner and jerk it out horizontally from under the spoon. With a little practice, the spoon will barely move. Put a $3'' \times 5''$ card on top of an empty glass and place a nickel directly over the center of the glass on the card. Now flick the card forward with a snap of a finger, and if all goes well, the nickel will drop into the glass.

3. The figure on p. 76 shows a girl throwing a ball straight up while in a uniformly moving train. Analyze the situation via the credo-of-the-three-laws, and then duplicate it. In other words, while walking at a constant speed in a straight line, throw a ball into the air straight up, and watch as it comes straight back down into your waiting hands. If you stop just after you heave the ball, where will it land? If you can't throw it straight up, then let it fall as you walk.

4. Find a heavy toy truck (the bigger the better), and tie a long rubber band to the front of it. Now attach a ruler or a piece of cardboard to the side so that it juts out in front of the truck. You're going to pull on the band while keeping its stretched length constant. That's what the ruler is for. Pull the truck along the floor with a *constant force* (i.e., with the rubber band at a constant stretched length). Observe how it moves.

5. We can duplicate what's happening in the figure on p. 93 and get a nice sense of the addition of forces, even without spring scales. Position four identical rubber bands over two pencils in place of the four scales. While holding a ruler behind them, observe the unstretched lengths of the bands, and then pull downward on the lower one. Stretch the pair 1 inch; the single bands should each elongate about 2 inches. Attach four or five identical rubber bands, one to the next, with paper clips. Pull on both ends. Do they all stretch equally? Play around with different combinations.

REVIEW

Section 4.1

1. What is meant by the word "fluxions"?
2. What did Newton mean when he said, "I feign no hypotheses"?
3. What was the dispute with Leibniz about?
4. What is the subject of the *Principia*?
5. *Clippers* were people who shaved the edges off gold coins before it became common practice to make them ridged. What does that have to do with Newton?
6. Was Newton recognized as a genius during his own lifetime?

Section 4.2

1. Name something that has a mass of about 1 gram.
2. State the *first law* (i.e., Newton's).
3. Has anyone ever proved the first law to be absolutely true?
4. Is there any difference between rest and uniform motion as far as the first law is concerned?
5. What is an *inertial reference frame*?
6. State the *second law* and write it symbolically.
7. Can an object ever move in a direction other than that of the net applied force? Can it accelerate in a direction different from that of the force?
8. What are the units of momentum in the SI system?
9. What pushes on a rocket to propel it when it is flying either on Earth or in *empty* space?
10. Can you shoot a hole through a silver dollar out in space, where bullets are weightless?
11. Is $F = ma$ the most fundamental statement of the second law?

12. How much harder would it be in free space to pull a 10-kg mass at a constant speed than a 1-kg mass at that same speed?
13. What is a *newton*?
14. Discuss the difference between *weight* and *mass*.
15. What is meant by *action-reaction*?
16. Can you push on a table with 100 lbs without the table's pushing back on you with 100 lbs?
17. Can you propel yourself through space by moving your legs or arms?

Section 4.3

1. What is the net force on a person standing still?
2. Why can't you grab your belt, pull upward, and fly?
3. What force propels you forward as you walk?
4. Which is the only force acting on you in free fall?
5. Are astronauts orbiting Earth actually weightless?
6. What do you do differently when you run in place, as opposed to normal running?
7. What is meant by the term *friction*?
8. To walk eastward you must _____ on the ground in a westward direction.
9. In the process of accelerating on a smooth floor, your feet will sometimes slip. Why?
10. An astronaut in space wearing magnetic boots starts walking along a freely floating iron beam. Describe both his motion and that of the beam.
11. What is Mach's definition of *mass*?
12. Why have we discarded the notion of quantity-of-matter, no longer taking it to be synonymous with mass?

QUESTIONS

Answers to starred questions are given at the end of the book.

Section 4.1

1. What were Newton's scientific accomplishments during the plague years?
2. Newton established a whole approach to the doing of science with his famous doctrine *"hypotheses non fingo."* What does it mean and what is its significance? How does it relate to, say, atomic theory?

Section 4.2

1. What did Newton mean by *quantity-of-motion*? This was a central concept for him.
2. An old adage says, "Everything that goes up must come down." Is it true? How would you resolve it with the first law?
3. If the first law really describes Nature, why does

"everything" around here come to rest sooner or later?
4. I once saw a little old man try to yank down a half-sawed tree by attaching a chain and pulling on it with a little old pickup truck. Wheels spun and dust flew, but neither tree nor truck moved. Explain in light of the first law. By the way, after a while he got out of the truck, cut a bit more, and finally won the tug.
5. A horse and a dog are running with the same momentum. Which is moving faster? Which will be harder to stop (provided that the dog won't bite you)?
6. Describe (in the imagery of momentum and the second law) how a baseball bat in the hands of a player at home plate manages to turn an incoming fast ball around and propel it over some distant fence. Explain the physics of the bunt. If you don't know what a bunt is, forget it.
* 7. Why do you have to pedal harder to get a bike started than to maintain it in constant motion?

8. A rocket ship firing one of its two identical engines is moving faster and faster. Neglecting changes in mass, what happens to its acceleration when the second engine ignites while the first is still on? What would happen if half the mass of the ship was jettisoned instead?

9. Can an astronaut in orbit tell which objects would be heavy or light on Earth if everything is effectively weightless? How?

10. Suppose that you're at work out in space. If you get hit on the head with an effectively weightless hammer, is it likely to hurt? Explain.

11. A rocket ferry with two identical constant-thrust engines is floating at rest in space next to its mother ship. The first motor fires and burns out. Then the second ignites and runs until it's exhausted. Compare the speeds just before the second motor lit and after it went out (neglect any change in mass).

12. Why does a good football kicker (golfer, baseball batter, pitcher, etc.) get maximum speed and range with a smooth follow-through? *Hint:* Think about contact time. Why are exaggerated follow-throughs a waste of effort?

13. Which stings the hand more, catching a fast ball or a slow ball? Catching a baseball with or without a glove? Catching a ball while running toward or away from it? Explain.

* 14. Which would be more difficult to throw at the same speed while hovering "weightless" in space, a tennis ball or a (formerly 15-lb) bowling ball?

15. It requires a horizontal force of about 20 lbs to drag my giggling son along a straight line on the living-room floor at a fairly constant speed. What's the friction force acting on him? What's his acceleration and what's the net horizontal force acting on him?

* 16. Godzilla and King Kong have a tug of war, each pulling with 10,000 lbs. What's the net force on the rope? Describe the motion. Now suppose that the same rope is attached to a ceiling hook, and Kong (weighing 10,000 lbs) hangs from its bottom end. Does this situation differ in any way from the previous one as far as the rope is concerned? What happens if he starts to climb up the rope?

17. Suppose you and your copilot were stranded hovering motionless in space 100 meters from your ship after your backpack rockets ran out of fuel. How might you use momentum conservation most effectively to save your lives?

18. What would happen to two stationary astronauts if, while floating in space, they decided to have a catch with a baseball?

19. As you walk to the back end of a rowboat, does the vessel move in the water? If so, in which direction will it glide, and what happens when you stop?

20. A railroad flat car is almost frictionlessly coasting along at a fairly constant speed. The last sixteen surviving hobos in the country jump onto the car as it moves beneath an overpass. What will happen to the motion of the car if its mass equals the mass of the drop-ins? Annoyed, they start pushing one another off. How does that affect the speed? What if they took a running jump off the back? Explain.

21. A truck carrying a load of caged canaries pulls onto the weighing scales at a highway toll station. While no one's looking, the driver bangs on the side of the cab so the birds fly around inside. If the canaries weigh a total of 200 lbs, what's the most weight the driver can get away without paying for?

* 22. Suppose you were standing still on a cardboard box that could just support your weight. What would happen if you decided to jump straight up off it?

Section 4.3

1. A 200-lb man holding a 20-lb keg of cold beer is standing on a scale in an amusement park. He heaves the keg straight upward, and while he is still pushing, a card with his weight and fortune pops out of a slot. Will it read 200 lb, 220 lb, less than or more than either of these? Explain.

2. If you jump onto a bathroom scale, will it ever read anything but your weight? Explain.

3. A weightlifter and a 200-lb barbell are both resting on a large floor scale. During the process of lifting and holding the barbell, will the scale reading ever momentarily exceed and/or be less than the combined weight of the two bodies? Explain.

* 4. Imagine a flat, very light, wheeled cart that is low to the ground. What would happen to it if you decided, while standing on it, to walk along its length? Would anything change if you got started on the ground and traversed the cart at a constant speed?

5. Imagine two kids, each in a separate canoe, each holding one end of a long taut rope. The older and heavier one decides to pull on the rope. What will happen? Would the motion have been any different if the lighter one did the pulling with the same force? Explain.

6. Suppose five or six marbles are set out in a straight line touching each other (e.g., along the groove of a plastic ruler). What happens if somebody shoots another marble straight into the end of the row? If someone shoots two in at once? Explain.

7. At one time or another everyone has shot a rubber band by stretching it over a thumb and letting it go. Analyze the process in terms of the credo-of-the-three-laws.

* 8. During a tennis serve, the racket hits the ball at about 120 ft/s, remains in contact with it for around 1/250 of a second and then continues forward at roughly 110 ft/s. Describe what is happening in terms of momentum.

9. Suppose you were out in space and you had an alien creature from the planet Mongo tied to your belt by a long string. What would happen if your captive suddenly pushed you straight away? How might things change if the string was elastic?

MATHEMATICAL PROBLEMS

Answers to starred problems are given at the end of the book.

Sections 4.2 and 4.3

1. Determine the momentum of a 100-kilogram wrestler sailing through the air at 2 meters per second.

2. A 200-lb Mafia Don, out on appeal, and his 400-lb bodyguard and rehabilitated killer take a moment away from family business to go jogging side-by-side in their new film *Godfather 19*. Which has the greater momentum and by how much?

* 3. The blue team in the annual Pentagon charity tug-of-war is slowly being dragged toward the line by a superior red team. The six blue players are each hauling away with 150 lbs horizontally, but the six red players each pull with an average of 160 lbs. How might you jump in and reverse what's happening; i.e., how much force needs to be applied and in what direction?

4. A baseball once used by the Brooklyn Dodgers has a mass of nearly 150 grams and can be thrown at a speed of up to 100 mi/hr. How much force must be applied in order to catch such a projectile, stopping it in 2/10 of a second?

* 5. A dried pea fired from a plastic drinking straw has a mass of $\frac{1}{2}$ gram. If the force exerted on the pea is 0.07 lb over the 1/10-second flight through the straw, at what speed does it emerge? (1 lb = 4.45 N.)

6. Imagine that a one-kilogram package wrapped in plain brown paper is placed in your outstretched hand by a man wearing dark glasses. What force (in newtons and pounds) must you exert so that the package remains perfectly still? So that the package accelerates upward at 9.8 m/s² (32 ft/s²) while you run away?

7. Calculate your weight in newtons, and then determine your mass in kilograms. What's the mass of $\frac{1}{2}$ pound of hard salami, 3 ounces of Swiss cheese, 1 ounce of mustard, all on a fresh 4-ounce roll?

8. The acceleration of gravity on the moon's surface is 1.6 m/s². What fraction of your Earth weight is your moon weight? What would your mass be on the moon?

9. A fifteen-pound bowling ball is to be accelerated from rest to 15 mi/hr in $\frac{1}{2}$ second. How much force is required?

* 10. A 20-kilogram statue of an elephant made out of chopped liver, with black olives for eyes, stands expectantly on the head table at a Republican fund-raiser. What force does it exert on the Earth?

11. Two astronauts floating in space decide to have a catch with a 500-gram pepperoni. Malcolm XX (whose mass is 100 kg) heaves the pepperoni at 20 m/s toward Flo, the captain (whose mass is 50 kg). She catches it and heaves it back at 20 m/s. After he catches it, how fast is each astronaut moving and in what direction? What are the initial and the final momentum of the system (astronauts and pepperoni)?

12. Malcolm XX and Flo (see previous problem) are tied to opposite ends of a very light (massless) rope, 60 meters long. She decides to see what he's doing and pulls herself toward him. How far will she have to move before they meet?

13. Captain Flo (having a mass of 50 kg) walks at 3 m/s (wearing magnetic boots) from one end of her space ship toward the other. Describe the resulting motion of *Chicken I*, the 5000-kg vehicle.

* 14. A golf ball with a mass of 46 g can be blasted from rest to a speed of 70 m/s during the impact with a clubhead. That impact lasts only about 0.5 milliseconds (one half of one-thousandth of a second, or 0.5×10^{-3} s). Calculate the change in momentum of the ball and the force applied to it. How does this force compare with the weight of a car? Determine the ball's acceleration and compare it with the acceleration of gravity.

We come now to a discussion of Newton's very important theory of gravity. Interestingly, Sir Isaac was far from the sole creator of the **universal law of gravitation,** and the first section of this chapter examines the actual origins of the key ideas. Newton's contribution came by way of showing that planetary motion, as embodied in Kepler's laws, could be understood as a consequence of gravity. That analysis is examined in the second segment along with the idea of **centripetal force;** the **relationship between weight and mass; satellite motion;** and the concept of a **field.** The last section is a brief look at the tremendous social and political impact of Newton's grand synthesis.

ON GRAVITY

If the attractive force of the moon reaches down to the earth, it follows that the attractive force of the earth, all the more, extends to the moon and even farther.

J. KEPLER

5.1 FORETHOUGHTS

The universal law of gravitation, the crowning achievement of Sir Isaac Newton, was surely his own creation just as it was surely created out of the ideas of the centuries preceding. Bishop Oresme had speculated 400 years before that if other "worlds" could exist in the universe, matter in the vicinity of any one of them would be drawn toward its center. And that marvelous seer, Kepler, envisioned the planets propelled in orbit by a sweeping solar force. He astutely maintained that the attractive action of sun and moon raised the Earth's tidal waters—a notion Galileo found revoltingly mystical and Descartes rejected out of hand. The idea that invisible tentacles of force could reach out over vast ranges of space was always a troublesome draught to swallow. *Action-at-a-distance,* it is scientifically called, though it has long had the aftertaste of magic. Kepler's soaring vision was greatly influenced by the contemporary writings of William Gilbert, scientist, physician, and pioneer in the study of magnetism. With the lodestone (a natural magnet) one could surely prove the fact of this ghostly force that draws without contact (Section 19.1).

Magnetism, gravity, action-at-a-distance—for a while it was one metaphor, one mystery of matter joining matter. Dr. Gilbert suggested the idea that the attractive force, transmitted by a subtle "effluvium," was proportional to the quantity-of-matter; the more massive a lodestone, the greater its ability to draw objects to it. Magnetic force decreases with distance; that is obvious to even the simplest hand, and Kepler concluded that the sun's driving force diminished as well. As the planetary orbits increase in range from the sun, their speeds decrease—to him a sure sign of a decreasing force. He maintained that "two stones . . . in space . . . outside the reach of force of a third" would gravitationally attract and "come together . . . each approaching the other in proportion to the other's mass."

Dr. William Gilbert
(1544–1603)

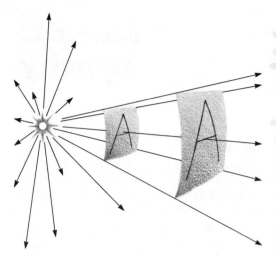

The uniform spreading out of emanations from a point source. Note that if the radius of the spherical segment is doubled, the corresponding area is four times larger.

Yᵉ suns conversion doth turn the planet out of this line framing its motion into a circular, but the former desire of yᵉ planet to move in a streight line hinders the full conquest of yᵉ Sun, and forces it into an Ellipticke figure.

JEREMIAH HORROCKS

ON GRAVITY

Gravity is the mutual bodily tendency between cognate [i.e., material] bodies toward unity or contact.

J. KEPLER

When anything diffuses uniformly in all directions from a point source, it spreads out thinner and thinner over increasingly larger surfaces as it advances. These surfaces can be envisioned as those of concentric spheres, each of area $4\pi r^2$. Thus the same quantity of emanation will initially sweep densely through a tiny spherical surface as will later emerge tenuously through a much larger one. Exactly as if you held something first near and then far from the nozzle of a paint spray, the flow becomes weaker, thinner; the quantity of paint arriving on each square of surface decreases with area, with distance squared ($1/r^2$). Kepler knew this was true with the intensity of light: the **inverse-square** emission pattern was already common knowledge. You can read by a candle flame only if you stand near it. Far away the light, spread out over a large region, is too dim to adequately illuminate the page—everyone knew that. Kepler came remarkably close to embracing this same imagery for his solar force, but other erroneous assumptions kept him from it.

The brilliant seventeenth-century Englishman Jeremiah Horrocks, who died too soon at only 23, was an ardent early disciple of Kepler. He established the wide applicability of Kepler's third law and showed the lunar orbit to be elliptical; both of these accomplishments would later prove essential to Newton, who relied heavily on Kepler's laws. Horrocks's notes hint at a perception of gravity acting as a mutual attraction between the planets, as well as with the sun.

Soon after Galileo's death, the French mathematician Gilles de Roberval took to the field (against Descartes) in favor of action-at-a-distance, accurately proposing that gravity pervaded the universe with a mutual attraction between *all* matter.

In 1645, Ismaël Bullialdus, a French astronomer, asserted (in the course of rightly criticizing Kepler's sweeping-force) that *the solar attraction acting on a planet was along the line from its center to that of the sun and dropped off inversely with distance squared.* He was a confirmed Copernican and one of the few who accepted the ellipticity of planetary orbits. Newton knew his work and knew it to be correct; in fact, he explicitly mentions the contributions of Bullialdus and the Neapolitan Borelli.

Giovanni Borelli was one of those rebellious types who followed the outlandish teachings of the "moderns," Copernicus and Galileo. He picked up the ancient Greek vision of the moon whirling about the Earth like a stone restrained by a sling. Long before, in 1585, G. B. Benedetti had discovered that *an object revolving in a circle and suddenly unleashed sails off in a straight line tangent to the curve at the point of release.* David presumably knew as much intuitively, as perhaps did Goliath.* For Borelli, the tendency of a planet to fly tangentially out of orbit was counterbalanced by its "natural instinct" to fall in toward the sun. Except for the "natural instinct" nonsense, that observation was quite true.

---

* Have you ever seen sparks flying off a spinning grinding wheel?

In 1665 gravity was envisioned in a muddled torrent of ideas, but all the key elements were there—scattered, but there. Descartes, already dead fifteen years, still ruled the mayhem. His universe, filled with aetherial matter swirling in vortices, in great planet-driving whirlpools, dominated natural philosophy. In 1665 young Newton worked in secret solitude off in the quiet countryside. Within a year he deduced the law of gravitation

> and thereby compared the force requisite to keep the Moon in her Orb with the force of gravity at the surface of the earth, and found them answer pretty nearly.

But he told no one. Perhaps "pretty nearly" was just not good enough; perhaps he felt the need to work out some of the details. He had intuitively treated the Earth and moon as point-masses, as if they acted with all their matter concentrated at their centers, and yet the proof that this could be done legitimately eluded him for a long while. In any event, he remained silent on the subject for well over a decade.

Since the early 1660s Robert Hooke had been vainly trying to empirically determine the form of the variation of gravity with distance. He conducted experiments within deep wells and "on Paul's steeple and Westminster Abbey, but none that were fully satisfac-

> When the body is freed from the sling, . . . it follows a trajectory from the point of release tangential to the last revolution of the sling . . . under the influence of the impetus already set up in it.
>
> **BENEDETTI** (1585)

Isaac Newton

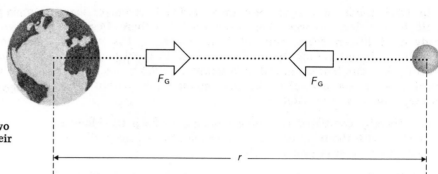

The gravitational attraction between two masses acts along a line connecting their centers.

tory." By 1670 he was an avowed proponent of **universal gravitation:** *gravity acting mutually on all matter,* every part attracting every other. He wrote:

> All Celestial Bodies whatsoever have an attraction or gravitation power towards their own Centres, whereby they attract not only their own parts, and keep them from flying from them, as we may observe the Earth to do, but that they do also attract all the other Celestial Bodies that are within the sphere of their activity.

"The sphere of their activity," indeed. Sometime before 1680 Hooke, too, decided that, like other emanations blossoming out spherically, the *force of gravity must be proportional to the inverse square of the distance.* Stated symbolically (although neither he nor Newton ever wrote it that way), this reads

$$F_G \propto \frac{1}{r^2}.$$

After 1676, Newton withdrew from science and spent the next several years in solitude contemplating biblical prophesy. In the exchange of letters that drew the pouting prodigy back to physics, Hooke wrote to Newton (Jan. 6, 1680): "My supposition is that the attraction always is in duplicate proportion to the distance from the centre reciprocal...." Others—Edmund Halley and Christopher Wren, in particular—would surmise as much, as well. When the three (Hooke, Wren, and Halley) met to discuss the problem in January 1684, they were agreed, as Hooke commented, "that upon that principle all the laws of the celestial motions were to be demonstrated." But not one of them had the mathematical prowess to actually do it.

In August of that year, Halley, then a bright and charming young man, went up from London to Cambridge to consult with the mathematical virtuoso in residence. John Conduitt, the man who married Newton's favorite niece, recounted the incredible incident years later.

PHILOSOPHIÆ

NATURALIS

PRINCIPIA

MATHEMATICA.

Autore *JS. NEWTON,* Trin. Coll. Cantab. Soc. Matheseos Professore *Lucasiano,* & Societatis Regalis Sodali.

IMPRIMATUR·

S. PEPYS, *Reg. Soc.* PRÆSES.

Julii 5. 1686.

LONDINI,

Jussu *Societatis Regiæ* ac Typis *Josephi Streater.* Prostat apud plures Bibliopolas. *Anno* MDCLXXXVII.

The title page from Newton's *Principia,* which he began in 1685 and published in 1687.

Without mentioning either his own speculations, or those of Hooke and Wren, he at once indicated the object of his visit by asking Newton what would be the curve described by the planets on the supposition that gravity diminished as the square of the distance. Newton immediately answered, *an Ellipse*. Struck with joy and amazement, Halley asked him how he knew it? Why, replied he, I have calculated it; and being asked for the calculation, he could not find it, but promised to send it to him.

The irony of that moment is marvelous. While others all around Europe were eagerly searching for the law of gravity, Newton had already lost it.*

Newton once again immersed himself in the problem, and in November, after being blocked for a while by a careless error, he sent Halley the calculation. So superlative was the work that Halley encouraged a lengthy exposition and personally took on the responsibility of seeing it into print. He soothed, supported, and cajoled the tempermental genius, calming Newton's steaming rage after Hooke claimed priority to his law of gravitation. And when the Royal Society, which was practically bankrupt, could not manage the cost of publishing the *Principia*, Halley paid for it from his own pocket although he was himself a man of very limited resources. Newton, who was fairly well off by then, curiously disdained to contribute financially to the project as if it were somehow beneath his dignity.

By midsummer 1687, the *Principia* was published, and for about ten shillings one could own the vision.

5.2 THE LAW

It is not enough to guess at the form of a law, however keen the guiding intuition. Law must be molded against Nature, hammered by observation; this was Newton's great accomplishment in the *Principia*. Kepler's three empirical laws stood at hand as a summary of innumerable bits of data strung out in an intricate pattern—if only that design could be shown consistent with the idea of a gravitational force acting on the planets. Stroke by stroke, Newton established that all movement, planetary and terrestrial, could be understood within the framework of his laws of motion and gravity.

Central Force

Early in Book I, Newton proved that Kepler's second law of equal-areas-swept-out-in-equal-times (p. 53) was in harmony with the image of each planet responding to a **central force,** *a force always directed toward the same point in space.* If the sun was commanding the planetary system with a drawing force pulling toward its center, it would surely reside at such a point. His arguments were totally geometrical in the classic tradition, avoiding calculus completely, although the Archemedian logic of limits is everywhere. The

---

* See the fascinating account in C. C. Gillispie's *The Edge of Objectivity*, Princeton University Press, 1960.

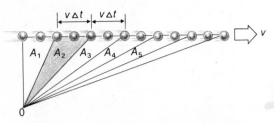

An object moving in a straight line past a point, O, sweeps out equal areas in equal intervals of time, Δt.

following gives something of the flavor of his treatment although without the detail and rigor it is only a suggestion of the original.

Imagine an object moving at a constant speed, v, along a straight path. Even in this simplest of all situations, a line drawn to the body from some fixed point, 0, sweeps out equal areas (A_1, A_2, A_3, etc.) in equal times, each triangle having the same base and altitude. Now, if the object is struck a series of sudden blows, all directed toward point 0, one at the end of each interval Δt, there will be a sequence of changes in the direction of motion. The line to the object again sweeps out triangular areas (A_1', A_2', A_3', etc.), which Newton, by the geometry of the situation, proved were equal. When the interval Δt is made vanishingly small, the flood of distinct impulses blends into a continuous **centripetal** (center-seeking) **force**, and the segmented path dissolves into a smooth curve. *Under the influence of a centripetal force, a body moves about a center of force, sweeping out equal areas in equal times.* The obscure pattern of Kepler's second law is now miraculously "understood."

Centripetal Acceleration

Borelli's slingshot vision of counterbalanced forces is the next ingredient. An object tends to move at a uniform speed in a straight line; that's the law of inertia, but the object can certainly be drawn off that course, presumably by a force. Velocity, as defined in physics, consists of both speed and direction. Even though the speed of a body may be constant, its direction of motion can still change. The *velocity* of a painted horse on a carousel varies from one moment to the next, facing one way, then another, gliding all the while round and around at *constant speed*.

The rate of change of velocity is acceleration; in this case, it is a variation only in orientation, known as **centripetal acceleration,** a_C. A ball of mass m whirling around with a uniform speed at the end of a string is constantly drawn inward by the tug on the cord. If the string breaks so that the restraining force, the centripetal tow, vanishes, the ball sails freely off in a straight line perpendicular to the string at that instant. Newton's second law tells us that the necessary centripetal force F_C must equal ma_C. We need a force to account for the deviation from straight-line motion, and there it is—even the logic is circular.

If a body is to revolve about in a circle of radius R at a constant speed, v, it can do so only under the influence of a specific centripetal force, whether that force is provided by the tension on a string, friction on tires, rails thrusting against wheels, or gravity. How does such a force depend on the physical circumstances? That problem was solved by Newton rather early, but to his chagrin, Huygens published the solution first. The particular force needed to yank a body off its straight-line inertial course, continuously tugging it into a perfect circle of radius R, is

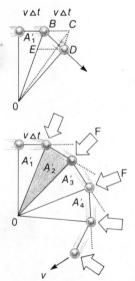

A planet moving in orbit under the influence of a central force sweeps out equal areas in equal times.

$$F_C = ma_C = m\,\frac{v^2}{R}.$$

A DERIVATION—CENTRIPETAL ACCELERATION

The object depicted in the figure is sailing along at a constant speed, v, in a circular orbit. Following Galileo's lead, we can envision its behavior as composed of two overlapping motions. In a given time interval, Δt, the body moves inertially in a straight line a distance, L, while dropping a distance, h. The Pythagorean theorem, applied to the right triangle whose sides are R, L, and $R + h$, is

$$(R + h)^2 = R^2 + L^2,$$
$$R^2 + 2Rh + h^2 = R^2 + L^2.$$

The R^2 terms cancel, yielding

$$2Rh + h^2 = L^2.$$

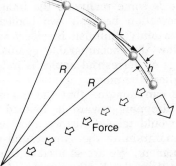

Now, we mean to let the time interval get vanishingly small so that the staccato motion blends into the curve, and we get a value for instantaneous acceleration. In that case, h^2 becomes exceedingly small as compared to $2Rh$ and may be neglected; whereupon

$$2Rh = L^2,$$

and dividing by $2R$ gives us

$$h = \frac{L^2}{2R}.$$

Since $L = v\Delta t$,

$$h = \frac{1}{2}\left(\frac{v^2}{R}\right)\Delta t^2.$$

As we saw earlier (p. 70), the term in parentheses corresponds to the constant acceleration by which the body drops through distance h; that is,

$$a_c = \frac{v^2}{R}.$$

The larger the circle, the less rapidly the body changes direction; the higher the speed, the more rapidly it changes and the greater the acceleration. In any event, this was the key Newton had to have since gravity itself was to serve as the centripetal force holding the planets in their near-circular orbits. Using that notion, Newton showed that Kepler's third law (p. 54) could obtain only if the source of the centripetal force itself varied with radial distance as $1/r^2$. Moreover, if a planet moved in an elliptical orbit (Kepler's first law) with a center of force at a focus, the attractive influence would again have to vary as $1/r^2$. All Kepler's jewels were no more than the natural manifestation of a planetary inverse-square centripetal force directed toward the sun—*gravity*.

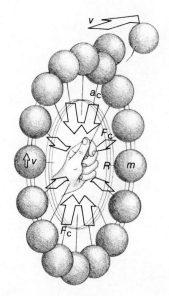

The presence of a center-seeking force, F_c, is required if an object is to move in a circle.

Note that Newton did not hesitate to apply his own three laws to celestial bodies even though there really was no experimental justification for doing so. It was a big assumption and a major step in establishing the universality of the credo-of-the-three-laws. Of course, the ultimate justification was simply that it worked.

Apple Force

If there is some reality to the imagery of gravitational force, then the interaction between two bodies, at least two celestial bodies, must certainly diminish inversely with the square of their separation. Now the incomparable genius moved to show that the force that tugs on terrestrial apples is indeed the selfsame celestial power that tethers the moon.

We can imagine that the moon in circling Earth is, in effect, always dropping toward it. Instead of flying off in a straight line, which would take it farther away each instant, the moon falls inward, maintaining the same radial distance from Earth as it drops each moment. We are so used to thinking about motion with respect to a fictitious flat Earth that the idea of falling and not getting any closer is disquieting at first.

The moon's centripetal acceleration, its acceleration of circular fall, is just v^2/R and can easily be determined numerically. The lunar orbital speed is simply the distance traveled in one revolution divided by the time it takes, the period. That distance is the circumference of the near circular orbit, $2\pi R$, that is, $2\pi \times (384,000,000$ m$)$. The lunar period is 27.32 days, and so the orbital speed (v) turns out to be 1022 m/s. The moon's centripetal acceleration then becomes

$$a_C = \frac{v^2}{R} = 0.00272 \text{ m/s}^2.$$

If this acceleration is actually caused by gravity, then any object at that range from Earth, whether initially orbiting or "motionless," should experience that same downward acceleration. Galileo proved as much near the surface with his projectiles. In other words, a body dropped at a distance of 384 million meters from the center of the Earth presumably accelerates toward it at a mere 0.00272 m/s^2. Newton next cleverly compared that value with the terrestrial rate of 9.8 m/s^2 at the surface, 6400 m from the center of the planet.

This scenario has the moon dropping under the influence of gravity, much as an apple drops from a tree. Indeed, we have all seen astronauts floating around within and without their *orbiting* space ships, and they are effectively weightless in free fall (just as a worm in a descending apple would be "weightless").

If the force of gravity is assumed to trail off as $1/r^2$, the acceleration should do the same. The moon is just about 60 Earth-radii away, so it "should" experience a gravitational acceleration

$v \leftarrow$

$a_G = 0.0027$ m/s^2

$R = 384,000,000$ m

$a_G = 9.8$ m/s^2

$(60)^2$, or 3600 times smaller than does an earthly apple; that is,

$$a_{\mathrm{G}} = \frac{9.8}{3600} = 0.00272 \ \mathrm{m/s^2}.$$

Amazing! The centripetal acceleration equals the gravitational acceleration: $a_{\mathrm{C}} = a_{\mathrm{G}}$! The whole calculation, so simple and yet so profound, is quite superb. How it must have shocked Newton's contemporaries! It's hard to believe, even now, that the hypothesis (i.e., the existence of a gravitational force) is not really proved true even though these predictions based on it are brilliantly borne out—yet that is certainly so. Einstein's theory of general relativity provides an alternative view; it is an even more precise though far more mathematically complex description of gravity. Yet the simple classical Newtonian theory is so powerful that we have blasted our way to the moon by its guidance.

All terrestrial objects experience a downward acceleration due to gravity (which we represented as a_{G}). At sea level this acceleration has a specific value equal to 9.8 m/s² or 32 ft/s² and is assigned a special symbol, g. Regardless of weight, that is, regardless of the gravitational force on the body, its acceleration will equal g. That implies that the weight of an object, F_{W}, is proportional to its mass since the acceleration of fall,

$$\frac{F_{\mathrm{W}}}{m} = g,$$

can be constant, independent of mass, only when m cancels from the left side. Accordingly, if weight is proportional to mass ($F_{\mathrm{W}} \propto m$), the bigger the mass, the bigger the weight, and so the resulting acceleration remains constantly equal to g for all objects. A piano is more massive than a pea and would naturally require a greater force to make it accelerate. But in free fall that force is its weight, and that weight is greater than the weight of the pea, and so both accelerate equally—the pea under a tiny force, the piano under a large one. We conclude, along with Sir Isaac, that since weight (at least at the Earth's surface) depends on mass, then perhaps quite generally the force of gravity on a body depends on its mass and thus has the form

$$F_{\mathrm{G}} \propto \frac{m}{r^2}.$$

THE APPLE MYTH

One day in the year 1666 Newton had gone to the country, and seeing the fall of an apple, as his niece told me, let himself be led into a deep meditation on the cause which thus draws every object along a line whose extension would pass almost through the center of the Earth.

VOLTAIRE (1738)

Weighty Matters

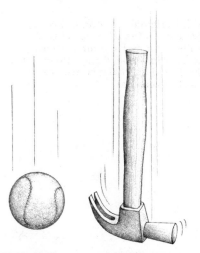

9.8 N

A 1-kg mass on Earth has an average weight of 9.8 N.

 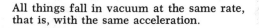

All things fall in vacuum at the same rate, that is, with the same acceleration.

$$\frac{\text{weight of key}}{\text{mass of key}} = \frac{\text{weight of ball}}{\text{mass of ball}} = \frac{\text{weight of hammer}}{\text{mass of hammer}} = g$$

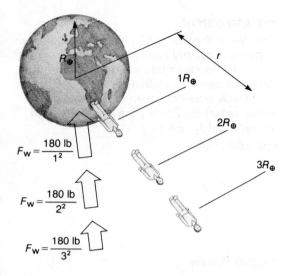

$F_W = \dfrac{180\ \text{lb}}{1^2}$

$F_W = \dfrac{180\ \text{lb}}{2^2}$

$F_W = \dfrac{180\ \text{lb}}{3^2}$

The variation of weight with distance from the Earth.

Cavendish's original drawing of his apparatus for determining the value of G. To prevent disturbance from air currents, he enclosed it in a sealed case. He observed the deflection of the balance rod from outside with telescopes.

Arguing from his third law, Newton suggested that since the Earth pulls on a small apple (with a force of about 1 newton), the apple must tug on the Earth with an equal oppositely directed force (of 1 newton). This implies that the gravitational interaction between *any two objects* of mass m and M separated by a distance r is expressable as

$$F_G \propto \frac{mM}{r^2}.$$

This is the force each body will experience pulling it toward the other. Realize that the notion of universality (that *all* objects attract each other) was a bold generalization for which there was precedent in speculation but no direct evidence whatever! It surely seemed reasonable that two apples should attract each other just as two planets should, but that was conjecture at most. For that matter, assuming an inverse dependence on exactly r^2 rather than, say, $r^{1.999}$ was just as unsubstantiated, though far prettier.

It was not until well over a century later that Henry Cavendish, using an exceedingly sensitive pendulum balance, measured the gravitational interaction between small lead spheres. (Nothing is quite as convincing as seeing the balls mounted on the suspended rod actually revolve toward the stationary balls.) As a result of his work, we can now write the force expression as an equality:

$$F_G = G\,\frac{mM}{r^2}.$$

Here G, which equals 0.0000000000667 (or 6.67×10^{-11}) SI units, is an experimentally determined constant, which presumably applies equally well in Cavendish's basement and in the Andromeda

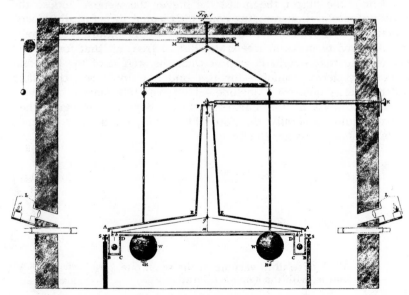

galaxy. Because G is so exceedingly small, only when at least one of the interacting bodies is extremely large will the force be significant. You and I at arm's length would gravitationally tug on each other with less than a millionth of a pound. In comparison, the good Earth pulls me down with a force of 180 lbs. Your terrestrial **weight**, F_W, simply equals the force of gravity, F_G. In other words, with your mass represented as $m_\odot$, and the Earth's mass as $M_\oplus$, your weight due to the planet at any center-to-center separation of r is

$$F_W = m_\odot \, a_G = G \, \frac{m_\odot M_\oplus}{r^2}$$

Notice that Earth weight decreases with distance; there is about a 0.012% weight loss as one ascends to the top of the Empire State Building (for me that's roughly 0.02 lb). The **gravitational acceleration,** a_G, also varies with distance from the Earth's center, so that $a_G = g = 9.8$ m/s^2 is only an average at the surface.

It must be said that Newton was not totally in the dark regarding the numerical value of G. After all, he had g and knew the Earth's radius (6400 km) and therefore could calculate its volume. He needed only the mass of the Earth to compute G, and that Sir Isaac could intelligently guess at. You see, that unknown mass could be imagined as some average density times the volume of the planet. The density of water is 1 g/cm^3, and most minerals are from 2 to around 10 times as dense. So if he reasonably assumed a density somewhere in that range, he would not have been far off the actual value of about 5.5 g/cm^3. In fact, unless the Earth were hollow, it would have been hard to be in error regarding G by any more than a factor of 4 or so—which is not at all bad as a guess.

Finally, realize that one is never actually weightless in the strict sense of being liberated from the Earth's tug. Even in free fall, gravity is the only force still acting. Unlike a car seat that pushes forward on your back as the vehicle accelerates forward, generating body tensions that produce the distinct feeling of motion, gravity acts uniformly on all parts of the body. In free fall, one feels nothing— no pulling force, no sense of continuously dropping, no perception of endless fall—just effortless floating, as if one is gliding under water.

"Weightless" acrobatics inside Skylab.

Satellites

Long before Sputnik, Newton prophetically observed that any projectile launched more or less horizontally is in a real sense an Earth *satellite* (a word coined by Kepler). Thus a rock thrown straight out of the window of a tall building sails for a while in a smooth yet modest orbit that soon intersects the Earth not far from the launch point. Galileo (Section 3.2) showed that this trajectory should be a parabola, and it certainly very closely resembles one, but those calculations neglected the curvature of the Earth itself. His projectiles, in theory, were accelerated straight down toward a horizontal plane by gravity and not radially toward the Earth's center. Generally the difference is negligible; nonetheless, there is a difference. A horizontally hurled baseball (neglecting air friction) would actually arc along an elliptical orbit with its far focus at the Earth's center. Of course, the orbital motion would be abruptly interrupted when the ball crashed into the ground. But if it was fired more swiftly still, it would sail farther; the ellipse would fatten out, becoming less elongated. Further increasing the speed would result in ever larger, rounder elliptical paths. Finally, at one particular launch speed, the baseball would glide just above the Earth's surface clear around to the other side without ever striking the ground. Like a tiny leather moon, it would wheel about the planet's center in circular flight. Flashing over the treetops, it could continue to revolve about the globe until practicality (air friction, collisions with birds, etc.) brought it down.

Throwing the ball still faster would loft it into an extended elliptical orbit, with the Earth's center now at the near focus. Note that this is identical to what would happen if an onboard rocket fired, speeding up a satellite already in circular orbit. Ever increasing launch speeds will finally stretch out the already eccentric ellipse, first into a parabola and then a hyperbola, both of which are open curves (and our baseball would soar away forever).

A low, tight, circular orbit essentially means that the satellite follows the Earth's curvature. Under the influence of gravity an object at sea level falls 16 feet in one second. Now if we constructed a plane tangent to the Earth at the point right where you are standing, how far would you have to walk along that plane before you found yourself 16 ft above the Earth, i.e., before the Earth dropped off 16 ft below the tangent? That's easily found (p. 121) to be about 4.9 miles. Thus, if our baseball was fired horizontally at 4.9 miles per second, it would also descend 16 ft each second and so continue to fall without changing altitude as the Earth's surface beneath it dropped away at the same rate. For this lowest of all possible orbits, the launch speed, roughly 18,000 mi/hr, is the highest.

Ordinarily a satellite-launching missile begins to roll over soon after vertical lift-off, in order to insert its payload into orbit somewhere beyond the atmosphere. At about 100 miles above Earth's surface, a launch vehicle must be moving horizontally at a bit more than 17,000 mi/hr to attain a circular orbit. Unlike the situation of elliptical paths, over which the speed varies, decreasing with distance from the planet (i.e., the center of force), the speed in circular flight is constant. In such cases, the centripetal force ($F_c = mv^2/r$) equals the satellite's weight, which, in turn, drops off as $1/r^2$, so the speed must decrease with altitude as well. *The larger the orbit, the slower the motion.* This is the mechanism behind Kepler's third law.

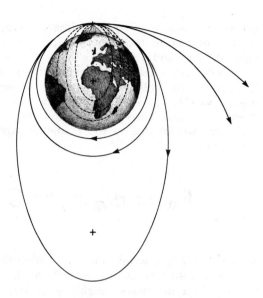

The orbits of projectiles fired horizontally at various speeds (based on a drawing by Newton).

Changing orbits.

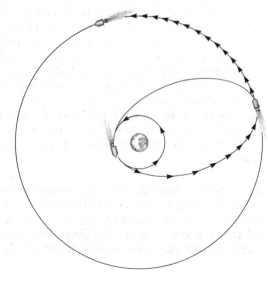

That makes life rather complicated for an astronaut who wants to go from a smaller to a larger and slower circular orbit. Paradoxically, the first step is to fire a rocket in order to move faster and thereby enter an extended elliptical transfer-orbit. Engines blast for a moment, the capsule accelerates over a predetermined time interval and off it goes at that top speed, hurled into an ellipse with the injection point nearest the planet. Sailing away, the craft decreases speed until it reaches its farthest orbital distance from Earth. There, moving at a minimum rate, just before the capsule would "fall" back toward the planet, the astronaut again fires a rocket tangentially. Accelerating out of the ellipse and into the new, larger circle, the vehicle finally ends up moving slower than originally, despite all the maneuvers.

To get back home, one needs only to slow down enough, via a retro-rocket, to descend into any small elliptical path that intersects the planet. By contrast, our astronaut could leave the Earth's domain entirely, roaming off in an open orbit, provided the vehicle could reach a speed of about 25,000 mi/hr before its fuel is exhausted. This is the *escape velocity* that an unpowered projectile must attain near the Earth's surface if, having gone up, it's not going to come back down again, despite the familiar adage. In other words, this must be the initial speed of a vehicle if it is to free-fall against the Earth's relentless pull, going upward slower and slower, all the way to infinity before it comes to a stop. Since chemical rockets burn out rather quickly, the escape velocity becomes the minimum practical launch speed for leaving the planet's vicinity. Of course, if we had a really long-firing engine, we could depart for the stars at a much more leisurely pace.

The Gravitational Field

All right, then, an apple loosed from its branch plunges earthward under the influence of gravity. And we even know how that force varies with mass and distance; but how, pray tell, does the *apple* "know" it? How does it find out which way is down and how hard it is supposedly being pulled? Does some imperceptible tentacle reach up and grab it? Does an invisible medium, filling the void, mechanically transmit the tug? Newton himself wrote:

> I have not been able to discover the cause of those properties of gravity from phenomena, and I feign no hypothesis. . . . To us it is enough that gravity does really exist, and act according to the laws which we have explained, and abundantly serves to account for all the motions of the celestial bodies and of our sea.

Sir Isaac seems—privately, at least—to have favored the notion of force transmission by some as yet undiscovered material agency. We are content with *direct contact*: two things smashing into each other, a hand pressed hard against the wall—force transmitted by

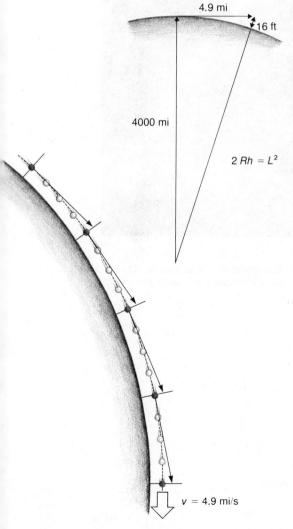

Falling in orbit around the Earth.

4.9 mi

16 ft

4000 mi

$2\,Rh = L^2$

$v = 4.9$ mi/s

touching. It used to be thought that the void was filled with a substance known as aether. If that were so, the pull of gravity could
be sent out in continuous mechanical contact from sun to Earth,
and vice versa. Alas, the aether has long since vanished and with it
that nice neat imagery. But when we begin to think of matter composed of separate, spaced atoms, the idea of *direct contact* becomes
meaningless. Atoms do interact, yet they seem never to touch in the
familiar way we like to think of macroscopic bodies as touching.
Even here, in the very essence of *contact,* there is the mysterious
action-at-a-distance. Granted, it is a small distance between atoms,
but even though they do push and pull on one another, they do
not touch belly-to-belly, and direct contact in the usual sense is an
illusion.

Our modern picture, known as the **field** conception, is a subtle
contrivance to make comprehensible the magic of action-at-a-distance—comprehensible if not entirely satisfying.

Euler, in the mid-eighteenth century, developed a mathematical

The flight of a Minuteman III
missile across the California sky.

description of fluid flow, assigning a velocity to each and every point in a moving liquid. This was the origin of field theory, in that particular case a field of motion. In much the same way, we could associate a temperature with each point in this room and speak about the local temperature field. The concept became an indispensable part of physics in the late 1800s, when it was applied to the theory of electricity and magnetism (Section 17.3).

Gravity is usually conceived of as a *force field*. We can detect and measure it (if only indirectly) by simply hanging a convenient mass, a *test mass*, on a scale and reading off its weight. We can thus deduce the gravitational field of some object by moving our probe from one spot to another throughout the space surrounding the body. By assigning both a numerical value and a direction to the force on our test mass at each point in the region, we can theoretically map the field.

Every object that has mass affects the surrounding space, permeating it with a gravitational field that extends out indefinitely, dropping off as $1/r^2$. Any entity possessing mass brought into this domain will interact directly with the field, experiencing an attractive force. Earth is embedded in and interacts with the sun's gravity field, and vice versa. Each of us is immersed in the field of everyone and everything else around, of the cars and trees, and pencils and spiders. If there were no friction whatever, all the people sitting in a classroom would slowly be drawn to the region of the greatest mass, of the highest concentration of bodies, where the field is strongest. Delicate instruments known as gravitometers can even detect high concentrations of mass (such as mountains, buildings, buried meteorites, etc.) from local variations in the Earth's gravitational field.

The invisible tentacles, the imperceptible effluvium, the mechanical aether are all now replaced by yet another construct, an unseen field spread out in the void of space. In early conceptions, space was a vessel and the field was strung across it, but the newer vision of Einstein's general relativity has space-time itself as the medium of gravity. Matter in this picture is seen as a localized deformation of the space, distortions of which propagate out into the surroundings as the field.

Nowadays it is generally believed that the gravitational field of an object moves out into space at a speed equal to that of light. This finite rate of propagation (a contemporary notion) leads to an apparent conflict with the third law. Suppose you walked up a flight of stairs. In so doing, you would be moving from one point in the sun's gravity field to another, and the solar tug on you would change accordingly. It does not matter by how much; our concern is that the change would occur immediately as you moved. Yet your own field, in which the sun is immersed and with which *it* reacts, will be unaltered in the vicinity of that star for about eight minutes. This is so because, spreading out at the speed of light, the ripples

The gravitational field of the Earth.

A Frenchman who arrives in London finds a great change in philosophy as in everything else. He left the world full, he finds it empty. In Paris one sees the Universe composed of vortices of subtle matter. In London one sees nothing of this. In Paris it is the pressure of the moon that causes the flux of the sea; in England it is the sea which gravitates toward the moon. With your Cartesians, everything is done by an impulsion that nobody understands; with Mr. Newton, it is by an attraction, the cause of which is not better known.

VOLTAIRE

of your changing field take eight minutes to cross the 93 million mile gap. For all that time, action-reaction directly between sun and you is presumably out of tune. If the sun vanished—*now*—for eight minutes we would see it hanging in the sky, all the while drawing on the Earth as if nothing were amiss. Indeed, however small the distance, there will be a lag of this sort.

Contemporary theory goes even further in maintaining that forces between objects result from the exchange of field particles. The "carrier" of gravitational force is a still hypothetical beastie known as the *graviton*. It is these particles streaming from Earth to apple and vice versa that presumably carry the requisite information from one to the other, that tell the apple "this way down." And with the assumption that gravitons possess momentum, it is then possible for conservation of momentum to hold exactly at all times. It is for this reason that the conservation principle takes on great fundamental importance.

In comparison with the other kinds of forces we will examine in later chapters, gravity is by far the weakest. Moreover, the observed interactions can be transmitted only by a tremendous torrent of gravitons, and so each one individually must exert only a minute effect and will therefore be exceedingly difficult to detect. Thus far, the graviton remains elusive. The evidence to date neither confirms nor denies its existence.

All of us, feeble though we may be, draw on everything in space within an immense expanding sphere as many light-years out as we are old, for at the moment "you" were born, *your* personal gravitational field began to spread out at the speed of light. Although the strength of interaction diminishes appreciably with distance, it never quite vanishes, and so, in time, everything in the cosmos will "know" that *you* have been.

5.3   NEWTON'S LEGACY

Successes of the grand Newtonian synthesis, the new understanding of the universe, swelled in a veritable torrent that swept away Aristotle as if he had been a schoolboy and raised in his stead the knight of the *Principia*. With mathematical precision one could "know" every subtle pattern of the planets and their moons; could comprehend the flight of projectiles and all the ancient mysteries of the swelling tides; could fathom the strange variation of pendulum periods with place and apprehend the bulging of spinning Earth and distant Jupiter alike. Halley took the Newtonian mechanism and wrung from it yet another popular triumph. He calculated the elongated elliptic orbit of the great comet of 1682, and it sailed back, blazing, to posthumously confirm his prophecy 75 years later. The majestic world-machine whirled in a steady rhythm of inexorable order, in a steady, perpetual rhythm of precise and numbered law.

It is amusing now to remember that almost fifty years after the publication of the *Principia*, John Bernoulli, the eminent Swiss mathematician, won a prize from the *Académie des Sciences* for deriving

MIRANT STELLA

HAROLD

Halley's comet, which returns every 76 years, was first recorded in 240 B.C. The Bayeux tapestry, a portion of which is shown here, was embroidered in 1070. It depicts the arrival of the comet in 1066, the year of the Norman invasion of England. King Harold of the Saxons is being warned of the ill omen. Not long afterward, he was defeated at Hastings by William the Conqueror. The comet will be back again in 1985–1986.

Kepler's third law from Descartes's appealing vortex nonsense. All of this suggests that nationalism may not be the handmaiden of objectivity, and given enough time, a good theoretician can "prove" almost anything.

Continental Europe was certainly slow to embrace the new mechanics, the new action-at-a-distance mystery of universal gravity. Despite its obvious power, its inexplicable imagery could hardly compete with the satisfying directness of swirling cosmic whirlpools driving everything, touching, pushing—particle against particle.

Voltaire, the French author, philosopher, and wit, after a brief stint in the Bastille, was exiled and went for several years to England, where he became an ardent Newtonian. His popularizations, along with many others, like Count Algarotti's *Sir Isaac Newton's philosophy explain'd for the use of the ladies*, were immensely successful in spreading the irascible knight's credo, the mathematical revelation of God's design.

Frederick the Great of Prussia and Catherine of Russia, imitating the English and French courts, surrounded themselves with scientists and mathematicians. Kings, queens, fine ladies, and genteel men chatted about the latest scientific advances and delighted in dabbling with beakers and balances and all the instruments of the new fashion.

The intelligentsia responded to the pervading sense of reason and design with a gradual change in their theology, a shift in belief from the orthodoxy of Christianity to the divinity of Nature. Spinoza saw God and the natural universe as one, and Hume, the ultimate skeptic, was doubtful about all accepted religious tenets of faith. The

When Newton saw an apple fall, he found
In that slight startle from his contemplation—
'Tis said (for I'll not answer above ground
For any sage's creed or calculation)—

A mode of proving that the earth turn'd
round
In a most natural whirl, called "gravitation";
And this is the sole mortal who could
grapple,
Since Adam, with a fall, or with an apple.

LORD BYRON (1788–1824)

Gravity leads us to God and brings us very near to Him.

COTTON MATHER (1663–1728)

American clergyman and author

Men's minds seem caught in a general movement towards natural history, anatomy, chemistry and experimental physics.

DIDEROT

Encyclopédie (1775)

eighteenth-century *deists* believed in a "natural religion," emphasizing morality and predicated on human reason rather than divine revelation. The Creator set the marvelous cosmic clockwork in motion, forever bound by His law, and then stepped aside no longer to drive it or to interfere in the piddling day-to-day affairs of humankind.

Pierre de Laplace, having brought Newtonian physics to a marvelous perfection, presented his masterwork *Mécanique céleste* to Napoleon Bonaparte, whereupon the emperor asked why there seemed to be no mention of the Creator anywhere in the treatise. "Sire," responded Laplace, "I had no need of that hypothesis."

The English philosopher John Locke was a close friend of Newton and an old classmate of Hooke and the poet Dryden. Believing that the concept of the "law of nature" could be applied to religion and government as well as to the physical world, he set about evolving a new political philosophy—one that would echo resoundingly in America years later.

The very first sentence of the American Declaration of Independence rings with the phrase "the Laws of Nature and of Nature's God." Divine law guided the universe—just so, divine law must guide humanity. The young Virginian Thomas Jefferson, philosopher-statesman and scientist, was the Declaration's principal author and a clear beneficiary of Lockean thought. Benjamin Franklin, who did some minor work on the Declaration, was quite familiar with the *Principia*. He wrote a piece on its philosophical implications entitled "On liberty and necessity; man in the Newtonian universe." The mechanistic system of checks and balances, the very clockwork of American government, was also a Lockean device born of the notion of action-reaction.

Just as the evolution of cosmic law and order brought a conclusion to the Copernican revolution, so did it contribute to the wellspring of ideas that boiled over in the American and French revolutions and in the English industrial revolution. For better or worse, the Time of Science was upon the land.

EXPERIMENTS

1. The period of a pendulum, T, is the time it takes to swing from one extreme position up to the other and back. Given the approximate relationship

$$T = 2\pi \sqrt{\frac{L}{g}},$$

where L is the length of the pendulum, see if you can determine a value for g, as Huygens did. Use several different weights as bobs. Do they have any effect on T if L is fixed? Vary L several times, and see if g is constant.

2. To study the nature of centripetal force, attach equal weights (a few large washers will do) to both ends of a string passing through a spool. Holding the spool, twirl the weight coming out of the top end so that it sails around in a horizontal circle. The other vertically hanging weight serves as a counterbalance. In this instance the centripetal force equals the weight. Vary the speed of the orbital motion while observing the corresponding radius. Now increase the hanging mass, and try it again.

REVIEW

Section 5.1

1. Was gravity Newton's personal creation?
2. To what did Kepler attribute the tides?
3. According to Kepler, the planets were propelled by a _____ _____.
4. What is meant by action-at-a-distance?
5. Discuss the meaning of the term *inverse-square*.
6. Why was Hooke experimenting on St. Paul's steeple?
7. Samuel Pepys, the famous diarist, was president of the Royal Society in 1686. How might he have had a tiny part in our story? See if you can find his name anywhere in this section to confirm your conclusion.

Section 5.2

1. Define the term *central force*.
2. What is meant by a *centripetal force*?
3. Write an expression for the centripetal acceleration of an object in circular motion.
4. For an object to be pulled off its straight-line inertial path and made to move in a circle, a centripetal force must be applied. What provides that force for an Earth satellite? For a rider on a merry-go-round?
5. What does it mean to say that the moon is falling toward the Earth?
6. If you double the mass of an object, what happens to its weight?
7. By how much is the weight of an astronaut reduced after traveling 4000 miles from the surface of Earth?
8. What, if anything, do you suppose happens to your weight as you descend a flight of stairs above ground level? Below ground level?
9. Earth bulges at the equator because of its rotation. Would you weigh more or less there? What happens to g there?
10. The quantity g changes with distance from Earth's center. Does G also change?
11. What is the *escape velocity* from the surface of the Earth?
12. Would you expect the escape velocity from the moon to be more or less than from Earth?
13. How would you determine the gravitational force field of an isolated object out in space?
14. What is a *graviton*?
15. How large is your sphere of gravitational influence beyond which you have as yet left no mark on the universe?

Section 5.3

1. Why is Halley's name associated with a comet?
2. Did continental Europe immediately embrace Newtonian theory?
3. What role did Voltaire play in establishing the new physics of Sir Isaac?
4. What has Newton to do with the Declaration of Independence?
5. Did Ben Franklin know anything about Newtonian mechanics?

QUESTIONS

Answers to starred questions are given at the end of the book.

Section 5.1

* 1. Explain why scholars long before Newton considered that gravity might be an *inverse-square* force.

2. Make a list of all of the properties possessed by gravity that were correctly proposed prior to Newton's work.

3. Some people have maintained that it was Newton who, on seeing an apple fall, *first* suggested that gravity might also extend all the way to the moon. Why is that historically inaccurate?

* 4. Suppose you held a small piece of paper one foot from and directly in front of the source of a spray of paint for one second. How would the amount of paint vary if you repeated the experiment with the paper two feet away?

5. Hooke was rather upset with Newton for claiming *his* law of gravity. Did Hooke have a basis for feeling cheated?

Section 5.2

* 1. By tugging on a string attached to a ball, one can swing the ball around in a circle overhead. How must the tug change if the speed of the ball is doubled? If the speed is constant, but half the mass flies off? If mass and speed are constant, but more string is let out to double the diameter of the circle?

2. How might a space station in the shape of a wheel be spun in order to simulate gravity? Explain in detail.

3. Two masses floating out in space attract each other gravitationally. If one is twice as massive as the other, what is the ratio of the force each experiences? If their center-to-center separation is halved, what happens to the force?

4. As you descend into a tunnel, what happens to your weight? Suppose the tunnel went down to the center of the Earth. What would your mass and weight be there? *Hint:* Consider each section of the planet to be attracting you.

* 5. How much would you weigh on a planet the same size as Earth but with twice its mass? What would you weigh 4000 miles above its surface?

6. Determine how the gravitational acceleration at the Earth, a_G, varies with mass, $M_\oplus$, distance, r, and G.

7. Explain why the gravitational force does not vary the speed of a satellite in circular orbit but does change it when the satellite is in an elliptical orbit.

* 8. How does Kepler's third law follow from the notion that gravity is the centripetal force holding the planets? In other words, why does planetary speed decrease with increasing orbital diameter?

9. Is there a point somewhere between Earth and the moon where their two gravitational forces cancel and you would be more or less weightless? How might you locate such a point theoretically? Incidentally, we sent a satellite into orbit around the corresponding point between Earth and the sun in 1978.

10. Why did the astronauts require a giant Saturn rocket to leave the Earth and only a tiny motor on the Eagle to leave the moon? The escape velocities are 11,200 m/s and 2400 m/s, respectively.

11. Why is it easier to launch satellites in an easterly direction than in a westerly direction? The J. F. Kennedy Space Center is on which coast of Florida? Why?

* 12. Suppose that an astronaut in a 500-mile circular orbit moving eastward (i.e., counterclockwise looking down on the North Pole) releases an apple into space. What will be its flight path? What if the apple is thrown directly in the forward direction (eastward)? In the backward direction?

13. Suppose that the Earth could be compressed, collapsing down to half its diameter. What would happen to the intensity of the gravitation force field at its surface?

Section 5.3

1. "In the thirteenth century the important words, words whose meaning were taken for granted, included God, sin, grace, salvation and heaven; in the eighteenth century these were replaced by nature, natural law, reason, humanity, and perfectibility." Discuss this statement by D. Schroeer as it relates to Newton's legacy.

MATHEMATICAL PROBLEMS

Answers to starred problems are given at the end of the book.

Section 5.2

1. What fraction of what you weigh now would you weigh in a rocket ship four Earth-radii away from the surface of our planet?

2. An ant sitting at the very edge of a phonograph record ($\frac{1}{2}$ ft out from center) rotates with the disc at $33\frac{1}{3}$ revolutions per minute. That corresponds to a speed in ft/min of $33\frac{1}{3} \times 2\pi R$, where $R = \frac{1}{2}$ ft. Calculate its centripetal acceleration.

* 3. A car driven by a TV cop during the traditional chase scene heads into a curve with a 100-m radius at 60 mi/hr. If the car's mass is 1500 kg, what centripetal force must be provided by friction on the tires if the car is not to skid out on the turn? Give the answer in both newtons and pounds (1 mi/hr = 0.447 m/s).

4. What happens to the weight of an object if its mass is

doubled and its distance from the center of the Earth is also doubled?

5. Presuming that you know the value of G, the *universal gravitational constant,* from measurements with a pendulum balance, how might you arrive at the mass of the Earth? Make a rough calculation of that value.

6. To get a feel for the force of gravity, calculate the attraction between two spheres of mass 100 kg, whose centers are separated by 1 m. That's roughly the mass of a football player (about 220 lbs).

7. The moon orbits the Earth in a nearly circular orbit of radius 384,000,000 m, taking about 27 days to go once around. Use that information to calculate the mass of the Earth.

* 8. Show that

$$v = \sqrt{Ra_{\text{G}}}$$

represents the speed of a satellite in circular orbit a distance R from the center of Earth, where a_{G} is the acceleration of gravity up there.

9. At what speed must an object be injected into circular orbit 500 miles above the Earth's surface? Take the mass of the planet to be 6×10^{24} kg (i.e., 6 followed by 24 zeros).

* 10. The *escape velocity* at a distance R from the center of a body of mass M is given by

$$v_{\text{esc}} = \sqrt{\frac{2GM}{R}}.$$

Using this equation, determine at what speed you could blast off from the moon and never fall back? (Moon mass is 1/81 that of Earth; radius is 1738 kilometers.)

11. Locate a point along the Earth-moon center line at which the tug of each of these bodies would cancel and you would actually be weightless.

12. Calculate the acceleration of gravity, a_{G}, 1000 miles above Earth's surface.

13. What would be the speed of a low-altitude lunar orbiter?

* 14. The smaller star of a binary pair orbits the larger at a distance of 1000 million kilometers once every 10 Earth-years. Find the mass of the larger star. Use the same technique to find the mass of the sun.

In midair they are weightless in free fall.

A. Einstein.

Aus

Zürich. 14.X.13.

Hoch geehrter Herr Kollege!

Eine einfache theoretische Über-
legung macht die Annahme plausibel,
dass Lichtstrahlen in einem Gravitations-
felde eine Deviation erfahren.

Grav. Feld → Lichtstrahl

Am Sonnenrande müsste diese Ablenkung
0,84" betragen und wie $\frac{1}{R}$ abnehmen
(R = Sonnenradius). Entfernung vom Sonnen-Mittelpunkt)

0,84"

Sonne

Es wäre deshalb von grösstem
Interesse, bis zu wie grosser Sonnen-
nähe helle Fixsterne bei Anwendung
der stärksten Vergrösserungen bei Tage
(ohne Sonnenfinsternis) gesehen werden
können.

A letter from Einstein to George E. Hale regarding "a simple theoretical
consideration" arising from his theory of gravity. (*The Hale Papers*.
Translation by Erwin Morkisch.)

The **special theory of relativity** is Einstein's penetrating vision of **space** and **time** and **motion.** This chapter is an attempt at making the basic elements of that vision understandable and its primary conclusions accessible. It begins with a brief history of the man and then turns to the paradox of his youth that set him rethinking the fundamental ideas of physics. The theory is built on two assumptions, or postulates, a discussion of which forms the second section of the chapter. From these postulates it develops that **events perceived to occur simultaneously by one observer do not necessarily occur simultaneously for all observers.** That is a central discovery because it pertains to the way we measure length in both time and space.

The last section establishes that **time itself is relative,** flowing at different rates for different observers; that **distance is likewise relative** to the viewer; that **the speed of light is an ultimate limit,** which cannot be attained or exceeded by moving matter; and finally that time can be thought of as one coordinate of a four-dimensional **space-time continuum** through which we sail, moment by moment.

Newtonian physics turns out to be the slow-end approximation of a relativistic universe. Only up near the speed of light does everything become wonderfully weird.

Albert Einstein, one of the greatest of all physicists, recast our understanding of the physical universe in a fundamental way, a monumental way. So extraordinarily far-reaching was his work, so rich and creative the construct, that we gain an entirely new vision through his insights. And yet, even now, few of the uninitiated know more of his triumphs than a meager chain of key words: space, time, four dimensions, relativity, etc.

Einstein's flair, his consummate strength, lay in a razor-sharp perception cutting away the extraneous to reach the very essence, a power to focus and concentrate, a power that took nothing for granted, least of all the obvious. A past master at the *thought experiment,* he asked "what if" of the simplest things and with unfailing intuition showed us answers never dreamed of, neither simple nor obvious.

Gentle, kind, in sagging sweater and carpet slippers, he was *stupor mundi,* the wonder of the world.

CHAPTER 6 Special Relativity

ALBERT EINSTEIN TO HALE
October 14, 1913

*My very dear Colleague:

A simple theoretical consideration makes plausible the assumption that light-rays experience a deviation within a gravitational field

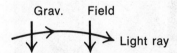

Grav.   Field

Light ray

On the edge of the sun the deflection should be 0.84″ and as $\frac{1}{R}$ diminish (R = Distance from the center of the sun)

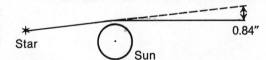

Star

Sun

0.84″

Therefore it would be of the greatest interest to know within how great a proximity to the sun bright stars could be seen by daylight [without a solar eclipse] with the help of the largest magnification.

*Addressed to Prof. George E. Hale, Pasadena, Calif., by Dr. Albert Einstein, Zürich, Switzerland.

6.1 A PARADOX

Munich just before the turn of the twentieth century was a bustling metropolis of more than a quarter of a million. The grand capital of Bavaria, ringing with church bells and frothed in beer, was the new home of Hermann and Pauline Einstein, their young son Albert, and infant daughter Maja. They were a cheerful family of essentially areligious "modern" Jews, interested more in life and assimilation than in tradition and "ancient superstition." The small electrochemical plant that happy-go-lucky Hermann and his brother Jakob operated did fairly well for a while, and in time, Pauline got her home in the suburbs.

When Albert was twelve, his parents befriended an impoverished medical student from Russia, Max Talmey. And it was Talmey, during Thursday visits for lunch, who introduced young Albert to several popular books on physics and mathematics. Though he despised the regimented discipline of the Luitpold Gymnasium (high school), where he was a totally undistinguished student, Albert soon, quite on his own, gained considerable facility with mathematics and an extensive knowledge of philosophy.

Managerial acumen not being Hermann's forte, the business ultimately folded, and all the family except Albert went off to Pavia near Milan in 1894. The solitary, introspective youngster was supposed to remain in Munich, living in a boardinghouse while finishing the necessary course work for a diploma from the Gymnasium. Within months, however, being on his own and feeling independent if not impudent, he began to concoct a scheme to hasten his departure. There was some business about a medical certificate attesting to a supposed nervous breakdown and other machinations as well, but before the precocious smart aleck could pull it off, the indignant taskmasters from Luitpold summarily drummed him out. Expelled without a diploma, a high school throw-out, he was sent packing with the reproach that his "presence in the class is disruptive and affects the other students."

Liberated from the weight of what he felt was a relentless German authoritarian education, Albert delightedly crossed the Alps to join the family in Milan. Hermann, envisioning his son an electrical engineer, decided to send the boy to the Swiss Federal Polytechnic School in Zurich. It was Spring 1895, and the lad was only sixteen, but already portents of an extraordinary future were there. With a letter to his uncle in Stuttgart, Albert enclosed an essay entitled "Concerning the Examination of the Aether State in Magnetic Fields." More a program of research than an exposition, it dealt with interrelationships of electricity, magnetism, and the supposed aetherial medium pervading space.

Long after the billowing shock of hair had grayed and the saintly face had taken on that sad, gentle look, Einstein recalled that it was then, at sixteen, that he discovered and began to struggle with a strange paradox. Light (regardless of its source) was understood to be an electromagnetic wave, an intricate oscillatory structure of electric and magnetic fields, troughs, and crests sweeping through space, spreading out like ripples in a pond. But if we could travel at the speed of light, flashing along at nearly 300 million meters per second, what would we see of light? If we were moving in imagination alongside a wave peak, the light would "appear" unchanging, motionless; yet the accepted theoretical formalism does not allow of such a situation. A stationary pulse of light cannot exist. It has no meaning; it must vanish. Light is totally unlike a stream of bullets that can be followed and examined as if frozen in flight. It is self-sustained by *change*, a thing of varying interwoven fields that, by alternation, generate each other in a rhythmic dance that has no rest.

Another way to appreciate the puzzling nature of the situation is to imagine a digital clock that lights the time for an instant and then goes black, only to light up again announcing the arrival of the next second, and so on. Now suppose that we moved straight away from the clock, gradually accelerating up to the speed of light (which is customarily designated c). If, just at the instant before attaining 300,000,000 m/s, we saw the clock read 11:59 and 59 seconds, we would *never* see it light up 12:00 while coasting along at c. The image of the clock face showing 59 seconds would reach our retinas, but once we had attained the speed of light, presumably no further signals could catch up to us. The clock would go black and stay that way. Indeed, it seems that if we slowed down at any time, say, two or three hours later, we would at last see the clock reading 12:00.

Einstein at the age of sixteen had put his hand on a profound dilemma, a paradoxical conflict between Newtonian mechanics that allowed of such speeds and electromagnetic theory that could not reasonably abide with them.

It took ten years before the paradox of Einstein's youth vanished in the new light of his **special theory of relativity.** In 1905, Albert Einstein was an unknown clerk in the Berne Patent Office, Switzerland. Newly married, shy but friendly, the young Ph.D. had plenty of time to think and create; reviewing applications at the Patent Office was hardly taxing. That was the year he published his marvelous first paper on relativity, "Zur Elektrodynamik bewegter Körper" ("On the Electrodynamics of Moving Bodies")—marching to a drummer so different that it would be years before many others heard the tune. At 26 he had created the *special* or *restricted* theory— restricted in the sense of specialized, since it pertained only to uniform motion.

Galileo and later Newton recognized that uniform motion had no perceivable effect on mechanical systems. One can play pool or Ping-Pong or anything else aboard ship and never know the vessel is moving, regardless of speed, so long as it is constant. The young man

The rays of light are small particles of matter emitted from the sun, or any luminous body, with a velocity so immense, as to enable them to move at the inconceivable rate of eleven millions of miles in a minute [183,333 miles per second].

THOMAS KIMBER
The American Class Book; or, A Collection of Instructive Reading Lessons: Adapted to the Use of Schools (1812)

6.2 THE TWO POSTULATES

The Principle of Relativity

juggling oranges in the lounge of a 747 can't tell from the response of the oranges in his hands whether he is cruising at 600 mi/hr relative to the ground or at rest on the runway. The apparent behavior of a mechanical system is quite independent of its uniform motion. This, then, is the **classical principle of relativity:** *The laws of mechanics are the same for all observers in uniform motion.*

Newton struggled with the distinction between absolute and relative motion, a distinction that is central to Einstein's analysis. Was there something somewhere in the vast universe that was totally stationary, something from which all motion could be reckoned absolutely? "For it may be," wrote Newton, "that there is no body really at rest, to which the places and motions of others may be referred." Still, space itself, unchanging and immovable, might serve as his fixed frame of reference. For him, absolute space was the motionless heart of the cosmos, and yet, paradoxically, it could have no special place in Newtonian mechanics, where all nonaccelerating (inertial) environments—like your living room or the lounge on that 747—were equivalent with regard to the laws of motion.

Einstein generalized the principle of relativity, broadening its scope. It was a splendid proposal but one already anticipated by the French mathematician, Jules Henri Poincaré (1854–1912). For that matter, Poincaré's intuitions anticipated much of the *special theory*, but unlike Einstein, he was too conservative to press ahead and formulate a completely innovative theoretical treatment.

Cruising in a blimp or rocket ship, on a fire engine or a sailboat, we expect no novel experience, no wondrous changes in the customary. Our battery-operated toothbrushes and vibrators presumably still hum along, pocket computers compute, popcorn pops, rubbed balloons stick to walls and still pick up tiny bits of toilet tissue, rubber bands stretch, and flashlights light; physics is undaunted by uniform motion, and life goes on quite normally at 2 mi/hr or 2000 mi/hr. This is the conjecture Einstein raised to the status of his **first postulate:** *All the laws of physics (not just those of mechanics) are the same for nonaccelerating reference frames.* Each and every observer, moving at constant velocity with respect to one another, experiences a harmony of universally established law. Consequently, *it is quite impossible to distinguish between inertial frames, and no experiment whatsoever can establish whether a particular platform is uniformly moving or at absolute rest.* Of course, the latter concept loses all significance if it can never be determined—motion is *relative*, not *absolute.*

Now this conclusion is really very profound although it is not new. Leibniz had long ago argued against the ideas of absolute rest and absolute motion, but Newton's conception prevailed. In any event, it is worth restating since the idea is subtle. Suppose that some special place in the universe was actually at rest so that all motion could be measured with respect to it and therefore motion was absolute. The **principle of relativity** makes the point that even if we somehow stumbled onto this wondrous spot and stood there absolutely at rest, we would never know it. The billiard balls and juggled oranges would behave no differently there than anywhere else, no differently at "rest" than in uniform motion. Well, then, what good is the concept if we cannot even recognize it when we find it? Clearly, it's of no practical use at all. In effect, the principle of relativity abolishes the concepts of **absolute rest** and **absolute motion.** All we can know empirically is that one object is in motion *relative* to another.

Light plays a central role in relativity theory, a role that can best be appreciated if we know a little about its physical characteristics. To that end, let's now anticipate some of the descriptive work we will examine in more detail in Chapter 20.

During the eighteenth century there were two distinctly different views of what light was. The dominant school envisioned it as a stream of *particles*. The other, less persuasive group saw it as a *wave*. Although Newton insisted that he "feigned no hypotheses," he hesitantly embraced a particle description that nonetheless had wave aspects to it. Others—Hooke and Huygens, for example—thought of light purely as a wave, a vibration waving through the all-pervading aether like a ripple on a pond. *A wave is a propagating disturbance of a medium,* a building up and a falling away of some aspect of that medium. For example, sound waves are a series of compressions and rarefactions, usually of air. The air is made to vary locally in density, being alternately compacted and thinned out, and these disturbances propagate beyond the source into that surrounding medium. Some regarded light in the same way, as a vibration of the invisible, elastic aetherial gel that was generally believed to fill all of space.

The eighteenth century, for the most part, was the era of the particle model of light, and just as rigorously, the nineteenth century —particularly the second half—became recommitted to the wave vision. The twentieth century, in turn, evolved a blending of both, a wave-particle duality (Section 11.2).

Compared with that of ordinary moving things, the tremendous speed of light is quite spectacular. Not surprisingly, it evaded direct measurement for centuries. Even the great Galileo, blinking a lantern at a distant compatriot, who blinked his in return, failed miserably at that attempt to measure it. Olaus Römer, in 1676, was successful in carrying out the first fairly accurate determination from studies of the duration of eclipses of Jupiter's moons. By 1849, when Armand H. L. Fizeau measured the speed of light flashing through the notches of a rapidly rotating toothed wheel, the wave picture was already well established. Michael Faraday forged the next link when he discovered that light propagating in a material was affected by magnetism, thereby implying a kinship of some kind.

The Speed of Light

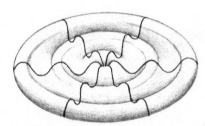

A water wave.

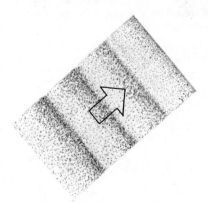

A sound wave.

The luminiferous ether, that is the only substance we are confident of in dynamics. . . . One thing we are sure of, and that is the reality and substantiality of the luminiferous ether.

LORD KELVIN

Popular Lectures and Addresses (1891)

HOW COMFORTABLE THE OLD IDEAS

The beginner will find it best to accept the ether theory, at least as a working hypothesis. . . . Even if future developments prove that the extreme relativists are right and that there is no ether, it is likely that the change will involve no serious readjustments so far as explanations of the ordinary phenomena are concerned.

A. A. KNOWLTON

Physics for College Students (1928)

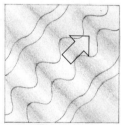

aether-filled space distorted into waves by electromagnetic field

electromagnetic field waves in empty space

Waves in the aether conceptually replaced by waves in the field.

Then James Clerk Maxwell stunningly showed that electromagnetic waves were theoretically possible and, moreover, would travel at a speed almost identical to that measured by Fizeau. In 1888, Heinrich Hertz succeeded in actually generating electromagnetic waves—radio waves. Using sparking electrical circuits, he demonstrated that the radiation behaved much as light did. The conclusion was inescapable: Light was an electromagnetic wave in the luminiferous aether (Section 20.1).

A balloon rubbed against cloth will pick up pieces of lint or tissue paper, obviously exerting a force. And just as with gravity, we speak about electric force fields (Section 17.3). Similarly, a magnet displays the presence of a magnetic force field when it picks up paper clips or twists a compass needle. The theoretical picture of light depends on the fact that a varying electric field generates a magnetic field that reaches beyond it, and moreover, that a varying magnetic field, in turn, gives rise to an electric field that extends beyond it. An electromagnetic wave consists of a web of varying electric and magnetic fields. Hand over hand, as it were, the two build and diminish, generate and regenerate each other, sweeping out into space, supposedly permeating, stretching the aether.

In the last year of his life (1879, the year Einstein was born), Maxwell wrote a letter to someone at the U.S. Nautical Almanac Office suggesting a scheme for measuring the speed at which the Earth plowed through the surrounding aether. A young naval instructor destined to become one of the most renowned of experimentalists, Albert Abraham Michelson, took up the challenge. By comparing two beams of light shining along perpendicular directions, he could detect differences in the underlying aether arising from the planet's motion. In 1881 Michelson reluctantly published the startling result that there was no apparent motion of the aether with respect to the Earth. A minor mathematical error unhappily marred the effort and lessened its impact. But that only made Michelson more determined in his search for the so-called **aether wind.** Joined in time by Edward W. Morley, Michelson redid the experiment even more precisely in 1887, but again the results were negative. Light traveled at the same rate, whether along or transverse to the supposed direction of the aether wind streaming past the planet. Generally accepted at the time as the stationary backdrop of the universe, aether served as the frame of absolute rest through which celestial bodies moved and against which absolute motion was measured. Despite all attempts, aether stubbornly evaded detection, and physicists, equally as stubbornly, rationalized its elusive behavior.

Poincaré was probably the first to grasp the philosophical significance of the enigmatic findings. In 1900 he wrote:

Our aether, does it really exist? I do not believe that more precise observations could ever reveal anything more than *relative* displacements.

It seems that Einstein was probably unaware, even in 1905, of the exceedingly important Michelson-Morley experiment although he, too, reached the same conclusion:

> The introduction of a "luminiferous aether" will prove to be superfluous, inasmuch as the view here to be developed will not require an "absolutely stationary space."

Light was now to be seen as a self-sustaining electromagnetic wave propagating in the electromagnetic field rather than in the aether—a substitution of one construct for another.

Whether aether actually exists or not is in a way irrelevant; whether we have seen the last of the notion, however, is not certain. Yet, if we cannot measure it directly or even indirectly, if it has no observable manifestations and provides no indispensable theoretical component in the understanding of phenomena, who needs it? That's not to say that modern physics completely shuns the unobservable. Incredibly, we have a few phantoms (like the graviton, the several quarks, and the virtual photon) that play a "necessary" role in the presently accepted theoretical structure. But aether, which had been around for 2000 years, was not indispensable, and it vanished from the list of useful theoretical entities. It was created out of the necessity of one vision and was dispensed with when it became irrelevant to a different vision.

Einstein's **second postulate** is known as *the principle of the constancy of the speed of light: Light propagates in free space with a speed (c ≃ 300,000,000 m/s) that is independent of the motion of the source (and of the observer as well)*. Thus we are liberated from the original paradox since light will always be seen to move at c, regardless of how fast the observer chases it. Yet, having rid ourselves of a paradox, are we not now confronted with worse—an illogical universe? How can it be that if I shine a flashlight at a rocket ship bearing down on me at several hundred thousand miles per hour, someone aboard will nonetheless measure the incoming beam as traveling at c, just as if that person were at rest standing next to me? What is in fact illogical is our tendency to presume that "strange" is wrong, that what is different from our familiar though limited understanding of experience must be erroneous.

In effect, Einstein said: Believe the postulates and press on to new discoveries, however bizarre they may seem at first. For surely, if the speed of light is always found to be constant, our ordinary notions of distance and time will need careful scrutiny. And scrutinize we must, for a whole range of experiments performed since 1905 are all consistent with a constant speed of light, irrespective of the motion of the source. These experiments go well beyond the original Michelson-Morley demonstration, from Dayton C. Miller's rerun of it using sunlight and a latter-day improved version incorporating infrared lasers to Willem de Sitter's (1913) study of double

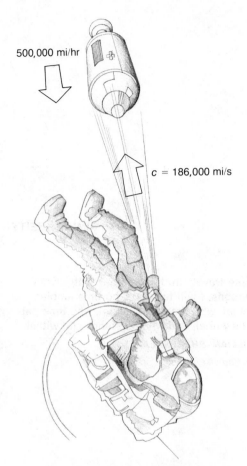

500,000 mi/hr

c = 186,000 mi/s

The speed of light in vacuum will be measured to be constant, regardless of the motion of either the source or the receiver.

AETHER PERSISTED

We have learned too that radiant heat energy is believed to be transmitted by a medium called the ether. At the present time, some scientists believe that other ether waves produce various other effects. . . . It is possible, then, that light waves are ether waves.

CHARLES E. DULL

Modern Physics (1939), a high school text

star systems. In a striking contemporary experiment, subatomic particles known as neutral pions were brought up to the tremendous speed of 99.98% of *c*, whereupon they naturally decayed, emitting electromagnetic radiation. Despite this extraordinary speed of the sources, the subsequent emissions were measured to be traveling at *c*, just as if the pions were motionless when they burst.

However puzzling, **the speed of light is the same for all observers.** This is the postulate, and moreover, it is the experimentally verified fact! It is a tribute to Einstein's powerful insight that he conceived that notion out of a sense of how Nature must behave, even without any experimental evidence at the time. In the decades that followed, the second postulate became an established law.

6.3 SIMULTANEITY

Time travels in divers paces with divers persons. I will tell you who time ambles withal, who time trots withal, who time gallops withal, and who he stands still withal.

WILLIAM SHAKESPEARE

As You Like It (Act 3, Scene 2)

There is matter, and there is radiant energy (electromagnetic radiation in the form of radio waves, microwaves, infrared, light, ultraviolet, X-rays, and gamma rays). Our most effective mode of interacting with the material world, of gathering information, is via electromagnetic emanations, typified by light. Beyond the mere fact of sight, light is the swiftest instrumentality at our disposal for communication and for the perception of events. Our observation of the universe on the most fundamental level is inextricably tied to the propagation of radiation, all of which occurs at *c* in vacuum. Yet this speed is finite, and that, in itself, shapes our understanding. If it were infinite, the distinction between special relativity and classical Newtonian theory would all but vanish. In fact, wherever *c* is so large in comparison with the relevant motions that it can be considered effectively infinite, Newtonian physics suffices very well. This is precisely why the discrepancies, the amazing "peculiarities," went unnoticed so long—we live, for the most part, in a comparatively slow-moving macroscopic world. Running, driving, even space-ship flying at thousands of miles per hour—all of it is meager crawling against 186,000 miles per second.

Light is the ultimate messenger, the edge of our perception, and thereby the edge of our reality. What we know of that reality comes by way of measurement, and that process is bounded by the finite rate with which events can be observed. The effect this limitation has on the concepts of distance and time, which for centuries were considered self-evident and absolute, is most fundamental.

The Bureau of Standards sends out radio signals with which we can set our clocks—"When you hear the tone . . . [beep]." But the farther you are from the transmitter, the later the beep will arrive and the more off your clock will be from those at the Bureau. An observer in the Andromeda galaxy would have to wait 2,200,000 years for the beep to arrive marking *now*. This isn't really a problem, though, since we can make the necessary corrections, knowing distance and *c*. Still, what does *now* mean for us in regard to our friend in Andromeda? That is, what does it mean to say that two

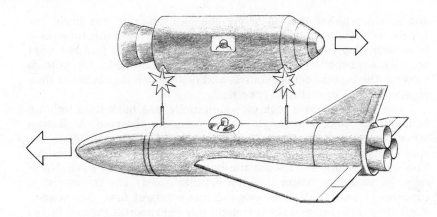

Two events seen to occur simultaneously by one midway observer may not appear simultaneous to another midway observer who happens to be moving with respect to the first.

events occur simultaneously, one here and one there? That, too, can be settled easily enough: *Two events are simultaneous if they are seen to occur at the same time by an observer located midway between the sites where the events happened.* All right, then, if there is no such thing as absolute rest, is there such a thing as absolute simultaneity? Do two events occurring simultaneously for one midway observer occur simultaneously for all midway observers, regardless of their relative motion?

Let's suppose that two events occurring at different places are seen to occur simultaneously by at least one midway witness. To set the scene, imagine a rocket ship and a space station gliding past each other in the dark void at a constant relative speed of, say, 20,000 mi/hr. The platform's commander, a stubborn but honest fellow, will be our chosen simultaneous viewer. Playing out his part, he switches on two rather erratic explosive flares at either end of the station, not knowing whether they will go off together or not. At the very moment they do explode (as determined later by him), he finds himself lined up nose-to-nose with the ship's pilot, a woman no less willful than himself. The platform commander is a midway observer, and the light from both flares, fore and aft, reaches him at the same instant, having traveled equal distances (at the speed c). So far as he is concerned, the two events—the explosive beacons flashing on—are unquestionably simultaneous.

Now the space-ship pilot is also a midway witness. Indeed, if her craft were long enough and close enough, the explosions might even leave some dents at either end as a permanent proof of her midway observership. Certainly if there were no relative motion, both viewers would see the flares flash on simultaneously, as they would if c were infinite. And yet, even in the short time during which the flashes are propagating at c to the moving ship, that vehicle is advancing toward the rear beacon and receding from the front one. The ship's pilot will actually overtake the light from the rear beacon,

What then is time? If no one asks me, I know: if I wish to explain it to one that asketh, I know not.

ST. AUGUSTINE

There is one single and invariable time, which flows in two movements in an identical and simultaneous manner. . . . Thus, in regard to movements which take place simultaneously, there is one and the same time, whether or no the movements are equal in rapidity. . . . The time is absolutely the same for both.

ARISTOTLE

and so she will see it flash *before* she sees the front one flash. For her the explosions are certainly not simultaneous. *And this time there is no adjusting for distances and the speed of light to find out what actually happened!* Neither pilot nor commander can say who is moving, the motion being relative, and neither can do any more than observe what is in reality happening.*

The pilot reports a lack of simultaneity, the back flare lighting first, but she understands why the commander observed the flashes occurring at the same instant. She sees the platform sail off to her rear and judges that the commander is moving away from the aft flare's light beam and toward the beam from the front flare. Naturally, he would see them flash on simultaneously (as far as she is concerned) only if the rear one did, indeed, light first. Her conclusions are consistent, and she is content that her report is correct. But so is he, and both are right. *There is no such thing as* **absolute simultaneity** *—events separated in space that are simultaneous for all midway observers in one uniformly moving frame are not necessarily simultaneous for midway observers in any other frame.* They all think they are correct; they all "understand" why the others record discrepancies, and there is no way to resolve the discrepancies. They are inherent in the nature of things; they are the reality. *Here-and-now* must replace *now—everywhere*.

6.4 A TEA PARTY

Buy the two postulates (which to date are abundantly backed by experiment) and receive along with them the relativity of simultaneity, as well as a free trip to Wonderland, where you will see the Hatter's watch and shrinking Alice.

The Hatter's Watch

A racehorse crosses the finish line—the *simultaneous* arrival of its nose at the wire and the corresponding reading on a clock face mark the moment. This is the way we reckon time. Yet if the front beacon on the space station in the previous example was replaced by a flashing digital clock, each observer would see the rear beacon lit at a different *time*. Since two observers in relative motion cannot agree on the simultaneous occurrence of events, they must surely differ in their measurement of time. In a word, **time is relative,** not *absolute*.

Imagine that we construct a light-clock, in which a pulse of light bounces back and forth between two mirrors, and some mechanism counts off the number of traversals, much like the familiar tick-tock. With such a clock mounted within our space ship, there would be no relative motion with respect to the pilot on board, and she would notice nothing extraordinary about it. Quite the contrary would be

---

* In this example the flares are fixed to the station, but that's not necessary. Any two events—two blinking stars, two blazing comets, whatever—that are observed by the midway commander as simultaneous will not be seen to occur simultaneously by the ship's pilot, and vice versa.

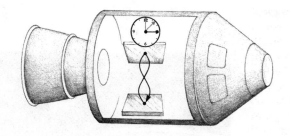

There was a young lady named Bright
Who traveled much faster than light;
She started out one day,
In the relative way,
And returned on the previous night.

A. H. REGINALD BULLER

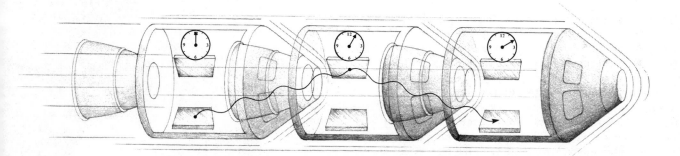

The passage of time when a light-clock is used, as seen by two observers in relative motion.

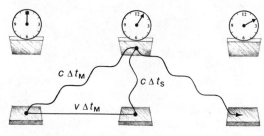

$$(c \, \Delta t_M)^2 = (v \, \Delta t_M)^2 + (c \, \Delta t_S)^2$$
$$c^2(\Delta t_M)^2 - v^2(\Delta t_M)^2 = c^2(\Delta t_S)^2$$
$$(\Delta t_M)^2 = \frac{c^2(\Delta t_S)^2}{c^2 - v^2} = \frac{(\Delta t_S)^2}{1 - v^2/c^2}$$
$$\Delta t_M = \Delta t_S / \sqrt{1 - v^2/c^2}$$

true for an observer on the space station as the ship whisked by. He would see the light pulse behave as if it advanced diagonally as both the vehicle and the clock's mirrors flew past. In other words, while someone moving with the clock sees the beam rattle to and fro between two stationary mirrors, the picture is markedly different as seen from outside. The pulse surely goes from one mirror to the other and back, but the mirrors themselves (as seen by an observer at rest with respect to the station) move along the flight path of the ship. The situation matches that of two boys having a catch across the aisle of a fantastically rapid train. One of them lobs the ball as the train streaks out of New York, and the other catches it as they enter Washington. To them it traveled two feet transverse to the car, but to the rest of us on the ground it sailed 200 miles diagonally. In the same way, the light beam travels a longer diagonal path. Bound by the second postulate to a constant speed, c, the beam must take longer to make the longer trip; that is, the interval between ticks and tocks will be larger.

There it is—*the moving clock, as seen by a stationary observer, runs slow!* For that matter, any clock (a wristwatch, a pendulum, a beating heart, a fertility cycle, or a dividing cell) must also slow down, must match the light-clock. Otherwise one could easily tell from the difference that they were actually moving, in violation of the first postulate. Everything, whether mechanical or biological, every process aboard ship slows down as viewed by an observer in relative motion peeping in the window. And since motion is totally relative, the space-ship crew, in turn, see all the station clocks running slow while their own appear to function quite normally. Let Δt_S represent a time interval recorded by someone stationary with respect to a clock, and let Δt_M be the correspondingly longer interval observed by someone moving with a speed v relative to it. A bit of geometry in the diagram and some algebraic manipulation relates both intervals via the expression

$$\Delta t_M = \frac{\Delta t_S}{\sqrt{1 - v^2/c^2}}.$$

In other words, an egg timer that when observed at rest takes three minutes ($\Delta t_S = 3$ min) to complete a cycle will take a longer time, Δt_M, to do its routine when viewed by someone in motion with respect to it. If v is small compared with c, their ratio is small and its square even smaller. Then $\Delta t_M \simeq \Delta t_S$, and the intervals are approximately equal, as they are ordinarily for us.

Suppose, now, that we sail past each other at 93,000 miles per second ($c/2$), clocks in hand. Don't even worry about who is doing the moving—it's relative. The equation then yields $\Delta t_M = 1.15\ \Delta t_S$. You see your own clock, with respect to which you are stationary, tick off $\Delta t_S = 1$ minute, as usual, while I see your clock, moving with respect to me, actually take $\Delta t_M = 1.15$ minutes for that same second hand to go once around. At the instant I observe your clock reaching exactly 1 minute, my own already reads 1 minute 9 seconds, and I naturally conclude that your time is running slow. Since these arguments are symmetrical, you, in turn, will see my tulips sprouting 1.15 times slower than they "ought," my candles burning longer, and my beard graying at a more charitable rate than you might otherwise have expected.

If you think all this is purely hypothetical and, like the Hatter's watch, someone has put butter in the works, let me reassure you that decades of experimentation bear out the actuality of the **time dilation** in every detail. As long ago as 1938, H. E. Ives of the Bell Telephone Laboratories first corroborated the relativistic retardation of time. He used the oscillatory rhythm of radiating hydrogen atoms as a natural clock. The faster the atoms were made to travel, the slower they vibrated, in precise accord with the prediction of special relativity.

More recently, experiments with unstable subatomic particles, like muons and pions, have further verified the theory. Pions at rest live for an average of about 0.026 millionths of a second before decaying. Large machines, like the cyclotron, can easily accelerate these particles up to, say, 0.75 c, whereupon they should survive long enough to travel about 5.9 m. Instead, they average paths of about

8.8 m, traveling farther and living longer than they "should," as seen by a laboratory observer. Yet if we calculate how long the pions "think" they lived (as might be measured by clocks moving along with them), using the time dilation equation, their lifetimes turn out once again to be 0.026 millionths of a second. As far as they are concerned, everything is marching on quite normally in their reference frame.

In October 1971, four exceedingly accurate cesium-beam atom clocks were flown around the world twice on regularly scheduled commercial jet flights. The idea was to compare the clocks with those at the U.S. Naval Observatory after they had circumnavigated the Earth. Because of the planet's spin, there were two trips around, first in an eastward and then in a westward direction. Things were admittedly complicated by the presence of gravity, which must be dealt with by the general theory of relativity. (The problem is that time speeds up as the gravity field weakens with altitude. This was in addition to the speed-dependent slowdown of time expected from the special theory.) In any event, the eastward clocks should have lost 40 ± 23 nanoseconds (a nanosecond is a thousandth of a millionth of a second, that is, 10^{-9} s) in the journey, and the westward-flying clocks should have gained 275 ± 21 nanoseconds. After $7600 had been dropped on airfare, it was found that, with respect to terrestrial reference standards, the eastward clocks had lost 59 ± 10 nanoseconds, and the westward clocks had gained 273 ± 7 nanoseconds—in breathtaking agreement! Ah, the Hatter would have loved that flying tea party.

Shrinking Alice

Neither time nor *space* is absolute in this universe where the speed of light is constant. Those pions in the last section travel a certain distance across the laboratory before committing suicide. But two observers, one moving with the particles and one standing in the lab, cannot possibly agree on that distance. The particle's relative speed, v, multiplied by its observed lifetime will certainly be the length traversed, but each viewer records a different lifetime. In that particular case, the pions think they have lived their usual 26 nanoseconds, and sailing along at $0.75c$, they believe they have traveled their customary 5.9 meters. The observer in the lab sees them live for 39 nanoseconds, during which they move from one end of the machine to the other, exactly 8.8 meters. "Well, then," asks Alice, "how far have they *really* traveled?" By now you know the answer, or at least you have been in Wonderland long enough to see it coming—they moved *both* distances, for **length is not absolute;** *it varies with the observer*. If the lab technician holds up a stationary rod that has a length of $L_S = 8.8$ m, the pion observer, moving with respect to it, will see the rod shrink to $L_M = 5.9$ m, and vice versa.

Time dilation leads directly to an expression for the contraction of length:

$$L_M = L_S \sqrt{1 - v^2/c^2} \ .$$

An object will appear to shrink along the direction of relative motion. Thus a meter stick sailing by at one-half the speed of light will seem to be only 0.866 m long to us stationary viewers. Of course, it doesn't matter which we imagine to be actually moving, us or the stick.

This remarkable equation is known as the **Lorentz-FitzGerald contraction,** after the two men who (just before the turn of the century) were responsible for its initial derivation. At the time it represented a last-ditch effort to save classical theory from the perplexities of the Michelson-Morley experiment. They simply made an assumption that the measuring device shrank as it moved through the aether by just the amount necessary to yield a null result. For them it was an ad hoc contrivance; for Einstein it came out of the theory as a reasonable consequence. We measure length by simultaneously lining up the ends of rods and rulers; simultaneity is relative, and so is length.

Someday we may have rocket ships that travel fast enough for length contraction to be significant. An astronaut, moving at 99.98 percent of *c*, will observe outside distances contracted to about two one-hundredths of their extent at rest. The distance to a star, measured to be 50 light-years* by an earthbound motionless astronomer, will appear as only one light-year to that intrepid soul. The astronaut will see the Earth and star moving at tremendous speed, with their separation shrinking accordingly, as if they were the end points on a gigantic yardstick. Strange as it may seem, flashing along at nearly the speed of light, our astronaut will arrive at the star in just about one year, rather than 50! That is, one year on his clock; we will see him, through a telescope, still young and vigorous, waving the flag in the vicinity of the star, a little more than 100 years after lift-off—50 of our years to get there and 50 more for the light from the flag-raising ceremony to get back. Should he decide to come right home, he would be wise not to expect a big welcome from his *old* friends. Unless they have gone star-trekking themselves, they will all have aged about 100 years to his two.

The question of what objects will actually look like to a rapidly moving observer because of the Lorentz-FitzGerald contraction is rather complicated. Unhappily, the semienlightened comic book renditions of people and things simply shrunk along the direction of relative motion is quite erroneous. In fact, if you are moving at close to the speed of light, where the contraction is applicable, things nearby will be flashing past so quickly you are not likely to *see* them

v/c	$\sqrt{1 - v^2/c^2}$	$1/\sqrt{1 - v^2/c^2}$
0.0	1.00	1.00
0.1	0.99	1.01
0.2	0.98	1.02
0.3	0.95	1.05
0.4	0.92	1.09
0.5	0.87	1.15
0.6	0.80	1.25
0.7	0.71	1.40
0.8	0.60	1.67
0.9	0.44	2.29
0.99	0.141	7.09
0.999	0.0447	22.37
0.9999	0.0141	70.71
0.99999	0.00447	223.6

---

* One light-year is the distance traveled by light in one year. It equals about 5,800,000,000,000 miles (5.8×10^{12} mi).

at all. Even so, further complications arise from the fact that light arriving at the retina (or on a piece of film) at some instant must have left different regions of the object at different times, depending on how far away they are. The result is that real three-dimensional objects will seem to have been rotated somewhat. For example, a passing sphere will still look spherical although the combined effect of the contraction and the different transit times will make it appear to have rotated a bit, with the far end coming into sight and the front end being obscured.

Fire up the rockets, press on to c, and let's see what happens; those equations hold more surprises. As v approaches c, time spreads out in increasing laziness until at $v = c$ the dilation equation loses its mind, and seconds dissolve into infinity, a tick into all of time. Nor is the Lorentz-FitzGerald saga comprehensible when $v = c$ and everything shrinks to nothingness. For speeds beyond c, all this becomes unreal, literally and mathematically. Only one conclusion is evident: *The speed of light is an upper limit on the rate of propagation* (see Question 6.4-13). There will be no accelerating to speeds far in excess of c and then whirling around to see yourself before you left—to pass and then catch the light from your own departure. Obviously, if we could move fast enough and see far enough, we could overtake and see the light that left Lincoln at Gettysburg or Kennedy at Dallas. But relativity benevolently saves us from watching our own history replayed, even though classical physics sets no such restriction.

A force applied long enough to a mass might be thought able to accelerate it up to any speed. Newton's second law, $F = \Delta(mv)/\Delta t$, allows of unlimited rates and in so doing seems to come into direct conflict with relativity. Still, the great "atom smashers" of today (such as the two-mile-long Stanford Linear Accelerator) can hurl tiny subatomic particles like the electron at incredible speeds of upwards of $0.9999999997\,c$, but strain though they may, c stands unattained, let alone surpassed. Einstein's way out of this apparent dilemma was the recognition that **mass is relative,** *not absolute!* The mass of an object depends on its relative motion with respect to the observer. Thus an object with a **rest-mass** of m_S (i.e., its mass as measured when stationary) moving at a speed v will attain an increased mass of m_M, such that

$$m_\text{M} = \frac{m_\text{S}}{\sqrt{1 - v^2/c^2}}.$$

Newton's second law is perfectly fine as is. It is the mass that increases, making it impossible to attain a speed of c. The harder an object is pushed, the greater will be its mass and the greater its inertia, its resistance to the further change in motion. And that resistance increases exceedingly fast as c is approached—all the king's horses and all the king's men cannot ever push Humpty Dumpty to c.

There was a young fencer named Fisk,
Whose thrust was exceedingly brisk.
So fast was his action,
The Lorentz-FitzGerald contraction
Reduced his rapier to a disk.

ANONYMOUS

Top-c-Turvy

In a large-screen TV set electrons flash along the picture tube at up to 90,000,000 m/s.

How cleverly Newton managed to dodge the question of mass and still reach the second law!

Alfred H. Bucherer, in 1909, first confirmed the relativity of mass with fast-moving electrons emitted from radium. Today the mass increase is so well established, so commonplace, that it's built into the very design of modern particle accelerators, which could not otherwise function.

The two-mile-long particle cannon at the Stanford Linear Accelerator Center (SLAC).

A moving particle has more mass and therefore even weighs more than when it was at rest. The unspoken *quantity-of-matter* conception of mass that seems so appealing is blown away with absolute rest and absolute time and absolute length. The constant, unchanging quantity-of-matter that we cling to intuitively is just not the mass that determines motion. A loaf of bread traveling at near c in a rocket ship may show a gigantic mass and weight as it whisks by

Earth, but it still tastes the same and looks the same to the hungry crew, and it makes the same number of sandwiches—manna it's not. Aboard that ship there would be no more of you, whatever your relative mass, than there was before lift-off.

The theory of relativity admits of three possibilities regarding the ultimate speed: (1) Ordinary matter, matter possessed of a rest-mass ($m_s \neq 0$), cannot attain a relative speed equal to or greater than c. (2) Entities having zero rest-mass (such as light itself, gravitons, and the particles known as neutrinos) are created with a speed c and cannot exist at any other speed. (3) Particles created with speeds in excess of c and for which c is a lower limit are theoretically allowed. These recently conceived hypothetical entities, known as *tachyons*, have never been observed and may well have no reality. Just because they are not theoretically forbidden, they don't necessarily exist. After all, we obviously don't now have all the theories we're ever going to have.

As it stands now, *c is the ultimate speed,* the limit for the propagation of matter and energy.

> Gentlemen! The views of space and time which I wish to develop before you have sprung from the soil of experimental physics, and therein lies their strength.

Standing at the podium, Herman Minkowski delivered his lecture to the Eightieth Assembly of German Natural Scientists and Physicians at Cologne, 1908. Then professor at Göttingen and a renowned mathematician, he had once taught Einstein at the Polytechnic in Zurich, but he was far from favorably impressed by young Albert. Ironically, here he stood years later presenting a paper entitled "Space and Time" that represented an elegant new reformulation of the special theory. He read on:

> Henceforth space by itself, and time by itself, are doomed to fade away into mere shadows, and only a kind of union of the two will preserve an independent reality.

Minkowski had recast relativity in a four-dimensional geometrical framework that would prove invaluable to Einstein's treatment of gravitation, his general theory of relativity.

"I'll meet you for lunch in the restaurant on the thirtieth floor of the hotel at the corner of 43rd and Park Avenue at exactly 1:30 P.M." We all know how to play the game—it takes three coordinates to locate something in space. Two street designations roughly fix the position of most buildings, and the floor number completes the location of our meeting point in space. Certainly, we could be far more precise by picking a reference point—say, a corner of your room—and then measuring along the three perpendicular axes where the walls and floor meet. Thus, if you go 14.71 miles along one axis and 3.32 miles along another, you will be stand-

DIMENSION

I have already said that it is impossible to conceive more than three dimensions. A learned man of my acquaintance, however, believes that one might regard duration as a fourth dimension. . . . The idea may not be admitted, but it seems to be not without merit, if it be only the merit of originality.

DIDEROT

Encyclopédie (1777)

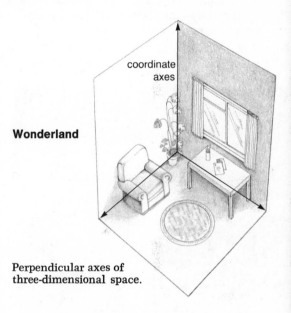

coordinate axes

Wonderland

Perpendicular axes of three-dimensional space.

Billy Pilgrim says that the Universe does not look like a lot of bright little dots to the creatures from Tralfamadore. The creatures can see where each star has been and where it is going, so that the heavens are filled with rarefied, luminous spaghetti. And Tralfamadorians don't see human beings as two-legged creatures, either. They see them as great millipedes—"with babies' legs at one end and old people's legs at the other," says Billy Pilgrim.

KURT VONNEGUT, JR.

Slaughterhouse-Five (1968)

"Clearly," the Time Traveller proceeded, "any real body must have extension in *four* directions: it must have Length, Breadth, Thickness and—Duration. But through a natural infirmity of the flesh . . . we incline to overlook this fact. There are really four dimensions, three which we call the three planes of Space, and a fourth, Time. . . . Some philosophical people have been asking why *three* dimensions particularly—why not another direction at right angles to the other three?—and have even tried to construct the Four-Dimension geometry."

H. G. WELLS

The Time Machine (1895)

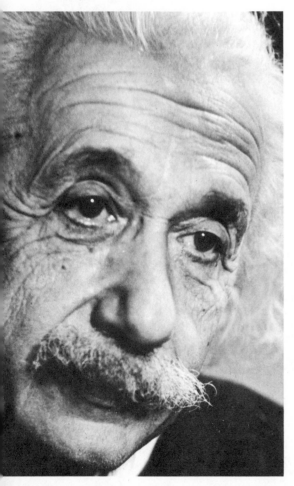

ing at the elevator door in the hotel. This is our familiar three-dimensional space. But in a real sense these three dimensions do not fix the event, our meeting, because that requires the specification of yet another coordinate, 1:30 P.M.

The unfolding of events requires four coordinates, three of space and one of time, if they are really to be located. And we can envision ourselves—the sequence of occurrences that mark our individual existence—as a progression of events in the *four-dimensional* realm of **space-time.** We are swimming, as it were, in a space-time continuum.

The product of c and time, having the units of distance, is interpreted as the fourth coordinate in space-time, a coordinate in every sense as significant as the other three. The history of our lives, of any object in the universe, can be imagined as a sequence of points sweeping out a smooth curve in four-dimensional space-time, known as a **world-line.**

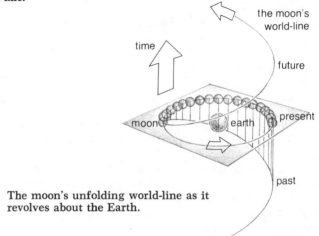

The moon's unfolding world-line as it revolves about the Earth.

It is not possible to envision motion through four-dimensional space-time pictorially; we are, after all, three-dimensional creatures locked into the imagery of a three-dimensional space. But we can get some sense of the concept of a world-line by considering a two-dimensional spatial motion and using time as the third axis (see the figure) along which events unfold. Moreover, you might imagine what life would look like if everything left afterimages that persisted in place. There you are now, bright and clear, the vision of the last moment at the end of an ever progressing luminous trail of faintly glowing images that winds downstairs, in and out of cars, back to the breakfast table, and from there, perhaps hours and miles away, to bed and yesterday and before. Each life is one meandering streak through space-time.

Although the imagery boggles the mind, the meaning is perfectly clear—may your world-line be a long one with as many wiggles as you find amusing.

REVIEW

Section 6.1

1. Is there any limit in Newtonian physics to the upper speed at which a thing can travel?
2. What is the speed of light?
3. Light is self-sustained by _____ .
4. What is so different about the motion of a bullet as compared with an electromagnetic wave?

Section 6.2

1. What is the classical principle of relativity?
2. How did Einstein generalize the principle of relativity?
3. Describe what is meant by *absolute rest*.
4. Is there any such thing as *absolute motion*?
5. Was the idea that all motion is relative first conceived by Einstein?
6. What role did Maxwell play in the development of the theory of light?
7. What did Poincaré have to say about the aether?
8. What was the Michelson-Morley experiment?
9. State Einstein's second postulate.
10. Why should the constancy of the speed of light be taken as an empirical law?

Section 6.3

1. Under what conditions can we expect that Newtonian mechanics will not substantially differ from relativity?
2. Why were the shortcomings of classical mechanics not apparent for more than 200 years?
3. What is a *midway observer*?
4. Define what is meant by *simultaneity*.
5. Rather than being absolute, simultaneity is _____ .
6. Discuss the role played by each of the two postulates in the analysis of the question of absolute simultaneity. How would the conclusions have differed if either postulate was erroneous?
7. What was Aristotle's view of simultaneity?

Section 6.4

1. What is the practical significance of the whole discussion of simultaneity?
2. Describe what is meant by a *light-clock*.
3. A moving clock seen by a stationary observer runs _____ .
4. Why do you think the effects of the slowing down of time were not observed before 1905?
5. If you and I were moving at a speed of $c/2$ with respect to each other, how long would it take for my clock to indicate the passage of one hour?
6. If you and I passed each other holding meter sticks in the direction of travel, what would each of us report about the other's ruler?
7. What is the Lorentz-FitzGerald contraction?
8. How could the equation for the contraction have been derived even before relativity was formulated? (Here is another instance of a correct result coming out of an incorrect perception.)
9. The faster an astronaut travels on the way to a star, the _____ will become the distance to be traversed as seen by that astronaut.
10. How long will a vertical meter stick appear to an observer moving horizontally with respect to it at 99% c?
11. What is the meaning of the term *rest-mass*?
12. Is Newton's second law applicable in relativity theory?
13. Has the variation of mass with relative speed ever been observed experimentally?
14. An observer at rest with respect to a meter stick will see it as having its maximum length. Will it also have its maximum mass?
15. Why does the notion of quantity-of-matter become useless as a synonym for mass in relativity theory?
16. What is a *tachyon*?
17. What is the limiting speed of any material object?
18. How many coordinates are needed to fix the location of a point in three-dimensional space?
19. Discuss the meaning of the term *world-line*.
20. What is the fourth coordinate of space-time?

QUESTIONS

Answers to starred questions are given at the end of the book.

Section 6.1

1. In terms of classical physics, what would you see if you rushed away at, say, a speed of 10 c, spun around, and stopped?
2. Is there any way you could see the digital clock (discussed earlier) light 12:00 and *then* 11:59 and 59 seconds?

3. Is there a way to see 12:00 light on the clock face as many times as you want within a matter of minutes (in accordance with classical theory)?
* 4. Keeping within the classical framework, suppose you travel at c for three hours, having seen 11:59 and 59 seconds

as the last clock reading. Now as you pass the planet Mongo, you slow down, and 12:00 finally lights up on the face of the distant Earth clock. What time will a Mongoian astronomer looking at Earth see on the face of that same clock at that moment?

Section 6.2

1. Explain how the first postulate of relativity bears on the concept of absolute rest.

2. Describe as thoroughly as you can at this point what is meant by an electromagnetic wave.

* 3. Relativity did away with the idea of aether. Why? What, then, became the medium for light?

4. Suppose you were traveling directly toward a woman at 50 mi/hr in a convertible Edsel and you tossed a ball in the forward direction at 20 mi/hr. How fast would she see it moving toward her? If you shined a light beam at her, how fast would she see it approaching?

5. How does relativity theory deal with the paradox Einstein perceived when he was a young man?

* 6. If two observers are separating from each other at 99.999% of *c*, and one shines a beam of light at the other, at what speed will each measure the light to be moving?

Section 6.3

1. How might the history of physics have been different if the speed of light were 1000 mi/hr rather than 186,000 mi/s?

2. Discuss the special role played by light in our perception of the universe.

* 3. Give an example of two midway observers in relative motion for whom two events would occur simultaneously.

4. What was meant when I said *here-and-now* must replace *now—everywhere*?

5. Suppose there were two observers, one at each end of a railroad car moving at 100 mi/hr, and someone switched on a lamp in the middle of the car. Would the light reach each of them at the same instant? What would someone outside at rest say about that?

Section 6.4

1. From the diagram on p. 147, is it obvious that the slowing down of time must increase with increasing speed? Discuss your answer.

2. The crucial point in the discussion of time running slow is that the dilation is not limited to light-clocks but must be true for *all* clocks. Why is this so?

* 3. It was noticed years ago that pions created in the upper atmosphere by cosmic rays were managing to travel all the way down to ground level. Although they presumably "should" not have lasted any longer than 0.026 millionths of a second before decaying, the journey was observed to take 160 millionths of a second. Explain how this could happen.

4. Our space-ship pilot, Jill, not far from Earth flies over a baseball stadium at 99.99% of the speed of light just in time to watch the national anthem light up on a large screen. She sees everyone in the stands singing and naturally joins in. Will she notice anything strange about the beat? Will she sit down before, after, or along with everyone else when somebody yells "Play ball" at the end of the song?

5. Review the discussion on p. 149. If someone moving along with the pion held up a rod 8.8 m long in the direction of motion, how long would it appear to the lab technician?

* 6. Return to Question 3. From the pion's perspective, it lives its usual 0.026 millionths of a second. And yet it travels about 48 km, as seen by an observer on Earth. Sailing at very nearly the speed of light, it can still have moved only about 7.8 m in this time, and yet it traverses the atmosphere. Explain.

7. Neglecting transit-time effects, how large would a 1-meter-long loaf of Italian bread look to a hungry observer with very sharp eyes if the two had a relative speed of 0.9 *c* in a direction parallel to the loaf?

8. A long, thin rocket takes off straight up from the ground and continues in that direction at 10,000 mi/hr. It is observed by three people: a guy selling ice cream on the beach, a reporter rising up in the gantry elevator at a constant speed, and the pilot aboard the missile. To whom will the ship look the longest? Who will see it as shortest? Who will see the gantry tower as shrunk?

9. A man named George held a meter stick horizontally out the window of a speeding car. It was seen and measured by five other people, some of whom were also in cars. When the six measurements were later compared, the two largest values were equal. One of these two was recorded by George's brother Frank. Knowing that George was driving the fastest car, place Frank during the experiment.

* 10. What will you reply the next time someone asks you something like "What is the distance from New York to London?" And let's assume here that you wish to be *absolutely* truthful.

11. What happens to the gravitational mass of a space ship as its speed increases relative to an observer?

12. If while flying past the Earth in a rocket ship at a very high speed you detected an extraordinarily large gravitational attraction acting on the ship, what would you conclude?

* 13. Astronaut Sam shined a flashlight beam on the front wall of his space ship as it passed near Earth. If the ship was traveling with a speed equal to or in excess of *c*, what would Sam see? What would you on Earth see? How does this violate the first postulate, and what can you conclude about such speeds?

14. Using the figure on p. 154 as a guide, draw a diagram showing part of the world-line of a book at rest on your desk.

15. Discuss Billy Pilgrim's remark on p. 153 as it applies to the concept of space-time.

MATHEMATICAL PROBLEMS

Answers to starred problems are given at the end of the book.

Section 6.4

* 1. A muon at rest ordinarily lives for an average of about 2 millionths of a second before decaying. How long will it appear to live to an observer moving with respect to it at 0.998 *c*?

2. The vaudeville team of Shultz and O'Hara, playing to a full house at Minsky's Burlesque, take exactly 60 seconds to tell their worst joke. How long will an observer in a rocket ship flying by at 0.995 *c* have to wait to lip-read the punch line? Will the observer start laughing *after* seeing the audience respond?

3. Suppose you looked into the window of a passing space ship and saw its clock running, such that 52 seconds went by on its face while 60 seconds elapsed on your clock. What was the ship's relative speed?

* 4. Calculate the approximate time dilation that would be observed by someone on the ground watching a clock in a supersonic jet flying at 1800 mi/hr. Use the approximation

$$\frac{1}{\sqrt{1 - \dfrac{v^2}{c^2}}} \simeq 1 + \frac{1}{2}\frac{v^2}{c^2},$$

which is OK for speeds less than around 0.3 *c*. Try doing it exactly to appreciate why the approximation is so useful.

5. Imagine yourself on a trip at a constant speed of 0.995 *c* to a distant star 10 light-years away from Earth. Roughly how long will the journey take, as seen by someone back on the planet? How long will it seem to you? What would be the Earth-star distance as seen by you? As seen by the Earth observer?

* 6. Suppose a long, narrow pepperoni is fired out of a new secret weapon developed by the Mediterranean arm of NATO. As it flashes past the viewing stands, its length is measured to be 44% of what it was before launch. How fast is it moving?

7. Calculate the length contraction for a high-speed jet fighter traveling at 3600 mi/hr. Try it exactly and then use the approximation (for speeds less than about 0.3 *c*):

$$\sqrt{1 - \frac{v^2}{c^2}} \simeq 1 - \frac{1}{2}\frac{v^2}{c^2}.$$

8. Suppose an electron is hurtling down the two-mile-long Stanford Linear Accelerator at 99.98% of *c*. How many feet long is the trip as seen by the electron?

9. A foreign power wishes to increase its food supply relativistically. At what speed must a ham and Swiss on white bread with mustard be fired if its mass is to double, as seen by the genius at the cannon? What do you think about the value of the scheme?

10. Construct a diagram of m_M/m_S versus v/c as speed increases approaching *c*.

* 11. An electron sailing down the picture tube of a TV set can hit a speed of 90,000,000 m/s. What is its mass compared with its rest-mass, as seen by a stationary viewer watching the twenty-fourth commercial during the late movie?

PART II Energy

My young son's Batmobile ground to a stop, its batteries finally drained. This technological dropping, a gift from the grandparents, is a dwarf incarnation of our twentieth-century power appetite, eating D cells as fast he would feed it—an *energy crisis* in a miniworld.

We learn about *power* from brown-outs and Batmobiles, from air conditioners with BTUs, sheiks with sunglasses, and the price of a gallon of gas. Running on fake chocolate bars and breakfast crunchies, it's the Pepsi generation, counting *calories* and *megatons*.

Although we blow it on electric toothbrushes, motorized cocktail stirrers, and "Lord only knows what else," we have tapped the very whirlwind of the universe—energy. Over the centuries that word has stood with many different meanings, for energy can manifest itself in numerous ways that seem distinct and quite unrelated. Most of its forms are familiar. We buy *electrical* energy "packaged" from the hardware store or "on tap" from the local power company. We can get *chemical* energy at any meat market, pub, or gas station. Rubber bands, girdles, and trampolines store *elastic* energy. *Radiant* energy pours in from the sun, light bulbs, and TV stations. We roast chestnuts with *thermal* energy and defend "the American way of life" with *nuclear* energy. Boulders roll down mountains because they have *gravitational* energy. Charging elephants always get the right-of-way because they have *kinetic* energy. And in the end all matter, because of its very existence, has *mass* energy. As we will see, several of these forms are actually identical, though disguised, and *all can be converted from one to the other!*

We measure energy through its effects on matter, and that means tangible *changes* in the condition or state of a material object (as represented, for example, by temperature, speed, position, mass, etc.). In that context *energy is the stuff of change*, and in the final analysis, change occurs through motion. We know too much now simply to say, as one used to say, "Matter is substance, and energy is the mover of substance," but there is truth to that phrase, too. We have found that there exists a chameleonlike entity called energy that can vary from one form to another. But *we do not, in fact, know what it is*—at least not to the extent of being able to provide a satisfactory conceptual definition. In other words, firemen can be changed into policemen, who can be changed into politicians; all are different examples of people, and we think we know what people are. Heat can be converted to electrical energy, and that can be turned into light—all energy, but we do not know what energy is in the sense of being able to describe the single underlying concept without having to resort to descriptions of each of its different manifestations. At best, we know energy by its many faces rather than by some essence that shows those faces.

When we get down to the basics, to the fundamental ideas—and energy is one of them—what we know comes of observation, of measurement. And the definitions that best serve our purposes are *how-do-you-measure-it* definitions, operational ones, not *what-is-it* definitions, which presume that the idea is describable in terms of other, still more basic imagery. The story will unfold more or less as it occurred historically, first by operationally defining each form of energy and then establishing that they are all equivalent, convertible one into the other.

The saga began as a whisper with Galileo, passed to Huygens, then Leibniz, and then, in the tumult of the second half of the eighteenth century, it burst into the roar of the Industrial Revolution. The emphasis shifted to power and work, the practical commodities of those whose new engines drove the mills and factories as never before. Nature became the fountainhead of surging power, power that could be harnessed to run the great machines of industry. Wind, rushing water, and steam, indeed Nature itself, was the wellspring, the source that energized the Industrial Revolution. Where once the force of Nature raged uncontrolled, beyond comprehension, it now miraculously became the docile giant in bondage at the mill, the tireless servant of the new technology.

"Nature's energy revealed" was an image that delighted the poets and painters of the time as much as it did the industrialists. But soon enough the inhuman machine turned ugly. The oppressive factories, the smoke-black skies, and the ash heaps were everywhere. Blake, Wordsworth, Byron, and Tennyson cried out in their poetry against the mechanistic view, against Newtonian rationalism, which they felt had brought the "dark Satanic mills." When Keats wrote, "Philosophy will clip an Angel's wings," he was talking about science and the human spirit as seen through a pall of smoke in an ominous age of machines.

Romanticism flourished in the late eighteenth and early nineteenth centuries. It welled up in revolt against the rigid intellectual bounds of unbending law. In the arts, free-wheeling fantasy replaced the clockwork, replaced the rules of music and literature. Turner, Delacroix, Mendelssohn, Chopin, Shelley, Poe, Coleridge, Schubert, Goethe, Wagner, and many, many others were the standard-bearers of Romanticism.

The nineteenth century, the century that began to the accompaniment of Bonaparte and Beethoven, was the first great age of energy. It was then that physics formalized the whirlwind, mathematizing *work*, *power*, and *energy*. Like a battery salesman in a modern toy store, William Blake could write "Energy Is Eternal Delight"— though I, for one, can't help thinking of Hiroshima whenever I read that line.

In its modern context, energy is a broad and subtle notion that, like all fundamental concepts, tends to wriggle away from easy definition—so let's take it slow, from the beginning, piece by piece. Chapter 7, for the most part, will examine mechanical considerations, and Chapter 8 will focus on the thermal aspects.

It is important to realize that in physics today, we have no knowledge of what energy *is*.
R. P. FEYNMAN
Contemporary physicist and Nobel laureate

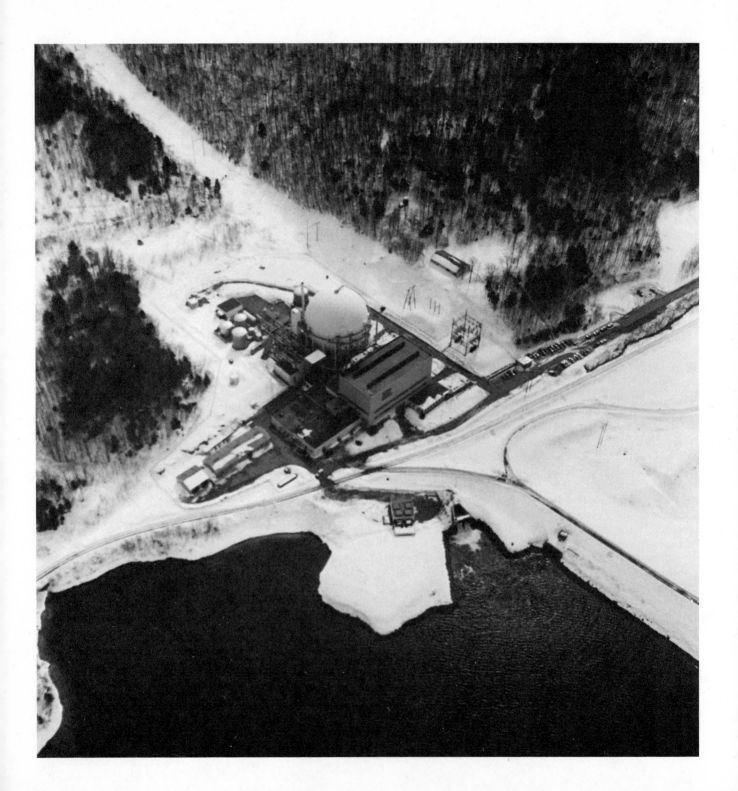

There came a new vision in the early nineteenth century, a vision born of the Industrial Revolution, a vision not of Newtonian forces but of **energy.** This chapter examines that vision. The formal notions that quantify **work** and **power** are considered in the first section since they are measures of energy change and so are central to the development. The two primary manifestations are **kinetic energy,** which arises out of mass in motion, and **potential energy,** which is energy stored as a result of moving against a force. The chapter ends with some preliminary thoughts on one of the great guiding principles of physics, the **law of conservation of energy.**

7.1 WORK, WORK, WORK

If we use the vision of Newton, change comes of pushes and pulls. Energy, without which there can be no change, seems somehow connected to force. In *The Two New Sciences,* Galileo already showed a grasp of the key ideas. He talked about the physics of pile drivers and recognized that the weight of the hammer and the distance through which it falls determine its effectiveness—force and distance relate in a crucial way. Earlier we learned that the product of *force* and the *time* over which it acts is a measure of the corresponding change in *momentum.* Now we will find that *the product of force and the distance over which it acts is a measure of the corresponding change in energy.*

Carry a refrigerator up three flights of stairs, or push a station wagon a few blocks, and you are certainly working as you exert force over distance. The engineers and scientists of the late eighteenth century referred to the product of force and distance as **work,** i.e.,

work = force × distance,

and we follow their lead, even though *change in energy* would be a better name.

Despite the beads of sweat on your brow after hours of laborious pushing on a reluctant elephant's rump, you have done no work at all *on* the beast if it doesn't move—at least as far as physics and the elephant are concerned. In the end, nothing about the pachyderm has changed from all your shoving. *For work to be done on an object, it must move along the line of force while the force is acting.* This is not to say that during all that pushing and shoving your muscles do no internal work. Indeed, whenever a tensed muscle changes length, it certainly does work, even if only against other muscles. Moreover, it takes energy just to maintain the tension in a muscle, whether or not there is any subsequent change in its length. In a somewhat similar way, my son's Batmobile will surely drain its batteries if left to push against an immovable wall, even though it does no work *on* the wall.

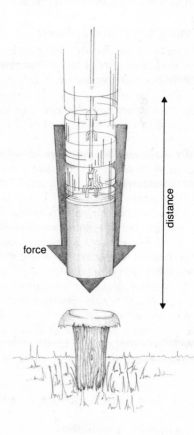

distance

force

Raising a 100-pound canary 5 feet straight up requires doing 500 **foot-pounds** of work on the bird, but holding it up there for an hour counts for nothing. You might just as well prop a ladder under it—that won't do any work on it either. Similarly, if you lift an apple whose weight in the SI system is 1 newton (about ¼ pound) up through 1 meter, you have done 1 **newton-meter** of work—also called a **joule.** In general, engineers in the United States still work in foot-pounds, whereas scientists prefer to talk in *joules,* but the two are interchangeable (1 joule = 0.74 ft-lbs).

Obviously we can do work, and so can our elephant; in fact, getting work out of fuel—*realizing energy*—is the name of the technological game. Dragging a plow through the earth, exerting hundreds of pounds of force over miles of land at the cost of a bushel of oats or a gallon of gasoline—that's the practical side of all this.

Wherever a force is in the direction of motion, work will be done. It requires a force to alter the speed of an object, to overcome friction, to compress a spring, to move against gravity; in each case, work must be done, and, in that sense, *work is the overcoming of resistance*. All the array of machines—from the can opener to the nuclear sub, machines that cut and crush, draw and hammer, lift or throw, turn or bend—all the mechanical movers do work.

Here the force (opposing the weight) is vertical, and so the work done depends only on the vertical distance traveled.

What the Whole World Wants—Power

Plumes of smoke were already rising high over the great ironworks of England by the latter part of the 1700s. The sky was lit with the fiery glow of the Industrial Revolution.* Ironmasters like John Wilkinson (of razor blade renown) made the cannons and ships and iron bridges, the very sinews of the era. But the manufacturers, particularly Matthew Boulton, the greatest of the lords of industry, set it all in motion.

Friendly, brilliant Boulton owned a fine factory in Birmingham that made snuff boxes, lovely buttons, and metal trinkets. Modestly powered by a waterwheel on a millpond, which unfortunately ran dry each year, the works slowed to a crawl in the summer. It was that annoying problem and the idea of using some sort of pump that he discussed with his friends Benjamin Franklin and Erasmus Darwin,† but to no avail. Yet another acquaintance with similar concerns, an industrialist named Roebuck, introduced Boulton to a young engineer, James Watt. This melancholy Scot, an inventor of considerable talent, had by then designed an improved steam engine, although it had not yet actually run. Boulton was immediately impressed by Watt's ingenuity, and Watt was delighted at the prospect of using the skilled craftsmen and tools at Boulton's factory. It was not very long before Watt's engine was raising water up from the pond to the top of the great wheel to drive the mill and make Boulton even richer.

100 lb

No motion, no work. (Art department's version of a 100-lb canary.)

zero   work

---

* A name given to the period by Friedrich Engels, colleague of Karl Marx.

† Charles Darwin's grandfather.

I was excited by two motives to offer you my assistance, which were love of you and love of a money-getting ingenious project.

M. BOULTON

From a letter to J. Watt

An early (1788) Boulton and Watt rotative-beam steam engine. This is a massive machine more than ten feet tall. To an observer on this side, the rest of the large black-iron flywheel is out of view below ground level.

In those days steam engines were still used only as pumps, a situation Watt would soon change. They had formed a partnership, Boulton and Watt, becoming the first and foremost manufacturers of efficient steam engines.* Boulton was so pleased with his new collaboration, with the mighty rush of the pistons, that he boasted: "I am selling what the whole world wants—power!"

It was the beginning of the age of power—wind, water, steam, whatever would drive the engines. Science had come to Nature not just to examine it but to wring energy from it, to borrow "the might of the elements."

---

* Wilkinson, using his new cannon-boring machine, made the precision cylinders for their engines.

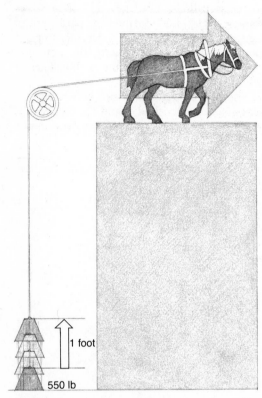

By definition, 1 horsepower is equal to 550 ft-lbs of work done in 1 second.

Now that is the wisdom of a man,
  in every instance of his labor,
to hitch his wagon to a star
  and see his chore done
by the gods themselves.

That is the way we are strong,
  by borrowing the might of the elements.

RALPH WALDO EMERSON (1803–1882)

American poet

Boulton and Watt brought their engines to Cornwall to pump out the seeping water that flooded the deep copper and tin mines. Shrewdly, they offered to install them free of charge. All they asked in return was a fraction of the difference between the cost of fuel for their engines and the cost of hay for a team doing the same amount of work each day. The mine owners began to grumble only as the years rolled on and they were still paying the power brokers.

When you contract by the job and you don't care how long it takes, you buy *work*. It doesn't matter whether a boulder is dragged up the mountain in an hour, a day, or even a year, as long as it gets up there. But if you are buying a day's labor, if you are concerned about getting things done fast, you want so much work every hour. That's the practical language of industry. **Power** is just that, *the rate of doing work*, the quantity of work done over the time it takes to do it:

$$\text{power} = \frac{\text{work}}{\text{time}}.$$

An engine that can do the same amount of work as another in half the time is twice as powerful. This early definition is fine, but it is rather restricted. Work is a particular sort of change in energy. In its modern sense, *power must be thought of more broadly as the rate at which energy varies, regardless of the manner.* We will see that the variation can occur in several ways other than just through work. For example, energy can be converted from one form to another, or it can be transmitted from one place to another, and that flow of energy is power as well.

Thomas Savery, an early worker in steam, had long before suggested that the measure of power be standardized by the rate at which a horse works, and Watt picked up the idea. He determined that a draft horse could exert a pull of about 150 pounds while walking at $2\frac{1}{2}$ mi/hr for a considerable time. That's equivalent to 33,000 foot-pounds each minute or *550 foot-pounds per second*, which became known as 1 **horsepower** (hp).

In the SI system of units, 1 joule of work performed each second is a power output of 1 **watt**, and 746 watts equal 1 hp. Almost all electrical devices are rated in watts to show their power consumption in the process of converting electrical energy into light, sound, heat, work, whatever.

By the way, a person can deliver work at the rate of roughly 1/10 hp, and a VW Beetle generates about 50 hp. Working steadily, the human body develops power in proportion to the amount of oxygen it can take into the lungs. A rate of consumption of about 1 *liter* (1000 cubic centimeters) per minute corresponds to a power output of about 1/10 hp or around 75 watts. An athlete in very good shape can raise that to roughly 5.5 liters per minute and so develop power continuously at close to 400 watts. When muscles are rested, they have a limited additional short-term supply of oxygen that may be called on for bursts of power. This reserve can be used at varying rates of, say, 450 watts for a minute or several kilowatts for a fraction of a second. But once that stored supply of fuel is exhausted, replenishing it will take some time. Athletes aware of the one-shot nature of the process attempt to use this rush of power only when it counts most.

We generally buy energy, not power. It doesn't matter, as far as cost is concerned, whether you burn one light all night or several for an hour; you pay for the total amount of energy used. The most common unit of electrical energy is an ugly contrivance known as the *kilowatt-hour*, which corresponds to 1000 watts of power delivered continuously for one hour. That is equivalent to 3.6 million joules and at present it costs roughly ten cents. Strangely enough, the cost per kilowatt-hour actually goes down for large-scale power consumers even though the logic of conservation suggests just the opposite.

When next your speakers blast the Rolling Stones at ten watts per channel, remember the modest Scotsman and the boastful button maker whose steam engines set the pounding tempo 200 years ago.

7.2 KINETIC ENERGY

To hurl a bowling ball, to have it accelerate from rest to some final speed, a force must be exerted ($F = ma$). Work is done in acting against inertia all the while the ball is in hand moving faster and faster. Let loose, it sails down the lane, crashes into the pins, sends them flying, does work on them and comes to rest. Work-motion-work. And if we simply define energy as *the ability to do work,** then the moving ball has some *energy of motion*. Similarly, the falling hammer of a pile driver rams a wooden pier into the ground, converting energy of motion into work and bringing us back to Galileo, where the analysis began.

Huygens never liked Descartes's idea of quantity-of-motion, momentum. To be meaningful, as we have already seen, momentum has to be formulated as a directional concept like force. A body at rest could explode into two pieces rapidly moving in opposite directions, and yet the total momentum would have to remain zero— one motion in the positive direction, one in the negative, and so they cancel. That does not bother us, but it bothered Huygens, and he sought to find a different measure of motion that was independent of direction, a measure that would not vanish unless the motion itself ceased. His studies of colliding bodies led him to the realization that there was something special about the product of the mass and the speed squared. Whether the speed was taken to be positive or negative, squaring it always yielded a positive nondirectional quantity.

Leibniz, Newton's bitter rival, picked up the idea, calling it the *living force*, the **vis viva.** He was able to show that vis viva was proportional to Galileo's product of weight and height. For Leibniz mv^2

---

* This definition has its shortcomings, too, as we will see in Section 8.5, where we consider energy that is unavailable for doing work.

became the crucial measure of motion rather than mv, and the great and meaningless vis viva controversy began between his followers and those of Descartes. It was not until the beginning of the 1800s, until the Industrial Revolution was well underway, that Thomas Young shifted the imagery and spoke of mv^2 not as living force but as *energy!* He went on to conclude rightly that "labour expended in producing any motion, is proportional . . . to the energy which is obtained." In other words, work done equals the resulting change in energy.

Nowadays we call energy of motion **kinetic energy,** a term introduced about 100 years ago by one of the great figures of the last century, Lord Kelvin. To see how work relates to it, let's consider a situation in which motion is changing. Imagine a body moving under the influence of a fairly constant force: a bullet decelerating as it penetrates wet clay; a rocket driven by a fixed-thrust motor; a ball freely falling of its own weight. In each case, work is force × distance, but we can restate that product so that it comes out only in terms of mass and speed (Question 7.2-6). By rewriting force as ma, using the definition of acceleration to replace a and substituting from the mean-speed theorem for distance, we can show that

work = change in kinetic energy,

that is,

$$W = \Delta(\text{KE}),$$

provided that kinetic energy is defined by the relationship

$$\text{KE} = \frac{1}{2} mv^2.$$

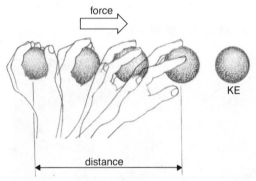

A force acting over a distance results in a change in kinetic energy.

This calculation (in which the factor of $\frac{1}{2}$ multiplying the vis viva arises from the theory of motion) was first carried out by Gaspard de Coriolis (1792–1843). In the imagery of Young, he described vis viva as the "transmission of work." Leibniz's living force was conceptually transformed into the modern butterfly of kinetic energy.

Force ruled the Newtonian era, but this was an age of work and power and now *energy*. The times affect the physics and the physics affects the times; vis viva was forgotten, drowned out by the roar of the engines, and in its place stood its lookalike, kinetic energy, one of the most fundamental notions of contemporary physics.

To get a feel for how numbers fit into all this, let's go back to our 12,000-lb elephant of Chapter 4. We found there that it would take a force of 825 lbs exerted for 10 seconds to slide the stranded beast across an ice pond, accelerating it at 2.2 ft/s² from rest up to 15 mi/hr (22 ft/s). The mean-speed theorem allows us to calculate the runway needed, and that turns out to be 110 ft. Now then, the work done in changing Jumbo's horizontal motion,

$$F \times \Delta d = 825 \text{ lb} \times 110 \text{ ft},$$

is 90,750 ft-lbs. The kinetic energy, in turn, increases from its initial value of zero to a final value of

$$\Delta(KE) = \frac{1}{2}mv_f^2 = \frac{1}{2}\frac{12000}{32}(22)^2,$$

which equals, you guessed it, 90,750 ft-lbs. That really is no surprise. They had to be equal; we defined them that way. Incidentally, someone's going to have to do 90,750 ft-lbs of work on Jumbo to bring him back to rest.

Note that doubling the speed of, say, an automobile quadruples the kinetic energy and also quadruples the work needed, both to get it up to that speed and to stop it. Since the braking force is fairly constant, once the brakes are slammed on, it will take four times the distance to stop a car moving at 40 mi/hr that it takes at 20 mi/hr.

Along with mass and speed, length and time, kinetic energy is a relative quantity. Your 32-lb suitcase is motionless next to you in a plane flying off to Chicago at 600 mi/hr. With respect to you it has no kinetic energy at all. But to someone on the ground it flashes past with a KE of almost 390,000 ft-lbs.

7.3 POTENTIAL ENERGY

Suppose that there is some sort of driving force constantly exerted on a body, like its weight. To move against that pull requires the application of a counterforce and the doing of work. This is so with a stretched spring, just as it is with the gravitational field. The crucial point is that the force—electrical, magnetic, gravitational, whatever—continues to act even after the displacement, when the object is held motionless. Once let loose, that force will drive the body back to where it came, imparting kinetic energy in the process.

Stretch out a slingshot working against its rubber band, and then let go. As the band snaps back, it will do work on the projectile, giving it kinetic energy. Raise a rock in the Earth's gravitational field, and then let go—down it comes, picking up speed and KE in the fall.

But now let's back up a bit. Work was ultimately converted into kinetic energy. That much is OK, but what was going on while the rock was held *motionless* up in the air? Similarly, work goes into stretching the rubber band, but not until it is fired will KE be generated. Clearly, it is possible to do work and not have it immediately appear as kinetic energy, and yet the potential for generating that energy is there. It's as if energy were *stored* in the system, waiting to be let loose. This retrievable stored energy, *energy by virtue of position in relation to a force*, is known as **potential energy** (PE), a name suggested by William Rankine (1820–1872).

A stretched garter holds *elastic* PE; a flagpole painter tries to forget his *gravitational* PE; a bright red stick of dynamite is gift-wrapped *chemical* PE; a nylon sweater just out of the dryer has *electrical* PE; two magnets pulled apart wait to unite with *magnetic*

The flat-spring heart of a windup toy car stores elastic potential energy.

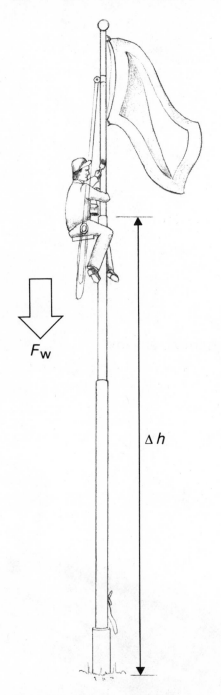

In a gravitational field the weight of an object multiplied by its height above some level is its potential energy with respect to that level.

PE; and all those neatly painted warheads buried in their silos threaten with *nuclear* PE.

Of course, this is all a construction; the very notion of energy is a contrivance of the human imagination. And yet it holds together a diversity of observations, providing a unity that makes them *understandable*. Having embraced the principle, we construct a fine logical web and lay it over Nature, just as the poet would—but still, the warheads are real enough.

Gravitational PE

When a painter climbs to the top of a flagpole, he does work in overcoming the downward pull of gravity. Work corresponds to a change in energy, and we now associate that change with potential energy since he is motionless on his perch:

$$W = \Delta(\text{PE}).$$

The work done by the painter in mounting the pole is the product of the force he exerts, i.e., his weight, F_W, and the vertical height through which he rises, Δh. That same product is the increase in his PE,

$$\Delta(\text{PE}) = F_\text{W}\,\Delta h.$$

Potential energy, like kinetic energy, is a relative quantity; in fact, the term "height" is itself relative. How high is your nose? Above what—the ground, the floor, your lip, or sea level? Imagine a three-pound potted hemp plant sitting on a two-foot-high window sill. With respect to the floor, the pot already possesses a gravitational PE of (2 ft × 3 lb), or 6 ft-lbs, as a gift from whoever put it there. Now if it's raised an additional foot up, it then possesses 3 ft-lbs of PE with respect to the sill, and 9 ft-lbs with respect to the floor. That just means that if it was released, it would be capable of delivering three times as much work to the floor on impact as to the sill. Of course, if all this is happening on the forty-second story of an apartment house, 500 feet up from the street, you are talking about 1509 ft-lbs of PE with respect to the sidewalk. In any event, the change in potential energy, when the pot was lifted a foot, was 3 ft-lbs, regardless of the reference level.

Potential energy arises from the interaction of two or more physical objects; in this case it is the Earth and the pot pulling on each other via gravity. When their separation increases, the Earth-pot system increases its gravitational PE. We must conclude that as the pot falls *down*, the Earth moves *up* to meet it. Certainly we have never measured that minuscule acceleration of the planet, but the logic demands it. *Potential energy is stored in the system of interacting objects;* neither Earth nor pot is by itself the proprietor of PE. If that bothers you, imagine the Earth shrunk down to the size of a very small asteroid, or a large grapefruit. Now you can see it move toward the pot in your mind's eye.

Electrical PE

Of the several fundamental forces that rule the universe (Chapter 15), none is more richly harvested than that of the *electromagnetic field*. Charges interact, they push and pull on one another, and that means work can be done by moving them with respect to one another against these forces. The result is energy stored in the system, *electrical potential energy*.

A spring or a rubber tire, distorted—stretched, compressed, or twisted—is a storage place of energy, *elastic* PE. But in the final analysis, that energy is accrued by doing work against electrical forces that exist between the atoms and molecules of the elastic material. The many seemingly different manifestations of elastic energy, ranging from stretched bubble gum and muscles to the twisted mainspring of a cuckoo clock, are all basically *electrical* PE.

In a quite similar way, *chemical* PE arises from interatomic and intermolecular attractions that are also electrical in nature (Section 13.2). Every molecule, by virtue of the arrangement of its electrons and protons, is a storehouse of electrical energy. The rearrangement of electrons that results from a chemical process, such as the burning of fuel, causes a change in their kinetic and electrical potential energies. Energy that was hidden deep in the molecular bonds is set loose to drive the lawn mowers and the rocket ships.

The combustion of any of the common fuels, i.e., their rapid combination with oxygen, releases roughly the same amount of energy per kilogram for each, and that is from 10 million to 50 million joules. Burning completely 1 kg of so humble a fuel as dried dung will provide 17 million joules (12.5 million ft-lbs). That much energy is equivalent to your carrying an 8600-lb truck up the 110 stories (1454 ft) of the Sears Tower! A tenth of a kilogram of butter (about ½ lb) spread liberally over a kilogram of bread (2.2 lb) and then set ablaze will liberate roughly 13 million joules. It is the conversion of chemical PE within the body, the slow "burning" or oxidation of food, that powers us all.

Living cells possess tiny globules, mitochondria, that in turn contain proteins known as enzymes. When there is plenty of oxygen available, these enzymes can take apart glucose (sugar) molecules, in effect "burning" them to yield carbon dioxide, water, and energy. Much of that released energy is then stored in the electrical bonds that hold together the compound adenosine triphosphate (ATP). This molecule, found in every life form, is the primary energy transmitter in biological systems. Energy-rich ATP leaves the mitochondria and moves within the body of the cell to serve as a ready source of power wherever needed. The energy liberated when its phosphate bonds break down can be used to contract muscles, operate nerve conduction, synthesize other complex compounds, etc. Indeed, whenever an organism does work, ATP participates in the process.

Just as American Indians once used "buffalo chips" as fuel, Asian Indians today use dried dung patties, which are displayed for sale all across the country.

7.4 CONSERVATION OF MECHANICAL ENERGY

The notion of *conservation* in physics alludes to immutability, to constancy—a particularly attractive theoretical idea. That some quantity remains unchanged as the system it relates to undergoes change implies that the quantity has a deeply fundamental character. Take a piece of material, chop it, bang on it, burn it, do whatever you like to it; yet in the end the mass of all the fragments will equal the original mass—*mass is conserved.**

Not surprisingly, physicists have long pursued conservation laws as if they were the very keys to the inner recesses of the universal clockwork. Descartes proclaimed the conservation of momentum (quantity-of-motion), and Huygens and Leibniz felt instead that they had the crucial conservation law for what the latter called vis viva. Actually, unbeknown to all of them, it is an easy matter to show that conservation of kinetic energy implies momentum conservation. Relativity theory established that the two are really the *time*-and-*space* parts of a single four-dimensional quantity and so are intimately connected. Indeed, conservation of mass is inseparably bound to the same relativistic notion.

To this day we still seek conservation laws with an almost mystical conviction that this is the true road to Oz. Of course, when it turns out that one of our conservation laws clearly is not in agreement with some new observation, we may be forced to pull it down from its high place and chuck it. As with all things in physics, we never prove them true; at best we fail to disprove them. Conservation of mechanical energy is one of those remarkable laws that have never yet been violated. Simply put, it maintains that

The total mechanical energy of an isolated system (PE plus KE) is always constant.

This is really a restricted version of a much broader and more profound conservation law (Section 8.4), which encompasses all forms of energy. As such, it applies only when there is no conversion of mechanical energy into any other form. Of course, that represents a tremendous restriction on real systems, where friction is generally unavoidable. Still, it gives us a beautiful new insight into the behavior of moving objects.

Christian Huygens

Law Abiding

A ball weighing 2 newtons held 100 meters above your outstretched hand has a gravitational PE with respect to you equal to its weight (F_W) times that initial height, or 200 joules. This energy is stored, ready to be converted to KE whenever the ball begins to drop. Initially $(\text{KE})_i = 0$, and $(\text{PE})_i = 200$ joules. As the ball descends, speeding up, the gain $\Delta(\text{KE})$ equals the loss $\Delta(\text{PE})$ at any instant, so that the total energy is always constant (neglecting friction). One quarter of the way down, PE will have decreased to 2 newtons $\times$ 75 meters, or 150 joules, and KE will have increased to 50 joules. At

---

* What is actually conserved, we now know, is mass-energy (Chapter 15).

the halfway point, PE = 100 joules = KE. The instant before the ball slaps into your hand, all the energy will have been transformed to kinetic (KE = 200 joules), with none any longer stored as potential. From the definition of KE = $\frac{1}{2} mv^2$, we can calculate the arrival speed to be about 44.3 m/s.

Note that if you hurled that same ball straight up with a speed of 44.3 m/s, it would again reach a peak altitude, where the ball would momentarily come to rest at 100 meters. Thereafter, it would fall back exactly as before, returning to your hand once more with a speed of 44.3 m/s. That's why police should never fire warning shots straight up.

A pendulum (or a swing hanging from a tree) is another nice example of energy conservation, as the bob transforms PE to KE and back to PE. Remember that potential energy is always taken with respect to some arbitrary reference. The most convenient place to fix the zero-PE level in the dropping-ball example was at the height of the outstretched hand; here it is at the very bottom of the pendulum's swing—again, *at the lowest point in the motion.*

The moving pendulum rises up to a maximum height at either extreme of its swing. There the bob is momentarily at rest (KE = 0), and all its energy is potential. As it falls, PE is converted to KE, the speed increases, and the height above the reference level decreases. At the instant when the pendulum is all the way down and hanging vertically, PE = 0, all the energy is kinetic, and the bob is moving with its maximum speed. Note that Δ(PE) equals the weight of the bob times *the height through which it has fallen*, regardless of the reference level. In other words, Δ(PE), the quantity of interest here, is independent of the zero level.

Along with Newton's laws, we now have the law of conservation of energy, an additional powerful tool with which to analyze problems. Indeed, the energy formulation can often provide a wide overview of a given situation unlike the usually more detailed force considerations of Newtonian theory. For example, we can expect that a planet in an elliptical orbit will slow down as it ranges farther from the sun. After all, that increase in distance means an increase in gravitational PE, which must be accompanied by a corresponding decrease in KE and therefore speed. Since it falls back when approaching the sun, a planet must speed up if energy is to be conserved.

The conversion of energy from one form to another in the swinging pendulum.

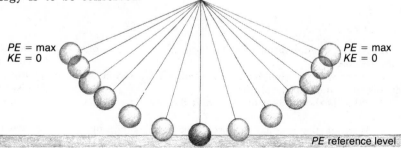

PE = max
KE = 0

PE = max
KE = 0

PE reference level

KE = max
PE = 0

The conservation of mechanical energy is one of the mainstays of modern science. So deeply is it rooted in the catechism of physics that many would doubt all else before it.

In the 1930s, physicists found themselves hard-pressed by an apparent violation of energy conservation. Electrons ejected from the nuclei of radioactive atoms (in what is called beta-decay) were flying off with "too little" KE. Rather than believe in the possibility of an exception to energy conservation, rather than accept the obvious facts that were there for all to see, the theoretician Wolfgang Pauli was steadfast. In 1931 he postulated that an invisible phantom particle emitted with each electron carried off exactly the amount of energy needed to maintain conservation. It had to be a remarkably strange little creature with no charge, essentially massless, and traveling at the speed of light. The *neutrino* ("little neutral one") could, on the average, sail straight through a thickness of well over a light-year of solid lead before colliding with a nucleus, thereby leaving some sign of its reality. Better to believe in this fantastic dwarf than to think the unthinkable. Twenty-five years later, in 1956, the incredibly elusive neutrino was finally detected. And although we cannot measure its energy to absolutely set right the balance sheets, its mere existence reaffirms our faith, as do countless other experiments. *Energy is conserved*—we believe.

EXPERIMENTS

1. It's easy to get at least a rough idea of the amount of power your body can generate either on a continuous basis or in spurts. Simply jog or run up a flight of stairs. Your weight times the vertical height risen equals the work done, and that divided by the elapsed time is the power. You can compare your performance to the record dash (set in 1978) up to the eighty-sixth floor of the Empire State Building in 12 min 32 sec.

2. Attach a rubber band to a toy truck as in Experiment 4, Chapter 4. Position a board to make an inclined plane from the floor up to a stool. Pull the truck very slowly up the incline at a constant speed, and note the length of the band. Try it with the board at several different inclinations. Now slowly raise the truck straight up from the floor to the stool. Presuming that the band's stretched length is roughly proportional to the force exerted, what can you say about the work done in raising the truck via the different routes? The best way to do this experiment would be with a small spring scale, but the rubber band makes the point adequately.

REVIEW

Section 7.1

1. All the various forms of energy are _____, one into the other.

2. What was the significance of wind, water, and steam during the Industrial Revolution?

3. What is the product of force and the time over which it acts?

4. What is the product of force and the distance over which it acts?

5. Define the concept of *work*.

6. How much work is done on the Statue of Liberty if a ten-pound force is exerted on it for one hour?

7. List two different units of work.

8. A *joule*, the SI unit of _____, is equal to 1 _____.

9. How many *foot-pounds* equal a joule?

10. Define the concept of *power* in terms of work.

11. Give a broad definition of power that encompasses all forms of energy.

12. What is the meaning of the term *horsepower*?

13. What is a *watt*?

14. How many watts equal 550 foot-pounds per second?

15. Discuss the meaning of the term *kilowatt-hour*.

Section 7.2

1. Define *energy* (as best you can).

2. Leibniz introduced the term *vis viva*. What did he mean by it?

3. Work done equals the resulting _____ in energy.

4. *Kinetic energy* is energy of _____.

5. Define KE in terms of m and v.

6. Does an object have kinetic energy whenever it is moving, even at a constant speed?

7. In what sense is it easier to stop a kid on a bike than to stop a bus coasting at the same speed?

8. Suppose you knew the mass and velocity of a ball as well as the time it would take to come to rest in the catcher's hands. How would you calculate the force needed to stop the ball and the distance over which it would come to rest in the hand?

Section 7.3

1. Why is it convenient to assume that energy can be stored in a system?

2. Define *potential energy*.

3. What kind of PE is involved when you bend a bow in the process of firing an arrow?

4. What kind of PE is involved when Woody Allen dan-

gles out a window on the tenth floor at the end of a computer tape?

5. Discuss the relationship between work and PE.

6. Write an expression for the change in the gravitational PE of an object in terms of its weight and height.

7. A stretched rubber band stores _____ PE.

8. We know that a rubber band stores energy because it has the ability to _____ _____.

9. A spring is capable of storing _____ PE. How can you verify that?

10. Ultimately *elastic* and *chemical* PE are in fact specific manifestations of _____.

11. What chemical compound stores energy in biological systems?

Section 7.4

1. What does it mean to say that mass is conserved?

2. State the law of conservation of mechanical energy.

3. In what sense is the law of conservation of mechanical energy really an idealization?

4. If you throw a rock straight up at 50 mi/hr (neglecting friction), at what speed will it drop back into your hand?

5. At what point in its swing does a pendulum have the most kinetic energy? The most potential energy?

6. What can you say about the potential energy of a planet moving in a circular orbit around the sun?

7. What can be said about the kinetic energy and the speed of a planet in a circular orbit (see previous question)?

8. Why was the notion of the *neutrino* introduced originally?

QUESTIONS

Answers to starred questions are given at the end of the book.

Section 7.1

1. Discuss the meaning of R. P. Feynman's remark about energy quoted on p. 159.

2. Make up a list of the different kinds of energy you caused to be "consumed" in the last three hours. Specify the method in each case.

* 3. A 100-pound, gold-embossed set of the great classics is carried up the stairs to the sixth floor by a young man working his way through college. Does he do work *on* the books? Would he have done the same amount of work if he had taken the elevator?

4. Suppose it takes 10 newtons of force tugging via a string on a ball moving around in a circle with a 4-meter

circumference. How much work is done on the ball per revolution? Be careful with this one!

5. A 10-kg asteroid floating in space is to be accelerated from rest to a speed of 10 m/s. Inasmuch as the asteroid is weightless, will work have to be done on it during the acceleration?

* 6. A ten-pound book floating in free space is brought to a speed of 10 ft/s and then maintained at that rate for 100 ft. Is work done on the book during that 100-ft portion of the trip?

7. A ten-pound book sitting on the ground is brought to a speed of 10 ft/s and then maintained at that rate for 100

ft by a horizontal force of 2.5 lbs. Is work done on the book during that 100-ft portion of the trip?

8. If a 24-hp electric motor will raise an elevator 6 floors in 16 seconds, how long will it take a 48 hp motor to do the same work?

9. If a 50 hp engine can raise a load 100 ft in 10 s, how long will it take to raise the same load 200 ft?

10. Two cars that are otherwise identical are fitted with different engines. If one can drive to the top of a hill in half the time it takes the other, what can you say about the engines?

* 11. Two roads lead to the top of a very steep mountain. One goes straight up the side; the other spirals gradually several times around the mountain. Does it take the same amount of work (neglecting friction) to reach the peak via either route? Assume you are driving my '69 Toyota. Why might it be wise to choose the winding road?

12. If Superman is really "more powerful than a speeding locomotive," what does that mean regarding his ability to pull an equal number of freight cars?

Section 7.2

1. The gas exhausting from an engine exerts a reaction force. Is any work done on the rocket while it's held in place as the engine fires? Once the missile is released, does the gas do work on it? Out in space beyond gravity, how does that work manifest itself?

* 2. Why must we consider kinetic energy to be a relative quantity? Does that mean that work is relative as well? Discuss your answer in detail, keeping in mind that $W = \Delta(\text{KE})$.

3. In the light of the previous question, suppose that while flying in a jet at 500 mi/hr someone next to you does 10 joules of work on a ball, imparting 10 joules of KE to it as it sails down the cabin. How will things look to someone on the ground?

4. A speeding black car is roaring down the highway straight for our dear Lois Lane, who is lying unconscious across the road. *But* just in time, Superman jumps in the way, exerts 100,000 lbs, and brings the car to a screeching stop in 1 ft. What was the car's kinetic energy the instant before it encountered the "Man-of-Steel"?

* 5. Presume that friction is negligible and that you can exert a 100-lb push on a stalled car over a level distance of 100 ft without collapsing. What else must you know in order to predict the speed attained by the car? Explain your answer.

6. We wish to show that $W = \Delta(\text{KE})$, provided that $\text{KE} = \frac{1}{2}mv^2$. First prove that constant force, F, times distance, Δd (that is, $F \times \Delta d$), equals

$$m\left(\frac{v_t - v_i}{\Delta t}\right)\left(\frac{v_t + v_i}{2}\right)\Delta t.$$

Then show that this reduces to

$$F \times \Delta d = \frac{m}{2}v_t^2 - \frac{m}{2}v_i^2.$$

Section 7.3

1. An athlete with pole in hand runs, vaults into the air, and lands. Describe the different forms of energy manifest at every moment leading up to the instant before hitting the ground.

2. Someone yanking on a rope attached to a pulley system does 10,000 joules of work raising a piano up to a second-floor window. Assuming friction to be negligible, state by how much the piano's gravitational PE is altered.

* 3. How much work can be delivered by a pile driver weighing 3000 lbs, 20 ft above ground level? What is its potential energy with respect to the ground? With respect to the bottom of a 10-ft-deep well?

4. Suppose a compressed spring fires a ball weighing 10 N straight up to a maximum height of 5 m. What is its gravitational PE at maximum altitude? How much energy was stored initially in the spring, assuming negligible friction losses?

5. Would you say that there is a lot of energy stored in ordinary chemical fuels? Justify your answer. What does this suggest about the efficiency with which we use food energy?

6. Imagine an object in a free-fall elevator. Discuss the relationship between work done by gravity and $\Delta(\text{PE})$ as considered by observers inside and outside watching an object float around within the elevator.

* 7. Setting the change in gravitational PE equal to $F_w \Delta h$ assumes something about F_w and consequently about g. Discuss the situation. Go slowly with this one—it's subtle.

Section 7.4

* 1. The car on a roller coaster is usually tugged up to the top of the first hill by a chain drive and then released to coast the rest of the way. Realistically, some small amount of energy must be lost in friction all along the journey. (a) Where is the gravitational PE a maximum? (b) Where is the KE a maximum? (c) Having once reached its maximum gravitational PE, can the car attain it again? (d) Having once reached its maximum KE, can the car attain it again? (e) Must each successive hill be lower than the one before it? (f) Must the second hill be lower than the first? (g) If the track closes on itself, will the car go around forever?

2. An arrow is fired from a bow straight up. Neglecting frictional losses, compare the elastic PE at the moment before release with the gravitational PE at maximum altitude, with the KE the instant before the arrow lands.

3. What is the purpose of jumping a few times on the end of a diving board to set it oscillating before leaping off?

4. Suppose you *throw* a Superball straight down at the ground from a height of, say, 4 ft. Assuming there are no losses, will the ball return to the height at which it was released? How much KE will it have at that instant? In what form is the energy at the moment when the ball is squashed motionless against the floor? What do you need to know in order to determine the maximum height the ball will reach? How can you find the speed at which the ball was thrown, having only its weight and a ruler?

5. Discuss Galileo's inclined-plane experiment concerning the law of inertia (p. 72) in terms of energy conservation. Analyze his pendulum demonstration (p. 73) via energy considerations.

MATHEMATICAL PROBLEMS

Answers to starred problems are given at the end of the book.

Section 7.1

1. A 20-foot-long ramp leads to a platform 5 feet above the ground. How much work is done in sliding a 100-lb box up the ramp, assuming no friction? How much work would have to be done in just lifting the box straight up to the platform?

* 2. A force of 10N exerted vertically lifts a 1-kg mass through 10 m. How much work is done by the force? A force of 10N exerted horizontally moves a 1-kg mass through 10 m over a frictionless floor. How much work is done by the force?

3. A 500-lb piano is pushed 10 ft across a smooth floor by a 250-lb piano mover. If while lifting vertically with 100 lbs he pushes horizontally with 100 lbs, how much work will he have done?

4. The Zambesi River rushes over Victoria Falls in Africa at a rate of 25 million gallons a minute. The falls are 355 feet high, 1 gallon = 0.1337 cubic feet, and each cubic foot of water weighs 62.4 pounds. Calculate the number of horsepower developed.

* 5. If you take all the power consumed in the United States in a year and divide by the number of people, you end up with a figure of about 10,000 joules each and every second per person. At what speed would you have to push a car exerting 1000 newtons of force (224.8 lb) all year, day in and day out, to consume your share?

6. The B-1 bomber has four engines, each capable of generating 30,000 lbs of thrust. Calculate the total maximum power in foot-pounds per second developed when the machine is bopping along at 1320 mi/hr.

7. The total energy consumption in the United States during 1973 was 77 million million million joules. What was the corresponding power? Assuming a population of 210 million, determine the per capita power expenditure in kilowatts.

* 8. One gigawatt (1 GW) is a thousand million watts. On the average, the United States uses about 220 GW of electrical power. Incidentally, a large modern power plant can generate around 1 GW of electricity. How many 100-W light bulbs, all on at once, is this equivalent to? How many joules of energy does this correspond to each and every second? What's the per-capita share of consumption of electrical power?

Section 7.2

1. Compute the kinetic energy of a 390,000-lb B-1 bomber flying at 1320 mi/hr. Compare this to the KE of a 1-gram paper clip moving at 1 m/s.

2. Determine the KE of a 4000-lb car traveling at a constant 40 mi/hr. (To convert mi/hr to ft/s, multiply by 1.467.) If the car gets 20 miles to the gallon at that speed, and each gallon of gas delivers around 130 million joules, how many joules per second does the car consume? Of this, only about 20% finds its way into the mechanical systems—pumps, transmission, etc. Just around 10,000 J/s actually ends up propelling the car; ultimately, half is dissipated in overcoming tire and road friction and half in air friction. How much horsepower actually propels the car? By the way, you lose about 1000 J/s in unburned gasoline that evaporates from the carburetor, never even getting to the engine!

* 3. How much power must be supplied to a 4000-lb car if it is to accelerate from 0 to 60 mi/hr in 10 seconds? This neglects friction of all kinds—tire, air, etc. (To convert ft-lb/s to W, multiply by 1.36.) Compare this to the propulsive power needed to maintain a steady 40 mi/hr (previous problem). Incidentally, air friction increases very rapidly with speed, becoming a major factor above roughly 40 mi/hr.

4. After a bit of a spat, astronaut Flo with her back up against the space ship playfully pushed her 150-lb copilot. Rather annoyed, she shoved with about 100 lb over 2 feet. At what speed did her surprised associate float off into space?

5. According to the record books, Aleksandr Zass—known to his admirers, naturally enough, as "Samson"—would on occasion (when he wasn't bending iron bars) catch a 104-lb girl fired out of a cannon at 45 mi/hr. Determine her kinetic energy, and assuming that old Samson brought her to rest in about 2.5 ft, determine what force he exerted on that nameless heroine.

* 6. One ton of Uranium-235 can supply about 74,000 million million joules of nuclear energy. That's enough to propel a 3.5 million-kg space ship (the size of a fully loaded Saturn V) up to a speed of _____ m/s.

Section 7.3

1. A 747 with a takeoff weight of about 500,000 lbs cruises at 600 mi/hr at an altitude of roughly 20,000 ft. Determine its KE and gravitational PE.

2. One barrel of oil has a chemical PE of about 6000 million joules. How many foot-pounds is that equal to? How high in the air could that much energy raise a 5000-ton ship, assuming it was all converted into gravitational PE?

3. A 1000-kg auto at rest at the base of a hill accelerates straight up a smooth incline, reaching a speed of 20 m/s at an altitude of 100 m. What was its total gain in energy at that point?

* 4. A 10-lb weight resting on the floor is attached to a light rope, which goes over a pulley 20 ft in the air and down so that its end touches the ground. A 10-lb monkey decides to slowly climb up the rope. How much work does it do in ascending 10 ft? How much rope ends up on the floor? Has the weight risen? By how much? What is the total increase in gravitational PE of the system?

5. Calculate a value for the gravitational acceleration on the surface of the moon, and then determine the change in gravitational PE when a boulder weighing 100 lb there is raised 100 ft up from the surface. Take the radius of the moon to be $\frac{1}{4}$ of Earth's and the mass to be about 100 times less.

Section 7.4

1. A 3200-lb car traveling at 60 mi/hr runs out of gas 10 miles from a gas station. Neglecting friction losses, if the station is 50 feet below the elevation where the car stalls, how fast will the vehicle be traveling when it gets there?

* 2. Show that a ball fired straight up with an initial speed v_i will rise to a height Δh such that

$$v_i = \sqrt{2\,g\Delta h},$$

provided that g can be assumed constant.

3. Two cars of weight 3200 lb and 1600 lb are moving horizontally at 60 mi/hr when they both run out of gas. There is a town in a valley not far off, but it's just beyond a

120-ft-high hill. Neglecting friction, will either car make it to town?

4. A 120-lb pole-vaulter just clears 18 ft. Roughly how fast was he running the instant before the jump?

* 5. A 120-lb woman steps off a 14-ft ladder and drops to a trampoline, stopping 4 ft above the ground. Assuming no losses, how much energy is stored in the trampoline at the instant she comes to rest? How fast will she be thrust into the air by the recoil, and how high will she sail?

6. A car racing up a winding mountain road runs out of gas while moving at 80 ft/s at a height of 100 ft above the base level. Assuming no losses, will our fearless driver clear the 195-ft peak? Would it help to throw out any extra weight or even jump out and run alongside the car? Not having any brakes, at what speed will the car be moving when it reaches the base of the mountain?

7. Imagine a hole running along a diameter through the center of the Earth and out the other side. Describe the motion of a ball dropped into the hole (neglecting any frictional losses).

8. Suppose there was a straight tunnel from New York to Paris and you entered it in a frictionless, motorless car. What would the trip (if any) be like?

This episode is entitled "thermo," which is the physicist's jargon for **thermodynamics, the study of heat and its relationship to the other forms of energy.** The constant random motion of the atoms of a substance produces a jumble of KE that is known formally as **thermal energy. Temperature is a measure of the concentration of that thermal energy in the same sense that saltiness speaks to the concentration of salt and not the total amount in the pot.** The third section examines the relationship between changes in heat and the accompanying variations in the temperature and state of a body. The next bit of business is a restatement of the law of conservation of energy to include thermal processes—a powerful generalization known as the **first law of thermodynamics.** The chapter ends with one of the great, potent insights of human thought, namely, the **second law of thermodynamics.** It deals with the unfolding of events, the flow of time, the inexorable grinding of order into disorder on the majestic scale of the entire universe.

It is rare that a hypothesis, once set aside, is ever resurrected, but that is precisely what happened to the **kinetic theory of heat,** now one of the mainstays of contemporary physics. That formalism treats matter as a structure composed of atoms and molecules that are in constant, chaotic motion. It was widely believed in the seventeenth century (by Bacon, Descartes, Boyle, Hooke, Newton, Locke, Leibniz, and others) that *heat was a manifestation of motion.* Galileo had said as much decades before. Certainly the concept that heat was related to the rapid jiggling of the tiny particles of matter composing all bodies had been around, more or less, for many centuries. It is a logical consequence of combining the ancient atomic hypothesis with the matter-in-motion view of the universe. In any event, the scholars of the seventeenth century did not do very much with the idea, providing neither convincing theoretical analysis nor strong experimental evidence in its behalf.

With the eighteenth century there came a gradual acceptance of a new theory of combustion. That conception was based on the presumed existence of a weightless, invisible "matter of fire" possessed by substances like wood, coal, and gunpowder—a fluid firestuff called **phlogiston.** Years later, in the 1760s, the same unfortunate invisible-fluid metaphor served yet another fiction. Heat was then recast as an imponderable, self-repellent, indestructible fluid, which the great chemist Lavoisier (around 1787) called **caloric.** The

On facing page, Count Rumford in 1801.

Fire is fast-moving, subtle, and sharp or penetrating. . . . It is fast-moving because right down to its simplest parts—that is, its atoms—it is in motion. . . . Fire is hot because, by its power, motion so penetrates any material, watery or earthy, that it moves and divides each of the parts of such matter and makes each part move rapidly.

ANONYMOUS

On The Elements (ca. 1160)

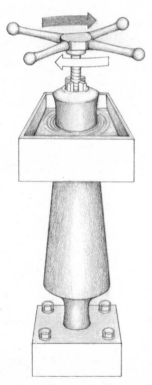

This artist's rendition was drawn vertical to fit in the margin. Rumford had no such problem, so he used the traditional horizontal boring technique, rotating the cannon barrels, not the cutting tools.

idea that heat was a form of matter was not new. Like so much else, it went back to the Greeks. And we really should not be surprised by the introduction of these intangible fluids. After all, they were no stranger beasts than the "spirits" (e.g., gravity, electricity, and magnetism) already living in the invisible forest.

Still, the imagery of a heat-fluid, a "matter of heat" flowing from hot objects to cold ones, was appealing, and the caloric theory proved quite useful, at least superficially—wrong hypotheses can explain things, too. Moreover, it promised to provide a quantitative approach to the study that seemed not to have been possible before.

The first important attack on the then widely accepted caloric theory was delivered in 1798 by one of the most fascinating figures ever involved in science. Benjamin Thompson was a professional soldier who saw little action yet rose, by luck and guile, to the rank of Major General. An elegant wheeler-dealer, he became Minister of War of Bavaria. Knighted by the King of England, he was later made Count Rumford of the Holy Roman Empire. A tall, handsome, philandering, blue-eyed rogue, he fathered three children, of whom only one was legitimate. He was a spy and a scoundrel and a beloved benefactor of the poor, a superb administrator and social reformer. But above all, he is remembered as a brilliant practical scientist.

Sir Benjamin was humbly born in Massachusetts in 1753. At 19 he moved to the little town of Rumford to be its schoolmaster but soon married a 33-year-old widow—rich of course—and retired. He had seen the "Boston massacre," and yet not long after that he became a major in the king's army. When the British General Gage moved on Lexington and Concord to start the Revolutionary War, it was on the strength of secret intelligence sent him in a letter written (in invisible ink) by none other than B. Thompson. Thompson even impressed General Washington, who almost offered him a commission, and years later he would impress the First Consul of France, Napoleon Bonaparte. Ingratiating himself with important people was his special talent. His second marriage (1805) to the wealthy, witty, and irrepressible widow of Lavoisier was a short-lived disaster. He was fond of growing roses, and she was fond of pouring boiling water on them.

After countless escapades and world renown, our semihero the Imperial Count was all but running the tiny state of Bavaria. In 1797, having newly outfitted the arsenal at Munich, he was watching cannons as they were being bored, and he contemplated the tremendous amount of heat liberated in the process. Determined to get to the heart of what was happening, he arranged to have a brass gun-barrel blank immersed in about two gallons of cold water and rotated against a blunt steel borer. Slowly the temperature rose until after two and a half hours, "the water *actually boiled!*" Heating by friction had certainly been known for a long time; people had always rubbed their hands together to warm them, and of course, aborigines (and Boy Scouts) presumably still start fires via friction.

The calorists would have said that caloric was "squeezed out" of the brass by the boring tool. But Rumford showed that heat could continue to be generated as long as work was done, as long as the horses turned the cannon barrel. It "appeared evidently to be *inexhaustible*" and could therefore not "possibly be a *material substance.*" Rumford accepted the alternative—heat was motion—but it would be several decades yet before caloric was blown away for good.

The mid-1800s revival of the wave theory of light, which had as its basis an undulating aether, suggested a similar picture for heat. Caloric gradually vanished, and in its place stood the quivering aetherial gel, a reincarnated cousin in the zoo of invisibles. Around the same time, the crucial recognition came that heat was a form of energy. Still, it was not until the end of the 1800s and even into the beginning of the 1900s that heat at last became universally recognized as *the energy of motion of the atoms constituting matter.*

> Heat is the *vis viva* resulting from the insensible movements of the molecules of a body. It is the sum of the products of the mass of each molecule by the square of its velocity.
>
> **LAVOISIER AND LAPLACE**
> *Mémoire sur la Chaleur* (1780)

8.1 THERMAL ENERGY

The previous chapter dealt with mechanical energy, the energy of an object behaving as a whole—in that context it could be considered organized energy. We talked about the KE, gravitational PE, and elastic PE of a body, understanding that all of its constituent atoms were acting together in an organized fashion. Nonetheless, it certainly is possible to impart to the individual atoms motion that is disorganized, motion not of the body but within the body.

Our contemporary approach is to call the internal energy of an object, which is associated purely with the random motion of its atoms and molecules, **thermal energy.** The atoms within everything are always more or less rattling around in a fairly disordered fashion. Thus thermal energy is kinetic energy in the micro-world.

Thermal energy, while in transit, is defined as **heat.** It is the energy of random motion transferred by way of contact from one object to another, from one group of atoms to another (exclusively as a result of a temperature difference). In other words, heat is the thermal energy gained, lost, or shifted via the cumulative effect of individual atomic collisions. A body contains, or stores, thermal energy, not heat; the latter is thermal energy that is transferred to, within, or from the body. Once transmitted, the energy is no longer called heat.* As before, we are again concerned more with how the energy of a system changes when something is done to it than with how much energy it happens to have at any time. An important agent of that change in thermal energy is heat, just as an agent of change in mechanical energy is work. Once transmitted, the mechanical energy of a system is no longer called work—a system stores KE or PE, not work.

---

* The concept of heat is like the notion of wind. Air in transit is wind. When a wind stops moving, it vanishes into thin air. Wind can blow up a balloon, but what gets stored in the balloon is air, not wind.

THE HEAT OF THE SUN

There are three things from which heat can be generated, namely, a hot object, motion, and a concentration of rays; nevertheless we should understand that the heat in these things is of a single kind. . . . In all of these three cases, the proximate cause of heat is scattering. And so when a hot object generates heat, it does so by the scattering of bits of matter.

ROBERT GROSSETESTE (ca. 1175–1253)

The language used so far has been rather careful because it is possible to increase or decrease the thermal energy of an object (make it hotter or colder) in several quite different ways. You can add to the thermal energy by pounding on an object with a hammer or by rubbing or bending it (that is, via work). You can do so by placing the object in contact with something at a higher temperature, like a flame, where it will pick up energy by way of atomic collisions (that is, via heat). Or as we will see, you can shine electromagnetic energy on it (that is, via radiation). Whatever gets the atoms more agitated will do the job, will increase the thermal energy.

Remember that uniform motion and the associated KE are each relative quantities—you can drive alongside a moving train, and to you it appears motionless. The orderly flow of matter and the corresponding ordered motion of its atoms and molecules do not constitute thermal energy; that arises only from *random* motion. If you throw an apple, all the atoms do, indeed, have more KE, but they are all moving together as a body. That is not random motion, and it contributes nothing to the ordinary agitation of the atoms and so changes neither the apple's thermal energy nor its temperature. You would hardly expect people in an airplane to feel hot just because they fly past a stationary observer who sees them all as having a tremendous amount of KE.*

8.2   TEMPERATURE

We creatures of the third planet have an intuitive sense of hot and cold—a body sense, a what-can-you-put-in-your-mouth-without-burning-it sense, a pull-the-covers-over-my-feet—they're-freezing sense. Hotness and coldness are the subjective boundaries on either side of our own warm bodies. Hot as hell is how hot? Cold as a fish is how cold? Physics must quantify, must stamp numbers on the metaphors, and that is what we are about.

Like other physical properties, *temperature* was measured long before it was understood, long before anyone quite knew what was actually being quantified by all those precise readings. Galileo seems to have invented (1593) the first temperature indicator although it was not a true thermometer, because its scale was arbitrarily graduated. He simply inverted a long narrow-necked flask containing colored water into a bowl of the same liquid. Air captured in the small bulb on top either expanded or contracted as it was heated or cooled, and the column fell or rose proportionately. You can see nearly the same thing happen by putting an inflated balloon in the freezer. The gas molecules, whose collisions with the balloon walls are what puff up the balloon, lose thermal energy in the freezer and do not bounce around as much, and the balloon begins to collapse.

Within sixty years of the introduction of the thermometer, sealed pocket-sized versions containing either alcohol or mercury were in use (both freeze at much lower temperatures than water). To reference the device to Nature, two standard points were determined,

Galileo's thermoscope.

---

* You cannot violate the first postulate of relativity with a thermometer (Section 6.2).

one high and one low, and the range between them was simply divided uniformly into what have come to be known as **degrees.** Although Huygens recommended using the temperatures of melting ice and boiling water for these references, as we do today, the early workers were far more imaginative. The annals of thermometry are replete with all sorts of strange reference standards, ranging from bowls of melted butter to cooperative stray cows!

The dreadful scale we still use too much in the United States is a gift of G. D. Fahrenheit (1686–1736). There remains some confusion on the matter, but he seems to have initially set the 0° mark at the coldest temperature he could reach, that of a mixture of water, ice, and sea salt. Following Newton's lead of thirteen years earlier, the upper reference was taken as normal body temperature. Fahrenheit fixed that point at 96° (it should have been closer to 98.6°). Water froze at what turned out to be 32°, and water boiled at around 212°. Fortunately, A. Celsius (1701–1744) soon devised a more pleasing scale, a variation of which is today used universally in scientific work and generally, as well, in most countries of the world. The freezing and boiling points of water are taken as 0° and 100°, respectively, and the range between is just marked in 100 equal divisions.

All right, now that we can measure it, what actually is temperature, beyond what you read off a thermometer? And here, once again, there is no totally satisfactory simple answer. For one thing, it is certainly different from *heat*, which is thermal energy in transit; **temperature** *is a property of matter* corresponding to the degree of "hotness." It relates to the *concentration of thermal energy*. At least in the simplest case of a gas, we can say that *temperature is a measure of the average kinetic energy of the molecules*. If most of the molecules are randomly flying around rapidly and have a lot of KE, then the gas will have a correspondingly high temperature, and vice versa. Yet even this conception has its difficulties in the domain of the exceedingly cold, where there are good quantum-mechanical reasons for rejecting it. There are those who maintain that, however unsatisfying, only operational definitions, ones based on measurements, are meaningful. Then, *temperature is what you measure with a thermometer*; the rest is irrelevant.

Human beings function well only within a tiny range of temperatures—much like the modern automobile. But the universe sustains an incredible sweep of hot and cold, which is completely alien to our lukewarm intuition. The highest temperatures we thus far conceive of are to be found deep within certain stars—roughly four thousand million degrees Celsius seems the theoretical peak. A hydrogen bomb ignites at about 40 million °C (70 million °F) and reaches nearly ten times that, whereas the innards of the sun perk away at a mere 15 million °C (27 million °F). Among the hottest things you are likely to have around at home is an ordinary light-bulb filament, which operates at about 2500°C (4600°F), and only once, for an instant as it burns out, flashing blue-white, will it reach the

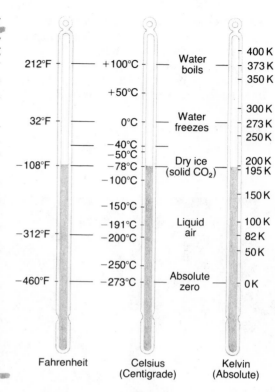

5/9 (degrees Fahrenheit −32) = degrees Celsius.

To Some Degree

It appears that the temperature concept is not a clear cut thing, which can be made to apply to all experience, but that it is more or less arbitrary, involving the scale of our measuring instruments.

P. W. BRIDGMAN (1927)

At present we know of no purely mechanical quantity—that is, one expressible in terms of mass, length and time only—which can be used, however inconveniently, in place of temperature. We are inclined to conclude that temperature probably is itself a basic concept.

A. G. WORTHING (1940)

melting point of tungsten, 3410°C (6170°F). Lead melts at 327°C (621°F), paper burns roughly at 230°C (450°F), and now we are down to the oven temperatures for roast beef and apple pie. Earthly air temperatures have reached the extremes of 136.4°F in the Libyan shade and −126.9°F in the Antarctic nightmare, and that small span is more than we are fit for.

Ice, of course, melts at 0°C (32°F), but if salt (NaCl) is added to icy slush, the temperature of the mix will drop as low as −21°C

A sprig of pachysandra frozen brittle in a bath of liquid nitrogen.

(−5.8°F). That effect has been used for quite a while to cool home-made ice cream during churning. To go much below that temperature with ease, we would have to use a liquified gas, such as nitrogen. It's a clear, colorless fluid that looks like water and gently boils away in an open thermos bottle at about −196°C (−320°F). Easy to handle and relatively inexpensive (somewhere between the price of milk and that of beer) liquid nitrogen is the workhorse of cryogenics (low-temperature physics). That's the stuff physics professors like to demonstrate by freezing, rock-hard, bananas and rubber balls and anything else that will fit in the bottle. Still it has to be treated with care; a droplet on the skin stings like a bead of molten solder.

The most exotic of all materials is wondrous liquid helium. Clear, colorless, and incredibly cold, it boils (in an open container) at −268.9°C (−452°F). Exceedingly difficult to handle and somewhat dangerous, it must be stored in a special thermos-bottle arrangement surrounded by a constant bath of liquid nitrogen, and even then it boils off more rapidly than one would like. With only the slightest provocation, a few liters of liquid helium will rapidly vaporize, wasting itself into a frigid, blasting plume of gas. Since it sells at the price of a decent champagne, mishaps can be both costly and painful.

That, then, is about the limit of coldness attainable without any complex pumping systems or gadgetry. But be content, nothing has ever reached a temperature more than about 4.2°C lower than that and probably never will!

Absolute Zero

Different liquids expand and contract at different rates with variations in temperature, and so even the best alcohol and mercury thermometers agree exactly only at the two calibration points (0°C and 100°C). But gases, because of their inherent physical simplicity, behave much more nearly alike and are ideally suited for precise thermometry (another gold star for Galileo). A very interesting pattern starts to emerge as we lower the temperature of any gas: The volume always decreases proportionately. All gases show the same effect; they seem to collapse straight toward zero volume at a temperature of −273°C (−460°F). Of course, no gas actually gets there and vanishes. All gases liquify before that, and the game is over. But −273°C is a very special value; it's the lower limit—the **absolute zero**—beyond which we cannot extract any more energy from the substance, and the temperature will presumably drop no further.

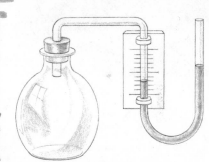

A gas thermometer.

William Thomson, Lord Kelvin (1824–1907), following a theoretical line of thought that was quite different, introduced an **absolute temperature** scale, which turned out to agree with that of the gas thermometer. Known nowadays as the *Kelvin* scale (K), it retains the Celsius degree but begins with 0 K as absolute zero (−273°C). This is in contrast with the arbitrarily selected zero for all previous scales. Thus water freezes at 273 K and boils at 373 K, and any Kelvin temperature is numerically simply 273° greater than the corresponding Celsius value. That puts liquid helium at a mere 4.2 K. In the quest for absolute zero no one has thus far come any closer than a frigid 0.00000005 K, and it does seem as though 0 K may well be unattainable.

8.3 QUANTITY OF HEAT

Four men, four friends dominated the intellectual life of Scotland in the latter half of the eighteenth century. They were Adam Smith, the eminent political economist; David Hume, one of the greatest of the modern philosophers; his personal physician, Dr. Joseph Black, professor of chemistry at Glasgow; and that university's resident

Lord Kelvin

Steve Allen once remarked on TV that during World War II he "served in a British Thermal Unit."

ENERGY CONTENT

Food	Kilocalories
Martini	145
Peanut butter sandwich	330
Buttered popcorn (1 cup)	55
Vanilla ice cream (⅓ cup)	145
Whiskey (1 shot)	105
Brownies (1)	140
Jelly doughnut	225
Dry red wine (1 glass)	75

scientific-instrument maker, James (Jamy) Watt. The founder of modern analytic chemistry and the incomparable engineer maintained a lifelong scientific dialogue and a truly affectionate relationship. That Black and Watt were both advocates of the fluid theory of heat was reasonable enough; the notion of caloric was a practical contrivance, and these were practical men, more concerned with using heat to their own ends than with abstractions.

Black was the great pioneer in the study and probably the first to distinguish between "*quantity of heat*" and *temperature*, which he often called the "degree of heat." The latter he saw as reflecting the "intensity" or concentration of what we would call thermal energy (and what he spoke of as "the matter of heat"). A large pot of water will require the addition of considerable heat in order to reach the same temperature that can be attained by supplying a small amount of heat to a little bit of water. A fraction of a teaspoonful can be boiled with a cigarette lighter, but the same amount of heat will obviously have a minuscule effect on the Atlantic Ocean.

A stew tastes salty because of the concentration of salt, and it feels hot because of the concentration of thermal energy. The bigger the pot of stew, the more salt must be sprinkled on and the more heat must be pumped in. The Atlantic Ocean, cold as it is, contains far more thermal energy than a scalding hot pot of stew, and no matter how salty the dish, the ocean will contain far more salt as well. In the same way, a large block of ice at 0°C has more thermal energy than a hot cup of tea at 100°C—the fact that it is spread out thin, not concentrated, accounts for the low temperature, but it is there nonetheless.

Roughly around the year 1760, Black introduced a scheme for quantifying heat that, because it is operational, is quite independent of the actual nature of heat—whether caloric or kinetic. Following his lead, we nowadays define the **calorie** as *the amount of heat that must be added or extracted in order to change the temperature of 1 gram of water by 1°C, either up or down.* This quantity turns out to be inconveniently small for weight watchers, so the world of dietetics uses the large calorie, or **kilocalorie** (cleverly spelled **Calorie**), with 1 Cal = 1000 cal. Engineers and aspiring air-conditioner salesmen (mostly in the United States) hold fast to another fading anachronism, the *BTU*, or **British Thermal Unit**. *One BTU is the amount of heat needed to change 1 lb of water by 1°F.* A 6000 BTU air conditioner will pump that quantity of heat out of a hot room *each hour*. Though no one knew it then, heat is energy, and we ought not to have this ridiculous collection of redundant units, but we still do:

$$1000 \text{ cal} = 1 \text{ Cal} \simeq 4 \text{ BTU} \simeq 4200 \text{ joules}.$$

A hamburger (330 Cal) and a can of beer (170 Cal) quietly burned in the belly will provide roughly 2 million joules of energy, as compared with, say, a quart of gasoline exploding in an engine, which will yield around 32 million joules.

One Calorie supplied will raise 1 kg of water 1°C, but there is no reason to assume it will do the same thing to 1 kg of iron or peanut butter or pizza. To the contrary, Dr. Black found that every substance changed temperature by a characteristic amount when infused with a Calorie of heat. He spoke of it as a "capacity for heat," and accordingly, we define the **specific heat capacity** *of a substance* as the *number of Calories of heat that must be added to raise the temperature of 1 kg of the material 1°C.*

Water, which has a relatively high heat capacity, is molecularly light. Therefore 1 kg will contain a great many molecules (three times as many as there are atoms in 1 kg of iron). Since temperature relates to average kinetic energy, most of the molecules are going to have to be set in motion, and that will take a proportionately large amount of heat. Moreover, some of that heat will go into increasing the motions of the atoms within each water molecule, and that activity does not show up as a temperature change. All of this taken together is our kinetic-theory justification for expecting water to be difficult to heat up, in comparison with, say, brass or iron. Of course, Black "understood" the phenomenon in terms of caloric; water, it seemed, had a lot of space between its atoms, where it could accommodate the material particles of heat, and so it had a high heat capacity. That really is a rather simple and effective picture—totally wrong but quite useful.

An object with a large heat capacity will change temperature (up or down) only when a great deal of heat is either added or drawn off. The hot-water bag is the ultimate testament. A potato or a lima bean, fished out of our bubbling stew pot, can be eaten with a little caution, but a juicy tomato or an onion (mostly water) will have to lose much more heat to get down to the same temperature, and invariably either one will retain enough to burn your tongue.

When two bodies at different temperatures come together, heat flows from the higher to the lower until the temperatures equalize and they are said to be in **thermal equilibrium.** Two spoons of equal mass, one made of gold and the other of glass, immersed in a hot cup of coffee will reach some new equilibrium temperature along with the brew. But the glass spoon, because of its higher heat capacity, will take up about four times as much thermal energy as the gold in coming to that temperature.

Heat Capacity

SPECIFIC HEAT CAPACITY*

Substance	Specific Heat Capacity
Water	1.0
Dry soil	0.2
Glass	0.1–0.2
Air	0.17
Ice	0.5
Brass	0.09
Wood	0.3–0.6
Paper	0.3
Iron	0.11
Hydrogen	2.4
Gasoline	0.5
Gold	0.03

* All determined around room temperature. The value for water is nearly constant, varying only in the third decimal place.

APPROXIMATE CALORIES CONSUMED PER HOUR

Body Weight	lbs	100	150	200	250
Body Mass	kg	45	68	90	113
Sleeping it off		40	60	80	105
Sitting on the grass		65	95	130	165
Standing on line		70	100	140	170
Strolling through park		130	195	260	320
Running from a mugger		290	440	580	730

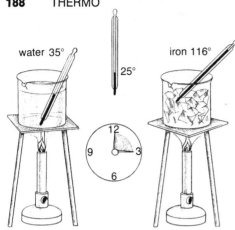

water 35°   25°   iron 116°

If a quantity of water (with a specific heat capacity of 1) changes temperature by 10°C, the same mass of iron, given the same amount of heat, will change temperature by about 91°C (because its heat capacity is only 0.11).

The number of incoming Calories versus the temperature change as 1 kilogram of H_2O is carried from −10°C to more than 100°C.

Return for a moment to the definition of *specific heat:* the amount of heat in Calories that must be added or removed (Δheat) from each 1 kg of mass (m) of some substance for each and every 1°C temperature change (ΔT) it is to undergo. In other words, specific heat is the Δheat per unit mass per degree Celsius, or

$$\text{specific heat} = \frac{\Delta\text{heat}}{m\Delta T}.$$

For example, 1 kg of water will increase in temperature by 10°C when 10 Calories of heat are poured in, since the specific heat of water is 1. Now suppose the same 10 Calories are added to 1 kg of iron, which has a specific heat of 0.11. By how much will its temperature change? That is, find ΔT such that

$$0.11 = \frac{10}{1 \times \Delta T}.$$

This is equivalent to writing

$$\Delta T = \frac{10}{0.11} = 91°C.$$

The iron changes temperature by 91°C. The 10 Calories will have a considerably greater effect on the temperature of the iron than on that of the water.

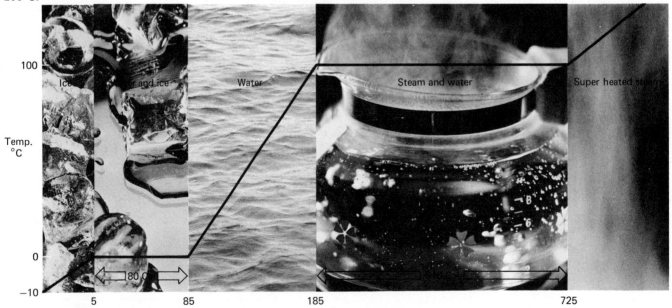

Ice   Water and ice   Water   Steam and water   Super heated steam

Temp. °C

100

0

−10

80 C

5   85   185   725

Calories

Some people find it strange to think of ice at a temperature different from 0°C, but it can be cooled like any other solid. Gold solidifies at 1063°C, and yet there it is on your finger, frozen well below that. The ice cubes in a home freezer should certainly be somewhat colder than 0°C. Suppose, then, that we take a 1-kg block of ice at, say, −10°C and trace what happens as it is heated up, melted, boiled, vaporized, and finally brought to a temperature in excess of 100°C, all at ordinary atmospheric pressure.

Inasmuch as the heat capacity of ice is roughly 0.5, we need to supply only ½ Calorie to raise each kilogram 1°C; to bring the block from −10°C to 0°C requires that just 5 Calories be provided. From here on, as heat trickles in, the ice begins to melt, but the temperature remains fixed at 0°C until all of it is liquified. Energy goes into unlocking the crystal bonds that hold the molecules together in the solid, and only after that is accomplished will there be any change in temperature. It takes 80 additional Calories just to melt one kilogram of ice. That quantity is called the **heat of fusion;** it's different for each material, and a value of 80 is actually quite high—a fact the old icebox relied on. Having thus far supplied 85 Calories, we now have 1 kg of water at 0°C. Realize that we could turn around and start removing heat from the water, whereupon ice would simply begin to reappear; in fact, the entire procedure is completely reversible.

A further input of 100 Calories will raise the water's temperature by 100°C, bringing it to the boiling point. Any addition of heat now goes into breaking the liquid bonds, allowing molecules to rip loose and fly off as steam. Bubbles of water vapor rise up and burst at the surface. Boiling water will perk away at a fairly constant 100°C until all of it vaporizes. (The fact that it never overheats makes it exceedingly convenient for cooking purposes.) This transformation to steam requires the addition of 540 Calories per kilogram, called the **heat of vaporization.** That represents a tremendous amount of energy stored in the steam, in the violent motion of its molecules. And that thermal energy can be transported, via steam, from a furnace to a piston or turbine or to a humble radiator, where the vapor condenses, liberating its energy cargo of 540 Calories for each kilogram of water formed!

Of course, if a water molecule just happens to be moving fast enough, it may fly off even a cool surface (like a wet bathing suit). Evaporation takes place at the boundary of a liquid at all temperatures. Below the boiling point only the occasional high-velocity molecule manages to escape, carrying off an above-average amount of energy. It leaves behind the rest of the molecules with a lower average KE and therefore a lower temperature. *Evaporation is a cooling process;* that's why we perspire and dogs pant. Each kilogram so lost carries away at least 540 Calories.

Change of State

°C	IMPORTANT TEMPERATURES	°F
−273.2	Absolute zero	−459.7
−269	Helium boils	−436
−196	Nitrogen boils	−320
−183	Oxygen boils	−297
−79	Dry ice (CO_2) freezes	−109
−39	Mercury freezes	−38
0	Water freezes	32
3.8	Heavy water freezes	38.9
~20	Room temperature	~68
31	Butter melts	88
~37	Body temperature	~98.6
~54	Paraffin melts	~129
78	Alcohol boils	172
100	Water boils	212
101.4	Heavy water boils	214.6
108	Saturated salt solution boils	226
232	Tin melts	449
327	Lead melts	621
445	Sulfur boils	833
657	Aluminum melts	1215
801	Salt (NaCl) melts	1473
961	Silver melts	1762
1063	Gold melts	1945
1083	Copper melts	1981
1000–1400	Glass melts	1830–2550
1300–1400	Steel melts	2370–2550
1530	Iron melts	2786
1620	Lead boils	2948
1774	Platinum melts	3225
1870	Bunsen burner	3398
2450	Iron boils	4442
3410	Tungsten melts	6170
3500	Oxyacetylene flame	6332
5500	Carbon arc	9932
6000	Surface of the sun	10832
6020	Iron welding arc	10868

The heat capacity of steam is roughly 0.5, and so if the vapor is contained in a boiler, we can continue to increase its temperature. In most modern applications, steam at temperatures in excess of 500°C (932°F) is used. *Superheated steam*, as it is called above 100°C, whirls great turbines that drive the vast majority of large ships and electrical generators.

The long road to efficient steam power was first trod by Dr. Black and his friend Jamy Watt; the former's discoveries guided the latter's inventions. Ironically, their work, though framed in the imagery of caloric, would serve as the stimulus for the development of thermodynamics and the ultimate overthrow of the fluid theory of heat.

Rumford at the cannon factory had come very near the crucial insight; he had sensed the *energy-circle* but did not take that final step to quantification. "More heat," he even remarked, "may be obtained by using more fodder for the support of the horses." *Chemical potential energy* in the hay, turned by the horses to *work* turning the cannon, turned into *heat;* round it went like a circle, like a chameleon walking round a circle, one form unfolding into the next. The chemical energy stored in the molecules of fodder powered the horses, which did work on the machinery, which moved the borer against the cannon, which heated up, gaining thermal energy via friction. But this was 1798. Almost all of Europe was content with caloric, and the word *energy* had not even found its way into the scientific vocabulary. Few saw the real link between the straining horses and the boiling water; no one knew that humanity teetered at the very edge of the whirlwind of energy.

The obscure young physician from Heilbronn, Germany, who would take the next giant step was 28 years old when his first essay appeared in 1842. Julius Robert Mayer, writing in a bold, speculative, almost medievally metaphysical style, proposed that the various forms of energy "are quantitatively *indestructible* and qualitatively *convertible*." *All manifestations of energy (PE, KE, work, and heat) are transformable one to the other, and energy as a whole is conserved*.

Not an experimentalist himself, Mayer reinterpreted some long-forgotten results on expanding gases to yield a rough value for what he called the "mechanical equivalent of heat." Surely a work of genius, his essay was nonetheless without accurate experimental confirmation and so was ridiculed as unprofessional, the mere dabblings of a would-be philosopher. Mayer published several other papers, but before the decade passed, he was overwhelmed by a deep depression fashioned of derision and made more intense by the deaths of two of his children. In 1849 he leaped from a second-story window in an unsuccessful suicide attempt. Within two years, brilliant, sensitive Mayer was a straitjacketed victim of the cruelty of an insane asylum. He was freed from that institution in 1853, but not until

As we cannot give a general definition of energy, the principle of the conservation of energy simply signifies that there is *something* which remains constant. Well, whatever new notions of the world future experiments may give us, we know beforehand that there will be something which remains constant and which we shall be able to call *energy*.

JULES HENRI POINCARÉ

8.4 THE ENERGY CIRCLE

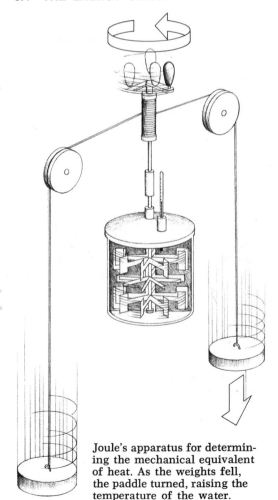

Joule's apparatus for determining the mechanical equivalent of heat. As the weights fell, the paddle turned, raising the temperature of the water.

late in the sixties, near the end of his life, was he freed from the despair that had taken him there. Only then did he finally receive the recognition he deserved.

As if the time were somehow ripe for change, within one year of Mayer's paper, there appeared the first report of the remarkable work of James Prescott Joule. This well-to-do brewer's son from Manchester, a pupil of the great John Dalton, had already spent three years in search of the numerical relationship between work and heat. Joule was a brilliant experimenter, who devised a whole series of arrangements using various means (e.g., electric currents, friction, compression of gas, etc.) to generate heat. His most famous was a paddle-wheel device driven by falling weights. As it spun, churning the water and heating it via friction, the temperature rose. By determining the mechanical work done as the weights fell and the amount of thermal energy imparted to the water, he could compute the needed relationship. Nowadays we say that *1 Calorie of heat equals 4200 joules of energy*. For more than fifteen years Joule, the eccentric brewer, the amateur scientist, struggled—much of the time against the current consensus. In the end (around the 1850s), the care and rigor of his research, coupled with the increasing acceptance of kinetic theory, finally established the equivalence of work and heat. By the latter part of the 1860s, caloric was a vanishing concept.

The discipline within physics that deals mainly with heat, its relationship to mechanical energy, and the conversion of one to the other is known as **thermodynamics.** The law of conservation of energy opened out to include thermal processes:

James Prescott Joule

> Energy can be neither created nor destroyed, but only transformed from one form to another.

That far-reaching notion is the **first law of thermodynamics.**

In the Eyes of the Law

We are now presumably in a position to understand all sorts of occurrences in terms of the imagery of energy. For example, suppose an apple weighing 1 newton falls 10 meters from a tree, down with a thump into your waiting hand. The original PE (1 newton × 10 meters) when up among the leaves is transformed totally into KE at the instant before the apple is caught. If we overlook a tiny loss of energy in the form of sound, the apple comes to rest sharing with your hand 10 joules of thermal energy, and both hand and apple increase slightly in temperature. A meteorite with a much greater KE may generate enough heat on impact for large portions of it to actually melt.

If you do 1/10 of a joule of work compressing a spring, putting energy into it, then carefully tie it up and drop it into an insulated bath of acid, what happens to the elastic PE as the spring dissolves away? From the mechanical equivalent of heat we can compute that

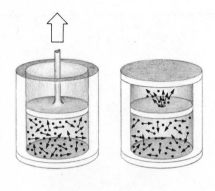

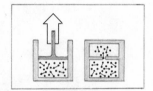

about 24-millionths of a Calorie was stored in the spring, and that energy will be liberated. Although the primary release of energy is via chemical PE, nonetheless we can expect the temperature of the bath to go up more while dissolving a compressed spring than one that was relaxed—but not much more. Likewise, if you submerge a wound pocket watch or one of those battery-operated clocks in an insulated cup of water and let it run down, the temperature of the water must go up.

A gas expanding, pushing up a piston, will do work in lifting the weight of the piston. Thermal energy of the moving molecules is converted into gravitational PE of the piston, and the gas temperature drops. It must drop because the gas has lost energy to the piston. By contrast, if the gas expands freely into a vacuum so that it does no work, there is no loss of energy and no appreciable change in temperature. (In fact, there is a minute increase in gravitational PE and an equally minute drop in temperature.) Reversing the process, push down on the piston, thereby compressing the gas and doing work on it. That is another way of saying that energy is transferred to the gas. Thermal energy is increased, and the temperature will rise. Try it with a bicycle pump; the shaft will become quite warm. In fact, the diesel engine operates without spark plugs by compressing the fuel in a cylinder until it gets hot enough to self-ignite. Similarly, a gasoline engine "knocks" because of preignition of the fuel. These are instances in which the thermal energy of the system increases because work is done on it, even though no heat is introduced from outside.

Some of the kinetic energy of a space ship reentering the atmosphere is deliberately converted to thermal energy in the heat shield. The ship slows down as the shield vaporizes. For that matter, my '69 Toyota cleverly converts KE, mine and its, into useless thermal energy in the brake linings and in the tires. In a related way, friction that occurs during the tidal action of the oceans, the washing back and forth of the waves, heats the water a bit. Where does that energy come from? The Earth, as a result of frictional drag, is slowing down in its rate of rotation and losing energy of motion, which shows up as heat.

A Superball would bounce up and down forever in a vacuum chamber if not for the internal heating that occurs each time it strikes the floor and compresses. It loses thermal energy (in part to the floor) and springs back less rapidly, rising not quite as high each time. A portion of the organized KE of the ball is converted during compression into the disorganized KE of its atoms.

An idealized machine that moves frictionlessly could, in principle, maintain its initial stock of motional energy forever, moving indefinitely. But if in the process it performs work or generates heat, the device must, to that degree, run down and invariably stop. The first law of thermodynamics demands a constant balance of the total energy: If you don't put any in, you can't take any out—at least not without altering the total. Insofar as our machine performs work, that work must be at the cost of the conversion of some fraction of the machine's own energy. Self-sustaining perpetual-motion machines that deliver work are impossible!

A red-hot object dropped into the ocean will cool rapidly, losing heat to the sea despite the fact that the ocean possesses far more thermal energy than it does. By contrast, an ice cube floating in the cool waters will draw heat from the sea and melt. The transmission of heat, it seems, has nothing to do with the total thermal energy of the objects involved but, rather, depends on the respective concentrations of energy, i.e., the temperatures. The process is local: There is no way for the ice cube to "know" whether it is floating in the ocean or in a pot of water; it reacts only to the temperature of its immediate surroundings.

Heat flows naturally from a higher- to a lower-temperature region. The reverse is certainly not prohibited by the first law or by any other law considered thus far; it just does not happen that way, and that in itself is a law. This simple statement predicting the direction of heat flow is one of several alternative forms of the **second law of thermodynamics.** That law is a completely independent physical notion drawn from observations of numerous phenomena. We can "understand" it rather nicely within the context of the kinetic theory: Molecules constituting a hot object are randomly moving, on the average, at high speeds with a good deal of KE. When the hot object is placed in contact with a body at a lower temperature (whose molecules, on the average, are moving more slowly), the ensuing collisions will transfer energy to the slower particles. They are, in effect, bombarded by the "hot" molecules. The net result is a flow of heat from the higher (T_H) to the lower temperature (T_L), and the average KE of each approaches that of the other.

Beneath the surface of this seemingly innocent, almost obvious law resides one of the most far-reaching and significant insights yet revealed about the very nature of our universe.

When Nicolas Léonard Sadi Carnot published his masterwork, *Reflections on the Motive Power of Heat* (a pamphlet of some 118 pages) in 1824, he was only 28 years old. His father, Lazare, had been the military genius of the French Revolution and Minister of War to Napoleon. The young Carnot brought his special ingenuity to bear on the great practical symbol of his age, the fire-chariot of the Industrial Revolution—the steam engine. Thermodynamics was born out of his inspired efforts to analyze the operation of that hissing, pounding wonder; to determine the theoretical limit to its efficiency and, indeed, to the efficiency of all heat engines (e.g., piston and rotary internal combustion auto engines, gas turbines, etc.).

Carnot envisioned the steam engine as if it were a watermill where the fluid caloric (heat) flowed down from the temperature

8.5 THE PRICE OF POWER

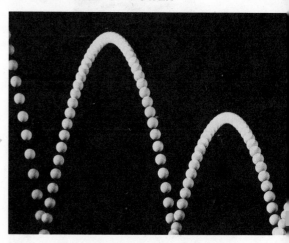

The loss of energy, mostly thermal, runs the system down—in this case a bouncing golf ball.

Sadi and the Second

Nicolas Léonard Sadi Carnot
(1796–1832)

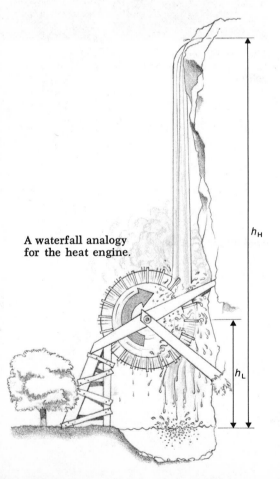

A waterfall analogy
for the heat engine.

heights (T_H) of the boiler through the engine, doing work, and on to the low point of the falls (T_L), the cool condenser. The *working substance*, the carrier of heat in this case, was steam. Generated in the boiler, it did work pushing on the piston. The steam was then cooled (the piston dropped back to its original position) and condensed, and water returned to begin the cycle again. Carnot's starting point was to imagine an ideal device, the ultimate heat engine, flawlessly insulated, completely without leaks, and totally devoid of friction. Such a perfect engine would be the best of all possible machines, operating—at least in theory—at the greatest attainable efficiency, that is, *delivering the most work for the smallest input of heat.* In very general terms, what you get out out divided by what you put in is known as **efficiency:**

$$\text{efficiency} = \frac{\text{work-out}}{\text{heat-in}}.$$

But what is the maximum theoretical efficiency—the best we can hope for with a heat engine?

"According to established principles at the present time," wrote Carnot, "we can compare with sufficient accuracy the motive power of heat to that of a waterfall." The idea was simple enough: Consider a fall and an appropriate ideal harness-the-rushing-water-machine located somewhere in the cataract. All heights are measured upward from a convenient zero reference, in this case the level of a motionless pool at the very bottom of the downpour. The height of the upper high reservoir is then h_H, and the altitude of the machine to be driven by the cascading water is a lower h_L.

The maximum amount of energy available to be converted to work by a mill (or a hydroelectric power plant) at that site is $\Delta(\mathrm{PE})$, equal to the weight of falling water times the change in its height ($h_\mathrm{H} - h_\mathrm{L}$). Of course, work equals $\Delta(\mathrm{PE})$, which is independent of the zero reference level. Clearly, the amount of energy actually available is only a fraction of the total that the fall could deliver if the mill was lowered to the zero level, the very bottom of the drop. That maximum energy is proportional to the maximum fall, h_H, and so the fractional amount of energy available to the mill is $(h_\mathrm{H} - h_\mathrm{L})/h_\mathrm{H}$. In fact, this is the maximum efficiency of the machine-waterfall system, and it obtains when the machine itself is perfectly lossless. In other words, it represents the ratio of the most useful energy we can get out of the system to the amount of energy flowing in.

Carnot argued that the *available energy* for a **heat engine** (a device that converts thermal energy into some other form) was similarly determined by the difference between the high and low temperatures across which it operated. The theoretical efficiency of his ideal engine, by analogy, was $(T_\mathrm{H} - T_\mathrm{L})/T_\mathrm{H}$. Note that the situation involving the waterfall depends on the reference level—at least, the efficiency does—even though the zero of gravitational PE is arbitrary.

For waterfalls the treatment is of little practical value anyway, but for steam engines it is perfectly applicable and very practical. That difference arises because the zero of temperature is not arbitrary at all. There is an absolute zero, and it is always the reference level for Carnot's efficiency equation. Those temperatures that reflect the cascading of heat are in *degrees Kelvin* on the absolute scale.

For a simple steam engine with a boiler temperature of 100°C, or 373 K, and a condenser cooled by water to, say, 4°C, or 277 K, the *maximum* possible efficiency is $(373 - 277)/373$, which equals about 26 percent. That figure is attained only by a perfect device; any real machine will waste energy and operate at a somewhat reduced efficiency. Clearly it would make sense in designing a heat engine to shoot for the largest possible temperature difference while trying to eliminate losses like friction. That's why superheated steam is widely used nowadays and why gasoline engines, with their high combustion temperatures, are fairly efficient.

Any heat engine carries a working substance (air, water, steam, combustion gases, whatever) from a high to a low temperature, extracting work in the process. The molecules of the working substance will carry energy away with them as they leave at T_L, energy that is effectively lost. That means that if a certain amount of energy goes into the engine as *heat-in* and a certain amount is exhausted as *heat-out*, then, by the first law

heat-in = work + heat-out.

Clearly, *it is impossible to convert all the heat-in into work as long as there is heat-out.* In other words,

heat-in = available energy + unavailable energy,

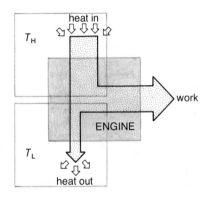

In a heat engine, not all the energy flowing in can be delivered as work; some fraction is inevitably lost as heat out.

and this limitation on the usefulness of second-hand or processed energy is an alternative statement of the *second law*. If the first law demands that we never get something for nothing, the second law demands that we never as much as come out even.

Only if the working substance leaves the cycle with no heat at all, only if it leaves and flows into an environment at absolute zero $(T_L = 0)$ will the engine be 100 percent efficient.* Since all real engines are less efficient than the ideal, none can ever operate at 100 percent efficiency, and all lose heat. No matter how clever we are, whenever we operate an engine, heat is lost; just put your hand on the hood of a car or on an electric motor after it has run for a while.

The total energy in an isolated system is constant, but unless that energy is distributed in such a way that two different temperature regions exist, heat will not flow and work cannot be done. *Energy is conserved, but it is not totally recyclable because some of it loses its usefulness, its availability to do work.* The great sea of

---

* This was essentially the route taken much later by Kelvin in establishing his absolute temperature scale.

thermal energy in the ocean is at a fairly low temperature and therefore, practically speaking, more or less useless. To what could we efficiently make it flow?

Before he died, at 36, of cholera in the epidemic that swept Paris in 1832, Sadi Carnot (his unpublished papers reveal) had rejected caloric and even anticipated the first law. The workers that followed him would in time establish that Carnot's conclusions were quite independent of the actual nature of heat and so perfectly applicable to a kinetic interpretation.

The crisis we face today does not exist because we somehow use up energy. The total amount of energy in the universe is presumably constant. The crisis exists because we dissipate energy, converting concentrated forms into useless, unavailable thermal energy. Nor will we ever get it back; the process in this universe at this moment is quite *irreversible*—so sayeth Sadi! *The inevitable continuing decrease in available energy is yet another statement of the second law.* The energy crisis here and now is probably temporary; new techniques for unlocking useful energy may save our electric toothbrushes for generations. But the second law will march on insatiably, despite humanity's ingenuity. It portends an ultimate insurmountable crisis.

The Heat Death

Heat, more or less unavailable heat, seems the most common by-product of all our engines—indeed, of all our processes, including life. Where has the energy gone from the innumerable gallons of gasoline burned in our cars; from the vast quantities of electricity, gas, and oil used in factories and homes? Few monuments are built by all the chocolate bars we eat; the rest evaporates as heat. In the end, most of the energy liberated from storage finds its way into random molecular motion, into low-temperature, fairly useless heat that spreads out over the planet and eventually leaks off into space. It's easy to convert mechanical energy to heat; friction is the perfect instrument for it. The difficulty comes in trying to go the other way, and for that there are no perfect means. Strike a match and its chemical PE erupts into heat, light, and sound. But in a moment, the latter two are usually absorbed, themselves vanished into heat. A wisp of thermal energy floats away, the match is gone forever, and you will only liberate more heat in the process of making the next match, let alone burning it. Turn on a TV set and watch electrical energy do its disappearing act. All our clever handiwork is flushed away as heat —even the songs.

It seems that the relentless flow of heat from high temperatures to low will ultimately cause all the hot to cool, all the cold to warm, and all the universe to end in uniform tepid boredom. The entire cosmos, the prophets of doom proclaim, will reach the exact same lukewarm temperature, and there it will remain, incapable of work, incapable of change, in homogeneous, lifeless chaos forever.

The second law, so innocuous at first, now sounds like an ominous funeral dirge wailing of a ghastly end, thousands of millions of years off in the future. Relax! By then the ninety-third law of thermodynamics will probably have given us a few more years. Or perhaps the universe, now expanding, may start to collapse and change its rules.*

The Arrow of Time

Of all the laws we have seen, only the second law of thermodynamics gives a direction to the unfolding of events, to the progression of processes. It points the arrow of time in the "forward" direction—downhill, in the wake of flowing heat. Time stops when nothing is happening. It is the succession of events that gives time meaning, and it is the second law of thermodynamics that gives it direction. Could the universe be run backward like a giant motion picture? Newton's physics is indifferent to time's pointing, and the first law of thermo would not object, but the second would!

Have you ever seen the gases rush back, the smoke descend, the light return, the heat pour in, the flame collapse, and a match reform, unburnt again and whole? Will a scrambled egg gather itself together, reshape its yolk, absorb energy, and jump back into its shell? Will the gray hairs darken, the skin grow tight and supple, the muscles strengthen, the sags unsag? Not likely! And yet energy would be conserved. The first law could stand by and watch as heat rushed back, turning into mechanical energy to fling a swimmer up out of the water and back to the diving board. Heat to KE to gravitational PE—that's allowed; but the second law insists that heat be lost, and our bouncing diver will never make it back up to the board. All by itself, the game runs downhill, and it is the one-way whine of the universe that we mark off in seconds.

We could watch a movie of a skater jumping, spinning, gliding forward and backward with equal grace, and it would be all but impossible to tell whether the film itself was being run forward or backward. Our clues to the march of time come from the natural progression of events, which more often than not lead to disorder. If you saw a motion picture of a burning cigarette in which the smoke settled down, the ash leaped back, and the weed reformed from a butt to an unlit whole, what would you think? Of course, the film ran backward. And what would you say if you saw it happen in reality before your very eyes? Well, perhaps *time* itself ran backward!

We are content to watch a cue ball slam into a wedge of fifteen neatly racked pool balls and see them all fly off, colliding and bouncing around. But if we see moving balls that have been crashing into each other and rebounding all at once converge and pull together into a perfect, motionless wedge with only the cue ball sailing out and

Ludwig Boltzmann

---

* Of course, we do not know whether the second law simply describes the universe as it presently is (i.e., expanding) or whether it actually commands the universe, determining how it will always be.

away, then something very strange has happened. Disorder has gone into order, and that somehow is not the normal way of things—not quite the way events seem to naturally occur in time.

Law and Disorder

The mysteries of the second law, the law of dissipation, were framed in yet another set of images by the German physicist Rudolf Julius Emmanuel Clausius. He was a young instructor at the Royal Artillery and Engineering School, Berlin, in 1850 when his first and perhaps his most important paper on heat was published. There Clausius, recognizing the value of Carnot's analysis, specifically set out the second law. The climax of his work came in 1865 with the introduction of the concept of **entropy*** as a measure of the transformation of energy from an available to an inaccessible form. This strange quantity, entropy, whatever it was, seemed somehow to be of fundamental importance. But it was not until 1878, when that restless, tormented genius, Ludwig Boltzmann, reformulated the vision, that entropy could be understood with a new clarity. Boltzmann said that *entropy was a measure of the disorder of the universe*. The neatly arranged molecular pattern represented by the chemical PE in a match is transformed into a mayhem of gas and heat when it is burned. The disorder of the universe increases, entropy increases, energy once concentrated is now spread out with an evenness that is the essence of disorder. *Maximum disorder corresponds to a uniform, undifferentiated evenness throughout.* A bull in a china shop will powder the pretty cups, crush the delicate vases and increase the uncompromising entropy of the universe.

The *second law*, framed in the rich colors of Boltzmann's construction, now reads: *The entropy of an isolated system will either increase in time or at best stay constant, while the entropy of the universe as a whole moves inexorably toward a maximum.*

A gallon of gasoline has its molecules arranged in complex relationships in a far more ordered state than does a cloud of gaseous exhaust thinly spread out and disassociated. Mixtures of any kind are more disordered than were their separate constituents. Two gases introduced into a chamber will "always" mix. Colored grains of sand shaken in a bag will "always" mix. Yet neither the gases nor the sands are likely ever to unmix spontaneously to become neatly separated, decreasing in entropy. For that matter, you can shake the bag of sand for centuries, and though it is possible, it is most improbable that the different colored specks will ever separate perfectly again.

True, you could reach into the system (then no longer isolated), do work on it, and sort out the different colored grains—you could decrease its entropy. But in the end, you would waste more heat in the effort, you would disorder your breakfast more than you would order the sand—and the universe would lose again. Wherever en-

---

* From the Greek for *transformation*. Clausius defined it as the heat transferred divided by the absolute temperature.

tropy is decreasing—in a factory where someone is stringing beads, with a dung beetle rolling its treasure into a ball, with you presumably learning the second law, among the cooling ice cubes in my freezer—order is rising out of disorder. But wherever this occurs, there will always be a hidden harvest of mayhem that overbalances the scales on the side of universal disorder. The freezer will make its ice all right, but at a considerable cost of electrical energy quietly blown off as heat.

Boltzmann found the mathematical fingerprint of disorder in a statistical relationship describing the likelihood that particular molecular arrangements will occur. If we take all the separate molecules that make up a match and put them in a box and shake them, the probability of looking in to find the molecules all bound together, neatly arranged as an actual match, is incredibly small. It is possible but, given the billions of possible combinations, exceedingly unlikely. By contrast, the random distribution of those molecules in the box can occur in so many different ways that to find them in one of these disordered configurations is by far the more probable.

Unwrap a deck of cards and observe the arrangement. That ordered state is one of 80 million million million million million million million million million million million possible arrangements of those 52 cards. Shuffle them a few times or toss them in the air, and your chance of seeing that special order again is one out of 80 followed by 66 zeros (80×10^{66})! Seems that order is less likely to occur naturally than disorder. As we have seen, once the molecules in an egg are scrambled, the chance that all of them will simultaneously find their way back to so unlikely an arrangement as the original ordered pattern is slim indeed. Every process irreversibly scrambles a bit of the cosmic egg; even your quiet breathing as you read this page increases universal entropy.

When Ludwig Boltzmann killed himself while on holiday near Trieste in 1906, he personalized disorder's inexorable mill—dust to dust. His monumental formula for entropy, carved in the headstone on his grave, is the universal epitaph.

I met a traveller from an antique land
Who said: "Two vast and trunkless legs of stone
Stand in the desert. Near them, on the sand,
Half sunk, a shattered visage lies, whose frown
And wrinkled lip, and sneer of cold command,
Tell that its sculptor well those passions read
Which yet survive, stampt on these lifeless things,
The hand that mockt them and the heart that fed;
And on the pedestal these words appear:
'My name is Ozymandias, king of kings:
Look on my works, ye Mighty, and despair!'
Nothing besides remains. Round the decay
Of that colossal wreck, boundless and bare
The lone and level sands stretch far away."

P. B. SHELLEY
Ozymandias

EXPERIMENTS

1. Fill three pots with water, one with some ice in it, another lukewarm, and the third hot tap water. Put your left hand in the cold bath and your right in the hot bath for a minute or so. Now place both hands in the lukewarm water. Considering the nature of heat flow, explain why both hands feel different sensations. What can you conclude about whether we feel temperature or heat flow?

2. If your refrigerator is pumping heat out of the food within it and wasting additional heat in the process, it must be dumping all that energy into the room. See if you can find its exhaust.

3. Press your finger quite hard on a smooth surface and rub it back and forth as fast as you can until your fingertip heats up. Bend and unbend a piece of thin metal or a length of wire rapidly until it heats up. Put a nail or a piece of wire on a hard surface, and pound on it a few times with a hammer. It, too, will quickly become quite hot. What do all these processes have in common as far as energy is concerned?

4. What happens to the temperature of a thick rubber band when it's quickly stretched? Use your lips as a sensitive temperature detector before and while the band is stretched.

REVIEW

Section 8.1

1. Distinguish between organized and disorganized energy.

2. What is *thermal energy*?

3. Define the term *heat*.

4. In what way are work and heat similar?

5. Does the temperature of a baseball increase as its KE increases when it is thrown?

6. Describe Rumford's contribution to the theory of heat.

7. What is the name of the theoretical formalism that pictures heat as a manifestation of the rapid random motion of atoms and molecules?

Section 8.2

1. How is the scale of a thermometer determined using standard reference points?

2. How many divisions are there on the Celsius scale between the freezing and boiling points of water?

3. Discuss the distinction between *thermal energy, heat,* and *temperature*.

4. What are the highest and the lowest temperatures you are likely to find at home?

5. What is the temperature of an open bath of liquid nitrogen?

6. Liquid helium boils in an open container at _____ °F.

7. What property of gases suggests the existence of an *absolute zero* of temperature?

8. At what temperature on the Kelvin scale does water freeze?

Section 8.3

1. If an iceberg contains more thermal energy than a cup of hot tea, why is its temperature so low?

2. What is a *calorie*? How is it different from a *Calorie*?

3. Define a *BTU*.

4. An 8000-BTU air conditioner removes _____ joules from a hot room every _____.

5. Explain the meaning of the term *specific heat capacity*.

6. When are two objects said to be in *thermal* equilibrium?

7. Is it possible to measure the temperature of a block of ice and find it different from 0°C (32°F)?

QUESTIONS

Answers to starred questions are given at the end of the book.

Section 8.1

1. Suppose we fired a bullet at a thick steel plate. What would happen to the temperature of the "slug" during impact? Explain.

8. What is meant by the term *heat of fusion*?

9. How many calories does it take to melt 1 g of ice?

10. Discuss the meaning of the term *heat of vaporization*.

11. Explain how evaporation functions as a cooling process.

12. Does turning up the flame under a pot of boiling water help to cook the egg within it any faster?

Section 8.4

1. 1 Cal of heat equals _____ joules of energy.

2. What is *thermodynamics*?

3. State the first law of thermodynamics.

4. How does the temperature of a meteorite change on impact with the Earth?

5. What happens to the KE of a moving car whenever you come to a stop on a level road?

6. Suppose that you are running and you decide to stop for a rest. How do you usually get rid of excess kinetic energy?

7. A large lead ball is dropped from a diving board into a pool. Both water and ball are initially at the same temperature. Will that temperature change after the drop?

Section 8.5

1. An object loses heat to its surroundings because those surroundings are lower in _____.

2. State the second law of thermodynamics.

3. Describe the transfer of heat between two bodies in contact from the perspective of the kinetic theory.

4. What is a *heat engine*?

5. What is a *working substance* in a heat engine?

6. Define the term *efficiency*.

7. State Carnot's efficiency equation for a heat engine.

8. In what respect does the second law of thermodynamics demand that we never break even in the use of energy?

9. Under what circumstances is the theoretical efficiency of a heat engine 100 percent?

10. What is meant by the statement that energy is not recyclable?

11. What ultimate crisis does the second law portend?

12. How is the second law related to the flow of time?

13. Entropy is a measure of the _____ of the universe.

* 2. Distinguish between heating an object by doing work on it and by supplying heat to it.

3. Describe what is happening when you warm your hands by exhaling on them.

4. What can you expect to feel if you insert a metal spoon into a flame while you are holding the handle?

* 5. The air blasting from a hair dryer can be either hot or cool. What is the difference physically?

6. Can you tell whether a frankfurter has been warmed by radiation in a microwave oven or by having heat pumped into it from a flame?

Section 8.2

1. Describe the operation of Galileo's temperature indicator. I had a sealed empty gallon can sitting on a table during the summer, and every time the air conditioner dropped the room temperature to a certain point, the can would make a loud clunking sound. Explain what was happening in light of the kinetic theory.

2. Suppose that we could reach into a material with some tiny probe and slam a few of its atoms. For example, suppose that we could fire some high-speed electrons at it. What would happen to the thermal energy and the temperature of the target?

* 3. A radioisotope thermal generator converts heat directly into electrical energy and is particularly convenient for use in long-lifetime satellites. The core of the device is a chunk of an appropriate radioactive material, which stays quite hot all by itself for decades. High-speed particles emitted by the radioactive atoms fly through the material, and most are absorbed within it. Explain how the core produces thermal energy.

4. Discuss the meaning of the term *temperature*.

5. What temperature Fahrenheit is equivalent to 0°C? To 100°C?

* 6. At what temperature will a Fahrenheit and a Celsius thermometer have the same numerical reading?

7. What is the lowest possible temperature in each of the three major systems—Fahrenheit, Celsius, and Kelvin?

8. Make a list of some of the uses of low-temperature liquified gases that have been in the news in recent times.

* 9. What effect, if any, does a fan have on the temperature of the air in a room?

Section 8.3

* 1. Can two objects possess the same amount of thermal energy and yet have different temperatures?

2. Suppose that you had two beakers, one containing 2 kg of water, the other 1 kg of water and a 1-kg block of brass. If both were at room temperature, which would reach 100°C first when both were placed over identical burners? Explain.

* 3. What will happen to a 1-kg mass of water at room temperature if we remove 2½ Calories from it?

4. Suppose that we have two equal-mass lumps, one of gold and the other of glass, both at 150°C. Will there be any difference in the final equilibrium temperatures if we drop both into identical quantities of water at room temperature?

5. Imagine that you have two equal-mass mugs, one made of glass and the other of brass. If you fill both with identical quantities of hot water and wait a little while for them to reach equilibrium, which will be hotter?

* 6. Is it possible to add heat to a body that is otherwise isolated from its environment and not cause its temperature to increase? Explain and give an example.

7. An icebox was just that—an insulated chest in which one put a large block of ice and whatever food had to be kept cool. Discuss how it worked.

8. Explain the mechanism by which you can actually manage to cool off on a hot day by drinking a hot cup of coffee (65°C).

* 9. Why do ice cubes dumped into a bucket separately tend to stick together after a little while?

Section 8.4

1. Suppose that you have a closed metal can filled with a gas and you squash the can a bit. What will happen to the temperature of the gas?

* 2. Imagine that two identical cars traveling toward each other at the same speed collide head-on and come to a dead stop. Discuss the collision from the point of view of momentum conservation. What happens to the KE of the two cars?

3. On p. 101 we talked about the momentum of a bunch of billiard balls in a closed box that was shaken up by an external force. If all this happened out in space, would the balls go on forever bouncing around inside the box? If not, why not?

4. A classical physics demonstration uses a bowling ball (4 kg) hung from the ceiling on a long rope. The fearless instructor stands with the back of her head against the wall, raises the ball so it touches her nose, and lets go. Why should she not push while releasing the ball? Will it return to the original height on the back swing? Will the pendulum keep swinging endlessly? Explain.

5. A person holding a ball throws it into the air and then catches it. Trace the various transformations of energy throughout the process.

6. One of the Amazing Spamoni Brothers jumps off a ten-meter diving board onto a trampoline and bounces up and down and up again. Will he ever stop? Explain.

* 7. One day in the summer of 1847, James Joule and his bride on their honeymoon were walking in the Alps, and quite by chance, they met William Thomson (Lord Kelvin), who just happened to be strolling around. But that's irrelevant to the problem. Joule was carrying a large thermometer on his way to the waterfall at Chamonix. It was his intention to measure the increase in temperature of the water at the

bottom as compared with that at the top of the fall. Why did he expect to find a difference? Incidentally, he was not successful, because the torrent of water was interrupted in several places by rocks, and there was a good deal of spray.

Section 8.5

1. My wife and I know that if we want the house to become a wreck, all we have to do is pay no attention to it for a few days, letting the three kids, one dog, two cats, and goldfish just go about their business. What fundamental law of the universe is at work?

2. Can an engine develop more work than the amount of heat flowing into it? For that matter, is it possible for an engine to convert into work all the heat flowing into it?

* 3. Why is it not possible practically to power a ship by drawing heat from the ocean to turn the propellers and then simply dumping the resulting ice overboard?

4. Is it correct to say that energy transforms naturally from mechanical to thermal and not the other way around? Explain your answer and give some examples.

5. From the perspective of entropy, discuss what happens when a bullet comes to rest on impact with a steel plate.

6. When a mountain climber reaches the peak, has he not increased the amount of energy now available to him in the form of gravitational PE? Does this mean that the entropy of the universe has decreased in the process? Explain.

7. When green plants store energy in the form of chemical PE via photosynthesis, does the entropy of the universe decrease since it certainly does decrease locally? *Hint:* Consider the source of almost all earthly energy, the sun.

8. Make a list of all of the processes you were involved in today that decreased the available energy of the universe.

* 9. Every steam-electric power plant produces "heat" that represents thermal pollution and is often dumped into nearby rivers to amuse the fish. Is this waste heat unavoidable?

10. How is entropy involved in distinguishing between the past and the future?

11. What can you expect to happen to the entropy of any closed system, that is, one completely isolated from the rest of the universe? Keep in mind that the *total* entropy of the universe can never decrease.

12. The human body takes in high-grade energy in the form of oxygen and food, and it emits almost exactly the same amount of energy. The latter is in the form of some chemical PE in excreta (low-grade), low-temperature heat given off, air exhaled, and work done. Despite the conversion of energy within the body and the resulting production of entropy, the body itself does not increase in entropy. Explain what's happening.

* 13. The human body ordinarily functions as an open system exchanging energy with the environment. In light of the two preceding questions, what would happen if it became a closed system isolated from the rest of the universe? *Hint:* Consider someone locked in an insulated vault.

14. Of all the laws examined thus far, which one applies most directly to the misfortune of Humpty Dumpty?

MATHEMATICAL PROBLEMS

Answers to starred problems are given at the end of the book.

Section 8.2

* 1. The temperature of the human body varies somewhat over the course of a day, and it is certainly not the same for everyone, but it's usually taken to be 98.6°F. How much is that on both the Celsius and Kelvin scales?

2. Room temperature is generally said to be 68°F. What does that correspond to on the Celsius scale?

* 3. Liquid oxygen normally boils at 90 K. What does that correspond to in degrees Celsius and in degrees Fahrenheit?

4. The interior of the sun is somewhere around 15 million degrees Kelvin. Estimate its Celsius and Fahrenheit temperatures.

Section 8.3

1. Using the definition of a calorie and a BTU, prove that 1 BTU = 252 cal.

* 2. A glass of nice dry red wine supplies the body with 75 Cal. What is the equivalent of that in joules?

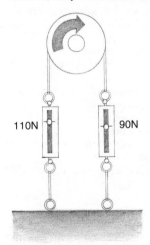

3. The power developed by an engine can be measured experimentally by attaching a disk to it and then running the disk against a belt, as shown in the figure. Before it's turned on, both spring balances read 100N. While it's turning against friction at 10 revolutions per second, the balances read 110N and 90N. The disk has a circumference of

0.5m. Determine the power delivered in the form of friction. How many Calories is that equivalent to per minute?

4. (a) How much heat must be added to raise the temperature of 1 kg of water from 20°C to 80°C? (b) How much heat must be added to raise the temperature of 1 kg of brass from 20°C to 80°C? (c) A 1-kg brass bowl containing 1 kg of water is at an equilibrium temperature of 20°C. How much heat must be added to bring the bowl and the water up to 80°C?

* 5. A large insulated container holds 500 g of cool water at 10°C. (a) What will be the final temperature if someone now adds 500 g of hot water at 90°C? (b) What final temperature results if someone pours in only 250 g of hot water? *Hint:* The heat that must be removed to lower the hot water to 10°C is then available to raise the temperature of all the water.

* 6. How much heat does it take to melt (a) a 10-kg block of ice? (b) a 10-g ice cube? (c) Now suppose that two such ice cubes are dropped into 200 g of water at room temperature (20°C). What will the final temperature be after the ice melts?

7. How much steam at 100°C is needed to melt a 1-kg block of ice at 0°C? How much is needed if the ice is at −150°C?

Section 8.4

* 1. Suppose that you could condense 10 kg of steam and use the emitted energy to fire the resulting water straight up into the air. At what speed would it be propelled and how high would it sail? Assume that g is constant, even though in this case that's a poor assumption.

2. You have a pond and a 10-g ice cube, both at 0°C, and you would like to melt half the cube by dropping it from some height into the water. What's the minimum drop needed to accomplish the business? (Assume g to be constant at 9.8 m/s².)

3. How high must a waterfall be if the temperature of the liquid is to rise half a degree Celsius on impact?

* 4. Lead has a specific heat of 0.03 and a heat of fusion of about 6, and it melts at 330°C. How fast must a lead bullet travel if it's to melt on impact with a hard wall (which we will assume absorbs no appreciable heat)? The temperature of the bullet prior to striking the target is 30°C.

Section 8.5

* 1. What is the efficiency of an engine that takes in 10 Calories for every 1000 joules of work it performs?

2. Calculate the best possible efficiency of a steam turbine in a power plant that uses superheated steam at 500°C, cooling it down to room temperature, 20°C. How much wasted heat must be expelled for every joule of chemical energy liberated by combustion in the plant?

* 3. A modern fossil-fuel power plant is typically 40 percent efficient. Suppose that it delivers 300 megawatts (300 × 10^6 W) of electrical power. How much power must be pumped in to operate the plant? How much power is lost in the form of waste heat?

4. Home furnaces are usually around 60 percent efficient. If it takes one million BTUs to heat a house on a winter day, how much power must be supplied by the chemical energy in the oil being burned? How many BTUs are lost up the chimney?

* 5. A dam on a river at a height of 250 meters above sea level supplies a hydroelectric plant located at an elevation of 50 meters above sea level. Gravitational PE is typically converted to electricity with an efficiency of about 85 percent. A river (like the Colorado) can provide about 700,000 kilograms of water each second to the turbines. How much gravitational PE is delivered each second? How much electrical power will be generated by such a plant?

6. The water in Niagara Falls first plunges down 50 ft through rapids, drops 167 ft over the falls, and then cascades another 98 ft down the lower rapids. This is the torrent the tourist sees, but all the while two giant pipes are carrying an average of 130,000 cubic feet of water per second down to the turbines of the power plant. Given that the Niagara Power Project generates 2.4 million kilowatts, calculate its efficiency (1 ft-lb/s = 1.36 × 10^{-3} kW).

PART III

Matter

The question "What is matter?" is of the kind that is asked by metaphysicians, and answered in vast books of incredible obscurity.

BERTRAND RUSSELL (1959)

British philosopher

It is probably as meaningless to discuss how much room an electron takes up as to discuss how much room a fear, an anxiety, or an uncertainty takes up.

SIR JAMES JEANS (1877–1946)

English astronomer

It should be easy to hold a thing in hand and know what matter is; it's stuff, substance, the material aspect of the universe. It's the constituent of butterflies and cigar butts—that which composes, which makes up all the tangible, all the sensible.

But in the end, language and imagery fail us if we try to go beyond our intuitive sense, if we press to answer "what *really* is it?" *Matter cannot be satisfactorily defined*, certainly not in terms of more fundamental notions. It is surely not enough to simply say matter is that which is perceivable (overlooking illusion); light is sensible, and yet it is something else again—not matter, but *energy*. Nor is matter merely that which possesses *mass* (or alternatively, that which possesses *inertia*). Light can be bent in its path by the gravitational influence of a star, and in that sense it behaves as if it had mass. Though it has no *rest-mass*, it can transfer mass from one place to another. "A beam of light carries energy," said Einstein, "and energy has mass." In fact (as we will see in Chapter 15), a hot apple pie has more rest-mass than a cool one and will even exert a greater gravitational attraction. Does that mean it has more matter?

Excluding "nothingness," all there is is either matter or energy, and we will find that both are related like the faces of a coin—one and different. That is a crucial point. The supposedly clear distinctions between matter and energy are not really so clear after all. Indeed, the neutrino, the tiny, chargeless, zero rest-mass particle, seems to walk the line between matter and energy. All the simple definitions simply fail.

Miss Buckley, my sixth-grade science teacher, used to say that matter was "that which was impenetrable and occupied space." And she always followed that bit of dogma with the ringing phrase "Matter can be neither created nor destroyed." She was not alone in those pronouncements; hers was the view of *classical physics*, the widely held belief prior to this century, which should have been swept away long ago but still lingers on. We know now that matter does not occupy space in any ordinary sense. The smallest particles are fluttery, elusive things that recoil from localization like wisps of smoke. The idea of ultimate impenetrable specks of solid matter is an ancient mirage. Indeed, if we accept the imagery of relativity theory, matter and space are one-and-the-same, and even the notion of "occupying" becomes meaningless. The theoretician can create a mathematical representation of the cosmos and within it intone that *matter is a local irregularity in four-dimensional space-time.* But that's a bit too far from chicken soup to give most people any real satisfaction.

Dear Miss Buckley should have known that the creation and annihilation of matter actually go on every day in laboratories and nuclear power plants all around the world (Section 15.1). Matter certainly vanished at Hiroshima and Nagasaki. Particles that supposedly occupy space can be made to dematerialize, to evaporate into puffs of energy. Bits of what dictionaries claim are *full space,* specks of matter, can be made to disappear, and with them into emptiness goes any hope of easy definition. As if to confound the situation even more, there is a stuff called *antimatter* (Section 15.3). When antimatter unites with matter in the ultimate embrace, both vanish—whatever they are.

Rather than struggle to define matter, however, let's deal with it as best we can. We can describe its properties and speculate about its structure; any more than that would be presumption. There will usually be no difficulties recognizing matter—certainly not in the quantities encountered in everyday life. If it is tangible, it is matter, regardless of the definition.

Bulk matter, whatever its composition, appears in four states: *solid, liquid, gas,* and *plasma* (Chapters 12–14). There exist roughly a hundred different primitive or basic materials known as *elements* (oxygen, iron, carbon, etc.), which combine to form all the various substances from DNA to TNT and LSD (Chapter 9). We believe that each of these elements is composed of tiny microscopic entities known as *atoms.* Each element is simply a collection of identical parts, atoms of one particular type, and there are roughly a hundred distinctly different kinds of atoms. These universal building blocks, in turn, are formed of various arrangements of subatomic matter—*electrons, protons,* and *neutrons* (Chapter 10). And even the last two of these minute particles may be fashioned of still more fundamental specks—the elusive *quarks.*

Our contemporary picture of the atom was fashioned in this century, in the ongoing intellectual flurry known as modern physics. The earliest of the new conceptions portrayed the atom as if it were a tiny solar system. A dense minute speck—the *nucleus*—composed of protons and neutrons commanded the very center of each atom. Bound by an attractive electrical force to this positively charged massive core were the negative atomic electrons, whirling around in lovely closed orbits.

The first and most far-reaching of the new revelations was that energy is not continuous, as had always been believed, but rather, like matter, it exists in lumps, or *quanta.* From this evolved an entire theoretical formalism known as quantum mechanics, which deals in particular with those phenomena in which the granularity of energy becomes highly significant—the domain of the atom.

Light is, in short, the most refined form of matter.

LOUIS DE BROGLIE
French physicist

To be sure, we say that an area of space is free of matter; we call it empty, if there is nothing present except a gravitational field. However, this is not found in reality, because even far out in the universe there is starlight, and that *is* matter.

ERWIN SCHRÖDINGER
Austrian physicist

C
Si
Ge
Sn
Pb

Carbon (C) exists in three natural elemental forms: charcoal, graphite, and diamond. Though not very abundant in the Earth's crust ($<0.1\%$), it constitutes about 17.5% of the human body. Carbon and hydrogen combine to form innumerable compounds, including those in all known life forms.

In time quantum theory (Chapter 11) produced a remarkably powerful mathematical description of atomic behavior. Along with relativity, it became one of the cornerstones of modern physics. Atomic electrons are still imagined circulating about a positive nucleus, but we no longer envision clusters of neat little orbits. Instead, we now calculate the atomic landscape in terms of the patterns of place in which the electrons are likely to reside. The concern shifted from discovering the details of individual electron orbits to evolving descriptions of the cloudlike distributions in which they spend their time—moving captives of the nucleus.

Quantum mechanics is a statistical theory, a theory that deals with the probability of occurrence of atomic events—not out of inadequacy, but because absolute certainty in Nature is itself an illusion. That, too, is a profound contemporary insight of far-reaching philosophical importance (Section 11.5). Quantum mechanics is our guide to every aspect of the internal functioning of matter. Out of that understanding has poured a technological wealth that includes everything from lasers to computer circuit chips.

We ourselves, like the very atoms that form us, are mostly emptiness, mostly "unfilled" void. Indeed, the central core of the atom, where practically all the mass resides, is a minuscule grain. Surrounding that core hovers the far-distant, tenuous, relatively massless electron cloud. On that scale, great expanses of empty space separate the tiny concentrations of mass from one another. Distances between the nuclei of adjacent atoms are comparatively immense, even within those dense objects we unwittingly call solid. If you could taste Alice's growing potion and swell up until the core of an atom in your body was the size of a cherry pit, you would stand about 800 million miles tall! And most of that huge volume of flesh and sinew and bone would be *void*, with hundreds of meters of emptiness separating the pit-sized concentrations of matter.

We are a collection of millions on millions on millions on millions of atoms, bound together as, very roughly, thirteen or so gallons of water, nearly thirty pounds of carbon, perhaps three additional pounds each of oxygen, hydrogen, and calcium, a bit more of nitrogen, a pound and a half of phosphorus, about a half pound of potassium and sulfur, three or four ounces of sodium and chlorine, an ounce of magnesium, a dash of iron, a pinch of iodine, silicon, and fluorine, and faint traces of many other elements picked up in our travels, for better or worse.

We are the constantly changing aggregate of nearly ageless atoms, themselves created thousands of millions of years ago! The living cells are recent enough, though almost without exception the atoms that form your hand are at least as old as the solar system and probably much older. They have circulated through fish and fowl and trees and dung and Whopper-burgers and back to the soil and now, for the moment, to you. Breathe in, and a wisp of exhaust from some passing car will fill your lungs with the distillate of dinosaur,* 100 million years gone. Probably formed in the thermonuclear fires of a star, the atoms you "ate" at breakfast extend your lineage back to the dawn of Creation. We *are* stardust, borrowers in the ancient ritual of "ashes to ashes."

---

* Gasoline derives from petroleum, which we think is the residue of buried organic matter ranging from dinosaurs to seaweed.

Hamlet: To what base uses we may return, Horatio! Why may not imagination trace the noble dust of Alexander till he find it stoping a bung-hole?

Horatio: 'Twere to consider too curiously to consider so.

Hamlet: No, faith, not a jot; but to follow him thither with modesty enough, and likelihood to lead it: as thus; Alexander died, Alexander was buried, Alexander returneth into dust; the dust is earth; of earth we make loam; and why of that loam whereto he was converted might they not stop a beer-barrel?

Imperious Caesar, dead and turn'd to clay,
Might stop a hole to keep the wind away:
O, that that earth which kept the world in awe
Should patch a wall to expel the winter's flaw!

WILLIAM SHAKESPEARE

Hamlet (in the graveyard)

DEBENT   IGNARI   RES   FERRE   ET   POST   OPERARI   QVATVOR   INSERTA   NATVRIS   IN   NVBE   REFERT
IVS   LAPIDIS   CARI   VILIS   SED   DENIQ3   RARI   NVLLA   MINERALIS   RES   EST   VBI   PRINCIPALIS
VNICA   RES   CERTA   VILIS   SED   VBIQ3   REPERTA   SED   TALIS   QVALIS   REPERITVR   VBIQ3   LOCALIS.

"The Alchemist," engraved by P. Galle from a drawing by the Flemish master Pieter Brueghel (1558). The shaggy fellow on the left is the amateur alchemist seeding the brew with his last gold coin. Mama bewails her empty purse, and their three kids are examining the barren cupboard. Out the window is a vision of the future—the family on the way to the poorhouse.

The main purpose of this chapter is to develop the modern notion of the **chemical atom** as it arose in the beginnings of the nineteenth century. The first section elaborates the origins of the concept of the **elements.** The second segment traces the discovery of many of the 88 naturally occurring elements, with a particular eye to establishing the patterns of similarity among them. It is the similarities that are the outward signs of a hidden inner order, an order arising from the invisible **atomic structure** of matter. Finally, in the last section the whole array of elements comes together in the grand scheme of the **periodic table.** At that point the pattern is revealed if not understood. That will have to wait for a theory of atomic structure.

Fire has always been the primal edge of mystery, and to the alchemist it was also the power that transformed. The Great Work of alchemy went on, for almost 2000 years, in the red glow of the vaulted furnace, in the rhythm of solemn prayer and incantation. In part, the Work was the **transmutation** of the base and common (iron, lead, etc.) into the pure and rare (gold); in part, it was the practitioner's own spiritual preparation. Gold could be melted and remelted, and unlike the mundane metals, it did not tarnish and corrode in the fire. It was the very symbol of purity, and only the pious could ever hope to carry out the sacred work, to make gold.

Alchemists labored in vain to create the *philosopher's stone,* the interim catalyst with which commonplace matter could be rendered perfect, the ultimate instrument of transmutation. The stone, the perfecter of the body as well, would surely be the cure of all disease, the balm of eternal youth, the *elixir of life:* medicine and chemistry hand-in-hand at the crucible.

After the armies of Alexander the Great had conquered Egypt, there came about a curious blending of Greek theoretical ideas on matter with the practical Egyptian mastery of metallurgy, cosmetics, and the dark embalming arts. The resulting Greco-Egyptian *khemeia* had among its practitioners those bent on transmutation almost from the very beginning.

The rising fortunes of Christianity, in time, brought a harsh suppression of pagan learning, and *khemeia* all but vanished into the dust of collapsing Rome. When the followers of Islam poured out of Arabia in the seventh century, they eagerly took up the ancient sciences, and *khemeia* was reborn as *al-kimiya.** In their hands **alchemy** thrived for 400 years.

---

* The prefix *al*'is the Arabic for "the," and words like alcohol, alkali, and algebra have that common origin.

9.1 FROM ALCHEMIST TO NUCLEAR PUFFER

If you would publish your infatuation
Come on and try your hand at transmutation;
If one of you has money in his fist
Step up and make yourself an alchemist.
Perhaps you think the trade is easily learnt?
Why then, come on and get your fingers
  burnt.

GEOFFREY CHAUCER (1340–1400)
"Canon's Yeoman's Tale" in *Canterbury Tales*

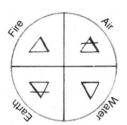

The defeat of the Muslims and the subsequent twelfth-century scientific renaissance brought alchemy back to Europe. During the Middle Ages, a few true adepts practiced the rigorous, methodical search for the stone, a quest that tolerated no element of experimentation, no deviation from the mystical though misty path. By contrast, the amateur, the get-rich-quick bungler, the casual devotee, the devoted but undisciplined—all pursued whatever course struck their fancy. These were the scorned **"puffers,"** whose derisive name was drawn from the interminable whooshing sound of the bellows feeding the flames. It was the puffer, as much concerned with discovery as with the stone, who isolated arsenic, bismuth, zinc, and phosphorus; who prepared muriatic acid (hydrochloric acid), aqua regia ("royal water"—a mixture of nitric and hydrochloric acids that dissolves gold), and oil of vitriol (sulfuric acid); who concocted such substances as potash, sodium carbonate, and ammonium sulfate; who manufactured laxatives (sodium sulfate) and peddled wonder drugs. It was the puffer who, at the end of the eighteenth century, at last freed of the weight of the stone, became the chemist.

Only now, only in this century, has the true nuclear puffer finally emerged. Ironically, although we at last have the power to transform lead into gold, it turns out to be too costly a game to play for fun or profit.

Earth, Water, Air, and Fire

A branch held in the fire will turn to ash; a shining iron rod will crumble to rust. Things transform. Then perhaps every material is merely a different manifestation of some more fundamental substance, some elemental matter that simply changes its appearance. The Greek philosopher Thales (ca. 624–546 B.C.) thought as much, proposing that *water* was the primitive essence of all things. After all, water could exist as solid, liquid, or gas. Before long, others suggested alternatively that *air* or even *fire* was the root substance. Empedocles the Sicilian (ca. 492–435 B.C.) postulated a pluralistic compromise; not one, but four elements—earth, water, air, and fire—constituted the universe. Aristotle added to the terrestrial list the *quinta essentia*, the fifth element, the eternal unchanging essence of the distant heavens—aether.

Along with Aristotle's other erroneous physical notions, the *theory of the four earthly elements*, which he embraced and embellished, dominated Muslim science as it did the medieval muddlings of the alchemists of Christendom. "Obviously" a burning branch (smoking, blackened, with sap bubbling) betrays its primeval constituents: fire, air, earth, and water. Many supposed that all things were but differing mixtures of the four elements, and that the wise alchemist who would transmute lead to gold needed only to alter and rearrange the balance.

Four-element theory, interpreted more or less symbolically (water/liquid, air/gas, earth/solid, fire/energy), reigned in the dark

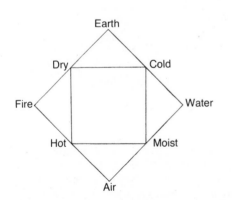

AN ELEMENT IS
the last point which analysis is capable of reaching.

A. L. LAVOISIER

halls of early science until the seventeenth century, when Robert Boyle, moving with the new spirit of experimentation, replaced the abstract with the operational.

Robert Boyle, the fourteenth child and seventh son of the Earl of Cork, was born in 1627 at Lismore Castle in Ireland. Always tall and thin, in later years he grew pale, frail, and gaunt—probably from drinking the foul concoctions he was forever dosing himself with, out of an exaggerated concern for his fading health. Boyle was a tremendously wealthy and influential man, a long-time director of the famed East India Company, and a founding Fellow of the Royal Society. High-minded, deeply religious, a committed bachelor, brilliant and honorable, he stands at a crucial moment in the rise of science, a moment midway between Galileo and Newton.

Though an alchemist himself, a confirmed believer in transmutation, Boyle effectively transformed the black art of alchemy into the science of chemistry with the publication of *The Sceptical Chymist*. Here he rejected the abstract Greek and mystical medieval views of matter and, instead, defined the modern concept of an element operationally, that is, in terms of the results of experimental procedures. *An* **element** *is a homogeneous material substance that cannot be further simplified or broken down into another substance by any chemical or physical process* (short of nuclear meddling). By contrast, two or more of these basic forms of matter can combine to produce innumerable **compounds,** and these compounds, in turn, may be decomposed into their elemental constituents.

All the world, from apples to zeppelins, is fashioned out of 88 basic varieties of matter, 88 naturally occurring elements. The daily romp through technology's playland brings us in contact with most of them, even some of the very rare ones. For that matter, the humble touch-tone telephone is by itself a beeping collection of 42 elements, including a few of the *exotique* like vanadium, palladium, beryllium, indium, molybdenum, and krypton. Many more can be found on any bathroom shelf compounded into lotions, laxatives, and elixirs (e.g., selenium in a shampoo, mercury in a contraceptive cream, zirconium in a deodorant, fluorine in toothpaste, and of course, iodine).

The elements are generally classed in three rough groupings: **metals, nonmetals,** and a compromise between the two extremes, **metalloids.** About three-fourths of all the elements are metals, and many of them have names ending in *-ium*. Metals have a shiny luster

Selenium (Se), a dark, hard, metallic-looking material, is both metal and nonmetal. In fact, when exposed to light, it becomes a good conductor and thus serves in a variety of photoelectric devices. It's widely used as an additive in the glass industry to cancel the green color that arises from iron impurities.

9.2 THE EIGHTY-EIGHT

Mercury (Hg), named after the planet, can be found in dental fillings (mixed with silver), thermometers, silent electric switches, fluorescent bulbs, vapor street lamps—and too often in seafood. Here a penny floats in a cup of the dense, shiny liquid. Don't be careless with the stuff; it's poisonous.

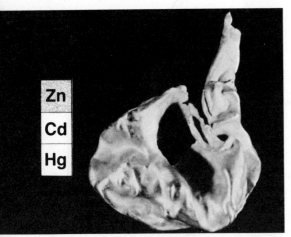

Zinc (Zn) has the familiar bright metallic appearance, but it's tinged with a very faint bluish cast. It has all the usual properties of a metal, although at room temperature it's surprisingly brittle. The white oxide is extensively used in ointments and as a paint pigment. Dipping iron into molten zinc coats it in a process known as galvanizing. You can see the pattern of flat Zn crystals on the surface of a metal trash can.

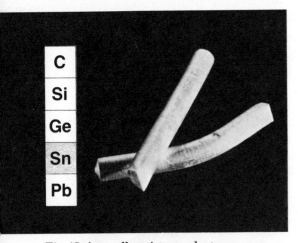

Tin (Sn) usually exists as a lustrous, very soft, silver-white metal, though a nonmetallic variation appears at lower temperatures. The latter was first observed as "tin disease," spreading on organ pipes in the cathedrals of Europe. Since tin resists corrosion, it is extensively used as a coating on steel, particularly in the making of "tin" cans.

and are good thermal and electrical conductors. On the other hand, nonmetals are usually comparatively poor conductors. With only one exception, metals are solids under ordinary conditions of temperature and pressure. That exception is *mercury,* which is also one of only two liquid elements.* The other is *bromine,* a reddish-brown foul-smelling nonmetal.

Nine substances, now recognized as elements, were well known even in remotest antiquity. Of these, *gold* (Au), *silver* (Ag), and *copper* (Cu) can be found free in nature as fairly pure grains and nuggets, a fact that probably accounted for their early use. *Iron* (Fe) meteorites with purities of around 95 percent were highly treasured by the ancients although it wasn't long before iron, along with *tin* (Sn), *lead* (Pb), and *mercury* (Hg), was being smelted from its ore in the fire. Interestingly, American Indians used meteoric iron but never did learn to refine iron from ore. Along with these seven metals, two nonmetals, *carbon* (C), as charcoal, and *sulfur* (S), available in the form of bright yellow crystals, have been used since prehistoric times.

Often the designations for the elements are far from obvious. The symbol *Hg,* representing mercury, takes its justification from *hydrargyrum,* the Greek for "water-silver," and *Pb* comes from the Latin *plumbum.* That one isn't any less esoteric unless you happen to know that the word "plumber" derives from the lead pipes used in ancient Rome. However obscure their origins, these are now the internationally accepted symbols.

Medieval alchemists added to the list of elements by isolating *arsenic* (As), *antimony* (Sb), and *bismuth* (Bi), three metalloids whose chemical behavior is quite similar. *Zinc* (Zn), too, was not recognized until the sixteenth century although brass, its alloy with copper, was well known to the ancients.

Hennig Brand, while in search of the philosopher's stone, unwittingly discovered *phosphorus* (P), the "light bearer," in 1669. To his utter amazement, a waxy, white residue in a flask glowed in the dark as night came on. When the old puffer scooped out a piece of the strange material, it burst into flames as he raised it into the air. For a price, Brand passed his secret on to one Johann Krafft, who sailed off to England to impress King Charles II. There Krafft met Robert Boyle, to whom he would only cryptically hint that the source of this remarkable substance "belonged to the body of man." In 1680, working quite on his own, Boyle prepared phosphorus from, of all things, dried urine.

By the mid-1700s the major search had turned to the analysis of gases. The Honorable Henry Cavendish, a truly eccentric English gentleman whom we've already encountered (p. 124), took the next step.

* The metals gallium and cesium are ordinarily solids, but their melting points (29.8°C or 85.6°F and 28.5°C or 83.3°F, respectively) are so low that they will liquify in a closed hand.

By the way, he was morbidly shy, particularly with women, and rarely if ever spoke to his housekeeper. He left notes around for her, and she in turn kept totally out of his sight on pain of dismissal. In any event, Cavendish collected the tiny bubbles that rose out of acid baths when certain metals were immersed. When it was later discovered that burning that particular gas produced droplets of water, the gas was named *hydrogen* (H), the "water maker." Hydrogen is the most abundant of all the basic substances *in the universe;* it is star stuff and building block, the primitive matter from which all else derives.

Alchemy gave way to science, to chemistry, especially in the work of such outstanding figures as Priestley, Scheele, and Lavoisier. Joseph Priestley was something of a radical, whose open support of the rebellious American colonists won him little love in England. Expressing outspoken sympathy for the French revolutionaries, he so infuriated some of his countrymen that they moved as a mob, burning his house and laboratory in Birmingham. Priestley, then in his sixties, resettled in the United States, welcomed by an old comrade, Benjamin Franklin, and befriended by President Thomas Jefferson.

He was already famous; his great discovery—that *air was not, after all, an elementary substance*—had been made twenty years earlier in 1774. Following centuries-old alchemical ritual, Priestley heated a well-known reddish-brown powder (mercuric oxide) until it was transformed into a shining bead of pure mercury. Unlike his predecessors, he collected the gas that was released in the process and found it to possess two extraordinary properties: It caused anything burning to blaze up more brilliantly, and it supported animal life. To his friend Franklin he wrote: "Hitherto only two mice and myself have had the privilege of breathing it." Priestley remarked that he felt "light and easy" inhaling pure *oxygen* (O), the colorless, odorless gas that constitutes roughly one-fifth of the mixture known as air. Oxygen is the most abundant of all elements *on Earth.* By weight it makes up 89 percent of water and about two-thirds of you. It constitutes roughly 50 percent of the planet's crust and is present in almost every common rock from sandstone to marble.

The Scotsman Daniel Rutherford investigated what remained of air in a chamber after a mouse had taken its last gasp, a candle flame had flickered out, and even phosphorous could no longer burn. For that work he is credited with discovering *nitrogen* (N), by far the largest constituent (about 78 percent) of air.

Although only twenty or so elements were known at the time of the American Revolution, the pace was quickening, and within only a quarter century eleven more would be found, including *chromium* (Cr), *titanium* (Ti), and *uranium* (U). New electrical techniques became available (with the discovery of the battery in the early nineteenth century), and with them Humphrey Davy was able to isolate *potassium* (K) and *sodium* (Na). Soon after, in 1825, *aluminum* (Al), the third most abundant element and by far the most common metal on Earth, was discovered. After iron, it is modern technology's

Antimony (Sb), though shiny and silver-gray, is in that middle ground between metal and nonmetal. Its principal uses are in the preparation of alloys with lead to make pewter-type metal, battery plates, and the like.

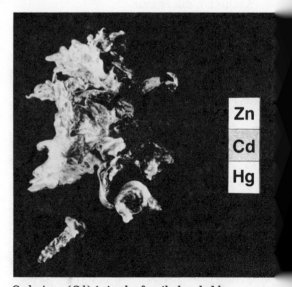

Cadmium (Cd) is in the family headed by zinc, an element with which it naturally occurs. Compounded with sulfur, it forms the brilliant pigment cadmium yellow. It's widely used nowadays in nickel-cadmium batteries. The photo shows a clump of the gray, shiny metal.

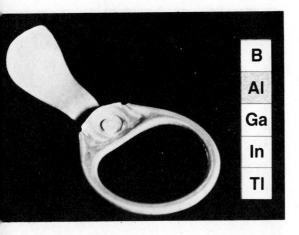

metallic darling—more than 11 million tons of it a year finds its way from ore to airplanes and beer cans.

By the end of the American Civil War, the number of recognized elementary substances had swelled to 63, and at the beginning of the twentieth century it was up to 82. The discovery of *rhenium* (Re) in 1925 finally completed the list of the 88 natural elements that can be found all around us on the third planet.

◄ Aluminum (Al) was named for alum, the compound in which it was discovered. It was first prepared by the physicist Oersted in 1825. Though aluminum was once exceedingly rare and valuable, the development of efficient refining techniques has made it one of the mainstays of modern industry.

9.3   ATOMISM

I go to Davy's lectures to increase my stock of metaphors.

SAMUEL T. COLERIDGE
English poet

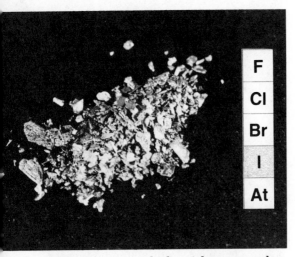

Iodine (I) exists as both a violet vapor and a bluish-black crystal with a metallic luster. Although its metallic properties are feeble, they are the most pronounced of the halogen family. In alcohol solution iodine is a common household antiseptic.

The concept of the *atom* is an old, though changing, notion, which probably dates back to Leucippus (ca. 450 B.C.), a student of Zeno, well over 2000 years ago. As first conceived, the atom was purely a creation of reason with no more reality than only human logic can afford; but the logic was straightforward enough and appealing. If a quantity of some substance was divided over and over again into ever smaller portions, could the process go on indefinitely, or would some last indivisible fragment be reached, the smallest particle of the substance, the *atomos (that which cannot be further cut)*? Leucippus and his eminent pupil, Democritus, argued that matter was composed of these ultimate, indestructible specks. For Democritus, who brilliantly developed the theory, nothing existed but *atoms and the void*—differently shaped invisible grains of matter and the emptiness in which they continuously moved.

In the long run, the most influential opponent of atomism was none other than Aristotle, whose own theory of motion could not abide with the void and who therefore rejected the atom. Epicurus (341–270 B.C.), the kindly moderate, the moralist, was the last great Greek exponent of atomism. His teachings remained quite popular, even into the Roman era; Cicero and Lucretius were ardent Epicureans.

Ironically, latter-day Roman Epicureans became self-indulgent sensualists in excess—the grade-B movie types with grapes and orgies. By the time Christianity ruled Rome, lechery, lust, atheism, and atomism were all synonymous with paganism. When the righteous mob burned the great library at Alexandria (ca. 390 A.D.) the will of this new piety was set, and atomism moved into the shadows, not to emerge for more than a thousand years.

The Greeks in antiquity had framed philosophical images of the universe, speculations that ranged the gamut from simple-minded nonsense to remarkably prophetic insight; atomism was one of their triumphs. Despite the similarity of the words and the likeness of the visions, there is a gap of centuries, a chasm of power, between the musings of the ancients and the hypotheses of the moderns, between the *atomos* of Democritus and the *atoms* of Dalton.

The shroud that obscured atomism throughout the Middle Ages fell away as Aristotelian physics began to collapse at the beginning of the seventeenth century. Torricelli, a student of Galileo in the last months of the master's life, created a fairly decent vacuum within a glass tube above a column of mercury, and in 1650 von Guericke constructed the first mechanical vacuum pump (Section 14.3). Gases could be compressed and expanded as if they were formed of discrete particles separated in the void by great distances that were subject to change. The springiness of air brought Boyle to embrace **atomism** as a viable explanation of the results of his "physicomechanical" experiments (p. 320).

Others found evidence in favor of the atomic hypothesis in the regularity of crystal formations, which implied a hidden pattern below the visible surface. Hooke proposed in 1665 that certain crystalline structures resulted from the close packing of tiny spherically shaped constituent particles—atoms stacked up, level upon level, like cannonballs.

Still others were convinced by the nature of the mixing process. The fact that salt grains dissolve and vanish into water suggests that the liquid, instead of being continuous, is riddled with empty spaces. A glass of alcohol combined with a glass of water will snuggle in to form something less than the expected two glasses of "good cheer"—again, the atoms seem to be sliding in between, filling holes. A droplet of ink or dye released into a cup of water will slowly spread out uniformly all by itself, suggesting a mingling of the particles of one with the other, a mixing of moving, colliding atoms. Similarly, the smell of frankincense (or frankfurters) can fill a house in minutes, and that, too, convinced the scholars of the seventeenth century.

The believers included Galileo, Gassendi, Bacon, Boyle, Hooke, Newton, and Leibniz, to name only a few, but the quantitative evidence was far less impressive than the list of exponents. Although atomism was an appealing picture, it seemed to be beyond verification, and that restricted its usefulness, if not its believability. The tiny specks had not yet left a mark that could be quantified.

Atomism didn't fare very much better in the eighteenth century, despite all the marvelous activity of the new chemistry. Practitioners like Priestley had no need of it, and the great Lavoisier, who fancied himself "the Newton of chemistry," would follow the master and *feign no hypothesis*, avoiding any "conclusions which are not immediately derived from facts." It hardly mattered to them what invisible scenario played itself out beyond reach in the mysterious microworld, provided all went as usual in the brew pot.

Lavoisier revived Boyle's conception of the elements, but disdained relating it to "indivisible atoms." For him at that moment, doing so would have been quite unscientific. Of course, since *fact is*

The Second Coming

Ink diffusing through water.

Antoine Laurent Lavoisier with his wife and assistant, Marie-Anne, as recorded by the French painter and revolutionary Jacques Louis David (1788).

made within some habit of belief (Section 1.3) it's not surprising that he could hold the tail and yet not smell the elephant.

After years of meticulous quantitative observation, Lavoisier set out the **law of conservation of matter,** an insight that would have a deep effect on all the work to follow:

An equal quantity of matter exists both before and after the experiment.

Matter is neither created nor destroyed while undergoing chemical change. The fact that this statement isn't absolutely accurate is of no practical concern, and we will not worry about it until later (Section 15.1). Suffice it to say that no deviation from this law has ever been observed, primarily because those deviations must be minute.

In 1794, standing convicted, incredibly enough, of having "added water to the people's tobacco," Antoine Laurent Lavoisier was beheaded in the name of reason and the Republic. More to the point, like many of the wealthy, he had invested in the hated *Ferme Générale,* a private tax-collecting corporation that gouged and bullied the common people. That and a few well-placed enemies, in particular the journalist Jean-Paul Marat, seem to have assured this superbly brilliant, arrogant aristocrat an abruptly terminated stay in revolutionary France.

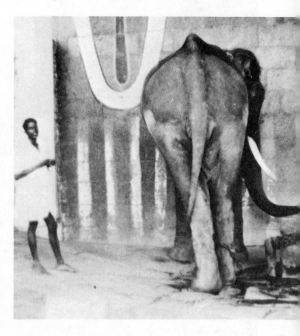

The Chemical Atom

For 2000 years atomism stood tantalizingly just beyond the reach of experimentation, beyond the grasp of all those great minds that vainly struggled to quantify the concept. And then miraculously the breakthrough came. As if to underscore the humanity of the moment, it came not in the laboratory of some dashing demigod but at the hands of an unknown country schoolmaster, a crude, gruff, awkward, graceless man named John Dalton. A dreadful speaker and a marginal "if not slovenly" experimenter, he would have seemed the least likely candidate to lead atomism out of the wilderness of metaphysics.

It had already been shown that *elements combine to form compounds in certain definite proportions.* That was **Proust's law.** One gram of hydrogen burned with eight grams of oxygen will combine to form nine grams of water. Put in more hydrogen than that, and it will simply be left over—1 to 8 is the only way to make water! The existence of a specific weight ratio is true of all compounds; for example, *carbon dioxide,* which we exhale and which is familiar as the bubbles in seltzer or soda water (an elixir Priestley invented), is a compound of three parts by weight of carbon to eight parts of oxygen.

It seemed to Dalton that these definite proportions were just the observable manifestations of an invisible atomic pattern. He suggested that *an* **element** *is a substance composed of only one kind of atom* and that *each type of atom, corresponding to a particular element, has a particular characteristic mass.* Furthermore, he assumed —and here he was wrong—that when two or more atoms combine to

John Dalton (engraving by Worthington from a painting by Allen).

On facing page, Dalton's symbolic representations of the elements (1808), or at least the substances he thought were elemental.

form a *molecule* of some compound, they do so with the smallest number of atoms. According to this rule, water would be composed of one oxygen atom and one hydrogen atom, instead of the two hydrogens that are actually present.

All this hardly seems startling, but Dalton pressed ahead. If water is formed of equal numbers of hydrogen and oxygen atoms in a weight ratio of 1 to 8, then each O atom must weigh eight times as much as an H atom! With the weight of a hydrogen atom arbitrarily set equal to 1, the weight of an oxygen atom becomes 8, a carbon atom 3, and so on down the line. Incidentally, he picked hydrogen as the reference because he suspected, and rightly so, that it was the lightest of the elements, that is, the lightest atom. Through reasoning very much like this, Dalton drew up the first table of *relative atomic weights*.

Of course, Dalton's error concerning the composition of the water molecule and the poor weight data available led to a good many wrong values. But even as this work was progressing, water was being electrically broken down, and for each volume of oxygen liberated, there were always two of hydrogen. In time, water was recognized as H—O—H or H_2O, and the table of weights was recalculated accordingly. Whatever was happening, it remained true that for each ounce, or pound, or whatever, of hydrogen there must be eight times

ELEMENTS

	Element	W.t		Element	W.t
⊙	Hydrogen.	1	⊕	Strontian	46
⊖	Azote	5	✺	Barytes	68
●	Carbon	54	Ⓘ	Iron	50
○	Oxygen	7	Ⓩ	Zinc	56
◉	Phosphorus.	9	Ⓒ	Copper	56
⊕	Sulphur	13	Ⓛ	Lead	90
◑	Magnesia	20	Ⓢ	Silver	190
◒	Lime	24	Ⓖ	Gold	190
⦶	Soda	28	Ⓟ	Platina	190
⦷	Potash	42	✸	Mercury	167

as much oxygen by weight. Thus one oxygen atom rightly weighs eight times as much as the *two* hydrogen atoms with which it combines, that is, 8×2, or 16.

One of the earliest and most influential presentations of the new theory came when Dalton delivered the 1803 Christmas lectures at the Royal Institution, newly established by the colorful Count Rumford. With typical insight and good fortune, Rumford had hired a young Cornishman as the director, one Humphry Davy, and it was he who invited Dalton. The theory was soon published, but acceptance of the reality of the Daltonian atom was slow in coming, despite the fact that the notion was widely used as a tool for interpreting chemical phenomena. Not even Davy, though he appreciated the power of the theory, ever believed that atoms *actually existed!*

Molecular Mayhem

The one gnawing flaw in Dalton's picture was his assertion that atoms unite 1 to 1, that the naturally occurring bits of any element participating in a chemical reaction are always the individual, separate atoms. As we've seen already, *a* **molecule** *is an aggregate of two or more atoms bound together tightly enough to behave as a single entity: In* **compounds,** *the molecule is the smallest unit that retains the chemical properties of the substance.*

We now know that a number of gaseous elements actually exist in the natural state in molecular form, usually with atoms pairing off in twos. For example, the air we breathe is, for the most part, a simple mixture of molecular nitrogen (N_2) and oxygen (O_2) with traces of hydrogen (H_2). But of course, none of this was known at the turn of the nineteenth century, when Dalton's theory was newly introduced and when it soon hit its first major snag.

The water molecule.

The interaction of nitrogen and oxygen to form nitric oxide.

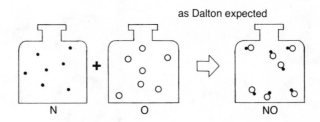

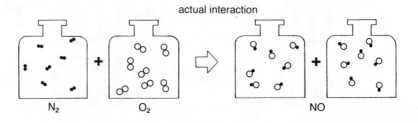

It seems that one volume of nitrogen and one volume of oxygen, both being gaseous and therefore composed of *widely spaced* atoms, should combine to form one volume of the gas nitric oxide (once again, with widely spaced particles, in this case NO molecules). But that doesn't happen! Joseph Gay-Lussac found, instead, that two volumes are created. A shocked Dalton, indignant and incredulous, chose to accuse the Frenchman of carelessness rather than to go back and question his own assumptions.

The resolution of these difficulties and indeed the whole future course of chemistry rested in a proposal set forth in 1811 by Amedeo Avogadro, Count of Quaregna—a brilliant proposal greeted with profound indifference. Avogadro made two bold but insightful guesses: (1) *The gaseous elements can exist in molecular form; that is, they can have as their smallest unit a cluster of two or perhaps more identical atoms.* (2) *Equal volumes of any and all gases (under the same conditions of temperature and pressure) contain the same number of molecules.* In other words, one bottle full of nitrogen will combine with one bottle of oxygen to form two bottles of nitric oxide (as in the figure). Since the volumes are identical, the number of molecules in the bottles is the same. The bottles of O_2 and N_2 actually contain double or **diatomic** molecules with twice the number of *atoms* that Dalton had assumed, and thus Gay-Lussac's results are quite understandable.

Perhaps it was a lack of clarity in his use of the word "molecule," but whatever the reason, Avogadro's ideas were almost totally ignored. It would be another fifty years before Cannizzaro would dust them off, tidy them up, and bring them back to life in 1860—four years too late for Avogadro to witness his triumph.

Sir Humphry Davy

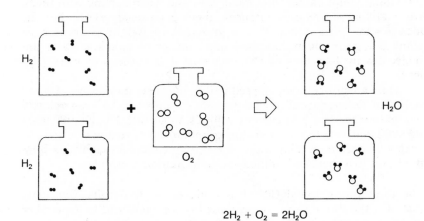

$$2H_2 + O_2 = 2H_2O$$

The combination of hydrogen and oxygen to form water.

H		H		H		N		Ammonia vapor	

The Winning Number

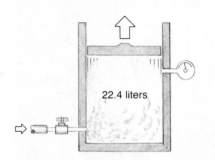

22.4 liters

22.4 liters of gas at standard temperature and pressure.

Molecular hydrogen (H_2) has a relative weight of 2. Suppose we let two grams of the stuff flow into an expandable chamber that can be maintained at ordinary room pressure. The gas would then occupy a volume of just about 22.4 liters.* According to Avogadro's hypothesis, 22.4 liters of oxygen (O_2)—or of any other gas—would contain the same number of molecules (provided the temperature and pressure were the same). Since each O_2 molecule has a weight of 2×16, or 32, we can expect any volume of oxygen to be 16 times more massive than the same volume of hydrogen. The 22.4 liters of oxygen must therefore have a mass of 32 grams. The same reasoning would lead us to conclude that 22.4 liters of N_2, with a molecular weight of 2×14 or 28, would have a mass of 28 grams. In fact, *22.4 liters of any gas at standard temperature and pressure has a mass in grams numerically equal to its molecular weight.* This rather useful measure is unimaginatively, if not confusingly, called a **gram-molecular weight.** For example, one molecule of carbon dioxide (CO_2) has a relative weight of $12 + (2 \times 16)$ or 44; therefore one gram-molecular weight of it has a mass of 44 grams and, as always, occupies 22.4 liters.

Actually, one gram-molecular weight of any substance (solid, liquid, or gas) regardless of the volume, contains the same number of molecules—a quantity known nowadays as **Avogadro's number.** Of course, we could take 22.4 liters of any gas and cool it down to a liquid or freeze it into a solid and still have one gram-molecular weight. By the way, most elements ordinarily exist as a collection of individual atoms rather than molecules, and when dealing with them, we ought more precisely to speak about **gram-atomic weight,** but the idea is exactly the same. Thus a gram-atomic weight of uranium, a shiny 238-gram block, has the same number of atoms as there are molecules in 44 grams of CO_2, whether it is gaseous or solid (dry ice).

Half a century more elapsed before the theoretical work of Maxwell and Boltzmann allowed Avogadro's number to be even roughly approximated. The best current value is about 602,200,000,000,000,-000,000,000, or 6.022×10^{23}. One gram-molecular weight of water, 18 grams (a little more than a tablespoonful), contains roughly six hundred and two thousand million, million, million molecules.

The Periodic Table

The First International Chemical Congress at Karlsruhe, Germany, just over the Rhine from easternmost France, convened in September of 1860 amidst a troubled mood of confusion and disagreement re-

---

* One liter equals 1000 cubic centimeters, or about a quart, and 22.4 liters is a volume almost precisely equal to that of a box 11 inches on a side, or very roughly, 1 cubic foot.

garding atomic and molecular weights, their true meaning, their actual values, their very reality. The whole field of chemistry was in a dark muddle.

The high point of this otherwise uneventful meeting came when a fiery Italian political revolutionary, Stanislao Cannizzaro, presented his superb defense of the chemical atom. That moment was a turning point, and many in that audience of renowned chemists sensed it at once. Professor Lothar Meyer later remarked, "It was as though scales fell from my eyes, doubt vanished, and was replaced by a feeling of peaceful certainty." Another among those present and impressed was a young Russian graduate student who had gone to work with the famous Professor Bunsen at Heidelberg a year before, one Dmitri Ivanovich Mendeléev, the hero of this small tale.

Dmitri Ivanovich Mendeléev and an early version of his periodic table of the elements.

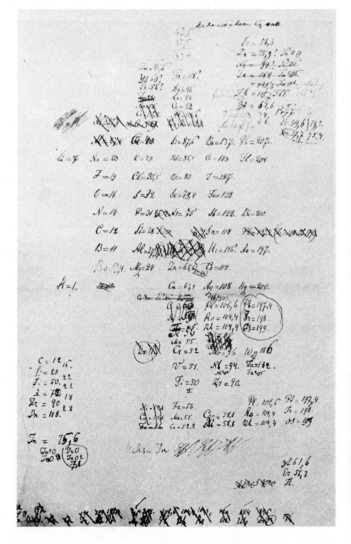

Mendeléev was the youngest of fourteen children, the darling of his energetic devoted mother, Marie, long widowed and poor. When he was fifteen, they left their home in Siberia to go west to the great cities of Russia, where he could pursue his education. With his mother pulling strings and with old friends in the background, Dmitri was finally admitted to college at St. Petersburg. Young Mendeléev was graduated at the head of his class in 1855, and he then went off to the Crimea to teach. Marie's dying words had been "Refrain from illusions. . . . Search patiently divine and scientific truth."

Unlike Dalton, who never found "time to marry," Dmitri at 42 took a second bride, a young art student, after divorcing his first wife. But Russian Orthodoxy, which doesn't recognize legalistic quibbling, saw that unholy trio unquestionably involved in blasphemous bigamy. Still, the shaggy-bearded genius was world-famous by then, and story has it that the Czar quipped: "Mendeléev has two wives, yes, but I have only one Mendeléev."

It was on returning to Russia after the Karlsruhe Congress that Mendeléev, like so many before him, began to seek some pattern, some intelligible order among the 58 elements then known. The first person to discover a seeming relationship between the elements had been Johann Döbereiner, a German chemist (and close friend of the poet Goethe). As early as 1817, he had found that there were certain groups of three elements each that displayed remarkably similar properties. *Calcium* (Ca), *strontium* (Sr), and *barium* (Ba) formed such a "triad," with Sr midway between the other two in its physical characteristics—melting point, chemical activity, solubility, and *atomic weight*. The **alkali metals** *lithium* (Li), *sodium* (Na), and *potassium* (K) made up a similar grouping. So did the three **halogens**, or "salt-formers": the gas *chlorine* (Cl), the liquid *bromine* (Br), and the solid *iodine* (I).

In the years that followed, others recognized similar arrangements, groups of four or five elements ordered in succession by virtue of their physical properties. Nonetheless, many dismissed the relationships as just the meaningless products of chance, indicative of nothing substantial. Then in 1864 John Newlands constructed a remarkable though rudimentary table of the elements, arrayed by atomic weight in columns of seven, wherein Döbereiner's triads appeared quite naturally. Unhappily, Cannizzaro's earlier arguments had not convinced everyone; the nonbelievers still prevailed. Newlands was ridiculed, and his work was denied publication. So matters lay blindfolded for another five years until Lothar Meyer and Dmitri Mendeléev, working independently, saw clearly the true pattern of the elements. Although Meyer's work was conceived at the same time, it wasn't published until 1870, a year after Mendeléev's. Still, the latter not only gets but probably deserves the lion's share of the credit—as much for the more useful form of his presentation as for the incredible predictions he drew from it.

Having listed the elements with their atomic weights on cards, Mendeléev dealt them out, arranged and rearranged them, shifted

Ca	Li
Sr	Na
Ba	K

and mulled, and rearranged again. It was a game of dedication and insight, predicated on a total familiarity with the elements, with every aspect of their behavior. And one aspect in particular, beyond atomic weight, seemed to guide his hand.

Years before, it had become clear that each element behaves as though it had some characteristic combining power, or **valence** (from the Latin for power). It's as if the atoms had "hooks" with which they could latch on to each other to form molecules, each kind of atom presumably having a specific number of such "hooks." For example, hydrogen (H), sodium (Na), and chlorine (Cl) all have a valence of 1, that is, one "hook" apiece, a single *bond* for each atom. When they interact to form table salt (NaCl) or hydrochloric acid (HCl), it's a matter of one atom's grabbing one and only one other atom. By contrast, oxygen (O) and sulfur (S) each have two "hooks" and therefore a valence of 2. An oxygen atom can combine with two hydrogens to make water; or it may unite with one sodium and one hydrogen to form NaOH, sodium hydroxide (commonly known as caustic soda or lye). We needn't actually think of all this in terms of individual atoms; not many chemists of the nineteenth century did. In other words, since sulfur is double-bonded, one gram-atomic weight of it will combine with two gram-atomic weights of single-valence hydrogen to make H_2S, the wretched-smelling hydrogen sulfide that wafts from hard-boiled eggs and organic wastes. And so, the notion can equally well be set in terms of weights for those who don't believe in atoms; 32 grams of sulfur combine with 2 grams of hydrogen.

Each element has a valence: For aluminum and iron it's 3; carbon, silicon, and lead have a valence of 4; and so on. Of course, it's not really that simple. There are substances—nitrogen is one—that display several possible valences. Ironically, out of Mendeléev's grasp, decades ahead in the twentieth century, was the deeper understanding that valence is a manifestation of the hidden structure of the atom, that it reflects the configuration of the atom's electron cloud hovering about the nucleus. But no one even suspected that then.

Mendeléev picked up his cards and at first astutely put hydrogen aside; it didn't seem to fit with the others and was best left by itself temporarily. Helium was next in atomic weight. It had been found (in 1868) to exist on the sun, or at least there was something on that star that gave out light matched by no known terrestrial element, and the name given to it came from *helios*, the Greek word for sun. Fortunately for Mendeléev, helium had not yet been isolated on Earth, so he simply neglected it. Like hydrogen, it would have been quite difficult to place without the rest of its still undiscovered family, the **noble gases.** After helium came lithium, which he placed at the start of the first horizontal row, a row that progressed in increasing atomic weight, one card after the next: beryllium, boron, carbon, nitrogen, oxygen, and fluorine. The next card in the stack according to weight was sodium, which is very similar in its properties to lithium, including a valence of 1, and so he began a second horizontal row under

Elements with a valence of 1.

Oxygen and sulfur each have a valence of 2, and carbon has a valence of 4.

Mendeléev's table.

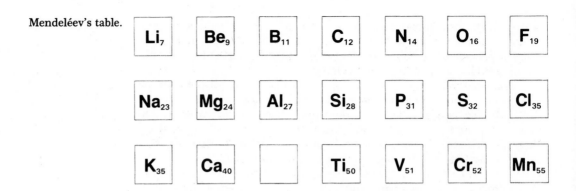

the first with Na. Across it ran another seven cards, ending with chlorine sitting just beneath its sister, fluorine. Potassium began the next row—and there immediately down the first column was Li, Na, and K, Döbereiner's triad, each with a valence of 1! Calcium was next in weight and it, too, fitted into a vertical family of like elements. The next substance in turn was titanium, but to Mendeléev it fitted more reasonably below silicon and carbon than below aluminum, so he skipped a box and confidently concluded (brilliantly guessed) that "there is a missing element there." Surely there were many "missing elements"; only 63 were known at the time, so there had to be gaps. Yet those gaps were far from obvious. Mendeléev had to see them as holes in a pattern of order that he himself was creating.

Here Mendeléev's remarkable courage and conviction became evident; he boldly insisted (correctly) that if they were to fit into his system, the accepted atomic weights of beryllium, indium, and uranium had to be wrong, and he simply assigned them the weights they "ought to have." As if that were not audacious enough, he picked out three holes in his table and proceeded, as had never been done before in all the history of science, to predict in detail the physical and chemical properties of the missing elements.

All this and more, as it appeared in print in March of 1869, was at first greeted with a chorus of skeptical snickers and a groan of indignation. How could anyone have taken those pompous predictions seriously? Mendeléev settled back to wait, and the wait was short enough. In 1875 a new element, *gallium*, was discovered in and named after France (Gaul). Mendeléev immediately recognized its properties as identical to those anticipated for one of his three truants. Four years later (in Scandinavia) *scandium* was isolated, and that time it didn't take the special insight of the Russian master to see it as the missing element, filling the gap between calcium and titanium. The final bit of icing on the periodic cake, the complete triumph, came to the already bemedaled and honored scholar in 1886 when (in Germany) the last wayward element, *germanium*, was found. Three phantoms that precisely matched the prophecy of the patriarch had miraculously come out of a game of cards.

By 1925 all 88 naturally occurring elements had been found, and the modern periodic table spread them out in lovely order. The final arrangement was remarkably simple: a progression by **atomic number,** something we'll come back to later (Section 10.4) and something Mendeléev could not have anticipated, although the related notions of valence and atomic weight provided an adequate beginning. Still, he did not hesitate to place tellurium correctly before iodine, even though its weight was greater. That was the genius of the man.

When the noble gases, *argon* (from the Greek for "inert"), *helium*, *neon* ("new"), *krypton* ("hidden"), and *xenon* ("stranger"), were discovered just before the beginning of the twentieth century, they neatly formed a column of their own at the far right of the table. This is the zero-valence column, comprising elements so antisocial that they do not seem to interact chemically with anything else. Indeed, until 1962, when some few compounds of xenon were finally formed, they never did.

Long ago, abundantly occurring compounds, like magnesia (MgO) and lime (CaO), were called *earths*. In time, Mg, Ca, and four other elements, each with a valence of 2, came to be known as **alkaline earth metals.** When still other elements were found to form similar though less common metallic oxides, they were christened **rare earths.** Fourteen of these metals, all with a valence of 3, similar properties, and relatively closely spaced atomic weights, range from *lanthanum* (La) to *lutetium* (Lu). That whole group, now called the **lanthanides,** had to be squeezed into the sixth row of the table between barium (Ba) and hafnium (Hf). For a while this "wart" seemed to cast some small doubt on the entire scheme—especially since the periodic table was not yet supported by a strong foundation of atomic theory.

A Modern Version

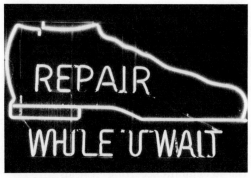

A neon sign.

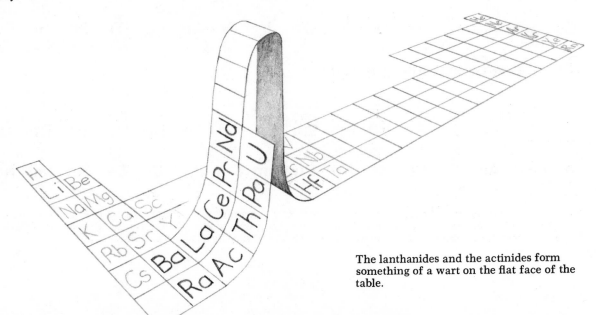

The lanthanides and the actinides form something of a wart on the flat face of the table.

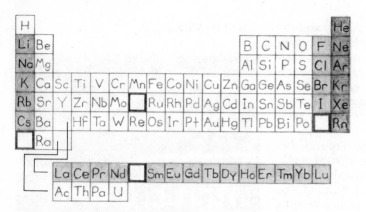

The 88 naturally occurring elements.

1 H Hydrogen 1.00797																	2 He Helium 4.0026

Metalloids **Nonmetals**

Metals

3 Li Lithium 6.939	4 Be Beryllium 9.0133											5 B Boron 10.811	6 C Carbon 12.01115	7 N Nitrogen 14.0067	8 O Oxygen 15.9994	9 F Fluorine 18.9984	10 Ne Neon 20.183
11 Na Sodium 22.9898	12 Mg Magnesium 24.312											13 Al Aluminum 26.9815	14 Si Silicon 28.086	15 P Phosphorus 30.9738	16 S Sulfur 32.064	17 Cl Chlorine 35.453	18 Ar Argon 39.948
19 K Potassium 39.102	20 Ca Calcium 40.08	21 Sc Scandium 44.956	22 Ti Titanium 47.90	23 V Vanadium 50.942	24 Cr Chromium 51.996	25 Mn Manganese 54.9380	26 Fe Iron 55.847	27 Co Cobalt 58.9332	28 Ni Nickel 58.71	29 Cu Copper 63.54	30 Zn Zinc 65.37	31 Ga Gallium 69.72	32 Ge Germanium 72.59	33 As Arsenic 74.9216	34 Se Selenium 78.96	35 Br Bromine 79.909	36 Kr Krypton 83.80
37 Rb Rubidium 85.47	38 Sr Strontium 87.62	39 Y Yttrium 88.905	40 Zr Zirconium 91.22	41 Nb Niobium 92.906	42 Mo Molybdenum 95.94	43 Tc Technetium	44 Ru Ruthenium 101.07	45 Rh Rhodium 102.905	46 Pd Palladium 106.4	47 Ag Silver 107.870	48 Cd Cadmium 112.40	49 In Indium 114.82	50 Sn Tin 118.69	51 Sb Antimony 121.75	52 Te Tellurium 127.60	53 I Iodine 126.9044	54 Xe Xenon 131.30
55 Cs Cesium 132.905	56 Ba Barium 137.34	51-71 La-Lu	72 Hf Hafnium 178.49	73 Ta Tantalum 180.948	74 W Tungsten 183.85	75 Re Rhenium 186.2	76 Os Osmium 190.2	77 Ir Iridium 192.2	78 Pt Platinum 195.09	79 Au Gold 196.967	80 Hg Mercury 200.59	81 Ti Thallium 204.37	82 Pb Lead 207.19	83 Bi Bismuth 208.980	84 Po Polonium (210)	85 At Astatine (210)	86 Rn Radon (222)
87 Fr Francium (223)	88 Ra Radium (226.03)	89-103 Ac-Lw	104 104	105 105	106 106	(107) 107	(108)	(109)	(110)	(111)	(112)	(113)	(114)	(115)	(116)	(117)	(118)

Lanthanide series	57 La Lanthanum 138.91	58 Ce Cerium 140.12	59 Pr Praseodymium 140.907	60 Nd Neodymium 144.24	61 Pm Promethium (147)	62 Sm Samarium 150.35	63 Eu Europium 151.96	64 Gd Gadolinium 157.25	65 Tb Terbium 158.924	66 Dy Dysprosium 162.50	67 Ho Holmium 164.930	68 Er Erbium 167.26	69 Tm Thulium 168.934	70 Yb Ytterbium 173.04	71 Lu Lutetium 174.97
Actinide series	89 Ac Actinium (227)	90 Th Thorium 232.038	91 Pa Protactinium (231)	92 U Uranium 238.03	93 Np Neptunium (237)	94 Pu Plutonium (244)	95 Am Americium (243)	96 Cm Curium (247)	97 Bk Berkelium (249)	98 Cf Californium (251)	99 Es Einsteinium (254)	100 Fm Fermium (257)	101 Mv Mendelevium (258)	102 No Nobelium (254)	103 Lw Lawrencium (257)

Note that the various rows of the periodic table have 2, 8, 8, 18, 18, and 32 members. Although no one knew it at the time of Mendeléev, this pattern is determined by the arrangement of atomic electrons. Brackets indicate the mass of a particular isotope. The weight of Carbon-12 is taken as exactly 12.

The last four and heaviest of the natural elements, *actinium* (Ac), *thorium* (Th), *protactinium* (Pa), and *uranium* (U), constitute still another group, known as the **actinide metals**. And they, in turn, pair up, sharing similar characteristics with the adjacent rare earths. This, then, was how matters stood in 1925, when *rhenium*, the last of the natural lot, was discovered. The table showed a continuous run of elements from the lightest to the heaviest, from hydrogen (number 1) to uranium (number 92).

Quite obviously, there are four—count 'em, folks—four glaring holes! These are the empty boxes left behind by four highly *unstable* absentee members of the club. All by themselves their nuclei break up, decaying into other long-lived elements. Though significant quantities of these four **radioactive** substances probably existed when the Earth was created some 4.5 thousand million years ago, that original motherlode has all but vanished.

Technetium (Tc) was the first of these lost spirits to be resurrected by nuclear alchemy. Tiny quantities of it were created in 1937, when *molybdenum* (Mo) was bombarded with subatomic particles that, when captured, would kick up the weight one box. And that's the gap Tc fills, the one right next to Mo. Nowadays it's manufactured in nuclear reactors at about $100 per gram.

Actinium is an exceedingly rare substance, primarily because, like all the actinides, it is radioactive—half of whatever amount exists disintegrates every 22 years. That period of time is known as its **half-life.** Very occasionally, actinium will split into two particular pieces, one of which is the short-lived phantom *francium* (Fa). Thus francium (which fills the spot in the table that is two boxes back from Ac) actually does briefly exist naturally, but in only 21 minutes, half of the minuscule amount just created will already have vanished. Identified in 1939 by its own radioactivity, francium has never been produced in quantities large enough to see.

Astatine (At), the third specter, was created in 1940, when bismuth was bombarded in order to bring its weight up two boxes. It also vanishes quickly and has never been accumulated in directly perceivable quantities.

The last of the missing four was first identified in 1945 and has since been produced in modest amounts in nuclear reactors. *Promethium* (Pm) is a lanthanide, the only rare earth not yet found in Nature. Once made, it is stable enough to hang around in appreciable quantities for dozens of years.

There we have them at last, the 92 elements, filling the periodic table, it seems, to the brim. By the way, all the known elements beyond lead are actually radioactive although bismuth decays so slowly it might just as well be considered stable. If we wait long enough and don't tamper with things, they will all decay away, and first bismuth and then eventually lead will be the heaviest natural element left in the remains of this planet (which will certainly have been destroyed long before by its own dying sun).

All that can be said upon the nature and number of the elements is confined to discussions entirely of a metaphysical nature.

LAVOISIER (1743–1794)

The glow-in-the-dark face of an old wristwatch painted with a radium-zinc sulphide mixture. Radium is radioactive, a potent emitter of alpha rays (p. 241). The young factory women who hand-painted these luminous dials in the early 1920s commonly kept their brushes neatly pointed by licking them. At least forty of those unsuspecting workers died as a result of ingesting radium.

Ironically, as the gaps in the table were being filled in, other researchers were already busy trying to create elements beyond uranium in weight. The first **transuranic** element, *neptunium*, number 93, was produced in California in 1940. By 1961 the actinides had been extended out to *lawrencium*, the 103rd element, the last in that series. Two of the actinides, *einsteinium* and *fermium*, were first found in the debris of the 1952 H-bomb test. By the end of the 1970s, elements 104, 105, 106, and 107 had added their minuscule presence to the list. All these substances are unstable, but some, like *plutonium*, number 94, are quite long-lived (it decays by half every 24,300 years).

EXPERIMENTS

1. Fill two beakers, one with ice water and the other with hot water. Gently pour in a few drops of ink and note the tremendous difference in the rate at which the ink diffuses throughout the water. Explain what's happening.

2. The same sort of diffusion experiment can be conducted in a gas. Gently introduce some smoke into a large inverted jar, and then place it with the opening down on a flat surface. Isolated from outside currents, the smoke slowly diffuses. Squirt a few shots of a scented air "freshener" in one corner of a room, and see how long it takes before you can smell it ten or fifteen feet away.

REVIEW

Section 9.1

1. What is the alchemist's meaning of the word *transmutation*?

2. Not all medieval alchemists literally believed that the philosopher's stone was necessarily rocklike. However, almost all understood its two main uses to be _____.

3. The acid aqua regia was of particular interest to the alchemist. Why?

4. What role did the Muslims play in the history of alchemy?

5. What was Boyle's definition of the word *element*?

Section 9.2

1. List some of the distinguishing characteristics of metals.

2. About what proportion of the elements are metals?

3. The ancients knew of *seven* metallic elements, which in time became related—at least symbolically—with the "seven planets." List those metallic elements.

4. How many naturally occurring elements are there?

5. Air is mostly composed of _____.

6. Which gas did Cavendish discover?

7. Which is the most common metal on our planet?

8. What is the most abundant element on Earth?

Section 9.3

1. State the law of conservation of matter.

2. Proust's law states that _____.

3. Who was the initiator of modern atomic theory? *Hint:* He was color-blind.

4. Define what is meant by the term *molecule*.

5. State Avogadro's two hypotheses.

6. Define the term *gram-molecular weight*, sometimes also known as a gram-mole.

7. What is Avogadro's number (which, of course, Avogadro never knew anything about)?

8. Make a list of several halogens, a few alkali metals, and a couple of actinides.

9. The chemical combining power of an element is known as its _____.

10. List at least five of the noble gases; for that matter, list all the elemental gases you can. Which ones would you not hesitate to inhale? Of course, you can die breathing large amounts of any of them (except O_2)—you'll simply asphyxiate.

11. Where does uranium stand in relation to the other 87 naturally occurring elements?

12. Discuss what is meant by a *radioactive substance*.

QUESTIONS

Answers to starred questions are given at the end of the book.

Section 9.1

* 1. How would you distinguish between the puffer and the traditional alchemist?

2. Who were the keepers of alchemical ritual around the time of the First Crusade (1096 A.D.) and for hundreds of years before?

3. Chaucer, the great English writer of the mid-1300s, and Brueghel, the master painter of the mid-1500s, give us some insights into the feelings of their day on alchemy. What were they?

* 4. What is the derivation of the word *quintessential*? What does the dictionary have to say about its meaning?

5. What was so significant about Boyle's approach to the elements? By the way, he called himself a "chymist" because the term *alchemist* was already somewhat synonymous with faker. It wasn't that people didn't believe in the possibility of transmutation, because most did. Rather, the association came as a result of all the imitation gold that was passed off on the unsuspecting.

6. Was the transition from alchemy to chemistry, which was started by Boyle, immediate or prolonged? Where did Newton stand on alchemy?

Section 9.2

* 1. Prepare a list of at least five nonmetallic elements. How many of these are you in direct contact with at this instant?

2. Air, of course, is not an element. What is it? Who took the first steps toward the realization that it wasn't elemental?

3. Make a search of your medicine chest for exotic elements, and formulate a list. For example, lipsticks (because of the pigments) contain such things as titanium, chromium, bismuth, aluminum, strontium, calcium, barium, etc. Maybe that's why they don't label all ingredients.

* 4. Considering the fact that it was Heraclitus who said, "All things flow, nothing abides," discuss why it was so reasonable for him to have suggested *fire* as the prime element.

5. There are only two elemental metals that are colored (*all* the rest are silvery gray in varying shades). Name the two.

6. Where in your home might you find each of the following?

aluminum	copper	oxygen
tungsten	sodium	chromium
carbon	iron	nickel
mercury	lead	silicon
zinc	magnesium	sulfur
chlorine	cobalt	

7. Define the concept of *half-life*.

8. Why was atomic theory forced into the shadows by the early Christian church?

Section 9.3

1. The idea of quantification is essential to mathematized science. Where do you see it emerging in chemistry? What was its relationship to the early development of atomic theory?

2. Describe the most outstanding contribution of Dmitri Ivanovich Mendeléev.

* 3. Brahe, Kepler, and Newton each contributed a different component of our understanding of planetary order (p. 9). Which of these concepts is most akin in its function to the role played by Mendeléev's periodic table in an understanding of the order of the elements?

4. One often hears the remark that the time was "right" for a certain discovery to be made. Was this so with the periodic table?

5. Helium was discovered first in 1868 by Pierre Janssen and again in 1895 by William Ramsay, who detected it in a uranium-bearing mineral. Explain why a double discovery was necessary.

* 6. Name at least three common *molecular* gases, i.e., gases composed of molecules rather than individual separated atoms.

7. Dalton made one principal erroneous assumption. What was it?

* 8. How does the fact that sugar dissolves in a cup of tea suggest that matter is composed of atoms?

9. Which of the following are legitimate elements?

palladium	kryptonite	zirconium
neodymium	samarium	steel
terbium	herbium	gadolinium
brooklynium	cobalt	ruthenium
yttrium	berkelium	chicagoium
ytterbium	lutetium	samtetium
petetetium	praseodymium	jameyomium

* 10. Nowadays common carbon is taken as the standard, and its mass is fixed at 12 **atomic mass units** (12 u). All other atoms are then specified in terms of these units, so that hydrogen has a mass of 1 u, oxygen 16 u, etc. Why do scientists fool around with yet another unit when they have grams and kilograms? (See Mathematical Problem 9.3-7.)

11. In 1663 Magalotti reported the following experiment: "A hollow sphere cast in silver" was filled to the very brim with water and screwed closed. Deep dents were hammered into the vessel, and amazingly, "water drops were seen to ooze through the pores of the metal." Does this seem to sup-

port the view of Democritus or that of Aristotle?

12. Who invented the eraser? *Hint:* It was the person responsible for the creation of seltzer.

13. If two blocks, one of gold and the other of silver, each with a freshly cut surface, are forced against each other for a year or so and then parted, there'll be gold in the silver, and vice versa. Explain.

MATHEMATICAL PROBLEMS

Answers to starred problems are given at the end of the book.

Section 9.2

1. A cubic centimeter of water has a mass of 1 gram. How many *molecules* does it contain? How many *atoms* are there in a 1-gram ice cube?

* 2. If your body is roughly 70% water, how many H_2O molecules are you? The answer ought to come out around a thousand million million million million.

3. Given that a cubic foot of air weighs a total of about 0.08 pounds, what is the weight of the nitrogen present?

4. A cubic foot of water weighs 62.4 pounds. How many pounds of that is hydrogen?

* 5. Make a rough estimate of the number of atoms in the universe. *Hint:* Assume a typical atomic mass of about 10 u, or 1.66×10^{-26} kg, a star mass of 10^{33} grams, 10^{11} stars per galaxy, and 10^{11} galaxies in the universe.

Section 9.3

1. How many grams of salt are there in a gram-molecular weight of salt (NaCl)?

2. How many molecules are there in 22.4 liters of gaseous trichloromonofluoromethane at ordinary temperature and pressure?

3. If we totally decomposed 11 grams of carbon dioxide, how much oxygen would result?

* 4. What quantity of water will be formed when 5 grams of hydrogen are burned in the presence of 40 grams of oxygen?

5. How many molecules are there in 684 grams of pure sugar ($C_{12}H_{22}O_{11}$)?

6. A pure gold ring has a mass of 1.97 grams. Determine the number of atoms in the ring.

7. By the latest convention, common carbon is assigned a mass of exactly 12 units. One-twelfth part of that is called an atomic mass unit (u), such that

$$1 \text{ u} = 1.6606 \times 10^{-27} \text{ kg.}$$

What are the masses of carbon, oxygen, and hydrogen atoms in grams?

* 8. Benzoylmethylecgonine (cocaine) has the chemical formula $C_8H_{13}N(OOCC_6H_5)(COOCH_3)$; what is its atomic weight?

* 9. If we had a millionth of a gram of francium at the start of an experiment lasting one hour and three minutes, how much of it would remain by the end of the session?

10. Suppose that you had 10 grams of actinium in a box on the table, and you left to buy a newspaper and didn't come back until 22 years later. How many grams of Ac would remain in the box?

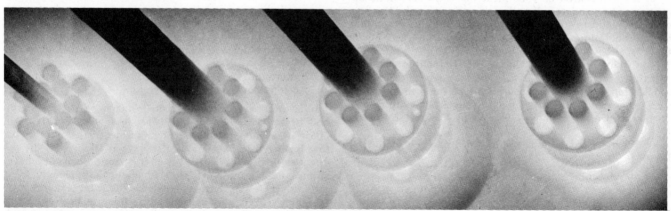

The glow of nuclear alchemy at the Brookhaven National Laboratory.

CHAPTER 10

The Complex Atom

Ever since the last decade of the nineteenth century, we have known that the atom was no one-piece Daltonian billiard ball but a complex structure of interacting parts. The beginning of this chapter deals with the first sub-atomic particle, the first constituent of the atom to be discovered, the **electron.**

Some kinds of atoms are unstable; they spontaneously shatter, hurling out fragments in a process spoken of as **radioactivity,** which we will consider here in the second section. **Alpha, beta** and **gamma rays** are the names that were given to these nuclear emissions early in the days of discovery. The third segment in the chapter traces the beginnings of our contemporary vision of the **nuclear atom;** as a **central, minute, positive core surrounded at some distance by a swarm of electrons.** The last bit of business focuses for a moment on the nucleus, its **proton** and **neutron** components, and the different forms of each element (with differing numbers of neutrons), known as **isotopes.**

Peeping out from under our "nuclear umbrella," we may find it incredible that atomic theory was still embryonic and struggling, that it still had its adamant detractors even into this century, our century. Physicist-philosopher Ernst Mach, scientist-historian Pierre Duhem, Nobel laureate Wilhelm Ostwald, chemist Marcelin Berthelot—all were among the more famous of the staunch opponents, the last of the lingering holdouts. To them, atomism was certainly a convenient construction, a potent working fiction, but one whose actuality could never be proved. How, indeed, can one prove beyond any possible doubt the existence of invisible entities that cannot be known directly?

As for us, we are thoroughly convinced of the reality of atoms, even though we should perhaps know better—even though we should know that theories, however powerful, *must* remain tentative, albeit only "a little tentative." They must be tentative even though we would rather they were not. The commanding evidence in favor of atomic theory that boiled up near the end of the 1800s became utterly overwhelming in the first half of the twentieth century. The universe certainly behaves as if it is composed of atoms; there is no doubt of that now. Amusingly enough, the creature that finally won the struggle was not anything like Dalton's solid, indivisible, billiard-ball atom. Instead, it was a remarkably complex structured form, a thing of interacting parts, many of whose intricacies evade us still.

But once for all be it understood that the atomic hypothesis is only a mode of picturing to ourselves what we know of the behavior of substances. What the "real" nature of matter is, is to us a matter of complete ignorance as it is of complete indifference.
WILHELM OSTWALD (1890)
Russian-German chemist

The atomistic character of matter belongs to the most certain facts of our present knowledge. . . . We can speak of the existence of atoms with the same certainty as of the existence of the stars.
HANS REICHENBACH
Atom and Cosmos (1957)

ON ATOMIC THEORY
Thus atomism has proved infinitely fruitful. Yet the more one thinks of it, the less can one help wondering to what extent it is a *true* theory. Is it really founded exclusively on the actual objective of "the real world around us"? . . . It behooves us, so I believe, to preserve an extremely open mind towards the palpable proofs of the existence of individual single particles.
ERWIN SCHRÖDINGER (1954)
Nature and the Greeks

Nowadays it would be sheer heresy to suggest that the atom might be no more than a functioning model that matches the manifestations of some other hidden truth. We believe in atoms, even though we do not yet understand their structure completely; we believe in atoms, even though we keep changing the picture of what they are. And in the end, atomic theory, however flawed, however immature, is perhaps the greatest single creative accomplishment of all time, the ongoing masterwork of human thought.

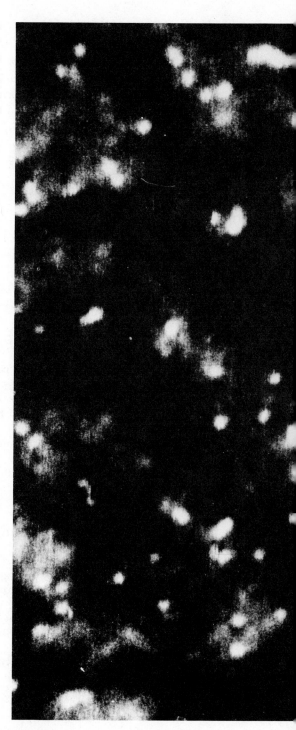

Individual uranium atoms on a thin carbon film. Each shows up as a ▶ bright spot. This one frame from a motion picture reveals considerable movement. Albert Crewe of the University of Chicago took it at a magnification of 10,000,000, using an electron microscope.

10.1 ELECTRONS

The first subatomic particle to show itself was a minute speck of charge and matter, whose coming culminated centuries of experimentation and image building. Although it was of course always there, physics was not ready for its arrival until 1897. Not until then were the visionaries mature enough to "see" what could not in fact be seen, to recognize the wee beastie only by its tracks.

Since the time of the ancient Greeks, we have known that when amber* is rubbed against a piece of cloth or fur, it attracts small bits of light material—pieces of straw, chaff from grain, twig fragments, tufts of lint or dust, etc. Yet that remarkable ability went all but unexplored until the sixteenth century, when the distinguished Elizabethan scholar William Gilbert took up the work. He greatly extended the number of materials known to possess attractive powers when rubbed, and he named them all "electrics" after the Greek word for amber, *elektron*. By the mid-seventeenth century, it was already being suggested that an *electric,* activated by rubbing, possessed some sort of "amber stuff," or *electricity*. And people spoke about a *charge* of electricity in the sense of an amount or load—much like a charge of gunpowder.

Inspired by Gilbert, Otto von Guericke, a physicist-inventor of considerable talent, devised a crank-driven rotating sphere of sulfur that could be stroked as it spun. Whirling within a cupped hand, it built up sizable quantities of "electric virtue" that sparked away impressively—a fact that von Guericke conveyed to the youthful Leibniz in 1672. Before long, all sorts of revolving rubbing machines were devised, and everything in sight from milk buckets to chickens was being charged by these *electrostatic* generators. The term "electric,"

---

* Amber is yellow-brown fossilized tree resin, long used for jewelry.

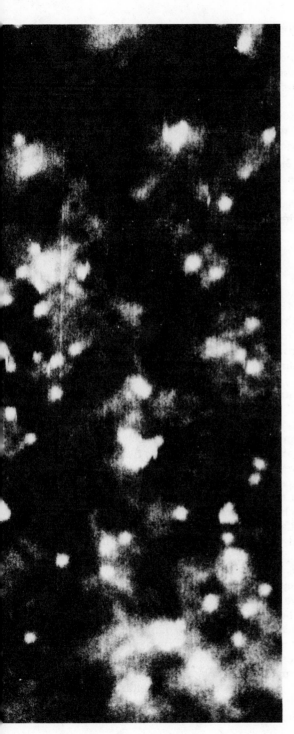

implying some special type of chargeable material, vanished from the technical vocabulary. All the great figures of science, including Boyle, his brilliant assistant Hooke, and even Sir Isaac himself, did their share of amused rubbing.

It was a charming Frenchman, one Charles du Fay, a botanist by profession, who first carefully studied the *repulsive* interactions of electricity. Du Fay found that objects of the same material electrified in the same way repelled one another. In other words, two pieces of glass stroked with silk will repel each other, just as two chunks of amber rubbed with fur will. And yet the charged glass will attract the charged amber, and vice versa. Sometime around 1734, du Fay concluded "that there are two distinct Electricities"—two kinds of electric charge, we would say. Simply stated in contemporary terms, *like charges repel, unlike charges attract.*

The notion of electricity had originated as a "virtue," a property of matter such as heaviness or rigidity, but by the middle of the eighteenth century, experimenters were already beginning to think of it as two invisible fluids. Alternatively, Benjamin Franklin proposed that there was only a single electric fluid. For want of a better name, he simply called it *positive*—an excess of that fluid results in positive electrification whereas a deficiency thereof corresponds to *negative* electrification. A lovely theory—somewhat in error, but lovely. Still, the terminology persists, and we speak of electrons as negative and protons as positive.

Under the powerful spell of Newtonian physics, it was natural to suspect that the force of interaction between charges "should" *vary inversely with distance squared,* as did gravity. The renowned mathematician Daniel Bernoulli around 1760 found that to be so experimentally, at least for his particular, rather limited arrangement. Of all people, those two totally different personalities, Priestley and Cavendish, working independently, observed that charged vessels behaved electrically in a fashion quite analogous to the way material bodies of the same shape would behave gravitationally—again *implying* an inverse-square law. The direct test of the force law (Section 17.2) was finally carried out with convincing precision in 1785 by Charles Coulomb. As anticipated, his efforts verified the inverse-square dependence of force on separation. Coulomb's electrical law is identical in form to Newton's gravitational law (p. 91); where one has the product of the two "quantities-of-matter" (masses), the other has "the product of the electrical masses" (charges). By measuring forces and the separations between the electrified bodies, one could at long last quantify charge. The scientist's ever present question "How much?" could be answered.

The pace of electrical discoveries picked up considerably as the technology of the Industrial Revolution plunged into the nineteenth century, an era the scientists themselves liked to call the Age of Electricity. Charges were made to flow as *currents;* the battery was invented; water was decomposed by *electrolysis* (passing a current

In amber there is a flammeous and spiritous nature, and this by rubbing on the surface is emitted by hidden passages, and does the same that the lodestone does.

PLUTARCH (ca. first century A.D.)

Neon sign.

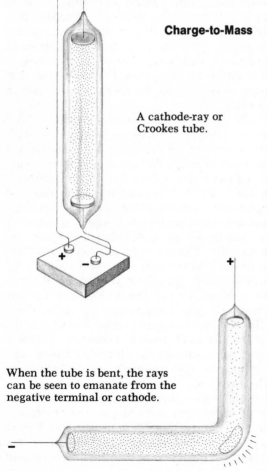

A cathode-ray or Crookes tube.

When the tube is bent, the rays can be seen to emanate from the negative terminal or cathode.

through it); and Humphry Davy, along with his assistant Michael Faraday, used electrolysis to explore chemical interactions among the elements. Faraday, the incomparable experimentalist, teetered on the very edge of electron theory, hinting that electricity was composed of particles of charge. His experiments on electrolysis suggested that the passage of a certain amount of charge was somehow related to the atomic mass and valence of an element; each atomic bond seemed to be associated with a specific charge. All the logic pointed toward particulate electricity, grains or "atoms" of charge, but Faraday remained surprisingly uncommitted to the notion. When G. Stoney revived Faraday's work in 1874, he coined the word **electron** to mean that particular quantity of charge. By then the world of science was just about ready to apprehend the invisible, to "see" the unseeable phantom of the electric force field as a shadowy granule of charge.

Charge-to-Mass Faraday had passed streams of charge through liquids. Later, in the second half of the 1800s, others studied the flow of electricity through gases. Outstanding among those early researchers was William Crookes. Sir William, a rather unorthodox fellow (particularly in his belief in the supernatural), was convinced that he could speak to the spirit world—a claim that never failed to rouse the disapproval of his colleagues (p. 18). Trained as a chemist, he seems to have shifted his interests toward physics after attending several of Faraday's lectures at the Royal Institution.

His experimentation with high vacuum began in the 1870s and ultimately led to the development of "Crookes tube." This was a glass bulb from which most of the air had been drawn and into which, at either end, were sealed two metal plates, or electrodes. (It's the forerunner of all those blazing bar and motel signs—the neon uglies.) When the wires that led from the two electrodes were connected to a source of high voltage, a glowing beam spread down the length of the tube. By displacing the positive plate, or **anode,** to the side, Crookes determined that the emanations were actually streaming from the negative plate, or **cathode;** they tended to move in straight lines, not always making it all the way to the anode. When these **cathode rays,** as they were soon called, struck the walls of the tube, the glass itself gave out a pale green fluorescent glow that betrayed the impact of the beam.

Objects that were inserted into and that partially obstructed the stream of rays (the Maltese cross was a favorite object in the experiments) cast sharp shadows in the glow at the far end of the tube. This again suggested straight-line propagation, much like light. Still, a spray of particles (e.g., buckshot) created a similar shadow effect. Crookes even put a little paddle wheel inside one of his tubes and spun it around with a bombarding beam. Later he was able to deflect the rays with a magnet, just as if they were negative charges. It had already been known for decades that charge moving in wires was affected by the presence of a magnetic field. (You can easily get a sense of the phenomenon by simply putting a strong magnet up against the face of a TV tube—but don't try it on a color set. The "cathode" rays will be displaced, creating distortions in the picture.) What gradually emerged from all of this was the "British" view of cathode rays, namely, that they were streams of microscopic, negatively charged particles.

The opposing "German" view was most vigorously pressed by Philipp Lenard, who had been interested in the work ever since he had read one of Crookes's papers as a teenager. Heinrich Hertz had discovered that cathode rays could pass through thin sheets of metal without leaving any holes behind, and this was enough to convince Lenard, his able assistant, that the rays must be nonmaterial. Nowadays we know that some subatomic stuff can sail through yards of concrete and steel as if they weren't there, but Lenard had no way of anticipating that, and he remained convinced. To him cathode rays were without substance; they were waves like light—"phenomena in the aether," he called them.

That's where the matter stood, locked in controversy and befogged by nationalism—corpuscle versus wave, British versus German—a controversy that would last at least twenty years.

When Joseph John Thomson began his definitive research into the problem, he was already forty years old and director of the Cavendish Laboratory at Cambridge University (a facility he brilliantly operated on a budget smaller than its present-day phone bill). In France, Jean Perrin had recently gone one step beyond Crookes to find that a metal obstruction located in the path of the beam actually acquired a negative charge. It remained for "J.J.," as his friends called Thomson, to show that the cathode rays and the charge were one and the same.

Assuming the rays to be composed of negative particles, Thomson shot the beam between a pair of parallel metal plates. The plates were themselves charged so that one repelled and the other attracted the rays, causing an overall upward deflection in their trajectory. By then applying just the right amount of magnetic field, forcing the beam downward and exactly canceling the initial deflection, he could determine from theoretical arguments the ratio of the **charge-to-mass** (q_e/m_e) of the particles, as well as their speeds (as much as 60,000 miles per second).*

Of course, "J.J." never actually saw the individual particles, i.e., electrons. No one ever has, nor could this experiment by itself yield

The electrical matter consists of particles extremely subtle since it can permeate common matter, even the densest, with such freedom and ease as not to receive any appreciable resistance.
BENJAMIN FRANKLIN (ca. 1750)

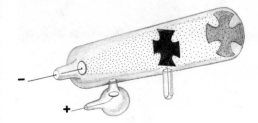

A shadow-casting cathode-ray tube.

A TV special—"Monty Hall meets the magnet."

---

* For obvious reasons, TV picture tubes were once commonly called cathode-ray tubes. Most picture tubes now use external magnetic coils to deflect the electron beam and so paint out the picture, line by line, on the tube face, which is coated with a fluorescent powder.

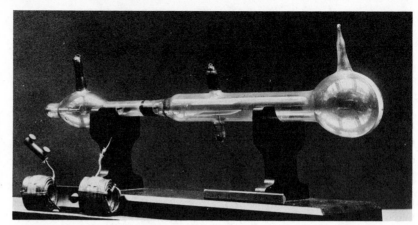

J. J. Thomson's original tube.

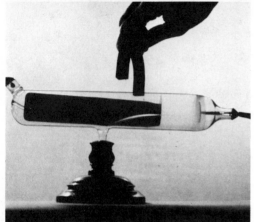

Bending a beam of electrons in an old cathode-ray tube, with a horseshoe magnet.

A Delicate Balance

ON ELECTRONS

At first there were very few who believed in the existence of these bodies smaller than atoms. I was even told long afterwards by a distinguished physicist who had been present at my lecture at the Royal Institution that he thought I had been "pulling their legs."

J. J. THOMSON

English physicist

Electricity is of two kinds, positive and negative. The difference is, I presume, that one comes a little more expensive, but is more durable; the other is a cheaper thing, but the moths get into it.

STEPHEN LEACOCK (1869–1944)

Canadian humorist

either their charge or mass separately. Yet in every respect the cathode rays behaved as anticipated, and so he wrote: "I can see no escape from the conclusion that they are charges of negative electricity carried by particles of matter." In sharp contrast, two years earlier in Germany, Röntgen had discovered X-rays, and those emanations, held to be "waves or pulses in the aether," were—in accordance with theory—quite unmoved by charged plates and magnets, just as light would be.

Although we properly credit Thomson with the discovery of the electron, some of his contemporaries in 1897 were not pleased with the notion of subatomic particles. Indeed, some thought he wasn't even serious in his conclusions, and for a while, old Lord Kelvin went on writing about electric fluids as if nothing had ever happened.

At this point, Thomson could have used the value of electron charge (q_e) deduced from Faraday's experiments, but instead he chose to measure that quantity for himself. This he did fairly well, using a technique devised by C.T.R. Wilson, a technique that the American Robert Millikan would bring to sheer perfection a decade later.

Millikan is responsible for one of those lovely classics of experimental physics that appear straightforward and deceptively easy. He began by squirting a fine oil mist from a perfume atomizer between two parallel horizontal metal plates. Lighted from the side by a bright lamp, each minute droplet shone like a tiny star in the field of a viewing microscope. The idea was simple enough: Most of the droplets would become negatively charged via friction on emerging from the nozzle, picking up some unknown number of electrons. As they fell, they would level off at a terminal speed (p. 67) that could be easily measured, and the measurement, from the theory of air resistance, would yield the mass of each particular drop. When the plates were subsequently charged, the top one positive and the bottom negative, the slow fall of a droplet could be interrupted, and that sphere of oil would be suspended motionless in a delicate balance between the electric and gravitational forces.

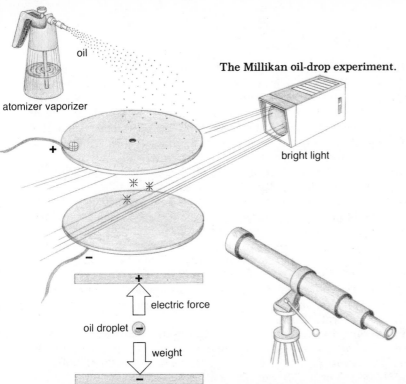

The Millikan oil-drop experiment.

oil

atomizer vaporizer

bright light

electric force

oil droplet

weight

Electricity is divided into definite elementary portions which behave like atoms of electricity.

HERMANN HELMHOLTZ (1881)

German physiologist and physicist

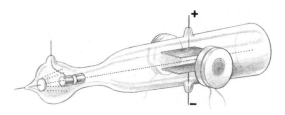

J. J. Thomson's electron-beam device showing the horizontal electrical deflecting plates and the vertical magnetic coils.

J. J. Thomson at the Cavendish Laboratory.

Crazy Horse dreamed and went into the world where there is nothing but the spirits of all things. That is the real world that is behind this one, and everything we see here is like a shadow from that world.

AMERICAN INDIAN

Black Elk Speaks

Having determined the mass of a given droplet, Millikan could calculate its weight, which at rest was precisely equal to the upward electric force. Knowing that, he could compute its total charge. He methodically found the net charge on thousands of droplets, one by one. What remained was a problem Millikan described as "similar to the one of finding the weight of a single egg, given the weights of a large number of paper bags each containing a different and unknown number of [identical] eggs." Each bag would simply differ by whole-number multiples (1,2,3, . . .) of the weight of an individual egg—and if you were lucky, you might even have a bag with only one egg in it. Thus the charge on each and every oil sphere was found to be a multiple of some minimum charge (1.6×10^{-19} coulomb, as discussed in Section 17.2), which was then assumed to equal q_e, the charge on the electron itself. This is an exceedingly small quantity; for example, a 100-watt light bulb draws a current equivalent to 6 million million million electrons flowing in every second, and a brief shuffling walk across some carpeting will build up a charge of several million million electrons.

Combining the value of q_e with Thomson's charge-to-mass ratio provides us with a remarkable number—the very mass of the invisible phantom—$m_e = 9 \times 10^{-31}$ kg; that's 0.000000000000000000-0000000000009 of a kilogram! A neutral atom (the atom as a whole always possesses zero *net* charge) stripped of one or more electrons, becomes a positively charged entity, a positive **ion.** It was not long before ions of several gases were made to run the course of a modified Crookes tube. The lightest of all, the hydrogen ion, was found to be 1840 times heavier than the dwarf electron. Clearly, the atom could no longer be seen as the minutest fragment of matter; clearly, it was no longer "that which cannot be further cut." The mythical Humpty had shattered.

10.2   RADIOACTIVITY

Röntgen's chance discovery of those mysterious X-rays flashing out from his Crookes tube sent much of the scientific community into a frenzy of activity. In basements and attics everywhere, researchers and cranks alike were busily at work in an effort to explore and exploit the new marvel and, indeed, to find their own miraculous emanations. And so it was that Henri Becquerel, in France, returned to his father's earlier work on fluorescent substances, which absorb and then reemit light, as does the pale green paint on the hands of a watch. Many people at the time wrongly believed that the emission of X-rays was somehow related to the glowing region of the Crookes tube, and Becquerel wondered whether his glowing materials might also send out X-rays. As fate would have it, he began to work with one of his father's compounds, potassium uranylsulfate—a *uranium* salt. Soon enough he had his answer. When the salt was exposed to sunlight, it actually did emit radiation that could penetrate the heavy paper wrapping on a photographic plate and fog the film.

On a sunless day in the winter of 1896, Becquerel placed the fluorescent compound atop a wrapped film plate as usual, but this time he put it aside in the darkness of a drawer to wait for clearer skies. A few days later, almost as an afterthought, he developed the plate; there, amazingly, was the blackened outline of the piece of uranium salt! Prior exposure to light was quite unnecessary; the radiation was emitted without it. By accident he had discovered what Madame Curie would later name—**radioactivity,** the property of certain substances to give out, all by themselves, penetrating radiation, to pour forth a seemingly inexhaustible torrent of energy.

Becquerel soon turned to other interests. Unlike X-rays, the new emanations couldn't produce those exciting pictures of bones, and they were generally greeted with indifference. This limbo of neglect persisted for about a year and a half until Becquerel brought the problem to a promising young student at the Sorbonne, proposing that she press on with the research. Marie Curie accepted the challenge and began her life's work. Becquerel had recognized in her the qualities of a scientist of extraordinary power. Years before, Mendeléev had said as much after seeing her as a little girl at work in her cousin's chemistry laboratory.

Madame Curie quickly determined that thorium, like uranium, was radioactive. She and her physicist husband, Pierre, discovered an element, until then unknown, about 400 times more active than uranium and called it *polonium.** Working for 45 months under the most dreadful conditions in an abandoned, leaking wooden shed, they finally isolated yet another radioactive element—*radium.* After four years of incredible labor, they had distilled a mere 1/30 ounce of the pure radium salt; 1/30 ounce laboriously wrung from more than a ton of the black uranium ore, pitchblende. But this new stuff was two million times more radioactive than uranium. Over the years, a single ounce of it would spew out the cumulative energy of ten tons of coal. That quietly blazing precious salt stayed warm of its own internal heat and glowed, self-luminous, like a magic firefly.

At last finding the time for herself, Marie Curie applied for her Ph.D. at the Sorbonne on the basis of this epochal work. The degree was granted, and within six months she, Pierre, and Becquerel shared the highest honor, the 1903 Nobel Prize. Three years later Pierre was dead—run down in the street by a horsecart. In 1934 after long illness, Marie Curie died of leukemia, a victim of the many years of overexposure to radiation—too close to the firefly. Even the pages of her lab notebook were later found to be contaminated with radioactive fingerprints.

Ernest Rutherford was just 24 when he arrived at the Cavendish Laboratory in 1895. Fresh from New Zealand, this unpolished, hearty

This lithograph, entitled "Radium," appeared in the December 22, 1904, issue of *Vanity Fair.* It depicts the Curies, Pierre holding a sample of the glowing element and Marie behind him.

Alpha, Beta, and Gamma

---

* The name derives from Marie's homeland, Poland, which at that time had vanished from the map, having been devoured by Russia, Austria, and Prussia.

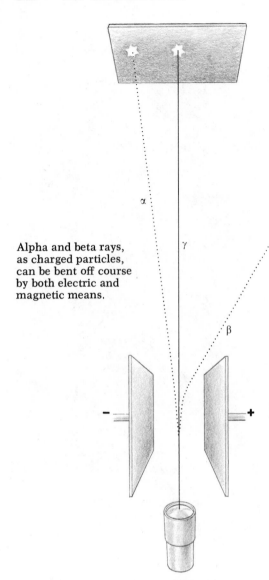

Alpha and beta rays, as charged particles, can be bent off course by both electric and magnetic means.

colonial was the first of the new research students at the lab and the first to work with "J.J." himself. So successful was their relationship that when Thomson was asked to recommend one of his graduating students for a professorship at McGill University in Montreal, he named the tireless, imaginative, dexterous Rutherford. It was in Canada that Rutherford began his study of the nature of what were still called "Becquerel rays." He found that the emanation emerging from radioactive substances "is complex, and that there are present at least two distinct types of radiation—one that is very readily absorbed, which will be termed for convenience α [**alpha**] radiation, and the other of a more penetrative character, which will be termed the β [**beta**] radiation." A year later in 1900, P. Villard discovered a third constituent, which, unlike the others, could not be deflected by a magnetic field. Rutherford established that these γ [**gamma**] rays, as they came to be called, were electromagnetic waves somewhat more energetic but otherwise identical to X-rays.

By far the most penetrating of the lot, gamma rays are completely absorbed only after passing through several feet of concrete or about one to five centimeters of lead. Usually more potent than X-rays, they have no trouble whisking through your body. Beta rays, which are actually electrons hurled out of the atom at fantastic speeds, up to 160,000 miles per second, can sail through from two to fifteen meters of air. A millimeter or so of aluminum will block them totally, and they will not burrow very deeply into people. Alpha rays are also particles, but they are positively charged and massive. Each has very roughly 8000 times the mass of an electron carrying twice its charge. They're fired out of radioactive atoms at formidable speeds, around 10,000 miles per second. Easily stopped, alphas will barely make it through a single sheet of paper or a few centimeters of air.

In a beautiful experiment (1909), Rutherford captured a stream of alpha particles within a glass chamber and showed that they were helium nuclei (helium stripped of its two electrons)! There's apparently something particularly stable about the configuration of the second lightest atom. Alphas are a common fragment of exploding atoms; they survive.*

10.3   RAISIN PUDDING

The raisin-pudding atom—negative electrons embedded within a large positive fluff. In its simplest form the electrons hover in circular patterns within fixed planes.

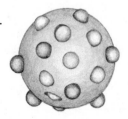

Moving back to England, Rutherford arrived at the University of Manchester in 1907 to take up his post as professor of physics and head of the research laboratory. Already working there was a congenial young Ph.D., Hans Geiger (later of counter fame), who had come to England for a few years to pick up some research experi-

---

* Most of the helium in the world comes from natural gas wells in the United States. When this government refused to sell the Nazis helium, they filled their zeppelins with inflammable hydrogen instead. The 1937 Hindenberg explosion put an end to that practice. In any event, it's likely that radioactive materials in the soil emitting alphas are the ultimate source of subsurface helium.

"Papa" had been awarded the 1908 Nobel Prize, inexplicably in chemistry, a fact he always rather resented, and was knighted in 1914. Sir Ernest Rutherford became the first of the nuclear puffers when he transmuted nitrogen into oxygen by bombarding the former with alpha particles. Of all his many contributions, perhaps the greatest was providing that first remote glimpse of the nuclear atom.

By the turn of the century, several different methods were available for getting at least a rough idea of the overall size of an atom. One of the more obvious schemes was just to allow a droplet of oil of known volume to spread out on the surface of water. If one assumed that the film would expand until it was only one molecule thick (and there were good reasons to suspect that it would do just that), the area of the oil slick immediately allowed the thickness and hence the atomic "diameter" to be calculated. This and all the various other methods that have since evolved have led to one result: Atoms typically are about one ten-thousandth of a millionth of a meter (10^{-10} m) across.

Rutherford's highly successful scattering calculations dealt with rebounding alpha particles off the positive atomic core, and they provided a rough measure of the size of the core itself. The confirming experiments (1913) bore out the suggestion that deep within the atom, at the center of the distribution of electrons, was a minuscule though massive nucleus, a mere 10^{-14} or so meters across—a speck less than one ten-thousandth of the overall size of the atom.

Rutherford did not deal with the surrounding swarm of electrons explicitly. He didn't have to, since they played no significant role in the scattering, but it does seem clear that he was envisioning a planetary system of some sort. He alluded to H. Nagaoka's earlier, purely speculative model (1904), which pictured the electrons hovering about an attractive center like the rings about Saturn. Yet nothing quantitative was done with the scheme beyond recognizing that since *atoms as a whole are neutral, there must be as much positive charge on the nucleus as there is negative charge orbiting it.* That was reasonable enough.

The year 1913 was a crucial one for the newborn nuclear atom. It was the year Bohr would attack the problem of the structure of the electron orbits, and it was the year Moseley would reveal the pattern of nuclear charge.

Atomic Number

Henry Moseley was the youngest and most brilliant of all of Rutherford's staff of clever young protégés at Manchester. His particular research problem was concerned with finding some relationship between X-rays and the atoms that emit them. When a high-speed beam of electrons (cathode rays) slams into a target, it liberates bursts of electromagnetic waves, energy in the form of X-rays. Each different element serving as a target emits its own **characteristic X-ray spectrum** —a paw print of sorts. After a tremendous solitary effort—inventing

10.4 NUCLEAR PHYSICS

H. G. J. Moseley (1887–1915)

A LETTER FROM THE ARMY
Let it suffice to say that your son died the death of a hero, sticking to his post to the last. He was shot through the head, and death must have been instantaneous. In him the brigade has lost a remarkably capable signalling officer and a good friend.

Regarding H. Moseley, Signalling Officer, 38th Brigade

new techniques, laboring day and night, surmounting one experimental problem after another—Moseley finally was able to devise an empirical equation describing each emitted pattern of X-rays in terms of the particular atomic structure of the target, that is, in terms of something he called the *atomic number*.

Each element, it seems, has an integer atomic number, and the numbers increase, one unit at a time, from hydrogen (1) to helium (2) to lithium (3) and beryllium (4) all the way up to uranium (92), just like a place number marking consecutive boxes in the periodic table. Moseley astutely concluded: "This quantity [atomic number] can only be the charge on the central positive nucleus." Hydrogen has one unit of nuclear charge, helium has two, lithium three, and so on. The **atomic number** *is the number of units of nuclear charge.* This of course suggests that there is something fundamental about the hydrogen nucleus, and in time it would indeed be recognized as the primary lump of positive charge, the proton. But that anticipates a bit of the story.

The idea that the elements of the periodic table were to be ordered not by weight but by nuclear charge had been in the air for several years already. But the bright young man at Manchester, the artist with tubes and glass, pumps and valves, the patient, insightful young man had made it true. He could observe the emitted X-rays and from them immediately tell whether or not the material under study was elemental and, if so, where in the table it belonged! For example, there was some controversy about argon (atomic weight 39.9) and potassium (atomic weight 39.1) since the former is an inert gas and should be located with the other inert gases. But that would have it strangely preceding the *lighter* alkali metal. Moseley's X-rays revealed argon to have an atomic number of 18 whereas that of potassium was 19. Argon clearly did belong before potassium, despite its greater weight.

It was not known then, but the atomic electrons that emit the energetic X-rays seen by Moseley are the ones that hover closest to the nucleus and are therefore most affected by the nuclear charge. So it was that Moseley's X-rays at last revealed the nuclear order of the elements, resolved the remaining discrepancies, and fixed for all time the pattern Mendeléev had come to by inspired intuition. From hydrogen to uranium, one element after the other, each atom carries a positive nuclear charge precisely one unit greater than the one before it.

Harry Moseley had "seen" a vision of the invisible, but he would undo no more of Nature's secrets; in 1915 at the age of 28, he was killed at Gallipoli in one of the most useless campaigns of World War I.

Protons and Neutrons

Well over 150 years ago, William Prout boldly (and anonymously) suggested that all the elements were built up from the lightest of atoms, hydrogen. Unfortunately, when the atomic weights of chlorine

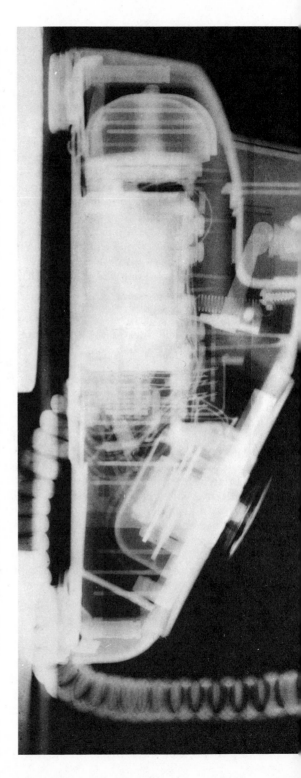

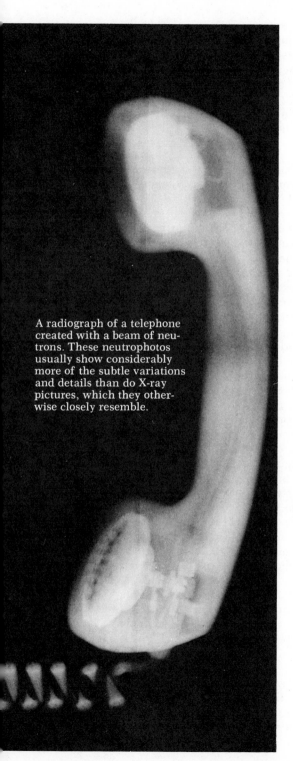

A radiograph of a telephone created with a beam of neutrons. These neutrophotos usually show considerably more of the subtle variations and details than do X-ray pictures, which they otherwise closely resemble.

(35.5), copper (63.5), and magnesium (24.3) were found not to be whole-number multiples of the weight of H (1), that hypothesis was put aside only to rise again a century later when the "facts" had changed.*

The Crookes tube positive-ion experiments of J. J. Thomson led him to conclude (1907) that the hydrogen ion, H+ (the atom stripped of its single orbital electron, i.e., the hydrogen nucleus), was indeed the building block of the elements. But the clincher came when Rutherford, pounding the nuclei of various elements with alpha particles, found hydrogen among the flying fragments. In 1920 he formally proposed the name **proton** (from the Greek *protos* for "first"), a word that had already been quietly in use for a dozen years. Physics was quite ready for the proton's debut.

With a mass of 1.7×10^{-27} kg, the proton is 1840 times heavier than the electron and certainly not the latter's positive twin.† The proton does however carry a positive charge that is equal to the negative charge of the electron, and so these subatomic specks logically seem, if not twins, at least mates.

The pieces were slowly coming together, and one could picture the **hydrogen atom** *as a single nuclear proton coupled by an attractive electric force to one orbital electron.* The next element, helium, should then consist of two protons and two surrounding electrons; lithium should have three and three; and so on, all the way up to uranium with 92 protons and 92 whirling electrons.

But there must be something else happening in the nucleus, since helium with an atomic number of 2 has an atomic weight of 4. Such differences between atomic number and atomic weight run throughout the table all the way to uranium, which has 92 protons and a weight of 238. There is a lot of mass unaccounted for if the uranium nucleus is to be thought of as a cluster of 92 protons. It should have a weight of 92, not 238! Rutherford proposed that the difference was due to the presence of additional *proton-electron pairs*, neutral units that would add to the mass but not to the charge. Thus the helium nucleus was to be conceived of as four protons and two neutralizing electrons, yielding a mass of 4 and a net charge of only 2. He even suggested the name **neutron** for the proton-electron unit.

Twelve years later, in 1932, James Chadwick, another of Rutherford's star pupils, discovered that when beryllium was bombarded with alpha particles, it was transmuted into carbon, and out streamed a flux of the long-sought neutrons.

We know today that once the neutron is removed from the atom, it is unstable, decaying into an electron and a proton‡ at a rate such that half of the neutrons present at any moment will decay within roughly the next 17 minutes. Though close, Rutherford was wrong.

---

* Here's another instance of the resurrection of a theory that was "proved" wrong in its initial form.

† That is the *positron* (Section 15.3).

‡ A massless neutral speck called a neutrino (Section 7.4) is also created during this minicataclysm.

We now believe that electrons do not exist inside the nucleus, and that the neutron is therefore a single particle slightly more massive than the proton. While nestled with the protons in the nucleus like grapes in a cluster, the neutrons interact with (and seem to be stabilized by) the others, so they ordinarily don't decay.

Thus uranium with 92 protons possesses an additional 146 neutrons, which together make a total nuclear weight of 238—and everything works out neatly, at last.

Isotopes—Birds of a Feather

Sir William Crookes as depicted in an illustration from the magazine *Vanity Fair*.

It is clear that, in some sense, there are electrons and protons, and we cannot well doubt the substantial accuracy of their estimated masses and electric charge. That is to say, these constants evidently represent something of importance in the physical world, though it would be rash to say that they represent exactly what is at present supposed.

BERTRAND RUSSELL

The Analysis of Matter (1954)

All the atoms of a particular element are supposed to be identical. That was one of the uncompromising assertions of Dalton's theory, but like the indestructible atomic billiard ball, it would be blown away in the first decades of the twentieth century. Incredibly, there *are* different kinds of hydrogen and helium, potassium and argon, different kinds of every one of the elements—often six or seven distinct variations on a theme.

People had speculated that perhaps the measured atomic weights of elements were not integers because each was actually the average value of a mix of nonidentical atoms. William Crookes was one of these prophets without proof. But the situation that would demand a whole new comprehension boiled over in Rutherford's lab at McGill. He and his assistant, Frederick Soddy, had traced the decay of several heavy, naturally occurring radioactive elements, like uranium and thorium, as they broke apart and formed new radioactive byproducts. These decayed in turn, cascading downward in atomic number, one to the next, until the series ended ultimately in stable lead. Strangely enough, along the way elements appeared that were identical to others chemically but that had obviously different radioactive characteristics. T. W. Richards had already noticed the then astounding fact that although ordinary lead has an atomic weight of 207.20, the lead present in uranium ore from Norway had a weight of only 206.05. Soddy later christened these variations of a given element **isotopes,** from the Greek *isos* meaning "same" and *topos* for "place"—having the same place in the periodic table.

In 1913 J. J. Thomson and one of his fledgling "saints of Cambridge," F. W. Aston, were the first to actually separate the isotopes of an element, but they did it quite unwittingly. They had put neon gas (atomic weight 20.2) into one of their positive ion tubes and deflected the beam with charged plates and magnets in order to measure its atomic weight. And there at the face of the tube were *two* separate spots. The conclusion was unmistakable: "Neon is not a single gas, but a mixture of two gases, one of which has an atomic weight of about 20, and the other of about 22."

Neon atoms are not all identical, even if the variations are chemically inseparable. It's the outer portion of the electron cloud surrounding the nucleus that apparently determines the chemistry, and the number of electrons goes hand in hand with the nuclear charge, the atomic number. The name of an element (hydrogen, uranium, neon, whatever), is associated with a specific atomic number, a specific place in the table. Knock out or add a proton, and you have transmuted one element into another. Thus any atom having 10 protons is neon, no matter what its total weight. It follows that 10 protons and 10 neutrons make neon ($^{20}_{10}$Ne) just as 10 protons and 12 neutrons make a different form of neon ($^{22}_{10}$Ne); each is an isotope of neon. Since the first and lighter form is far more abundant than the second, the naturally occurring mixture averages out to a weight of 20.2.

Isotopes of a given element differ only in their number of nuclear neutrons. Some of these are stable configurations that presumably will last for all time, but others are radioactive and transient. Some elements, such as xenon and iodine, have more than a dozen known isotopes each. Nine of the xenon isotopes are stable but iodine has only one stable form, as do aluminum and gold.

The three isotopes of hydrogen are so different from one another and so important that they have even come to have their own names. Ordinary *hydrogen* with a single proton (charge 1, weight 1) is the lightest and most common of the lot (99.98%). **Deuterium** with an added neutron (charge 1, weight 2) has a two-particle nucleus and is quite rare (0.02%). Radioactive **tritium** with still another neutron (charge 1, weight 3) is by far the least abundant. Since each isotope has only one electron, it's not surprising that the chemistry of these three isotopes (which differ so enormously in atomic weight) is distinctive. For example, living organisms respond quite differently to water formed of oxygen and deuterium (charmingly known as **heavy water**) than they do to the ordinary lighter brew, which differs as well in its freezing and boiling points. Heavy-water ice cubes, though they look and taste ordinary enough, will sink to the bottom in a glass of tap water. Except for minor variations, isotopes of each of the other elements behave almost the same chemically. The existence of one neutron, more or less, in the nucleus of a uranium atom will hardly affect the outer orbital electrons, which themselves rule the chemistry.

As we scramble to drain energy from Nature, hydrogen and uranium, elements at the two ends of the table, hold out the promise and the threat. The rare isotopes of each are the seeds of bombs and power plants.

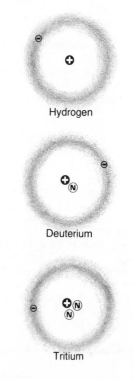

Hydrogen

Deuterium

Tritium

The three hydrogen isotopes.

EXPERIMENTS

1. Prop a piece of cardboard up and have someone place a fairly massive object beneath it, just out of sight. It could be a book, a tool, or perhaps a stone. Now see if you can determine what the hidden object is by studying the scatter pattern as you shoot marbles or checkers at the target. What determines the amount of detail you can "see"? Note how important it is that the object not be moved around in the process.

2. Place a *strong* magnet on the face of a black and white TV tube. (Color sets have metal masks that can become magnetized and so should *not* be used.) If your magnet is powerful enough, you can also see some *slight* effects waving it near a fluorescent lamp.

REVIEW

Section 10.1

1. Which was the first subatomic particle to be discovered?

2. Who invented the first mechanical electrostatic generator?

3. Two similar charges will exert a _____ force on each other, and two dissimilar charges will _____ one another.

4. The notion of positive and negative electricity was introduced by none other than the man in the coonskin cap, the great "ladies man" at the French court, _____ _____.

5. The force between two charges varies _____ with distance _____.

6. How does the force acting between two charged bodies depend on the charge on each?

7. Describe the main feature of a Crookes tube. Which electrode is the cathode?

8. How did Helmholtz envision electricity as early as 1881?

9. Most scientists do their great work before their thirties. What about "J.J."? What was his great work?

10. What was the purpose of the Millikan oil-drop experiment?

11. What is a *positive ion*? Incidentally, there is some evidence to indicate that the ion content of the air might affect human mood swings. Have you ever seen the ads peddling ion machines for home and office?

12. What is the experimental sequence that led to the determination of the mass of an electron?

Section 10.2

1. Who discovered *radioactivity*?

2. Describe Madame Curie's scientific accomplishments.

3. What are alpha, beta, and gamma rays?

4. X-rays and gamma rays can be distinguished by their _____; otherwise they're the same.

5. How are helium and alpha particles related?

Section 10.3

1. Describe the "Thomson atom."

2. Who was Hans Geiger?

3. Describe Rutherford's alpha-particle gun.

Section 10.4

1. Roughly, what is the diameter of an atom?

2. The nucleus is a minute grain approximately _____ meters in diameter.

3. Nagaoka's atomic model was surprisingly modern. Describe it.

4. What did Moseley mean by "atomic number"?

5. The bare nucleus of a hydrogen atom is a rather special positive ion known as a _____.

6. What's the mass of a proton, and how does it compare with that of an electron?

7. Are there any electrons residing within the nuclei of atoms?

8. Describe the creature known as a *neutron*.

9. Define the word *isotope*.

10. Describe the three isotopes of hydrogen.

QUESTIONS

Answers to starred questions are given at the end of the book.

Section 10.1

* 1. What was so important about Coulomb's contribution? Two identical conductors charged while in contact would carry equal charges after separation. If you then measured the repulsive force between them, you could define a unit of charge (the coulomb).

2. Describe the arguments for the position that cathode rays are charged particles (i.e., the British view). Note the tendency toward "either or" arguments that allow only familiar alternatives.

3. Did "J.J." prove that electrons were charged particles or simply that they behaved in his experiment as such particles were supposed to?

* 4. What were the implications, with respect to the atom, of the conclusion that electrons were tiny specks of charge?

5. Why is it reasonable to assume that all electrons are identical? Obviously we haven't looked at them all.

Section 10.2

1. How does the notion of serendipity enter into Becquerel's famous work?

* 2. When did scientists first become aware of the incredible amounts of energy locked up within the atom? Even then, it was almost always spoken of as "atomic energy."

3. Compare the higher speeds of beta rays with the speed of light.

4. Why do you think helium was discovered on Earth in a mineral containing uranium?

Section 10.3

1. Becquerel was touched by the wand of serendipity. Should we add Rutherford's name to the list, as well? If so, why?

* 2. Why was Rutherford so surprised by the results of Marsden's alpha-scattering experiment?

3. How was Rutherford led to his imagery of the nuclear atom?

4. Discuss the data–law–theory–prediction–test sequence as it pertained to the conception of the nuclear atom.

Section 10.4

1. Explain why we can expect any atom to contain equal quantities of positive and negative charge.

2. Discuss the ordering of elements in the periodic table in light of Moseley's discovery.

3. If we remove a proton from a cobalt nucleus, it transmutates into _____.

4. Why was Moseley able to find the secret of the periodic table when Mendeléev couldn't quite?

5. What fate did Prout's theory—that all the elements are made of hydrogen—have in common with the kinetic theory of heat? By the way, William Prout was a contemporary of Dalton.

6. What was the logic that led Rutherford to expect the existence of a *neutron* even before it was found?

* 7. When a uranium atom with an atomic weight (total number of neutrons and protons) of 238 emits an alpha particle, it transmutates into _____, with a weight of 234.

8. How many protons and neutrons are there in the nucleus of each of the following: ^{235}U, ^{14}C, ^{90}Sr, and ^{214}Pb?

* 9. Discuss what might happen to an atom if one of its neutrons decayed into an electron and a proton, assuming the former flew off as a beta ray. Lead-214 would thus become _____-214.

* 10. Tritium (^3_1H) decays via beta emission into _____.

* 11. In 1906 at Yale, B. B. Boltwood discovered a new element he called *ionium*. Its atomic weight was definitely different from but very nearly that of thorium. Indeed, it behaved in all respects like thorium, particularly in its chemistry. Why will you not find ionium in the periodic table?

12. Which element has a nucleus containing six protons and eight neutrons?

MATHEMATICAL PROBLEMS

Answers to starred problems are given at the end of the book.

Section 10.1

* 1. How many electrons would you get if you could buy one gram of them? How does that compare with the number of stars in the entire universe ($\sim 10^{22}$)?

2. Compute the electron's mass in atomic units, keeping in mind that a hydrogen atom corresponds to 1.0 u.

3. What is the total mass of all the electrons in 2 grams of H_2 gas at standard temperature and pressure (assume 1 electron per atom)?

Section 10.2

1. Ounce for ounce, radium puts out _____ times more energy than coal.

2. What is the approximate mass of an alpha particle?

* 3. Compute the kinetic energy of a beta particle moving at 200,000 meters per second.

4. Compare the speed of a typical alpha particle with the 1200 mile-per-hour speed of a jet plane.

Section 10.3

1. How many hydrogen atoms lined up end to end would it take to extend a distance of one inch? If you decided to make a drawing of this chain, sketching two per second, how long would it take to finish your masterpiece if you worked nonstop?

2. What volume of gaseous H_2 at standard temperature and pressure would contain the number of hydrogen atoms in the previous problem?

3. A proton has a mean radius (R) of very roughly 10^{-15} m. How many protons tightly packed would occupy one cubic centimeter, overlooking empty spaces between them? *Hint:* the volume of a sphere is $4/3\,(\pi R^3)$.

* 4. With the previous problem in mind, if atoms could be crushed so that nuclei were in contact, how much would a cubic centimeter of the substance weigh? What would its mass be? Is it possible that similar stuff may actually exist at the centers of certain stars?

5. Radium-226 decays by alpha emission into a gas that is itself radioactive. Ernst Dorn in 1900 placed it in its proper location in the periodic table, and William Ramsay determined its atomic weight to be _____. What is that element?

One of the great intellectual upheavals of this century came out of the gradual realization that classical physics, the physics of Newton and Maxwell (mechanics and electromagnetic theory), was surprisingly inept in the domain of the atom. The classical vision, which had worked so well on the large scale, failed abysmally in the miniworld, where its shortcomings were acutely apparent.

This chapter traces the development of those new ideas that culminated in our present-day conception of the universe known as **quantum theory.** This twentieth-century formalism (to which classical theory is a satisfactory approximation in the macroworld) shines a powerful searchlight into the previously secret miniworld of the atom. The first section of this chapter treats the rather obscure phenomenon of **blackbody radiation,** because it was there that Planck hesitantly proposed the key notion that **energy might not always be continuous.** The next segment examines the **photoelectric effect,** another crucial experiment that defied classical understanding. In order to explain this baffling mystery, Einstein introduced the idea that **light itself was granular.** Bohr's bold use of these revolutionary insights in his own rudimentary description of the atom was the first really successful atomic theory. Treated in the third section, the **Bohr theory,** now merely a historical relic, nonetheless provided a rush of superb insights. Section four explores the next revelation, namely, that **"particles" of matter have wavelike characteristics,** just as **"waves" of light have particlelike characteristics.** The chapter ends with a brief discussion of the statistical nature of **wave mechanics** and the ultimate limitation to knowledge imposed by the **uncertainty principle.**

CHAPTER 11
Quantum Theory

On facing page, a 1930 photo of two of the founders of modern physics: Niels Bohr on the left and Max Planck on the right.

As the atom was blossoming into a mature conception in the early days of this century, there began to develop, quite tangentially, a profound new vision of the microcosmos: quantum theory, an insight so powerful it would come to dominate modern physics.

Max Karl Ernst Ludwig Planck, thin-faced, wearing wire-rimmed glasses, balding, and mustachioed, at 42 was the reluctant father of the new imagery in the last cold months of 1900. This proper Ger-

11.1 A FORMAL ASSUMPTION

man professor was hardly a revolutionary figure, and yet, like the work of that other conservative, Copernicus, his efforts would revolutionize despite him.

When a body is heated, it radiates electromagnetic energy, and if the temperature is high enough, some of that energy takes the form of visible light. We see the cherry-red glow of a coal in the fire and feel the warming (invisible) infrared. Black objects (like that chunk of coal) when heated are most effective at radiating energy, and Planck, a theoretician, concerned himself with the idealization, the perfect radiator, known in the trade as a **blackbody.***

Extensive experiments had made it clear that a blackbody (or almost any solid body for that matter) raised to and held at some temperature gives out a whole range of electromagnetic emanations, from radio waves and infrared (IR) through all the colors of light, to ultraviolet (UV). The surface of the sun itself is a fair example of the sort of hot, glowing object we are talking about, and so is the filament in a light bulb or even a hot apple pie although it radiates far more IR than anything else. The perplexing thing at Planck's time was that the physics of the nineteenth century stubbornly insisted, out of the necessities of its own established principles, that there be a predominance of ultraviolet in the spectrum when, in fact, there was always comparatively little.

At a given temperature the radiant energy is richest in a particular emanation; it peaks. For example, the sun, with its surface at 6000 K, emits yellow light most strongly and so appears yellow-white. A paper clip in the flame of a stove is much cooler than that; peaking in the IR, it emits a good deal of red light as well, appearing to the eye "red-hot."

The higher the temperature of the body, the brighter the emanations, and the more the peak is shifted toward the blue. When an ordinary yellow-white tungsten bulb burns out, the temperature rises for an instant and the bulb brilliantly flashes blue-white. If you blow oxygen onto an ember in the fire, it will rise in temperature, blazing up from a dull red to bright yellow.

We can conveniently summarize the experimental data with a family of curves of brightness versus the kind of emanation from UV to IR—that much is easy enough. But to "understand" what is happening, to explain the shape of the curves on the basis of more fundamental processes occurring within the glowing object—that was a defiant problem, which gave no ground to traditional analysis. And it was particularly annoying since the turn of the century was a time when physics was most pleased with itself, most arrogant in the power of Newtonian mechanics and Maxwellian electromagnetic the-

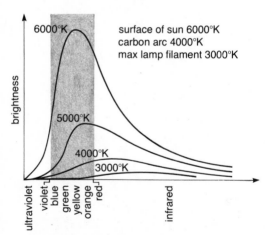

surface of sun 6000°K
carbon arc 4000°K
max lamp filament 3000°K

The amount of radiant energy emitted by a hot object at various wavelengths.

---

* Planck had already been working for several years trying to establish the superiority of Rudolf Clausius's as yet unrecognized entropy formulation of the second law of thermodynamics (p. 198). Indeed, it was with that purpose in mind that Planck began the study of blackbody radiation.

ory. It seemed capable of anything except, perhaps, this wretched, obscure little difficulty.

Max Planck approached the whole problem with a practical brilliance. He came at it from the rear—backwards, as it were. First he concocted "by happy guesswork" a mathematical expression that resembled all the members of the family of experimental data curves. Built into that equation were two constants, which he altered by trial and error until the expression slowly came to match the data curves rather nicely. Though merely a contrivance, a fudge-factor, in his initial treatment, one of those constants (h) would ultimately be recognized as a new fundamental quantity (like the gravitational constant G or the speed of light c).* Well, that much by itself was quite a success, even if it provided no understanding at all of what was going on.

Next Planck set out to construct a physical scheme that would logically lead to the equation he had already cooked up. Like others before him, he assumed that the radiation was emitted by microscopic oscillators of some sort, which vibrate away on the surface of the glowing body, generating waves in the process. Planck at that time was a devotee of Mach, who had little regard for the reality of atoms, and yet the obstinate insolubility of the problem ultimately led him to "an act of desperation." He hesitantly turned to Boltzmann's "distasteful" statistical method. Boltzmann, the great proponent of the atomistic view, and Planck were rather bitter adversaries for a while, primarily because Planck could not abide Boltzmann's probabilistic interpretation of entropy. And now he was forced to use his rival's formalism, which—ironically—he would actually misapply. One technical detail necessitated by the method was that energy had to be thought of, at least temporarily, as atomized or apportioned so that it could, in effect, be counted. This was a statistical analysis, and counting was central. Still, when the method was applied as Boltzmann intended, it naturally smoothed out energy, making it continuous as usual. But Planck realized that he could arrive at his sought-after equation only by keeping energy atomized. In essence, only certain vibrational energies could therefore be allowed. Instead of having any value whatever, an oscillator could have only certain discrete energies—a little like the gravitational PE of someone walking up the steps of a flight of stairs.

The rate at which a thing oscillates back and forth or, indeed, the repetition rate of any motion that occurs over and over again is called the **frequency,** f. The frequency at which a phonograph record whirls around and around, returning to the same position to finish a cycle and then sailing about once more is, of course, 33⅓ rpm (rotations per minute). The number of *cycles per second* or *cycles per minute*—i.e., the number of *cycles per interval of time*—is frequency.

---

* The other of his two constants has come to be known as Boltzmann's constant, although Planck was the first to evaluate it.

LUMINOUS BODIES

Light is now believed to originate in extremely minute and rapid vibrations of the atoms of matter. These vary in rapidity from about 400 million million to about 760 million million a second. The atoms of all luminous bodies are supposed to be vibrating at this enormous rate.

J. A. GILLET AND W. J. ROLFE

Natural Philosophy for the Use of Schools and Academies (1882)

The Unicorn in captivity, a fifteenth-century tapestry.

11.2 THE UNICORN AND THE PHOTOELECTRIC EFFECT

The SI unit of frequency is the **hertz,** with 1 Hz = 1 cycle per second. Planck concluded that the vibrational energy of an oscillator that was emitting or absorbing radiation depended on its frequency and was, in fact, a whole number multiple of the product of f and his empirical constant, h. In other words, the energy of a given frequency oscillator, at any instant, could be 0, hf, $2hf$, $3hf$, $4hf$, or.... Like a staircase, the levels mounted in steps of hf, which itself came to be known as the **quantum of energy** (a term introduced by Einstein in 1905 from the Latin meaning "how much").

"That energy is forced, at the outset, to remain together in certain quanta...," Planck later wrote, "was purely a formal assumption and I really did not give it much thought." Nonetheless, physically the quantum was an exceedingly peculiar beast. A homely analogy might be a wooden horse on a merry-go-round that could travel at 0, 10, 20, or 30 miles per hour but nowhere in between. Its motion would not be continuous, gradually changing, but instead would somehow jump from one allowed speed to the next in spastic bursts. Well, for a merry-go-round that certainly seems weird enough, and it was not any more appealing for these atomic-level goings on, either.

Although this complex and cloudy concoction finally led to the desired equation, the whole business was quite unsettling.* Planck himself, an exceedingly cautious man, at first seemed, if not downright disbelieving, certainly hesitant—as if having given birth to a unicorn while expecting an ass. In any event, few physicists outside the discipline paid any attention to the mythical creature, and those who did were more concerned with the equation that worked than with the quantum, whose reality was dubious and whose assured oblivion was tearlessly anticipated.

The quantum languished unnamed and unnoticed for almost five years, a veritable eternity in that bustling period of activity. It languished until a powerful seer arrived, a visionary who could perceive beyond the "reality" of that moment and know the unicorn for what it was. Albert the Marvelous was 26 in 1905; 26 and a lowly "technical expert, third class" at the little Swiss patent office in Bern and thankful for that, too; 26 and already one of the greatest of creative geniuses, although he could not even get a university position at the time. Sitting at his desk during the long, empty moments in the office secretly rewriting the universe, he would hear approaching footsteps and hide that incredible treasure in a drawer, dutifully returning to the tedium of patents and the somber look of business. The incomparable revolutionary, Albert Einstein, had glimpsed the quantal unicorn and come to believe in its reality as no one before him, not even Planck, had dared.

---

* In fact, Einstein would soon point out that Planck's analysis was logically inconsistent. But the important point was the discovery (however strangely made) of the granular nature of energy. The theory would be put right later.

PLANCK'S CONSTANT
$h = 6.63 \times 10^{-34}$ joule-seconds

Ironically it was Heinrich Hertz who had unwittingly set the stage for the new drama eighteen years before. While busily and quite convincingly "proving" that radiant energy (light, IR, etc.) existed in the form of electromagnetic waves, he happened on a *minor* observation that would ultimately (like a snake swallowing its own tail) attack the very basic understanding of such waves. He had constructed an apparatus whereby sparks, jumping the gap in a transmitting circuit, would generate invisible radio waves. These emanations, flashing out across the lab, revealed their passing presence by inducing sparks in a distant receiving circuit. The rest of the experiment, which is of little concern to us right now, went on to show that these waves behaved very much like light in all respects, differing only in that they were considerably longer. The conclusion was obvious: All the various manifestations of radiant energy were electromagnetic waves, ripples in the aether.* Classical theory had again prevailed, and all was joy in Physicsland.

One day Hertz happened to notice that the receiver's sparks induced by the radio waves became even thicker, stronger, when the polished brass knob that formed a terminal of the gap was simultaneously illuminated by a beam rich in ultraviolet. The important thing here was that ultraviolet, bathing the brass, was somehow increasing the electrical flow; in fact, for our present purposes, you can think of the entire setup (gaps, sparks, radio waves, all of it) as nothing more than the first crude detector of what later became known as the **photoelectric effect.** Hertz paid it little further attention, but others pressed on, improving the experimental technique by making metal target plates and placing them in vacuum tubes. Remember, this was the era of the cathode ray, and there were pumps evacuating glass bottles everywhere. By the end of the nineteenth century, it had become clear that *radiant energy (in the form of X-rays, ultraviolet, or light) impinging on various metals could eject electrons from their surfaces.*

That much was reasonable. What was not reasonable was the fact that many of the observed aspects of the phenomenon (as with blackbody radiation) stubbornly defied theoretical interpretation. The problem was that the imagery of incident waves (and light was universally believed to be a typical wave in all respects) just could not account for what was happening. For example, one should expect that the stronger the incoming light wave—that is, the greater its intensity or brightness—the more violent would be the wrenching of electrons free from the metal and the higher would be their emerging speeds. Certainly the height of an ocean wave determines how vigorously it will fling about a row of boats moored on a beach. And yet, that's not what happens; the speeds of the ejected photoelectrons inexplicably depended not on intensity but on, of all things, the frequency or color of the light! Making the source brighter increased the

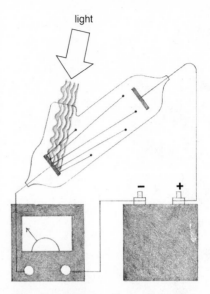

light

The photoelectric effect—light liberating electrons from a metal.

---

* Both the aether and the pure wave model seemed all but unassailable until they fell before Einstein.

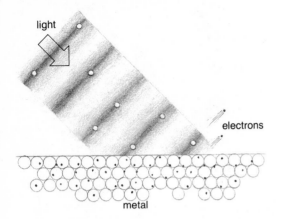

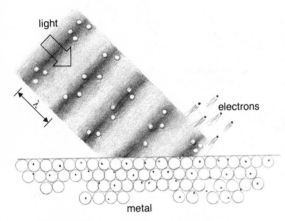

As regards the photoelectric effect, we can imagine that an increase in the intensity of the light beam (at some fixed wavelength) will supply *more* photons and therefore result in an increase in the number of ejected electrons.

number of electrons released, and yet, surprisingly, it had no effect whatever on their speeds.

Stranger still was the observation that switching on even the dimmest source of the proper frequency would *instantly* release a stream of electrons.* Now, that behavior was simply impossible to bring into harmony with orthodox wave theory. The prevalent conception clearly insisted that each incoming wave would spread uniformly over the surface of the illuminated metal, imparting only a minute amount of its energy to each of the millions upon millions of atoms in its path. It would therefore take wave after wave after wave for an electron to somehow absorb and accumulate enough energy to be ripped loose and flung out at the observed high speeds.

A rigorous application of existing principles unambiguously predicted that it might take months of irradiation before any electrons could possibly be ejected—and yet there they were, flying off instantaneously. Our honeymoon with the universe was brief and deluded. Something fundamental was obviously wrong with the accepted "understanding" of electromagnetic radiation. Classical theory teetered helplessly on the cutting edge of experiment.

The traditional interpretation likened the situation to that of an ocean wave rushing in on a line of boats along the beach—a wave that somehow could come down with concentrated fury, battering only one boat among the many, ripping it free on the first roll, and leaving the others unmoved. But that never happens. It would be unthinkable, at least for ocean waves, whose energy is spread out along the entire wavefront.

Earlier, while delivering the Silliman Lectures at Yale in 1903, J. J. Thomson suggested that electromagnetic waves might very well have configurations radically different from those of other waves; perhaps a sort of concentration of radiant energy actually did exist. After all, if one shone X-rays on a gas, only certain of the atoms, here and there, would be ionized, as if the beam had "hot spots" rather than being uniform. The notion was prophetic, but after pointing out its implications for the photoelectric effect, "J.J." took it no further.

Photons   Recall that Planck had earlier concluded that the transfer of energy between radiation and matter in a blackbody occurred in discrete steps. Regardless of the inconsistencies, he nonetheless supposed, as was usual, that radiant energy was continuous and, once emitted, behaved like a classical wave. Now Einstein, in 1905, stepped into the muddle. Audaciously breaking with the long-standing traditional view, he proposed that **light itself is granular, that it is actually composed of discrete bursts,** lumps of energy, or **photons,**† as they came to be

---

* The delay time, we now know, is less than three thousandths of a millionth of a second (10^{-9} s).

† The word *quantum* meaning *quantity* has been in the language for well over three hundred years, whereas the word *photon* was coined by G. N. Lewis in 1926.

called. *Of course,* the interaction between electromagnetic radiation and matter occurs in steps, as Planck maintained. It has to—radiant energy is quantized; it is absorbed and emitted discontinuously because it is itself discontinuous! In exactly the same way, the charge on a rubbed comb can change by only a whole-number multiple of the electron charge, because charge is quantized; it comes in lumps. Certainly we have known for a long time that matter, too, is particulate. In the microworld, granularity is more the rule than the exception.

Einstein, guided by Planck's work, postulated that *every electromagnetic wave of frequency f has its energy localized in a large number of individual photons, each one of which has an* **energy,** *E, given by the equation*

$$E = hf,$$

where *h* is again Planck's constant. In effect, this equation says that a photon of light that has a relatively low frequency has less energy than an ultraviolet photon of moderate frequency and still less than a high-frequency X-ray photon.

The theory then went on to simply maintain that a photon colliding with an electron vanishes, imparting all its energy to the charged particle. If the frequency and therefore the energy is just sufficient, the electron will be eased out of the metal; if the photon energy is still greater, the excess will go into speeding up the ejected electron. With this as the main scenario, the 1905 paper provided a complete quantitative description of every known aspect of the photoelectric effect. Moreover, it predicted a few details that had not yet been observed or even suspected.

The new theory was anything but an instant success. In fact, within a few years most people came to think that the young genius from the patent office, who was so successful in his other efforts, had gone rather too far out with this one. The necessary experiments, those that would be convincing, were exceedingly difficult to perform, and for a long while there were conflicting results that only clouded the issue even more. Few would go along with so revolutionary a theory without the coercion of hard data. By 1910 Planck could count only four prominent adherents of the photon hypothesis (Einstein, Stark, Larmor, and Thomson), and he himself, as the chief critic, was still opposing it as late as 1913.

Not until 1914–1915 would Robert Millikan finally and conclusively determine that Einstein's photoelectric equation was in complete agreement with experiment. He had spent ten years of meticulous labor intent on showing that Einstein was totally wrong, and in the end he had established "the exact validity" of the theory.* By plotting one set of measured parameters against another, Millikan

On the view we have taken of a wave of light the wave itself must have a structure, and the front of the wave, instead of being, as it were, uniformly illuminated, will be represented by a series of bright specks on a dark ground.

J. J. THOMSON (1903)

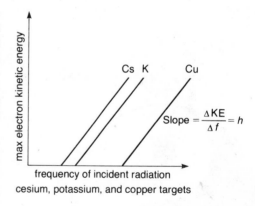

cesium, potassium, and copper targets

---

* Millikan still declined to accept the reality of the photon even while accepting the 1923 Nobel Prize "for his work on the elementary charge of the electron and on the photoelectric effect."

The more important fundamental laws and facts of physical science have all been discovered, and these are now so firmly established that the possibility of their ever being supplanted in consequence of new discoveries is exceedingly remote. Our future discoveries must be looked for in the sixth place of decimals.

ALBERT MICHELSON (1899)
American physicist

was able to determine the numerical value of Planck's constant, just as Einstein had predicted.* Amazing! What had first arisen as a mere mathematical contrivance in the analysis of the blackbody problem was now a fundamental constant linked to the corpuscular nature of radiant energy.

Light propagates from one place to another as if it were a wave. Several centuries of work (Section 20.1) had established *that* with pressing credibility, and yet now it became equally as clear that *light interacts with matter in the processes of absorption and emission as if it were a stream of particles.* This, then, is the **wave-particle duality,** the schizophrenia of light (and, as we will see, of matter as well). Radiant energy is neither wave nor particle. Those are concepts of the macroscopic world, the world of bricks and oranges and rolling ocean waves, where countless billions of atoms combine to make the imagery. A single atom no more resembles a shrunken billiard ball than light resembles a tiny sea. With both we are there at the invisibly minute and unimaginable, where models drawn from life seem destined to fail because nothing looks like the familiar.

Light manifests itself as particle *or* wave, and which we "see" depends on how we look. We certainly can ask how light behaves, but it makes little sense to worry much about what the truly invisible "looks" like. It looks like nothing else.

11.3   THE BOHR ATOM

The ancient Chinese "Diagram of the Supreme Ultimate" depicts the unity of the two cosmic opposites, *yin* and *yang*. Modern physics, too, has found a unity in the seeming opposites of motion and rest, of space and time, of energy and matter, of particle and wave. (No wonder Bohr chose this symbol as the centerpiece for his coat of arms.)

Naturally it was exciting for a new Ph.D. to come to work at one of the great centers of atomic research. Having done his thesis on the electron theory of metals, Bohr was doubly pleased at the prospect of collaborating with "J.J." himself, the discoverer of the electron. Niels Henrik David Bohr, a soccer hero of some considerable fame back in his native Denmark, was 26 when he arrived at Cambridge in the autumn of 1911.

As fate would have it, the guest speaker that winter at the annual Cavendish dinner was Ernest Rutherford, who, only a few months before, had published his theory of the nuclear atom. Bohr, sitting in back on a bench, was utterly captivated by this great, hulking, hearty genius. In Rutherford he had found a new prophet. Besides, he and "J.J." had not really hit it off well. A few weeks later Bohr took a trip to Manchester, where he was warmly welcomed by the great man. Having recently returned from the First Solvay Congress in Brussels, Rutherford was delighted to discuss the new physics with this eager young theoretician, and they sat and talked for hours. That international conference, which was held on the problem of the quantum, was the first of its kind, and it had a considerable influence on Rutherford, not to mention the fact that it provided

---

* Einstein received the 1921 Nobel Prize "for his services to Theoretical Physics, and especially for his discovery of the law of the photoelectric effect." He never did enjoy the prize money, having assigned it to his first wife at her request as part of their divorce settlement three years before he actually won it.

Some of the participants in the first Solvay Conference of 1911. Among the standees are Planck (on the far left), Einstein (second from right), and Rutherford (fourth from right). The young man next to him in the casual suit is Jeans. The only woman is of course Madame Curie. Thomson missed the picture altogether.

his introductory meeting with Planck and Einstein. The seeds of the quantum theory were finally blowing in the wind, sailing beyond the bounds of the German-speaking world, where it had been conceived, to Britain and France, where it had been neglected.

By the spring of 1912, Bohr had left Cambridge and was happily immersed in the whirlwind of atomic research that churned at Rutherford's lab as it did nowhere else in the world. "This young Dane," Rutherford once remarked, "is the most intelligent chap I've ever met." Bohr's modest, friendly nature had endeared him to the "Crocodile." But he would stay only four months in Manchester before returning to Copenhagen to marry and take up the duties of assistant professor at the university—four stimulating months during which he became convinced of the validity of the nuclear atom and began to evolve the preliminary notions of a theory that would ultimately spearhead the progress of the quantum revolution.

Images and Illusions

From the very outset everyone, including Rutherford, knew there was something inexplicably strange about the planetary model of the atom. The idea was simple enough: An electron held by electrical attraction orbits a proton, just as a planet under the influence of gravity revolves about the sun. But according to orthodox electromagnetic theory, an accelerating charged particle (and circular motion involves acceleration) must emit radiant energy. Charges whirling around in modern machines like the cyclotron most certainly do spew out electromagnetic radiation. It seems, then, that a circulating atomic electron must also continuously radiate. Losing energy, it should wind inward, finally crashing into the nucleus. And one can calculate that the death spiral would take only about a hundred-millionth of a second. Our very existence is therefore an embarrassment to such a theory, which ridiculously insists that all the atoms in the universe should have long ago collapsed!

Bohr began his creation by guessing—postulating—that, unlike a planet, which can revolve permanently at *any* distance from the sun, *atomic electrons have only certain stable, lasting orbits about the nucleus.* Why don't the electrons collapse? Here Bohr was refreshingly outrageous. He simply insisted that they don't, and that ended that! *Electrons while in any of these stable orbits do not radiate.*

The young Dane was not the only one groping toward an orbital model. J. W. Nicholson, an English astrophysicist, was perhaps his chief rival in the quest. Nicholson very astutely raised the question of there being a possible relationship between the structure of an atom and the kind of light it presumably emits. Bohr, believing that any such interdependence must be exceedingly complex and subtle, at first avoided the notion entirely. But when his former classmate H. M. Hansen returned to Copenhagen after a year and a half of studying spectroscopy at Göttingen, and the two began to discuss Bohr's work, the issue of the spectral colors immediately came up

Professor Niels Bohr standing in front of some equipment in the basement of the institute that bears his name on the occasion of its twenty-fifth anniversary (March 1946).

again. This time Hansen suggested that the spectra were not really so complex after all and showed his old friend **Balmer's formula.**

Any gas sealed in a tube and excited by an electrical discharge across it glows (like a modern-day neon sign) with a light that is characteristic of that gas. In particular, hydrogen gives out light that, when passed through a prism, splits apart into four vivid, narrow bands of pure color (red, blue-green, blue-violet, and violet) called **spectral lines.** Johann Jakob Balmer, a Swiss schoolteacher, had published a paper in 1885 (the year Bohr was born) that revealed a definite pattern among the measured frequencies of those lines. At first glance the frequencies $(4.57 \times 10^{14}, 6.17 \times 10^{14}, 6.91 \times 10^{14},$ and 7.31×10^{14} cycles per second) seem quite unrelated, but Balmer discovered that he could obtain each of them numerically from the single formula

$$f = 3.29 \times 10^{15} \left(\frac{1}{2^2} - \frac{1}{N^2} \right)$$

by setting N equal, in turn, to 3, 4, 5, and 6. He had arrived at this lovely little relationship by laboriously playing with the numbers in the tradition of Kepler. He had found the pattern, but without a theory he could not hope to "understand" it.

"As soon as I saw Balmer's formula," Bohr recalled, "the whole thing was immediately clear to me." Beginning with the simplest configuration, he assumed that the hydrogen atom consisted of a single electron revolving about a proton. Although it's true that there is one proton and one electron, not everyone at the time accepted even that much. He then maintained that *the electron ordinarily resides in the smallest allowable orbit, or* **ground state.** When an

The excitation of an atom and its subsequent emission of a photon, as viewed via the Bohr theory.

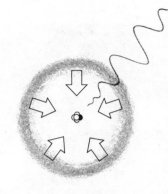
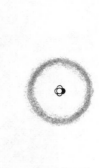

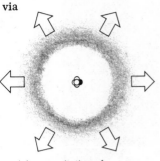

(a) excitation of
the ground state

(b) de-excitation with
emission of a photon

(c) ground state
10^{-9} seconds later

The transitions between atomic states determine the energy and frequency of the emission.

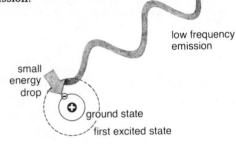

low frequency emission

small energy drop

ground state

first excited state

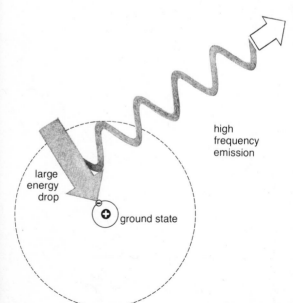

high frequency emission

large energy drop

ground state

atom is appropriately stimulated (perhaps thermally, electrically, or even by light), the electron is excited into a higher energy orbit that is more distant from the nucleus. There it resides, ordinarily for about a thousand-millionth (10^{-9}) of a second or so, before spontaneously descending to some inner orbit, ultimately dropping all the way back to home base, the ground state. During each drop (a moment when classical physics, with eyes closed, goes impotent) the electron emits its excess energy as a burst of electromagnetic radiation—a photon. With this jumping dance multiplied a million-millionfold throughout the gas, we see the familiar overall glow.

The picture is a little like a stepped circular sports stadium or Greek amphitheater, in which a ball revolving about the lowest tier is boosted up to some higher ring, where it orbits for a while, only to drop back to ground level in either one large or many small plunges. The sequence of Balmer's frequencies became a ladder of orbits, of **energy levels.** When the electron in an excited atom drops from an initial (E_i) to a final (E_f) energy level, it emits the difference as a quantum hf; that is,

$$E_i - E_f = hf.$$

The pieces all fall together. Using a simplistic mixture of Newtonian mechanics, Coulomb's law and the Planck-Einstein quantum, Bohr was able to derive expressions for the electron's allowed orbital radii, their speeds, and even the possible frequencies of the emitted radiation. Amazingly, his equation for f looked identical to Balmer's, except that where Balmer had the number 3.29×10^{15}, Bohr had the quantity

$$\frac{2\pi^2 m_e q_e^4\, K^2}{h^3}.$$

Running through this mess term by term: You know what 2 is; π^2 is pi (3.1416) squared; m_e and q_e are the mass and charge of the electron, and the latter is raised to the fourth power; K is the constant in Coulomb's law (which corresponds to G in the gravity law);

and h^3 is, of course, Planck's constant cubed. And now, wonder of wonders, if you substitute all of those numbers into that relation, it actually equals 3.29×10^{15}!

Bohr had actually derived Balmer's equation. He had quantified, mathematized, the atom, just as Newton two centuries before had given law to the planets. But Bohr knew too well that the whirling orbits were creatures of his own imagination. "When it comes to atoms," he remarked, "language can be used only as in poetry. The poet, too, is not nearly so concerned with describing facts as with creating images."

Sweet Success

Several striking confirmations of the new theory were immediately at hand. The calculated ground-state orbital radius of hydrogen (0.53×10^{-10} m) fitted nicely with everything known about atomic dimensions. Moreover, the model suggested that for heavier atoms the increased nuclear charge would draw more strongly on the surrounding electrons, shrinking their orbits. Atomic diameters are therefore understandably all close to the same size, about 10^{-10} m. Uranium, which is 238 times more massive than hydrogen, has an atomic dimension only about three times as great.

In addition to the Balmer series, two other invisible spectral-line series were known, one in the infrared and one in the ultraviolet. However, the latter was not fully established until 1914. When applied to those series, the model for hydrogen again brilliantly passed the test. New lines, predicted but never before seen, were actually observed.

Bohr conjectured that as the nuclear charge increased, the number of orbital electrons would also increase. One by one they would build all across the periodic table. Unhappily, however, the heavier atoms were beyond treating analytically with his simple techniques. Still, Bohr proposed that if an inner electron of one such atom was somehow removed from its orbit (e.g., by bombarding it with cathode rays), all the other whirling electrons would cascade downward until that vacancy was filled. He supposed that in the process of those long drops, characteristic X-rays would be emitted. Bohr discussed all this with his friend Henry Moseley at Manchester, and of course, it was Moseley in 1913 who brilliantly confirmed the imagery.

This was a high point for the new theory, the culmination of a stunning series of early successes. But the work had almost reached its limit. It surely had some semblance of truth, but it surely was not *the* truth. Although the theory would soon run out of power as a probing device, it had at least revealed the outlines of what was there, and that much was marvelous.

In a moment it was August 1914, and the world, with plumes and colors, prancing horses and confidence, marched off to a war that would end in a blind horror of poison gas and mud-filled trenches and death. The great laboratories emptied; their young men, smiling, tramped away to the battlefields; science slowed, paused, and turned

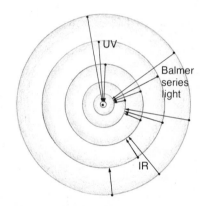

The Bohr hydrogen atom, showing various transitions and the corresponding emissions that result when the atom drops from any one of the first five excited states.

helium

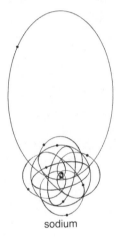

neon

Bohr's representations of several atoms.

sodium

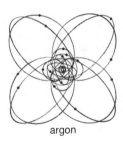

argon

to the practical. By the time the war ended in 1918, the first developmental stage, the *old quantum theory*, had just about run its course. Arnold Sommerfeld had already (1916) generalized the original model, making the orbits elliptical in an effort to broaden the scope of the theory. So appealing were those interweaving ellipses that the picture persists even now. Just ask any kid to draw an atom, and you'll get the elliptic illusion.

Valence

One of the last lovely accomplishments of the Bohr conception was an explanation of the chemical activity, or **valence,** of the elements. It evolved just after the war, principally at the hands of the American chemist Irving Langmuir. Taking a clue from the inert gases, Langmuir suggested that the electrons associated with each different kind of atom reside in a series of concentric shells about the nucleus, and only in the rare cases where the shells are totally occupied will an atom be content *not* to combine with other atoms.

Hydrogen has one electron; helium, an inert gas and therefore chemically inactive, has two electrons, and so Langmuir concluded that its configuration corresponded to a filled first shell. The first two electrons of any atom will occupy its innermost shell, and the remainder will spill over into successively larger shells. Lithium, the lightest of the alkali metals, has three electrons, two in the first closed shell and the third rather far off in a second shell. By the time stable neon is constructed, its ten electrons are assumed to complete the first two shells, with two and eight electrons, respectively. And so on down the periodic table, shells within shells, until a remarkable pattern emerges. With few exceptions, *all the atoms in any one column of the table have the same number of outer electrons!* The very active alkali metals all have only one such electron; the less active alkaline earths, headed by beryllium, have two; the boron, carbon, nitrogen, and oxygen families have three, four, five, and six, respectively; and the highly reactive halogens each contain an almost complete last shell of seven electrons.

The pattern alone was beautiful, but Langmuir went further, proposing that atoms with only a few occupants weakly bound in an outer shell tend to slough them off, to lose them, whereas atoms needing only a few electrons to fill a shell are eager grabbers. Thus sodium, whose eleventh electron sits alone in an outer shell, is an active lender, but chlorine, missing only one electron to fill its last shell, is an energetic borrower.

The electronic configurations of the first ten atoms in the periodic table.

H   He   Li   Be   B

Sodium, an alkali metal, will combine with the yellow poisonous halogen gas chlorine to form—of all things—table salt, NaCl. The metal *transfers* its extra electron to the nonmetal, and both, no longer neutral, become ions. The sodium, with 11 protons and 10 electrons, is on the whole positive, whereas the chlorine ion, now with 17 protons and 18 electrons, is negative. The two oppositely charged ions attract each other electrically and so bond into the salt molecule.

It's also possible for atoms to *share* electrons without actually giving them up. Each of the two hydrogen atoms in an H_2 molecule shares its electron with the other so that each effectively has two and therefore a filled shell. In exactly the same way, two chlorine atoms share a pair of electrons to fill their outermost shells and form the molecular gas Cl_2, and two hydrogens share their single electrons with an oxygen atom to form water (H_2O).

The pace of things intellectual picked up after the First World War—caution seemed less fitting in a battered world. Novelists like Sinclair Lewis, Ernest Hemingway, and John Dos Passos wrote of the generation's pained disillusion with traditional values. Gertrude Stein and James Joyce, among others, experimented with form, putting aside the bonds of old rules. Duchamp and his Dada artist comrades ridiculed the folly of the times with their own outrageous creations, which went beyond all bounds of orthodoxy, beyond all the canons of "good taste." The "gay decade," the "jazz era" was largely a period of revolt and irreverence, and that selfsame turbulent mood gave form to the new physics, the **quantum mechanics,** which rose almost full-blown out of the "roaring twenties."

It was in the summer of 1923 that the French aristocrat Prince Louis Victor de Broglie* proposed that the wave-particle duality manifested by light might be a fundamental characteristic of *all* entities (electrons, protons, oranges, fire engines, neutrons, etc.). Einstein had already shown (Section 15.1) that matter and energy were different aspects of the same thing. Why should they not display similar properties? In particular, might not matter have some sort of wave aspect associated with it?

If you jiggle a taut string smoothly and regularly, up and down, it will distort into a series of graceful curves that sail along the string. This is a familiar sort of thing and, as with all waves, its *peak-to-peak distance* is called the **wavelength** (usually symbolized by the

* Pronounced to rhyme with Troy.

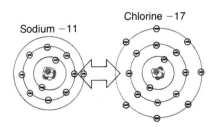

The interaction of sodium and chlorine, which binds them together to form table salt.

11.4 MATTER WAVES

Sharing of electrons in the molecules of hydrogen and chlorine.

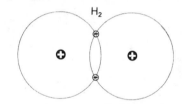

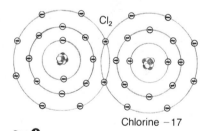

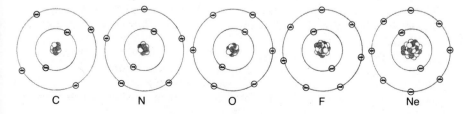

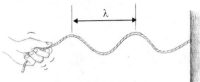

A wave on a string.

When the great innovation appears, it will almost certainly be in a muddled, incomplete and confusing form. To the discoverer himself, it will be only half understood; to everybody else, it will be a mystery. For any speculation which does not at first glance look crazy, there is no hope.

FREEMAN DYSON

Professor of physics

Greek letter lambda, λ). Well, de Broglie, beginning with the photon and reasoning from the notions of relativity, suggested that *material particles in motion have a wavelength,* too, given by the relationship

$$\text{wavelength} = \frac{\text{Planck's constant}}{\text{momentum}}.$$

Not that he knew how to visualize such waves or even what was doing the waving (if anything). Still, the idea that electrons, protons, neutrons—i.e., all matter—behaved like photons in that they displayed both particle and wave properties, was appealing to those who joy in the neatness of Nature, in its even-handedness.

In the summer of 1923, the American C. J. Davisson was scattering electrons off a polycrystalline nickel target at the Western Electric (A.T.&T.) lab in New York City, a project he had begun in 1919. Davisson would become world-renowned for showing that electrons have a de Broglie wavelength, even though that great discovery was serendipitous—he certainly did not set out to do the deed on purpose.

In April of 1925 (while he was collaborating with L. Germer), an accidental explosion rocked the lab, and by the time the experiment was back together and working, something very strange had happened; the data had "completely changed." Unbeknown to them, while they were cleaning up the target, the nickel, through prolonged heating, had recrystallized, changing from a structure containing many minute grains into one with just a few large crystals. The regular array of atoms in a single crystal will diffract the electron beam into a very precise pattern, exactly as it diffracts electromagnetic waves, X-rays. If electrons were purely corpuscular, as most people believed, they would surely not fly off in the complex way that was theoretically appropriate only for waves. Another year had to pass before anyone finally recognized that Davisson and Germer had actually verified the wave nature of matter, that de Broglie's equation actually worked in every detail!

You see, an electron of mass $m_e = 9.1 \times 10^{-31}$ kg moving at a speed of about $v = 4.3 \times 10^6$ m/s has a momentum that, you recall, equals the product of these, namely, $m_e v = 3.9 \times 10^{-24}$. Therefore, according to de Broglie, it has a wavelength of

$$\lambda = \frac{h}{m_e v} = \frac{6.63 \times 10^{-34}}{3.9 \times 10^{-24}} = 1.7 \times 10^{-10} \text{ m}.$$

But that is right within the usual wavelength range of X-rays, which is just about the size of the atoms themselves. Thus we can expect both electrons and X-rays to be scattered in the same special way— a way that was already well known for the latter but until then quite unthinkable for the former.

Davisson shared the 1937 Nobel Prize with G. P. Thomson, J. J.'s son, and therein lies yet another irony. The younger Thomson independently and deliberately set out to test the concept of matter-waves. By studying the scatter pattern that results when an electron

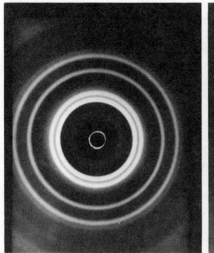

The almost identical concentric ring patterns produced when a beam of X-rays (left) and a beam of electrons (right) passes through the same thin aluminum foil.

beam is fired through a thin metal foil, he was able to confirm their wavelike behavior. And so, while J. J. Thomson "proved" that electrons were *particles*, G. P. Thomson "proved" that they were *also waves*.

De Broglie's wave-particle theory offered yet another picturesque way to visualize the electron orbits of the Bohr atom. When the wavelength of the electron in the ground state of a hydrogen atom is calculated, λ turns out to exactly equal the circumference of the innermost orbit. Indeed, *all the allowed orbits correspond to situations where the circumference equals a whole number of electron wavelengths;* the waves exactly fit, as it were. The picture is like that of a metal hoop that, when slammed, vibrates in one of any number of ways but always with an integral number of waves around the ring.

One of the pillars of the nineteenth-century wave theory of light, among the most convincing of all its convincing arguments, was the analysis of diffraction. As we will see in Chapter 21, when light propagates beyond an obstruction or through a hole, it casts an intricate shadow pattern that can be analyzed in all its detail only if light is assumed to be a wave. What is indeed remarkable is that researchers using all sorts of beams of matter, ranging from electrons and neutrons to potassium atoms, have now observed the same complex patterns. In each case the wave theory of the nineteenth century works beautifully, provided the particles can be thought of as having the appropriate de Broglie wavelengths. From each particle's mass and speed in a given situation, λ is calculated, and with that every dot and blotch in the pattern is predictable. We are clearly no longer dealing with "particles" in the sense in which that word was understood in classical physics. The shrunken billiard-ball specks, well defined and localized in space, have now somehow spread out into wavy vibratory wisps.

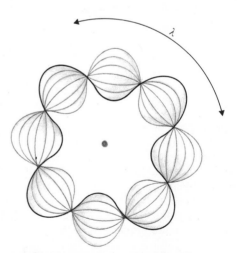

The hydrogen atom in the de Broglie theory. In this particular example four electron wavelengths exactly fit into an allowed orbit.

11.5   WAVE MECHANICS

Prince Louis de Broglie

The rabbi spoke three times; the first talk was brilliant: clear and simple. I understood every word. The second was even better: deep and subtle. I didn't understand much, but the rabbi understood all of it. The third talk was a great and unforgettable experience. I understood nothing and the rabbi himself didn't understand much either.

A story told by Bohr about a young man reporting on his visit to a great rabbi as related by

VICTOR WEISSKOPF

When he read de Broglie's memoirs in 1924, Einstein immediately recognized the merit of the work and incorporated some of its ideas into his own work. That affirmation of the as-yet unproved concept of matter-waves by the great Einstein had a profound effect on a young Austrian physicist who was to take the next giant step, Erwin Schrödinger, then of the University of Zurich. This was the turning point (1925–1926), the pivotal period that marked the arrival of the modern conception of *quantum* or *wave mechanics*.

Schrödinger generalized de Broglie's limited speculations into an all-encompassing mathematical formulation. He took off from and went beyond those notions to create a kind of master equation, summarizing the criteria to be met by all wave aspects of the microworld. Precedent for this sort of approach already existed in the form of the classical differential wave equation, which sets the mathematical requirements for all the various types of macroscopic waves, from those on a string to ripples in a glass of beer. Even more suggestive, though, was the superlative work of an Irish genius of the mid-1800s, William Hamilton. He had developed a very subtle and esoteric mathematical synthesis of Newtonian mechanics and the theory of light rays that *almost* anticipated quantum mechanics. But without the prodding of experiment, that great achievement remained a curious intellectual cameo of little practical value in a world that had not yet accepted the atom.

Schrödinger, guided by the inventions of Hamilton and de Broglie, conjured up his own form of differential equation. The **Schrödinger wave equation** cannot be derived step-by-step from previous classical ideas; it is an imaginative creation, a postulate, that must in the end draw its justification from its own ability to predict. But this sort of creative leap should not surprise—theory has always been *make-believe that works*.

So ripe was the moment for invention that a few months before the appearance of Schrödinger's first paper, a young German physicist named Heisenberg had reported an entirely different and powerful formulation. Werner Heisenberg, at the University of Göttingen, had devised a curious scheme that put aside all the nonobservable fictions, the models and pictures of orbits and jumping electrons, and dealt only with measurables—specifically with the experimental data on spectra. At first the two new competing theories vied for the crown, but Schrödinger soon showed that although they seemed quite different, they were in fact mathematically equivalent.

Quantum mechanics was immediately and eminently successful at treating the hydrogen atom, as well as a whole range of far more complex atomic systems. The Schrödinger equation for hydrogen has solutions only for specific values of energy, which turn out to exactly equal those of the Bohr orbits. But the new theory also describes the intensities of the spectral lines, something its planetary predecessor could not touch. Now there was no longer any reason to envision orbiting electrons whirling about the nucleus; it seemed that the atomic electron was instead smeared out somehow into a wave—whatever that meant.

Schrödinger's equation is an equation of motion of de Broglie waves. It is a mathematical statement written in terms of space, time, mass, energy, and Planck's constant. The solution of the equation is a function designated as Ψ (Greek letter *psi*) and referred to as the **wave function.** New theories are rarely, if ever, complete at the moment of formulation, and Schrödinger's was no exception. Although it was remarkably effective, several aspects were quite troubling from the outset. In particular, what was the physical significance of the Ψ wave functions? In the early days Schrödinger assumed that the negative electrical charge of the atom was actually spread out in space around the nucleus and that Ψ was related to the density of that charge-cloud. But this description soon proved untenable. Curiously, although Schrödinger's theory became an immense success, this spread-out-charge interpretation of essence-of-Ψ was a disappointing failure.

In the spring of 1926, not long after the publication of Schrödinger's papers, Max Born, then director of the Theoretical Institute at Göttingen, proposed what has now come to be the orthodox interpretation of Ψ. Taking a clue from Einstein, Born suggested that *the electron was truly a particle and that the Ψ wave associated with it was simply a probability function,* an information wave that would tell us *not* where the electron is but only where it is most likely to be. Thus, if we calculate the likelihood of finding an electron somewhere beyond the nucleus of the hydrogen atom, we get a whole range of potential locations, a cloud of possibilities. The densest part of the cloud, the most likely place to find the electron, miraculously, is out at a distance equal to the Bohr orbit. There is no longer any reason to think of atomic electrons flying around in little planetary orbits, though what they are doing from moment to moment is a mystery.

When dealing with a system of particles (e.g., a group of atoms or a beam of electrons), the theory cannot tell us exactly what will happen next to each individual participant. For any one particle it can at most provide a picture of the likelihood of occurrence of all the various possibilities. But it can provide superbly precise and reliable information about how the system as a whole will perform. Large numbers of particles respond as a group whose overall behavior can be predicted statistically quite accurately. Even if the individuals are lost in the crowd, the crowd *en masse* does its thing in a well-defined fashion that can be measured, understood, and anticipated. If we shine a beam of electrons at some obstruction, we can calculate rather precisely the resulting shadow pattern that will appear on a distant screen, despite the fact that we cannot predict the behavior of any one of the individual electrons separately. Similarly, an actuarial table can certainly tell you the probability of dying for any person in your age group. That much can be computed

$$-\frac{\hbar^2}{2m}\nabla^2\Psi + V\Psi = i\hbar\frac{\partial\Psi}{\partial t}$$

A newspaper clipping of Erwin Schrödinger.

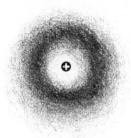

The distribution of possible locations of the electron in the ground state of the hydrogen atom. This is what's called the "electron cloud," even though there's really only one electron.

The quantum mechanics is very imposing. But an inner voice tells me that it is still not the final truth. The theory yields much, but it hardly brings us nearer to the secret of the Old One. In any case, I am convinced that He does not play dice.

ALBERT EINSTEIN (1926)

reliably. But it obviously cannot reveal how long you, a particular individual, will live.

This probabilistic interpretation of quantum mechanics is not the only possible one, but it is now the most widely held interpretation. Surely not everyone is happy, even today, with the drastic change in philosophy demanded by a theory that deals in *likelihoods* instead of *certainties,* and insists that Nature will simply not yield more. Still, quantum mechanics rules the theoretical roost; after half a century of extensive testing, it prevails. As a generalization of classical theory, it subsumes that formalism, and therefore, wherever Planck's constant can be taken as negligibly small, as in the human-sized world of every day, quantum mechanics reverts to Newtonian theory.

There was a time when scientists like famed Laplace idly fancied that if they knew the locations and speeds of all the atoms in the universe, they could, theoretically at least, foretell the future and retell the past—despite free will and passion—as if your next warm kiss had been set in motion some 20 billion years ago, when all the atoms were flung out to do their predetermined mischief. That is called *determinism,* and however beyond our actual computational ability, the logic still disturbs, particularly if you are not a believer in fate. Though Newtonian theory weaves a deterministic plot where every sigh results from the inexorable confluence of physical law, quantum mechanics happily puts a bit of butter, a touch of chance, into the Hatter's clockwork. The future of even one lone atom is beyond precise prediction, so kiss whomever you like, change your mind however you will, and rest assured that science, at least, knows *it* cannot hope to calculate your future from the bustling atomic mayhem.

Uncertainty

Born's statistical interpretation suggests that we can compute only the likelihood of events. Given a beam of protons or electrons or anything else, we can determine the probability that one of the particles will strike a point on a target, but we cannot know that *that particular proton will hit exactly there.* Now, this blurriness of vision, which seems inherent in the formulation, is quite peculiar. After all, if we know precisely where a proton is to begin with, as well as how fast and in what direction it's moving, why can't we follow it *exactly?* Presumably we could, but perhaps that's not the right question; perhaps we should be even more open-minded and ask whether or not we really can measure both the position and velocity precisely. These were the sorts of questions Heisenberg must have been asking himself in 1927 before he came upon the crucial insight known as the **uncertainty principle,** which gives some reason to the indeterminacy of Atomland.

Classical physics has always assumed that *there is the observer and the observed,* that in the process of measuring, one can poke

Despite its enormous practical success, quantum theory is so contrary to intuition that, even after 45 years, the experts themselves still do not agree what to make of it.

BRYCE DEWITT

Physics Today (September 1970)

around, albeit carefully, and not substantially alter what is being measured. You can stick a thermometer in a pail of water without worrying that the thermometer itself will noticeably change the water's temperature. But you cannot be quite so cavalier if you want to measure the temperature of a tiny droplet. Obviously it is desirable that the measuring probe have an inconsequential effect on the system being measured, but is that *always* possible?

In the final analysis we determine the position and/or velocity of any object by simply bouncing something else off it: radar waves off a car; sound (sonar) off a whale; light off your nose; or the end of a cane off a curb. A stream of sunshine photons bouncing off a flying baseball (which enables you to see it) has no perceptible effect on the ball's motion, but one photon slamming into a minute electron can alter the electron's motion drastically and quite unpredictably.

And there is the heart of the matter. To "see" something in Atomland, you will still have to use a probe, and yet the finest, most delicate probes possible obtrude on the measurements; they change what is going on! For example, suppose we wish to determine the position and velocity of a single electron. A photon probe with a low energy, chosen to be as gentle as possible, will have a correspondingly long wavelength. Although it will only slightly disrupt the target electron's motion, the long wave will bend around the electron, yielding only a very coarse determination of the particle's position. The way to sharpen the position measurement is to make the probe's wavelength comparable in size to the electron. But that means using a high-energy photon, which will locate the electron well enough but will subsequently blast it away. By detecting the recoiling photon, we can determine fairly well where the electron was before it was hit, but we then have little idea of its new velocity. In fact, the more precisely we measure the electron's location, the less precisely we know its final velocity, and vice versa.

The **Heisenberg uncertainty principle,** which is a postulate gleaned from wave mechanics, quantifies this "giveth and taketh away" relationship. It states that *in a simultaneous measurement the imprecision or uncertainty in the position, multiplied by the uncertainty in momentum, is at best roughly equal to Planck's constant.* Since mass is often well known, we can restate the principle in a slightly more restricted but useful way: The uncertainty in a particle's position, Δx, times the simultaneous uncertainty in its speed, Δv, is *at best* roughly h divided by its mass, m;

$$\Delta x \ \Delta v \simeq \frac{h}{m}.$$

Planck's constant is of course an exceedingly small number (6.6×10^{-34}), so when m is large, as it is with a garbage truck or even a Ping-Pong ball, the right side of the equation is minute. Under such circumstances, both uncertainties will be unobservably small, and the human-sized world has no need to concern itself with the details

An intelligence which at a given moment knew all the forces that animate nature, and the respective positions of the beings that compose it, could condense into a single formula the movement of the greatest bodies of the Universe and that of the least atom: for such an intelligence nothing could be uncertain, the past and future would be before its eyes.

PIERRE SIMON, MARQUIS DE LAPLACE (1749–1827)

I think I can safely say that nobody understands quantum mechanics.

RICHARD P. FEYNMAN

The Character of Physical Law (1967)

All Nature is but Art, unknown to thee;
All Chance, Direction, which thou canst not see;
All Discord, Harmony not understood.

ALEXANDER POPE (1688–1744)

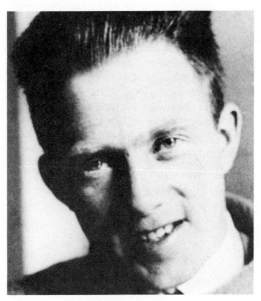

Werner Heisenberg.

of the Heisenberg principle. In effect, Planck's constant is nearly zero, and ordinary massive objects show no appreciable uncertainties; i.e., they behave in the usual deterministic way. The same is most definitely not so for atomic entities. The electron, for example, has a mass of only 9.1×10^{-31} kg, making the right side of the equation no longer negligible!

The special theory of relativity had already (1905) suggested that observer and observed were linked, that there was an unavoidable subjectivity to measurement. What one sees of time and space depends on the relative motion of the viewer and the viewed. Now quantum mechanics draws the observer and the observed even more closely together, especially in Atomland, where a single photon that couples the two wipes out the arbitrary subject-object distinction. The measurer's probe becomes an essential part of the system being measured. Ultimately we do not even know whether a particle "actually" has, simultaneously, a perfectly defined position and velocity. Perhaps it doesn't. In any case, we could not measure it if it did, not because we're momentarily without the proper method but because no such method is theoretically possible!

If we cannot know precisely an electron's present, we will have to rely on probabilities to tell its future.

EXPERIMENTS

1. Oscillatory patterns somewhat like the de Broglie waves for the hydrogen atom can be seen with an ordinary broad rubber band. All you need is something that will drive it into oscillation without getting in the way. Hold the band, somewhat stretched, out the window of a moving car. Keep it in a vertical plane perpendicular to the auto. At about 20 mi/hr the wind will set it humming. If the speed and tension are varied, the vibrational patterns will change.

2. Another amusing demonstration of oscillatory patterns resembling de Broglie waves in an atom uses vibrating globs of water. Heat a clean frying pan or griddle until it's quite hot. You've reached the right temperature when water drops bead up and dance around on a layer of vapor. Experiment with different-sized globules of water up to about a half teaspoon or so. Most will oscillate by themselves. The larger ones can be affected by blowing on them.

REVIEW

Section 11.1

1. Who was the father of quantum theory?
2. What is a blackbody?
3. In the process of developing his theory of blackbody radiation Planck first concocted a(n) _____ to fit the known data (just as Kepler had done with his elliptical orbits).
4. Define the term *frequency*.
5. Planck reached the conclusion that the _____ of his oscillators could have only certain values.
6. What is meant by the term *a quantum of energy*?
7. Was the notion of quantized energy an instant success?
8. Why is the section called "A Formal Assumption"?

Section 11.2

1. The photoelectric effect was discovered by _____.
2. List several different forms of radiant energy.
3. What is the *photoelectric effect*?
4. On what aspect of the incoming radiation were the speeds of the ejected photoelectrons dependent?
5. Are photoelectrons emitted immediately on exposure to light?
6. What was J. J. Thomson's notion of "hot spots" in the radiation?
7. Describe Einstein's view on the nature of light.
8. What is a *photon*?
9. In what respect can photons differ from one another?

Section 11.3

1. Classical theory predicts that the planetary atom will collapse in about _____ of a second.
2. What are *spectral lines*?
3. Define *ground state*.
4. How long does an atom typically remain in an excited state before emission of radiation?
5. What is an atomic *energy level*?
6. The ground-state orbital radius of hydrogen is _____ meter.
7. In terms of orbital electrons, how does sodium combine with chlorine to form table salt?
8. How does the single electron of hydrogen manage to emit so many different colors of light (i.e., spectral lines)?

Section 11.4

1. Explain the meaning of the term *wavelength*.
2. What was de Broglie's contribution to quantum mechanics?
3. Does a particle at rest have a de Broglie wavelength?
4. What was the conclusion that came out of the Davisson-Germer experiment?

5. What was ironic about G. P. Thomson's getting the Nobel Prize for his work on the wave aspect of matter?
6. Does a cruising garbage truck have a wavelength?

Section 11.5

1. Is the Schrödinger wave equation a postulate?
2. The allowed values of energy for the hydrogen atom in the Schrödinger theory exactly match those of _____ theory.
3. Is it correct to say that wave mechanics put aside the imagery of orbiting electrons?
4. What is a *wave function* in quantum mechanics?
5. Describe Born's interpretation of the Ψ-wave.
6. Under what circumstances does quantum mechanics revert to classical physics?
7. How do we generally measure the position and/or velocity of an object?
8. State the Heisenberg uncertainty principle.
9. Is it correct to say that the more massive a single object is, the more accurately we can predict its motion into the future?

QUESTIONS

Answers to starred questions are given at the end of the book.

Section 11.1

1. Which would you expect to be hotter, a bluish star like Rigel or a reddish one like Betelgeuse? (Both are in the Orion constellation.)
* 2. Discuss the major insight that came out of Planck's work.
3. Describe what is meant by a family of blackbody radiation curves. What role does temperature play in the size and shape of the curves?
4. What is the frequency of the Earth's spin? How does frequency enter into Planck's work?
5. Why did I refer to Planck as the "reluctant" father of the new imagery?

Section 11.2

1. What was Einstein's role in the development of the quantum concept?
2. What besides radiant energy is quantized?
* 3. Why are ultraviolet photons capable of causing a large range of chemical effects (from sunburn to the decomposition of wine) whereas light photons are far less so?
4. Discuss the part played by Millikan in establishing the photon as a reality. What were his own expectations regarding Einstein's work?

5. Discuss the accuracy of the Michelson remark quoted on p. 260.
6. Explain what is meant by the *wave-particle* duality. Discuss some differences between a wave and a particle.
* 7. The photoelectric effect is used in some electric eye systems and in gadgets like the one that reads sound tracks (in the form of light-intensity variations) on movie film. Describe the function served by the effect in such devices.
8. Describe the photoelectric effect in terms of conservation of energy.
9. Discuss the meaning of the statement "The photon is the quantum of the electromagnetic field." What would you call the graviton?

Section 11.3

1. Discuss the obvious problem that existed regarding the stability of the planetary model of the atom. What was Bohr's solution?
* 2. Describe the Bohr picture of atomic excitation and emission of radiation. The crucial point is that the orbits are quantized; only certain orbits can exist.
3. Explain why all atoms, regardless of mass, are very roughly the same size.
4. What holds the Bohr atom together? Why doesn't the electron fly off?

5. Describe the electron-shell model of the atom. How did the inert gases serve as a guide in developing the scheme?

* 6. Draw a pictoral representation of the shell structure of sodium fluoride, NaF, and explain how the molecule is held together. Do the same for hydrogen chloride, HCl.

7. Earlier (p. 169) we talked about chemical PE as being fundamentally electrical in nature. Expound on this point in light of Bohr's work.

8. Would you expect a singly charged negative chlorine ion, Cl^-, to be more or less active chemically than a chlorine atom? Explain.

Section 11.4

1. Discuss the wave-particle duality as it applies from photons to protons.

2. The work of Davisson and Germer was of tremendous importance in establishing the wave aspects of matter. Describe their experiment.

3. Expound on the differences between and similarities of a photon and an electron.

* 4. Draw a picture of the de Broglie electron wave pattern corresponding to the third Bohr orbit of hydrogen, and discuss what all of that means.

Section 11.5

1. Why do you think I put Bohr's story about the rabbi together with the discussion of Schrödinger's theory?

2. Describe the nature of the information quantum mechanics provides regarding the behavior of individual particles and systems of many particles.

3. What was Einstein alluding to when he said "I am convinced that He does not play dice."?

* 4. Discuss the relationship of the observer to the observed in both classical and quantum physics.

5. Use the uncertainty principle to explain how Planck's constant determines whether a system will appear to behave classically or quantum-mechanically.

6. Our most powerful devices for peering into matter, like the electron microscope, still come up with blurry images of individual atoms. Why?

* 7. Long before the advent of quantum theory, physicists knew that there were always experimental errors, that any measurement (from the location of a star to the mass of a shoe) was indeterminate to within some small experimental error. Moreover, they believed that better equipment and technique could *always* decrease the size of that error. How has this view been changed in this century?

8. Can you think of a measurement you can perform on a system that will not alter the system *in any way*? Remember that just turning on a light so you can see a thing will affect it. Touching it with fingers or instruments will usually transfer atoms and/or electrons to it, as well as a certain amount of energy via work.

* 9. Now that we have come all this way from the Rutherford atom, is there anything peculiar about the fact that the fundamental scattering calculations were performed in 1911? Though the results were correct, could the analysis that led to them have been correct?

MATHEMATICAL PROBLEMS

Answers to starred problems are given at the end of the book.

Section 11.2

1. Yellow light has a frequency of about 5×10^{14} Hz. (Remember, 1 Hz = 1 cycle per second.) Calculate the energy of a yellow photon. Compare this with the energy of an X-ray photon of frequency 1.2×10^{19} Hz.

* 2. The red light emitted by hydrogen gas consists of photons, each with an energy of 3×10^{-19} joules. Determine the frequency of the light.

* 3. Einstein's treatment of the photoelectric effect maintains that when a photon hits an electron on the surface of a metal, some of its energy goes into ripping the charge out of the metal, and the rest is transformed into the electron's KE. The measured amount of energy needed to free an electron from the surface of a piece of copper is 7.3×10^{-19} joules. Calculate the KE of a photoelectron liberated at the surface of copper by an ultraviolet photon with a frequency of 1.5×10^{15} Hz.

Section 11.3

1. The difference in energy between the ground state and the first excited state of a hydrogen atom is 1.6×10^{-18} joules. Calculate the frequency of the photon emitted when the hydrogen atom drops from the first excited state to its ground state.

* 2. Keeping in mind the previous problem, how much energy must an incoming photon have if it is to be absorbed by a hydrogen atom, kicking the atom up from its ground state to its first excited state?

3. The average kinetic energy of a molecule of hydrogen gas at room temperature is about 6×10^{-21} joules. Explain why the gas does not glow, giving off light at that temperature.

* 4. It takes about 2.2×10^{-18} joules of energy to rip a ground-state electron out of a hydrogen atom, thereby ionizing it. What does this correspond to in the Bohr model? Considering the previous problem, is it likely that colliding hydrogen atoms at room temperature will become ionized?

Section 11.4

* 1. Calculate the de Broglie wavelength of a 2000-kg (4400-lb) car traveling at 20 m/s (45 mi/hr). Compare that to the size of an atom, which is around 10^{-10} m.

2. The orbital radius of the ground state of the hydrogen atom is 5.3×10^{-11} m, and the electron's speed is 2.2×10^6 m/s. Calculate the de Broglie wavelength of that electron, and compare it to the circumference of the orbit ($m_e = 9.1 \times 10^{-31}$ kg).

3. A baseball with a mass of roughly 0.2 kg is hurled at a speed of 27 m/s. Calculate its de Broglie wavelength. How does this compare with the size of an atomic nucleus, which is roughly 10^{-14} m?

Section 11.5

* 1. Examine a typical electron sailing down the beam in a TV picture tube. It may have an energy of about 2×10^{-15} J. Now confine it to a region the size of an atom, such that it has a position uncertainty of 10^{-10} m, and calculate the corresponding uncertainty in its speed, Δv. Determine the fractional uncertainty, $\Delta v/v$.

2. Imagine that you have a 5-gram object traveling at 10 centimeters per second—something quite ordinary. Compute Δv and $\Delta v/v$, assuming that the position of the object is known to within a millionth of a meter. Compare the results with that of the previous problem. Note how the uncertainties in the macroscopic domain are trivially small.

Œuvres de Lavoisier _ Tom. III _ Pl. IX.

A Grande Lentille à liqueur.
B Petite Lentille pour rassembler les raïons plus près.
C Centre de mouvement horisontal de toute la Machine.
D Manivelle servant à imprimer le mouvement horisontal.
E Manivelle servant à imprimer le mouvement vertical par le moïen des Vis 1 et 2.
F Vis de rappel pour éloigner de la grande Loupe la petite Lentille ou la rapprocher.
G Porte objet aïant le mouvement de haut en bas et de bas en haut celui d'avancer et reculer parallellement à la plate-forme et de s'incliner au degré du Soleil et de s'avancer parallellement aux raïons.
H Chariot ou Plate-forme portant toute la Machine et les Opérateurs.
I Roues du Chariot tendantes au Centre de mouvement par leurs Axes et roulantes sur des bandes de fer incrustées circulairement sur une plate-forme de pierre.
K Escalier pour parvenir sur le Chariot il est soutenu de deux rouleaux excentriques.

DESSEIN en Perspective d'une Grande Loupe formée par 2 Glaces de 52 po. de diam. chacune coulées à la Manufacture Royale de St Gobin, courbées et travaillées sur une portion de Sphère de 16 pieds de diam. par Mr de Berniere, Controlleur des Ponts et Chaussées, et ensuite opposées l'une à l'autre par la concavité. L'espace lenticulaire qu'elles laissent entr'elles a été rempli d'esprit de vin il a quatre pieds de diam. et plus de 6 pouc. d'épaisseur au centre. Cette Loupe a été construite d'après le désir de L'ACADÉMIE Roiale des Sciences, aux frais et par les soins de Monsieur DE TRUDAINE, Honoraire de cette Académie, sous les yeux de Messieurs de Montigny, Macquer, Brisson, Cadet et Lavoisier, nommés Commissaires par L'Académie. La Monture a été construite d'apres les idées de Mr de Berniere, perfectionnée et exécutée par Mr Charpentier, Mécanicien au Vieux Louvre.

A Monsieur De Trudaine
Par son très humble et très obéissant Serviteur, Charpentier.

The great burning glass constructed by Lavoisier for the Royal Academy of Sciences.

The next several chapters treat the four major **states of matter,** that is, the four principal ways large numbers of atoms group together, acting more or less in concert. This chapter is concerned primarily with the tightly bound configurations commonly known as **solids.** When the atoms of a solid are arrayed row upon row in a regular, systematic, recurring pattern with a precise, long-range order, the substance is said to be **crystalline;** when no such order exists, a solid is designated as **amorphous.** These two classifications form the subject matter of the first two sections of this chapter. The final segment describes the four different types of interatomic "glue," the attractive forces that bind a solid together and determine many of its physical properties.

Crystal building blocks

type	geometry	example
Cubic	all 90° angles all edges equal	gold, lead, silver, diamond, salt, pyrite, carnet, alum, iron
Tetragonal	all 90° angles 2 of 3 edges at a corner are equal	zircon, tin, rutile, urea, indium, chlorine
Orthorhombic	all 90° angles no 2 edges of a corner are equal	sulphur, iodine, codeine, morphine, dextrose, topaz, epsom salt
Monoclinic	no 2 edges at a corner are equal 8 angles not 90°	cane sugar, gypsum, mica, borak, cocaine, saccharin, baking soda, meerschaum, talc
Triclinic	no 2 edges at a corner are equal no 90° angles	boric acid, copper sulphate, feldspar
Trigonal	all edges equal no 90° angles	ruby, bismuth, sapphire, hematite, quartz, tourmaline
Hexagonal	hexagon ends 90° sides	zinc, nickel, magnesium, calcite, mercury, hydrogen, osmium, ice, graphite

The stuff, the matter that constitutes our universe is composed of roughly 100 different kinds of atoms, all more or less in motion—an incredibly huge number of tiny, complex forms in an endless, unseen dance. So vast are their numbers that if we took the molecules in about one tablespoon of water (602,000,000,000,000,000,000,000 of them) and put them end to end, the column would reach from here to the sun and back 500 times!

The numberless atoms, molecules, and ions of the cosmos are not spread out in a uniform cloud; they are somehow clumped as stars and oceans and flower pots and cool breezes. Hung together under the influence of electrical forces, these individual submicroscopic specks combine in countless billions to produce the familiar collections we perceive en masse as substance. A chunk of frozen argon is composed of atoms that are all of one kind, whereas water is a mass of interacting molecules and table salt of ions—atoms, molecules, and ions.

The kinds of atoms that group together determine the strengths of their interactions, and these, in turn, define the ultimate structure of the substance in any particular environment. If the interparticle forces are strong enough, the collection of atoms as a whole *will maintain its shape and volume* (the amount of space it occupies). That is the distinguishing characteristic of a **solid**—aside from the fact that it may be dense or hard, or that you can hold it in your hand. A **liquid** is characterized by weaker binding forces, so it will flow, *assuming the shape of its container, although keeping a constant volume regardless of that shape.* If the forces are still weaker, the material exists as a **gas;** the atoms or molecules tend to disperse, and the substance *assumes both the shape and volume of its container.*

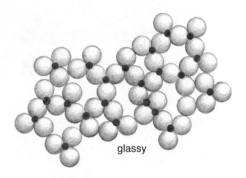

glassy

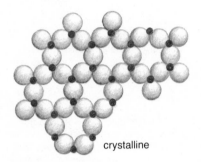

crystalline

Glassy and crystalline solids—short- and long-range order.

Some materials can exist as solid, liquid, or gas and readily shift from one form, or *phase*, to another and back, depending on the prevailing temperature and pressure. While at room temperature iron is solid. It is molten in the core of the Earth, vaporized to a gas by a nuclear explosion, and even whirled as an *electrified gas*, or **plasma**, in a star. Of course, H_2O is a less exotic example of the type, and many others occur, as well. There are also solids whose molecules exist only in that one state, not having any liquid or gaseous counterparts. Instead of melting or vaporizing, they simply decompose chemically. TNT is a loud example and coal a quieter one.

The phases of matter, as we will see, are distinguishable on an atomic level in terms of the movement and distribution of the atoms making up the substance. Internal *motion* and *disorder* both progressively increase as a sample transforms first from the solid to the liquid and then to the gaseous state.

There are many more materials in our environment in the solid state than in either the liquid or the gaseous state. Solids fall into two major classifications, *crystalline* and *amorphous*—the former containing crystals, the latter not.

The atoms, ions, or molecules of a solid *are* in motion, but it's a small, jiggling, vibratory motion about fairly fixed, closely spaced positions within the substance. When the atoms are arranged in *an orderly three-dimensional pattern that repeats itself, over and over again*, the solid is said to be **crystalline**. Most minerals and all metals and salts exist in at least one crystalline state.

On the other hand, if a substance is composed of atoms, ions, or molecules that are *not arranged in an orderly and repetitive array*, the solid is said to be *noncrystalline*, or **amorphous**. Rubber, pitch, resin, plastics, and various glassy materials are ordinarily of this kind. The atoms of glasses are strung out in an irregular mesh that lacks any long-range orderliness. What little order there is, is patchy, localized, and nothing like the neat repetitive array of row-upon-row of atoms that defines a crystal. By contrast, the long chain molecules of rubber and plastics resemble the entwined but still not total chaos of a bowl of spaghetti.

12.1 CRYSTALS   The regularity that is the ultimate stamp of the crystalline state is a long-range phenomenon, that is, one extending over many, many atoms. Still, a crystal is not necessarily large. After all, atoms are minute, and even millions of them arrayed in a perfect pattern would form a crystalline solid so small that it would be nearly invisible. Many materials are made up of countless single crystal grains, each of which may be no more than a fraction of a millimeter in size. Indeed, ordinary metal objects (like knives and forks, pots, pans, nails, wire, and hubcaps) are all **polycrystalline** (many-crystalled). The zinc coating on any galvanized steel garbage can shows an easily visible polycrystalline structure of flat, jagged, interlocking single crystals, just like the lovely patterns of ice that sometimes form on a window in the cold of winter.

Large single crystals are less commonplace in the home, but they exist in abundance naturally, and many are routinely "grown" commercially. Diamond, ruby, emerald, zircon, and topaz are among the more familiar single crystals we customarily mount in polycrystalline gold or silver and hang on ourselves. Quartz (SiO_2) can be found in Nature as large, clear, beautiful single crystals, but it is perhaps more familiar underfoot as seashore sand or those white rocks that seem to be everywhere. It's a rare gypsy who can still afford a real quartz crystal ball; alas, nowadays they're glass.

The growth of a crystal requires a continuous supply of fresh atoms from which to build outward, layer upon layer. These atoms must be mobile enough to find—rather, be drawn to—appropriate sites on the crystal's surface. This is usually accomplished when the crystal begins to form (to nucleate) in a liquid solution or in a molten mass. Ordinarily it will take a good deal of time to grow a large single crystal. More often than not, a molten glob will cool quickly, freezing its atoms into separate, randomly oriented grains. That's why common objects fashioned of metal are polycrystalline. Mobility and time are the crucial parameters, and anything that cuts down on either will limit the size of the crystals formed.

One of the distinguishing properties of most crystalline solids is their rather sharp transition to the liquid state at a well-defined melting temperature (omitting those that chemically decompose first). Crystals melt suddenly; their well-ordered atoms move apart, and the precise arrangement of the solid collapses into the disorder of the liquid state. In other words, the interatomic bonds are all about equal in strength, and all rupture almost simultaneously. The fact that a crystalline solid totally transforms into a liquid if maintained at its melting temperature makes its behavior quite unlike that of non-crystalline materials, which change state gradually over a broad range of increasing temperatures.

The lovely, smooth, flat, angular faces of crystals are the external manifestation of a superb internal order, an order that the crystallographer tries to appreciate in terms of geometry, of recognizable patterns. The atoms within a crystal are seen as if arranged in minute, identical building blocks, stacked one on another, filling the space of the sample and forming its overall geometrical design. But not every conceivable regularly shaped building block actually corresponds to real crystals. In fact, there are only seven such basic structures (see figure) out of which crystals can be constructed. The key point is simply that the imaginary blocks must completely fill space when they are stacked, and not every nice geometrical shape does. The problem is easy to see in two dimensions by just considering a tiled floor. Without leaving empty spaces between adjacent tiles, we can cover a floor with little triangles or squares or rectangles or even hexagons (six-sided thingies), but it simply cannot be done with regular pentagons (five-sided figures). There isn't a washroom in the universe with a flat floor tiled solid in little white pentagons.

Only certain symmetries work to fill space, and only certain

Sugar crystals grown on string.

A cluster of quartz crystals.

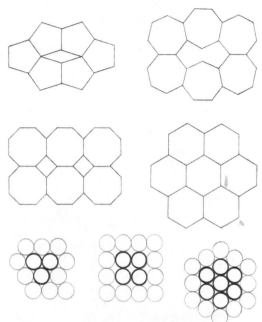

The packing of regular objects with 3, 4, 5, 6, 7, and 8 sides in two dimensions.

Regular hexagons fill the flat two-dimensional space of a bathroom floor in this self-portrait of your author.

building blocks will form crystals. Remember that crystals grow one layer at a time. A new bunch of atoms cannot leave great empty gaps; they nuzzle right in with their friends as tightly as they can, occupying the space of the crystal. Common salt, NaCl, is a rather nice example. Its basic building block is a tiny cube (roughly 0.5×10^{-9} m on edge) with one sodium or chlorine ion alternately at each of the four corners of each face. The external shape of a crystal need not be identical to the configuration of its building blocks (just think of all the different-shaped structures—pyramids, chimneys, buildings, etc.—made of brick), but for NaCl it is: Most brands of table salt come as perfect little cubes. Fed a constant supply of atoms, a salt crystal cube will continue to grow endlessly.

The building blocks are imagined packed tightly to form a crystal, but there must be some space between the atoms themselves in each elementary block, and that space is different for each of the seven blocks. To see this, you might stack marbles or cannon balls, or more simply, you can just arrange some pennies (two-dimensional atoms) in both a square and a hexagonal flat array. The hexagonal penny pattern is much more tightly packed. Thus both the structure and the mass of the individual atoms determine the **density** of any substance, i.e., the mass per unit volume,

$$\text{density} = \frac{\text{mass}}{\text{volume}}.$$

The densest earthly solid is the pungent-smelling metal osmium, whose heavy atoms (though assuredly not the heaviest) are so closely packed that a cube of it one foot on a side would weigh about 1400 lbs—22.5 times more than the same volume of water.

By comparison, atomic nuclei have densities of around 10^{15} g/cm³, or about a thousand million million times that of water. It has been suggested that certain collapsed stars are composed of nuclei whose former electron clouds have been crushed into them, leav-

The atomic configuration of ordinary table salt, NaCl.

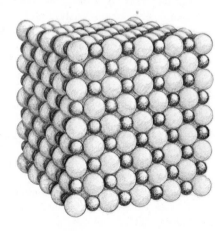

DENSITIES* OF A FEW SUBSTANCES

Material	mass/volume g/cm³	weight/volume lb/ft³
Air	0.0013	0.08
Oxygen	0.0014	0.087
Carbon dioxide	0.002	0.12
Wood (pine)	0.4	25
Gasoline	0.7	42
Alcohol (ethyl)	0.8	48
Ice	0.9	57
Water	1.0	62.4
Rubber (hard)	1.2	74
Sugar	1.6	99
Magnesium	1.7	109
Concrete	2.3	140
Aluminum	2.7	169
Diamond	3.2	200
Titanium	4.5	281
Germanium	5.5	341
Tin	7.2	449
Iron	7.9	490
Copper	8.9	555
Silver	10.5	655
Lead	11.0	687
Mercury	13.6	849
Uranium	18.7	1167
Gold	19.3	1204
Platinum	21.4	1335
Osmium	22.5	1405

* Both the mass and weight densities are listed for 0°C and 1 atm pressure.

Single crystal of salt.

ing only an incredibly dense ball of neutrons. Such a *neutron star* would be in a real sense a single immense nucleus 10^{15} times the density of earthly matter. A pea-sized chunk of the stuff, if brought back to our planet, would weigh in excess of 150 million tons.

12.2 AMORPHOUS SOLIDS

There are relatively few amorphous solids occurring naturally in the Earth's crust. Materials like opal and volcanic glass (obsidian) are the prevalent exceptions. Big irregular molecules in a molten mass have a difficult time aligning themselves in the precise patterns demanded of crystals. They tend to take longer at it, and more often than not, as the temperature drops, they are frozen into a fairly disordered tangle, a noncrystalline mass. Only in very rare circumstances of prolonged high temperature will glass have the needed time and mobility to come together in any sort of crystal.*

* Common silicate glass is made by simply melting silica sand (silicon dioxide, SiO_2) and cooling it too rapidly to allow crystallization as quartz. Ordinary glasses and glazes are made by fusing silica with oxides of aluminum, sodium, calcium, or boron—the latter yielding Pyrex.

The formation of glassy solids can easily be demonstrated with some crystals of ordinary sugar, whose molecules are appropriately large and irregular. Gently heat a spoonful of sugar until it melts into a clear liquid (heating it too much will chemically decompose it into a brown syrup called caramel). Now if you cool the molten mass *rapidly* so that it does not have time to recrystallize, it will become a glassy solid—a clear lollipop. Naturally occurring glasses, like the little beads found on the moon, imply rapid cooling, as would occur after material has been melted by a meteorite impact.

Since the molecules of amorphous solids are already fairly disordered, the melting process, which increases that disorder, is gradual. The intermolecular bonds vary in strength from place to place within the material, and the addition of heat ruptures the weakest of them first. As the temperature increases, more and more bonds break, and the material simply becomes increasingly fluid—an occurrence familiar to anyone who has ever worked taffylike hot glass in a flame or melted certain plastics. In a sense, glass is something of a liquid in that its molecules can move past one another even at room temperature. It is an exceedingly slow process, however, as witnessed by the fact that over the centuries the window panes in old cathedrals have, indeed, become slightly thicker at the bottom. It seems that glass is actually on the borderline between solid and liquid. As usual, we are having trouble with all-encompassing definitions that presume on Nature, but that's par for the course.

The stuff that lives or lived on Earth is made up of giant molecules—**polymers** (*poly* = many, *mer* = a unit)—composed of repeating groups of atoms in long chains. The amorphous solids, such as your fingernails, are great tangles of these giant molecules, as are all the synthetic polymers, such as Teflon, nylon, vinyl, and Plexiglas. The giant molecules of rubber and Spandex are like intertwined, coiled strands that straighten out as they are stretched and then spring back. The ability to deform under the influence of a force and then return to the original shape when the force is removed is of course known as *elasticity*, a property possessed, more or less, by most solids (more by spring steel and less by mushy clay).

Though many polymers do appear as amorphous solids, others exist in a crystalline configuration. The DNA (deoxyribonucleic acid) molecule, the carrier of life's genetic code, is a giant spiral affair composed of millions of atoms. Like hemoglobin, vitamin C, hydrocortisone, penicillin, and many other organic polymers, DNA has been crystallized. The crystallization of such substances is performed for the purpose of studying their structures, particularly with X-rays.

The eight-sided (octahedral) fluorite crystal.

12.3   THE BONDS THAT BIND US

All the solid bits of matter from dogs to hydrants are formed of atoms more or less confined to their individual jiggling spots by electric forces—atoms pulling and being pulled, bound to the gentle in-place dance of the solid state. There are four main varieties of interparticle "glue" that link atoms, ions, and molecules with their neighbors, locking them together in an electrical mesh of attractive force: *ionic, covalent, metallic,* and a group of weak interactions bunched under the heading of *molecular.* The subtle differences between these bonds working separately and in concert determine the many properties of solids.

Ionic Solids

As we saw earlier (page 267), when sodium combines with chlorine to form a salt molecule, it does so by transferring its outermost, weakly held electron to the chlorine, which is in need of that one charge to fill its own outer shell. In so doing, the sodium becomes ionized (Na^+), as does the chlorine (Cl^-), and these oppositely charged ions attract each other electrically. The two come closer and closer together until the repulsion that now arises from the overlapping of their respective electron clouds balances the ionic attraction, and the molecule stabilizes at that point. Still, each of these tiny charged bodies will attract other oppositely charged ions to it. If the new arrivals are free to associate, all that will fit will snuggle right in until each Na^+ is surrounded by six clinging Cl^- ions, and vice versa. This is exactly what happens in the solid. Each ion is then bound electrically to all the other oppositely charged ions and especially to the six nearest ones, thereby generating a highly stable interlinked cubic structure—an **ionic crystal**—that resembles a gigantic molecule.

Because ionic bonds are fairly strong, salt crystals, which are characteristic of the species, are quite hard, though brittle. Not until they are red-hot at a formidable 800°C will the bonds rupture and the crystal melt. Still greater strengths and even higher melting temperatures occur when atoms transfer more than one electron, as in the oxides of aluminum (ruby, sapphire, and emerald), titanium (rutile) and magnesium (magnesia). The last—MgO—melts at 2800°C. There is a whole range of ionic solids with a considerable diversity of properties: the carbonates, like calcite (i.e., stalactites, chalk, or marble); the sulfates, like calcium sulfate (i.e., gypsum, which when burned and ground is plaster of paris) or the lovely octahedral alum crystals; the nitrates, like sodium nitrate (i.e., saltpeter); the silicates, like topaz, garnet, and mica; the bromates and bromides and chlorates and chlorides and on and on.

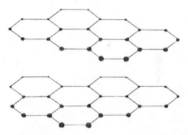

The parallel-layered structure of crystalline graphite.

Covalent Solids

Another kind of bond operating to form molecules (p. 267) involves not the outright transfer of electrons but a sharing. Two hydrogens share each other's single electron, effectively giving both a pair, and hence each atom has something resembling a closed first shell. The electron cloud then encompasses the two nuclei, gluing the system together as a single entity. The **covalent** bond, as it has been called ever since Langmuir named it, is also responsible for the formation of certain solids.

Graphite and diamond are the two remarkably different crystalline forms of pure carbon. Both are covalent solids, but diamond is

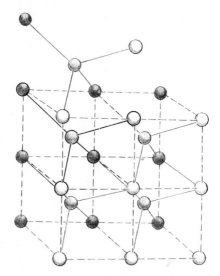

The atomic configuration of diamond.

Mica.

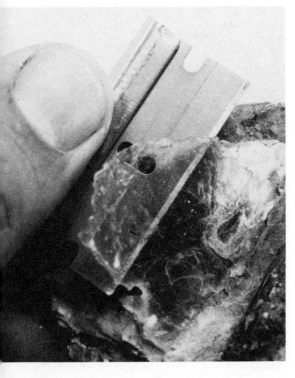

far more typical. Carbon, which has a total of six electrons (two in a closed first shell and four in a half-filled second shell), heads the column of the great electron sharers in the periodic table. It will enter into numerous relationships by sharing all four of its outer electrons to create four covalent bonds.

In diamond, an individual carbon atom forms a single covalent bond with each of four adjacent carbon atoms. These bonds extend, corner to corner, along the four diagonals of a cube, giving the crystal a cubic inner structure. Every atom in the crystal is thus locked in a four-way grip, creating a rigid three-dimensional solid, which is the hardest mineral known. So tough are these bonds that diamond must be heated above 3500°C before they will burst and the crystal will melt or, more likely, vaporize. With a few chemist friends and a helpful Parisian jeweler, Lavoisier demonstrated in 1772 that diamond, in the presence of oxygen, burns just like a charcoal briquette. Using sunlight focused by a large magnifying glass, they totally incinerated a crystal, causing it to vanish into a puff of carbon dioxide gas (p. 278).

Small black specks, a common imperfection or flaw in diamond, are important in determining the market value of a given specimen. In fact, heating a diamond can cause the whole crystal to turn black, to restructure itself into a form energetically more stable—*graphite*. Provided with enough thermal energy to rattle loose from their rigid criss-cross diamond bonds, the carbon atoms will settle down to the graphite pattern of widely spaced planes, each composed of an interlinked hexagon net. In this new guise, every carbon is tied to only three of its neighbors, *all in that same plane*, although one of those links is a strong double bond. Still, these separated parallel planes, which remind one of a deck of cards, are only weakly held together and can easily be pried apart (like sheets of mica).

Graphite is the soft, shiny, "greasy" material in pencil leads (to which a hardening binder is added); the slippery gray-black powder used in lubricants, particularly for locks; and the conducting rods, called brushes, that spark and burn out in vacuum cleaner motors.

Many of the compounds of carbon are covalent solids. For example, when sand (SiO_2) is heated with graphite, silicon carbide (SiC)—the common abrasive known as carborundum—is formed. Again, the ultrahardness of this material arises from the strength of the bonds.

Metallic Crystals

Metals (e.g., copper, gold, silver, iron, sodium, and aluminum) are characterized by the fact that their outer electrons, which are very weakly bound to the atoms, are also relatively few in number, so few per atom that in general they do not readily form covalent bonds. Instead, these electrons tend to be drawn off and exchanged with those of neighboring atoms in a sort of ongoing give and take. A solid metal, in effect, is a three-dimensional fixed array of positive ions permeated by a cloud, or "gas," of essentially free electrons. The electrons migrate randomly from atom to atom, not being bound to any one in particular. A metallic crystal is held together by the electrical attraction between the all-pervading negative cloud and the immersed assembly of positive charge.

The positive ions typically nestle into tightly packed simple crystal structures that tend to be quite dense. Since there is no atom-to-atom linkage, no well-defined fixed directionality to the bonds, parallel atom layers can be forced to glide over one another. Thus the material deforms rather than ruptures; it usually bends rather than breaks. Quite unlike brittle ionic and covalent crystals, metals generally can be hammered or drawn into shape.

The electron glue is relatively insensitive to the particularities of the ions present, and all sorts of metals can be mixed in the molten state to form all sorts of **alloys.** When soft copper is alloyed with about 15 percent of even softer tin, the larger tin atoms serve as a kind of grit, which then impedes the sliding of atomic planes in the composite solid. The resulting alloy is hard, durable bronze. A mix of bismuth, lead, tin, and cadmium, known as Wood's metal, is a bright, shiny, strong alloy with the feel of silver—yet it melts at a temperature (70°C) well below that of boiling water.

The presence of the electron gas makes it possible for free charges to be forced to migrate in any direction within the crystal. This motion, in turn, constitutes an *electrical current* and accounts for the fact that metals are good conductors of electricity. Similarly, if you stick a steel spoon in a flame, the ions and electrons will pick up kinetic energy via collisions with the fire's rapidly moving incandescent gas. But because they are mobile, the free electrons, once blasted, can quickly move down the spoon, and that constitutes a flow of thermal energy. Colliding with other electrons, over and over again, they soon distribute that energy throughout the spoon, burning your fingers at the far end. Metals, of course, conduct heat quite well. That's why we make pots out of them but not coffee mugs.

The fact that we cannot see through metals unless they are very, very thin (as in see-through mirrors) is due to the presence of the electron gas. All metal crystals (in fact, all metals, solid or liquid) exhibit a familiar shiny luster when they're smooth and clean. Free electrons scatter light in all directions exceedingly well, and before a beam can advance even a few dozen atom layers deep, most of the light is kicked back out of the surface. Peering into a mirror, comb in hand, you see your reflection as if the light had simply bounced off the metal coating.

Molecular Crystals

Thus far we have dealt with solids composed of arrays of either atoms or ions, but it's also possible for a crystal to be made up of interacting molecules. The molecules themselves are held together by

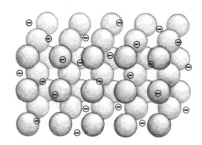

An electron "gas" within a metal.

Kennedy half dollars, like dimes and quarters, are now laminated; a copper core is sandwiched between two layers of a 75% copper–25% nickel alloy.

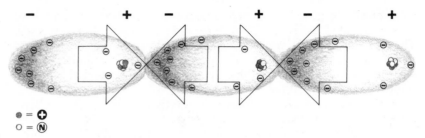

Three polarized atoms
attracting one another.

● = ⊕
○ = Ⓝ

Below, three views of a crystal model of
ice, revealing the six-sidedness of the
arrangement.

strong covalent bonds, but the forces linking them into a solid are usually much weaker. Carbon dioxide (CO_2), which ordinarily exists as a gas, freezes into the familiar solid known as dry ice at $-109.3°F$ ($-78.5°C$). So weak are its intermolecular bonds in the solid state that even when gently heated, dry ice gradually transforms directly into gaseous CO_2. Molecules at the surface break away, and the material evaporates; it *sublimes*. Crystalline iodine, composed of covalent I_2 molecules, performs the same vanishing act, going directly from solid to gas without ever being liquid.*

The mechanism whereby neutral atoms and molecules electrically attract one another is named after Johannes van der Waals (1837–1923), who first suggested the existence of such a force (without an explanation) to account for the behavior of gases. On the average, the distribution of moving electrons within an atom (or molecule) can be assumed to be uniform. At any instant, however, there may well be a slight random clumping of negative charge in one region or another that lasts for, say, a fraction of a microsecond. During that brief moment the atom is electrically negative at the end where the electrons are more numerous than usual and positive at the opposite end, where the nucleus is slightly more exposed; it is momentarily **polarized.** That nonsymmetrical atom will then temporarily polarize any neutral nearby atoms by pushing or pulling on their electron clouds. For example, a neighbor that happens to be close to the positive end of the original polarized atom will have its own electrons drawn toward that positive charge, bunching up on the near side, and so it will become polarized. For an instant, all the oppositely charged ends of these polarized atoms will be electrically attracted toward each other. These shifting fluctuations in the charge distribution have the overall effect of holding the substance together.

This weak van der Waals force *is always operative*, but in the few instances where it's the predominant cohesive interaction (as it

* Camphor and naphthalene (mothballs) also sublime; in fact, so does ice. Any ancient who remembers the time when clotheslines were not unfashionable would also remember that dungarees, frozen rigid, would still dry at temperatures below 0°F. Take a block of frozen coffee or ice cream, lower the pressure, and pump off the vapor as heat is slowly supplied. This will cause the cream to totally sublime, and the light porous material that remains is said to be *freeze-dried*.

is with iodine, dry ice, or the frozen noble gases, like neon and argon), the resulting solids are quite fragile, lack mechanical strength, are soft, and tend to melt and evaporate at low temperatures. Most substances that are ordinarily gaseous at room temperature condense into this sort of solid. For instance, neon, a stable atomic gas in its common state, becomes a solid only at $-248.7°C$. In most instances, van der Waals force plays a far more limited role, although it is quite significant in graphite, where it's the interaction that weakly ties the parallel planes of atoms to one another.

A molecule may be permanently polarized if its electron cloud is always distributed unevenly, being concentrated near one atom or the other. Hydrogen fluoride (HF) has its negative cloud drawn more toward the large positive nucleus of the fluorine atom than the small hydrogen atom. These polarized molecules attract each other to form HF crystals at temperatures below $-83°C$.

In much the same way, and far more important, water molecules are permanently polarized, with their hydrogen "Mickey Mouse ears" being the positive end (p.220). The nearly nude hydrogen nuclei enter into relatively strong intermolecular bonds, which are responsible for the extraordinary properties of ice. Water molecules in the solid state link into a fairly open hexagonal structure containing a good deal of empty space. That accounts for the six-sided symmetry of snowflakes and for the remarkable fact that, unlike most substances, which are more dense in the solid than in the liquid phase, *ice is less dense than water*. If it were not, ice would sink in lakes and ponds, which would freeze from the bottom up, destroying all the life that ordinarily lives beneath the frozen surface in winter.

When I was a kid, we used to keep bottles of milk out on the fire escape in the winter. Every now and then a cold snap would descend at night, and we would all wake up in the morning to find frozen milk hanging rigid out of the necks of burst bottles. Ice manages to be less dense than water simply via expansion; ice occupies more volume than the same weight of water. That is why rocks split and pipes crack when the water in them freezes.

Snow crystals.

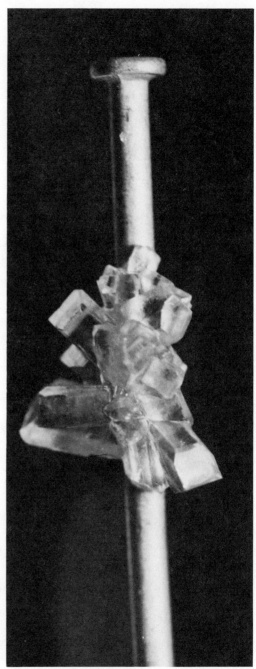

Sugar crystals I grew on a nail, using the procedure described on the next page.

EXPERIMENTS

1. Crystals can be grown quite easily; the process requires more patience than anything else. Boil a cup of water and stir in two to three cups of sugar; that is, keep adding sugar until no more can be dissolved. Now pour the liquid into a warm glass or jar. At this point you have two options. Either cover the jar with a piece of paper towel, or tie a small weight to a string and hang it in the liquid. The next step is not to disturb the brew for two or three weeks. Repeat the process with table salt, alum, borax, or epsom salt. Add a few drops of food coloring if you want colored crystals.

2. Melt some polycrystalline solder and notice how suddenly the transition to the liquid takes place. As crystalline ice melts, it obviously doesn't soften first—it's either frozen solid or watery liquid. Now melt some gelatin, wax, or paraffin. These amorphous solids soften and melt gradually.

REVIEW

Section 12.1

1. What are the two main classifications of solids?
2. When is a solid said to be *crystalline*?
3. Explain what is meant by an *amorphous* solid.
4. A crystal has *long-range* order. What does that mean?
5. Define the term *polycrystalline*.
6. What limits the size to which crystals will grow?
7. Why do crystals generally melt suddenly?
8. How many different basic building blocks are there in the formation of crystals?
9. Do regular pentagons fill a flat two-dimensional space?
10. Common salt crystals have building blocks that are _____ in shape with a _____ or _____ ion at each corner.

Section 12.2

1. What is obsidian? Describe its structure.
2. Why do large molecules tend to form amorphous structures?
3. Molten sugar cooled quickly forms a(n) _____ solid.
4. In what sense does glass resemble a liquid?
5. What is a *polymer*?
6. What is *elasticity*?
7. How is glass made?

8. Atom bomb tests turn the desert sand into _____.
9. Rubber is a(n) _____ solid composed of_____ molecules.

Section 12.3

1. List the four main kinds of interatomic bonds.
2. What kind of solid (in terms of bonds) is common table salt?
3. Are ionic bonds considered strong? Why?
4. The electrons in a covalently bonded hydrogen (H_2) molecule are _____ by each other.
5. Name two crystalline forms of carbon.
6. In diamond, each carbon atom is bound to each of _____ other adjacent carbon atoms.
7. Can a diamond be burned like a piece of charcoal?
8. What is a pencil lead made of?
9. Are metals in general fairly dense?
10. A metal may be thought of as an array of positive ions in a sea of free _____.
11. What causes metals to be shiny?
12. Define the verb *to sublime*.
13. Give an example of a material that sublimes.
14. What type of bonding is responsible for the solids with the lowest melting temperature?
15. What does it mean to say an atom is *polarized*?

QUESTIONS

Answers to starred questions are given at the end of the book.

Section 12.1

1. The atoms and molecules of a solid are in constant movement. Describe their motion and relate it to the ideas of hot and cold.
2. Discuss the difference between a glassy and a crystalline solid. Does the picture of the glassy substance (p. 280) seem totally without order?

3. Cast iron is hard and very brittle. When struck a hammer blow, a cast-iron pipe will shatter into fragments that reveal tiny twinkling facets. Assuming that the castings are cooled rapidly, discuss the likely structure of the material in relation to the information above.
* 4. Geodes are rocky balls that from the outside resemble stone potatoes. When a geode is split open, it reveals a marvelous glitter of crystals surrounding the walls, growing inward toward the center. How might such a structure have been formed?

Section 12.2

1. Explain what happened to the target in the Davisson-Germer experiment (Section 11.4) while it was being cleaned.

2. In an amorphous solid, the molecules are not all neatly arrayed. How does this influence the intermolecular bonds and the way the solid melts?

3. Make a list of several crystalline, polycrystalline, and amorphous solids that you have somewhere at home.

* 4. Is ice amorphous or crystalline? How about candle wax? How might you lend credence to your suspicions?

5. Why does glass shatter into jagged fragments? Are the facets on "crystal" cut-glass bowls actually cut as a diamond would be cut, or are they ground out with an abrasive?

Section 12.3

1. What is the crucial aspect of an ionic bond?

2. The figure below shows the electron probability cloud of a covalently bonded hydrogen molecule. How does the

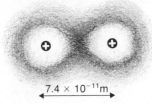

7.4×10^{-11}m

Electron probability cloud for a hydrogen molecule.

fact that the electrons are more likely to be found in the region between the nuclei account for the molecule's holding together?

* 3. How would you classify the individual constituent particles of an ionic solid? Of a covalent solid?

4. What is the ultimate source of all the different binding forces applicable to solids?

5. Describe the nature of metallic bonding, and discuss how it accounts for the conduction of heat in a solid piece of metal. Why are metals good electrical conductors and ionic crystals poor ones?

* 6. In addition to water, can you name a material (element or alloy) whose density decreases when it freezes (that is, one that expands on solidifying)? *Hint:* Such a trait would be very useful when making intricate castings.

7. What kind of bond holds solid xenon together?

8. Why are electrons far more easily liberated from metals than from nonmetals in the photoelectric effect?

9. Selenium has an electron shell structure of 2, 8, 18, and 6. Would you suspect it to be a metal or a nonmetal? Explain.

* 10. Although lithium and hydrogen are in the same column of the periodic table, the former is a solid metal and the latter is *ordinarily* a gaseous nonmetal. Discuss why this is so.

MATHEMATICAL PROBLEMS

Answers to starred problems are given at the end of the book.

Section 12.1

1. What is the weight of a cube of lead 2 ft on each side?

* 2. Express the mass density of water in kg/m³.

3. Compute the mass of 5 cm³ of lead. What is the mass of 5 m³ of lead?

4. How big is a 20-pound block of ice?

Geode.

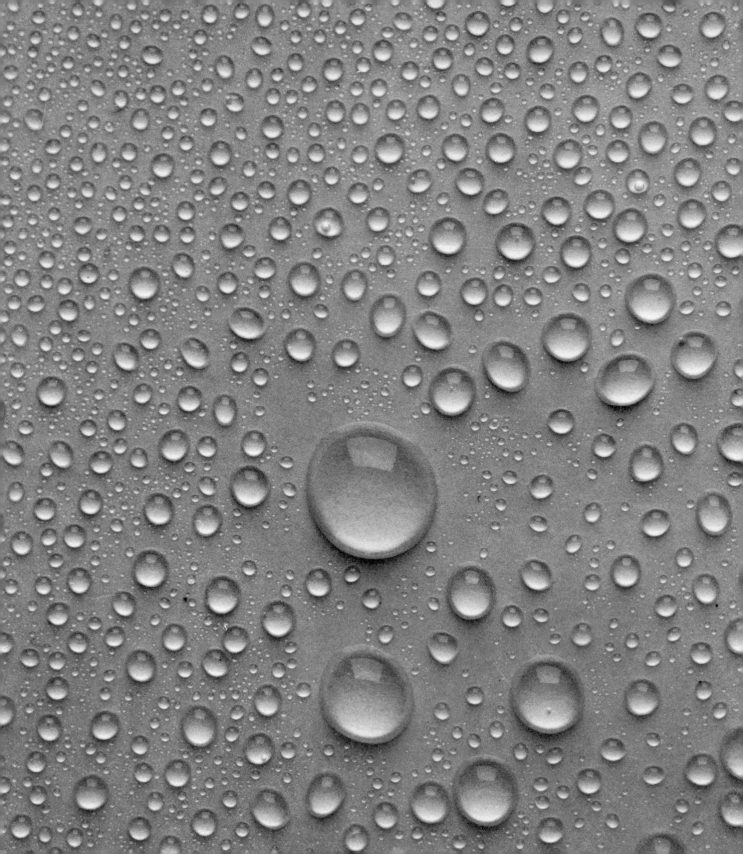

This chapter deals with the **liquid state,** with those remarkable conglomerations of atoms so weakly bound that they can flow past one another and yet held together strongly enough to maintain a fixed overall volume. **Surface tension,** the peculiar property manifested by the liquid at its interface with the surroundings, is the subject of the first section. The next deals with the interaction of different kinds of atoms within liquid **solutions. Pressure, hydraulics,** and **buoyancy** are the practical aspects of the large-scale behavior of liquids explored in the last section.

Typically the atoms of a solid, held close and strong, stand together as one, a thing of *fixed* shape—though nothing is really changeless, shape included. Solids expand and contract, sag, stretch, and wear; some decompose and even evaporate. They constantly change in minor ways. For example, an untouched block of steel sitting in air will draw foreign molecules (O_2, N_2, H_2O, etc.) into its surface. So solids are things that maintain their volume and shape (more or less) if you leave them alone. In other words, the internal binding forces are strong enough to overcome the tendency to deform under Earth's gravity.

When a solid is heated, its atoms, ions, or molecules pick up kinetic energy in the form of increased vibration; straining against the bonds that hold them, they oscillate more violently. When the melting point is reached in a particular substance, the atoms possess enough energy to overcome the rigid bonds, to slip out of their formerly fixed locations and tumble past one another. Patterns composed of small groupings of linked atoms still persist although they shift and regroup and shift again. There is still order, but it's a short-range changing order. There is still a web of interatomic cohesive force, but the energetic atoms can now move against it. Increasing the temperature toward the boiling point gradually breaks up the groupings, increasing the disorder. One can think of the liquid as a transition state between the random tumult of the gas and the calm precision of the solid.

There really are not many naturally occurring liquids in any great abundance. Other than water and petroleum, most of the familiar ones, like gasoline, turpentine, and benzene, are synthetic. Of the two elemental liquids, pure mercury is rare in Nature, and pure bromine is never found. At ordinary temperatures, the liquid state is most readily formed of highly polarized molecules, as in water, or large heavy molecules, as in petroleum, where the van der Waals force is substantial. As with Goldilocks's porridge, the temperature has to be "just right" or the interatomic forces will be too strong (resulting in a solid) or too weak (resulting in a gas).

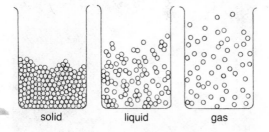

solid   liquid   gas

No physicist would have predicted the existence of a liquid state from our present knowledge of atomic properties.

VICTOR WEISSKOPF (1972)

Physicist

Liquid nitrogen at 77K (−196°C).

Usually the atoms or molecules are only slightly farther apart in the liquid than in the solid state, as suggested by the fact that most substances have very nearly the same density in both phases. Raising the temperature of a liquid ordinarily increases the molecular spacing, leaving holes between the ordered groupings and producing a very slight decrease in density. Like most solids, liquids are essentially **incompressible.** The electron clouds of the individual atoms and molecules resist being forced any nearer to one another. And, like solids, liquids therefore maintain their overall volume at a given temperature.

You cannot push your finger through the lid of a tin can, because the relatively weakly bound molecules in your hand will start to shift past each other (and hurt) long before the atoms will slide in the metal. However, you certainly can slam a hard steel blade through the lid, and there are a billion rusty "church keys" * to prove it. The atoms in a solid may be pushed aside as they can be in a liquid. It is often a lot harder to do, but we do cut, punch holes in, and bang nails into all sorts of solids—it's just a matter of degree.

The ability of a liquid to pour, to have its molecules slide over one another, varies with the cohesive force from one substance to another, from gasoline to water to motor oil to molasses to tar. **Viscosity** is what it's called: the stickier and thicker, the more viscous the fluid. And of course, heating a liquid—causing its molecules to move around faster and their groupings to shrink in complexity—reduces the viscosity just as it does in the extreme with a glassy solid: That's why tar is heated before it's spread and why motor oils have those (10–20–30) SAE numbers. A sheet of steel in a stamping press "pours" into the mold in much the same way that a liquid does, though the force must be much greater than gravity to cause the atoms to shift in the metal. For that matter, even diamond has been made to flow like wax, but it took a pressure of 25.2 million pounds per square inch.

13.1 SURFACE TENSION

The liquid level is the same in each, regardless of the shapes of the interconnected vessels.

Under the influence of gravity, each molecule in a liquid will simply fall until it can fall no more, gliding past the others until it settles in at the lowest possible potential energy. This is the reason rivers flow *down* to the sea and liquids assume the shapes of their containers (in the same way that dry sand would). But if there was no gravity, or equivalently, if the substance was somehow effectively weightless, a glob of liquid would draw itself into a sphere. Raindrops are roughly spherical—only roughly because of air friction—while they are in free fall and "weightless." Droplets of molten solder cool in midair into little spheres, and the astronauts have played at making perfect spheres of liquids while in orbit.

When you shake a little mercury in a jar, hundreds of tiny, nearly spherical beads appear and just sit there, quite unwilling to

---

* An ancient pointed instrument that—until displaced by the flip-top—was used to open beer cans.

assume the flat shape of the bottom of the jar! All these familiar liquid "tricks" are the outward manifestations of an internal tug-of-war that gives rise to something we call **surface tension.** A molecule within a liquid pulls and is pulled in all directions by the cohesive forces that bind it to the fluid. But a molecule at the surface, having no one above it, draws more strongly on its nearest neighbors to the side and below. The molecular surface sheet behaves as if it were an elastic membrane being pulled inward, assuming the least area it can. It was Thomas Young almost 200 years ago who suggested this membranous imagery, which is very helpful but should not be taken too literally.

A wire frame with one movable side (as in the figure) can be used to measure surface tension directly. When the frame is dipped into, say, a soap solution, a film will form across it. Pulling on the crossbar will stretch the film, and the corresponding force can be determined. Note that in this process work is done on the film, work against the internal cohesive forces of the molecules. *Energy is thereby stored as a kind of elastic PE in the surface.* Friedrich Gauss, the astronomer and mathematician, introduced this profound idea of **surface energy.**

Suppose we place a bead of water on a clean glass horizontal surface. If gravity alone were involved, the bead would collapse, spreading out uniformly so that its mass would drop as low on the surface as possible. In effect, the water would simply fall in the Earth's gravity field, resting as low as it could, minimizing the gravitational PE. Something resembling this does in fact happen, primarily because water molecules are attracted to glass molecules—water clings to glass and is said to *wet* it. The fairly exposed protons of the hydrogen in water attract the oxygen atoms on the surface of the glass (SiO_2).

However, if the glass surface is greasy, water will not wet it; instead of sheeting, it will bead up into globules. The large surface area and therefore the surface energy of the flat sheet will then be considerably reduced. The difference in energy will simply go into raising the mass of water in each droplet; that is, it will go into gravitational PE. In other words, when the water beads up, the elastic PE of the surface is lowered, and the gravitational PE is increased.

The smallest possible area that can enclose a given volume is a sphere, and that corresponds to the lowest surface energy configuration. The surface of a droplet of liquid in effect "falls," dropping in potential energy, as it approaches sphericity. That's why there are no cubic raindrops, why spray cans shoot out a mist of tiny near-spheres, and why "weightless" globs of liquid in space are almost perfectly spherical.

Surface tension is responsible for some familiar as well as some not entirely expected behavior displayed by liquids. The dry bristles of a fine camel's-hair brush ordinarily stand apart in a bushy tuft. If you dip the hairy end of one into a cup of water and then pull it out, the liquid film adhering to the bristles will pull them together

The distribution of forces exerted on an atom at the surface is quite different from the uniform pattern acting within the body of the liquid.

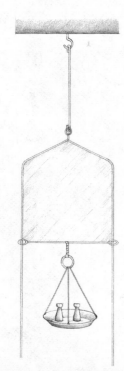

Measuring the surface tension of a liquid film.

in a tight mass. That's why your wet hair after a shower or swim is a matted, clumpy mess.

The membrane imagery is especially appealing when you consider that it is possible to form a cup out of fine wire mesh and fill it with water to a considerable depth before any liquid runs out the openings. The integrity of the surface film is crucial here, as it is when we "float" objects on it.

A steel needle or a double-edged razor blade (oiled slightly so they will not be wetted) can be supported on the surface layer of water even though they have a much greater density and should otherwise sink. Provided that the film remains intact, it sags beneath the needle, exerting an upward elastic restoring force that equals the needle's weight. So long as the weight is not too great to be balanced by the surface tension, the object remains suspended above the liquid. No part of the needle is within the fluid, and it is *not* floating in the usual sense of the word (Section 13.3). There are delicate insects that can tiptoe across a pond without ever breaking the surface. And just as often, there are bugs that have become wet and are so tightly bound to the surface of a pool that they become food for the fish, no matter how hard they struggle to break free.

Next time you fill a glass beyond the brim without having it spill or make soap bubbles or blow ripples on a cup of coffee or shed a tear or end up with a mustache of beer froth, remember surface tension and have another gulp.

A water strider resting on the surface of a pond. Its "feet," or tarsi, are covered with fine hairs, which increase the area in contact with the liquid and work something like snowshoes.

Surface tension arises from intermolecular forces at work at the surface of a liquid. Let's now examine how the same sort of forces act between different kinds of molecules to produce **solutions.** *A solution is a uniform mixture of two or more substances that displays characteristic properties of its own.* Although liquid solutions are the first to come to mind, gaseous and solid mixtures are also common. For example, the alloy of copper and silver known as sterling silver is a solid solution, as are brass, bronze, and steel, but not all alloys are solutions.

Ordinary household ammonia is a solution of the colorless gas ammonia in water, and it is the escaping NH_3 molecules that smell so terrible. Carbon dioxide gas dissolved in water produces the familiar solution we know as seltzer or club soda (and fish rely on dissolved oxygen). Shaking a bottle of beer provides enough kinetic energy to liberate some of the CO_2 from its bondage within the spaces between the liquid molecules in a rush of gas-filled bubbles.

Sugar crystals dissolved in water yield yet another commonplace solution, one that can provide us with some insights into the energetics of the process. When sugar dissolves, the ordered array in the solid is totally shattered, and the water molecules are pushed apart to accommodate the large intruders. Both of these occurrences require energy, require doing work against the attractive forces that exist between the molecules of the water (liquid-liquid) on the one hand and those of the sugar (solid-solid) on the other. The process can begin spontaneously when a fast-moving, energetic water molecule collides with a crystal and knocks out a sugar molecule. Heating the liquid and/or stirring it will increase the average KE of the water molecules and hurry things along.

Now, there also exists a sugar-water attraction, which quickly causes the freed sugar to become tightly surrounded by water molecules. This coming together of "solid" and "liquid" molecules under the influence of an attractive force effectively liberates electrical PE, just as a falling rock liberates gravitational PE. Molecular collisions carry off that energy, distributing it throughout the solution. In the case of sugar and water, this *source* of energy is almost—but not quite—enough to run the entire process. The needed additional energy comes from the overall kinetic energy of the molecules as they bang into one another. The temperature of the solution therefore drops somewhat. Such a process, in which thermal energy is absorbed, is said to be **endothermic.**

On the other hand, if a particular reaction converts an excess of potential energy into KE as the "solid" and "liquid" molecules come together, the solution as a whole will increase in temperature. Here thermal energy is liberated, and the process is **exothermic.**

As a rule, the presence of dissolved molecules will somewhat inhibit any solution from both solidifying and vaporizing: *The freezing point will be lowered, the boiling point elevated.* This is why substances like ethylene glycol ($C_2H_6O_2$) are added to the radiator water

13.2 SOLUTIONS

"Water."

There is nothing softer and weaker than water and yet there is nothing better for attacking hard and strong things.

LAO-TZE

The Tao

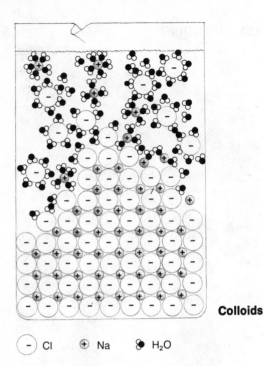

- Cl   + Na   H₂O

Salt dissolving in water.

in cars—they call it antifreeze. My grandmother used to put salt in water "to make it cook better"—to have it reach a higher-temperature boil and therefore cook faster.

The fact that its molecules are permanently polarized makes water particularly effective as a solvent. As any kid will testify, saliva will dissolve almost everything from jelly beans to aspirin. Water dissolves so many other substances that it is rarely if ever found in a pure state in Nature. A salt crystal immersed in water is, in effect, invaded by H_2O molecules. These surround the positive Na^+ ions at the crystal's surface, clustering about them, attracting them with their negative oxygen ends. At the same time the negative Cl^- ions are also ringed with water molecules, their positive hydrogen ends facing toward the chlorine. Encircled and shielded electrically, the individual ions float off into solution as the crystal dissolves away. So long as there are enough H_2O molecules to isolate the ions from one another, they will stay in solution, but if the water evaporates to the point where the ions are exposed and can reunite, they will simply recrystallize.

Colloids   Relatively large solid particles that do not dissolve can float around in a liquid for quite a while before ultimately settling out under the influence of gravity. Muddy water, orange juice, and calamine lotion are examples of **suspensions,** mixtures in which the solid does not enter into solution. The moving molecules of a liquid, constantly bombarding anything floating within it, can indefinitely support foreign particles, keeping them from settling, provided they are small enough. When particles in the size range of about 10^{-9}m to 10^{-6}m are dispersed in a liquid, they produce what is called a **colloid.** A colloid, such as milk of magnesia (magnesium hydroxide in water), in which tiny solid specks are suspended in a liquid, is referred to as a **sol.** Michael Faraday made a gold-water sol in 1857 that still shows no sign of any settling.

A colloid of liquid within liquid is known as an **emulsion.** Perhaps the best-known emulsion is homogenized milk, in which microscopic globules of butterfat are dispersed in a water solution. Oil and vinegar can be emulsified with the addition of a stabilizer, such as egg yolk. The result is mayonnaise.

A **gel** is a kind of cohesive colloid, in which the solid material forms a fine network that surrounds little droplets of the liquid (like an inside-out sol). Gelatin, jellies, certain hair goops, "canned heat," and napalm are all gels.

13.3 PRESSURE IN GENERAL   Village women in Asia still carry water in jugs gracefully poised on their heads. Having a density of 62.4 lb/ft³, water in any sizable quantity can be surprisingly heavy. Just two cubic feet of it, jugged and balanced, weighs 124.8 lbs. Certainly the considerable weight of the stuff pushing in on the human body (particularly on the eardrums) would be evident to anyone swimming several feet below the surface. The possibility of generating truly tremendous forces beneath the sea becomes apparent, especially in those nearly-crushed-submarine movies that Hollywood was once so fond of—"Dive! Dive! Dive!"

The force on each square foot of the ocean's floor, even in a region only one mile deep, is in excess of 300,000 pounds! That horrendous load arises, just as on a smaller scale with the jugs, simply from the weight of the supported water—in this case the mile-high column. A fluid (liquid or gas) in contact with the entire surface of a submerged object exerts a force acting everywhere on that surface. Thus far we have become accustomed to dealing with forces applied at specific points on a body. This sort of broadly distributed force is a little different. We now have to think in terms of both the force and the area over which it acts. The notion of **pressure** is particularly convenient here. It's defined as *the total force acting perpendicular to a surface divided by the area of that surface:*

$$\text{pressure} = \frac{\text{force}}{\text{area}},$$

or symbolically,

$$p = \frac{F}{A}.$$

The most common pressure unit in the United States is *pounds per square inch,* or psi for short. In the metric system *newtons per square meter* (N/m^2) is the appropriate measure.*

The force exerted by a fluid at rest on any rigid surface is perpendicular to that surface; it cannot be otherwise unless the substance doing the pushing has some degree of internal rigidity, and a fluid does not. You can certainly push on a wall with a stick in a direction that is not perpendicular to the wall, and you can even do the same with a moving stream of water from a hose, but a fluid at rest pushes only perpendicularly.

This idea is easy to appreciate in terms of the kinetic theory. The force arises out of millions upon millions of collisions between the moving molecules of the fluid and the rigid surface. Each impact exerts a minute force that is perpendicular to the wall if the wall itself is perfectly smooth but otherwise may or may not be. Nonetheless, since there are many collisions *randomly* occurring, the forces imparted parallel to a real surface tend, on the average, to cancel. A molecule moving in a generally upward direction will, on impact, push upward along the surface. But it is just as likely that somewhere else on the wall there will be a downward-moving molecule and a canceling downward force. Still, every collision will more or less exert a push perpendicularly and all these add up.

Of course, the generation of pressure is not restricted to fluids. The pressure exerted by this book lying on a table depends on its

---

* $1 \text{ lb/in}^2 = 6900 \text{ N/m}^2$.

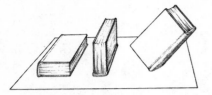

Pressure varies with the area over which the force acts.

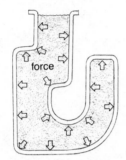

Force acting perpendicular to the surfaces in contact with the liquid.

In Fluids at Rest

weight, which is constant, and on the contact area, which can be changed. If the book is sitting flat on its broad cover, the pressure is relatively small. The pressure is a good deal less than it would be if the book rested on its end and still less than if the book was standing on a corner, where the area is tiny and the pressure relatively large.

One of the great faddist idiocies of a few decades ago was the bare steel-spiked high heel used on women's shoes. Ordinarily, a shoe may have roughly 20 square inches of contact area (heel and sole) with the floor. A 120-pound woman with all her weight shifted to one foot would then exert a meager pressure of 120 lb/20 in², or 6 pounds per square inch. But when that weight was concentrated on the heel of one shoe (as it might briefly be while the wearer was walking) with a spike area of about 0.12 square inches, the pressure would soar to 1000 psi! The resulting dents, holes, and chipped tiles prompted some shopkeepers to ban their wearers as a walking menace. (Actually it was a sitting menace as well, because the body weight shifts to the heels just before the wearer lands.) If these reminiscences are a bit too ancient for you, think of the humble thumbtack or the snowshoe, which summarize it all.

Imagine yourself submerged in a pool holding a postage stamp with an area A parallel to and well below the liquid's surface. The force on the stamp (the weight of the column of water above it) depends on its depth. The volume of that column (regardless of the shape of the stamp) is simply the area, A, times the depth, d. Weight-density (usually represented by the Greek letter rho, ρ) is weight per unit volume, weight divided by volume, so multiplying ρ by volume gives us the column's total weight, $A \times d \times \rho$. We get the pressure by dividing that weight of water by the area of the stamp, so A cancels out, yielding simply

$$\text{pressure} = \text{depth} \times \text{weight-density},$$

or

$$p = d \times \rho.$$

Note that pressure is quite independent of both the size of the stamp and the size of the body of liquid involved. The pressure at a point one foot down in a filled bathtub is the same as it would be one foot down in a lake or one foot down in a flower vase.

It follows that a *"weightless" fluid* (one in free fall or far out in space) *generates no internal pressure*. That much is true when the quantities of fluid involved are modest and internal gravitational effects are therefore negligible—conditions that do not hold within something big, like a star.

A rigid pillar of ice exerting pressure on the ground, like other solids, resists the flow of its own molecules and therefore undergoes no appreciable deformation. Gravity draws the atoms downward, and the strong internal binding forces keep them from any sizable relative motion. Yet if the column of ice were to melt instantly, some-

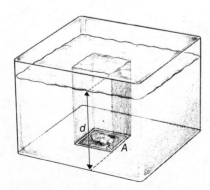

Calculation of pressure within a liquid.

thing quite different would happen. The shaft of water would quickly belly out, molecules would slide over one another and tumble outward, and the column would collapse.

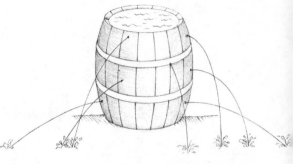

The effect of this outward rushing of fluids is commonly observable when water spurts from a hole in the side of a filled container—the lower down the hole, the greater the pressure and the faster it shoots out. Similarly, grains of dry sand cannot form a self-sustaining vertical cylinder; rather, they run off into a conical heap. Both water and sand require a container to resist the tendency to flow out, that is, to provide an inward counterpressure. The situation is much like trying to make a tall stack of marbles. Without a strong binding force, some glue, they will just roll away. To be sure, all of this assumes the presence of gravity—no gravity, no pressure, no atoms tumbling, no sideways spurting out of holes.

Water spouting from a few holes in a barrel.

The mobility of atoms in a fluid is the mechanism that transmits pressure equally in all directions. In other words, the pressure recorded at any point in a liquid is independent of the orientation of the measuring gauge. You can turn your head sideways (under water or in a jet plane), and your ears will still hurt. The existence of small spherical bubbles in water attests to the nondirectional character of pressure. Indeed, if the pressure at any point within a fluid was not equal in all directions, the molecules there would experience an unbalanced force, and they would move. The resulting flow of the fluid would continue until the pressure equalized. Your local TV weather forecaster is forever drawing maps of high- and low-pressure regions and the blowing winds.

The Dutch boy who stuck his finger in the dike and held back the Atlantic could actually have accomplished such a feat (provided the water level was not too high above the hole). The pressure he would have had to overcome equaled the liquid depth times the density and was independent of the size of the water body. The force exerted on his fingertip was of course perpendicular to its surface, tending to pop it back out of the hole like a cork in a champagne bottle.

Pressure depends on the depth of the hole below the liquid's surface and is otherwise independent of how much water is behind the dike.

The pressure anywhere on the flat bottom of a container of fluid, regardless of how weird the shape of the container, depends only on the height to the surface of the liquid. In the case of a long, narrow-necked vessel with flat shoulders, this is not at all obvious although it can be easily understood. *The pressure at every point at a given depth in a liquid is the same.* If it was not, the fluid would experience an unbalanced horizontal force and would simply flow until the pressure equalized. At the level where the bottom of the neck ends, the pressure in the fluid is the same, whether there is glass or water just above it. Thus the pressure of the liquid produces an upward force on the horizontal glass shoulder. By Newton's third law, the glass in turn exerts an equal downward force on the fluid. At that level, whether there is water or glass above and pushing down, the downward force is identical. A goldfish near the bottom could not tell whether it was swimming under the shoulder or the neck.

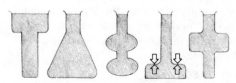

The pressure at a given depth is independent of the shape of the vessel.

By the way, it was Simon Stevinus in 1586 who proved the point experimentally. As he set the experiment up, each differently shaped vessel, clamped so that it was stationary, had a movable bottom that was actually one of the pans of a balance. The amount of weight that had to be placed on the other pan to keep the bottom from dropping out equaled the force exerted by the water.

Hydraulics

Two of Galileo's disciples, Torricelli and Viviani, gained widespread acclaim throughout Europe in the mid-1600s, principally for their development of the mercury barometer. Their work, which was something of a technological breakthrough, stimulated a resurgence of interest in the ancient mysteries of vacuum and pressure. Already a celebrated mathematician, the young Frenchman Blaise Pascal was one of those captivated by the new research. In or around 1651, when he was about 28, he wrote a major treatise on the subject entitled *On the Equilibrium of Liquids*.

That memoir contains the first precise statement of what has come to be known as **Pascal's principle:**

> An external pressure applied to a fluid which is confined within a container is transmitted undiminished throughout the entire fluid. It equalizes in all directions and acts via forces perpendicular to the retaining walls.

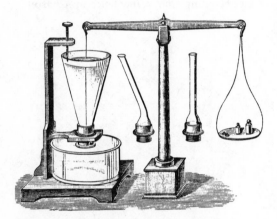

Stevinus's experiment to determine the pressure on the bottoms of differently shaped containers.

In other words, when pressure is put on some region of the surface of a confined liquid (for example, by a piston pressing down on it), there results a minute localized compression of the liquid, which then effectively spreads out until the pressure is rapidly equalized everywhere. This process has nothing to do with the internal pressure arising from gravity, and it would obtain equally well in a weightless environment. Pascal's syringe (see figure) illustrates the principle, as does the seltzer bottle or its cousin, the aerosol spray can, both of which use compressed gas to exert pressure on a liquid.

Note that the same amount of pressure can be generated within a fluid by two different-sized pistons, provided that the forces applied are proportionately different (see figure)—the larger the contact area, the larger the force required to produce the same pressure ($p = F/A$). At this point Pascal recognized a tremendously important practical application of his principle, which gave rise to a whole new class of force multipliers known as *hydraulic* machines (from the Greek for *water* and *pipe*).

If two chambers with different-sized pistons are attached so that they share a common working fluid, the pressure generated by one will be transmitted undiminished to the other. Most gas stations used to have service pits in which the mechanics stood while working on a car parked above, straddling the gap. Nowadays, cars are unceremoniously hoisted into the air on hydraulic lifts. Most often these lifts are activated by compressed air pressing on oil, but the simplest form is essentially a U-tube, narrow on one side, wide on the other,

ON WORK—EVEN BEFORE THE NOTION WAS FORMALIZED

And it is truly admirable that there is encountered in this new [hydraulic] machine the constant rule which appears in all the older machines, such as the lever, the wheel and axle, the endless screw, etc., which is, that the path is increased in the same proportion as the force.

BLAISE PASCAL
Traité de l'Equilibre des Liqueurs

with movable pistons sealing both ends. A downward force and corresponding *pressure* exerted on and by the narrow piston produces an equal *pressure* on the broad piston and a resulting upward force. Although the pressures are identical, the two forces clearly are not. If the face of the large piston under the car has 100 times the area of the small one, it will experience 100 times the force applied to the small piston.

Realize that when a machine (like a lever or a pulley) multiplies force, it does not multiply *work;* you never get out more energy than you put in. For the particular case at hand, the small piston will have to be pushed down 100 inches for every inch the large one rises, because the volume of liquid displaced by the small piston effectively spreads out thinly beneath the large one. Since for each piston that volume is its area multiplied by its displacement and inasmuch as the areas are in the ratio of 1 to 100, the displacements must be in the ratio of 100 to 1. Thus the input force times the distance over which it is applied (work-in) equals the output force times the distance over which it is applied (work-out). One person doing a lot of pumping could jack up an elephant provided *work-in equals work-out.*

Keep an eye open for hydraulic gadgets; you will find them on garbage compacters, missile platforms, plows, tractors, presses, airplane landing gear, forklifts, cherry pickers and powered shovels. They are even there in the family automobile linking the foot pedal to the brakes.

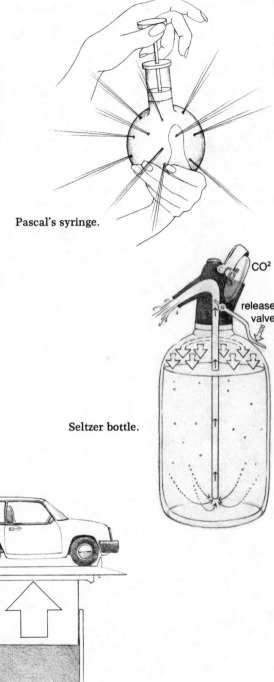

Pascal's syringe.

CO^2

release valve

Seltzer bottle.

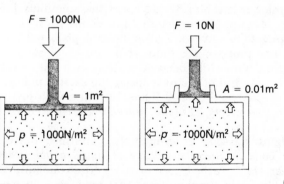

The force times the area of the piston determines the pressure.

$F = 1000N$

$A = 1m^2$

$p = 1000N/m^2$

$F = 10N$

$A = 0.01m^2$

$p = 1000N/m^2$

A hydraulic lift.

Eureka!

Determining the volume of a crown by displacing a liquid.

The sea had borne ships for several thousand years before the bright moment of Hellenistic Greece, and yet **hydrostatics** (the study of fluids at rest) truly begins there, in the third century B.C., with the incomparable Archimedes. Easily the greatest physicist of ancient times and one of the greatest mathematicians of all time, Archimedes was born and spent most of his long life in Syracuse, the major Greek settlement on Sicily's southeastern coast. He was of noble birth, a kinsman and friend of Hieron II, Tyrant of Syracuse, who was a most amiable and enlightened ruler. Like other well-to-do Greeks, Archimedes probably spent some time in Egypt as a young man, studying at the very hub of the scientific world, Alexandria.

Concerning Floating Bodies is one of several of his treatises that miraculously survived the long journey from Greece to Islam to Christendom to microfilm. In it he stated his buoyancy principle (one of the earliest of scientific laws), described the circumstances under which bodies float, and almost anticipated Pascal's principle— all in a remarkably modern tone. *"The surface of any liquid at rest,"* he rightly pointed out, *"is the surface of a sphere having the same center as the Earth."* Earth is a sphere, and the lakes and seas clinging to it curve about its center—a startling thought considering its antiquity.

Legend has it that King Hieron ordered a royal crown to be made for which he provided an exact measure of pure gold. When the piece was finally delivered, its weight, easily checked on a balance, was correct enough, but Hieron nonetheless was deeply disturbed by the suspicion that his jeweler had substituted silver for gold within the crown's hidden interior. Archimedes accepted the seemingly impossible challenge of determining the true nature of the royal headgear without damaging it in any way. After he had pondered and been perplexed by that problem for quite some time, its solution suddenly flashed before him while he was musing in a warm tub at the public baths. Unmindful of his nakedness, the excited philosopher leaped from the water and dashed home, shouting with joy as he ran nude through the streets, *"Eureka! Eureka!* I have found it! I have found it!"

Back at his lodging he soon experimentally confirmed the insight: *A submerged body displaces a volume of liquid equal to its own volume.* In other words, a body occupies a certain amount of space and, when totally sunk, simply pushes that amount of incompressible liquid out of its way. If the crown was immersed in a filled-to-the-brim container, the quantity of liquid that would overflow would equal the volume of the crown itself. Alternatively, in a less than full container, the rise in liquid level after submersion—that is, its apparent increase in volume—would again equal that of the crown. In either case, Archimedes now had a means of determining the crucial volume, which he could not otherwise accurately calculate because of the complex shape.

If the goldsmith was honest, a piece of pure gold of the same weight as the crown would have exactly the same volume. The presence of less dense silver would make the crown more bulky than the undiluted test piece of identical weight. Alas, most versions of this tale end with the anonymous goldsmith's untimely demise.

It is common knowledge that objects held under water are lighter than they are outside; the water buoys them up, pushes upward somehow. That is certainly clear to anyone who has tried to submerge an inflated tire tube or a beach ball, lifted a floating log, or carried a heavy scuba tank underwater. And it was equally clear to the sailors of antiquity, but it remained for Archimedes to quantify the process. His **buoyancy principle** maintains that *an object immersed in a fluid will be lighter (i.e., it will be buoyed up) by an amount equal to the weight of the fluid it displaces.* A ten-pound body that displaces two pounds of water will weigh only eight pounds while submerged.

Buoyant force arises from the pressure difference occurring between the top and bottom of an object, which always exists because pressure varies with depth. To see this more clearly, imagine a solid cube held beneath the surface of a liquid, as in the figure. The overall pressures on the four vertical sides are equal since the conditions are identical on each, but not so on the top and bottom. *Fluid pressure increases with depth,* so the pressure on the bottom of the cube is greater than the pressure on the top. Because those faces are at two different levels in the liquid, they must be at two different pressures. As for the *forces* acting, in each instance the force is equal to the pressure times the area of the respective face. Thus the upward force on the bottom face is greater than the downward force on the top face. Or equivalently, the force acting on either the upper or lower face equals the weight of a column of water extending up from that face to the surface of the liquid. The difference between those two forces, the **buoyant force,** is the difference in weight between the two fluid columns. That difference equals exactly the weight of a quantity of fluid identical in volume to the cube itself—*the weight of fluid displaced.*

That this is true as well for an oddly shaped object, like a human body, is easily seen. Any submerged object, regardless of what it's made of, displaces a volume of water equal to its own volume. Now imagine that we remove the object in question, in this case a submerged swimmer, and think only of the volume of water that then exactly fills the region the swimmer's body formerly occupied. Since the tank now contains only water, all of it motionless, we can assume that the body-shaped glob of liquid experiences no net force. Its weight downward must be canceled by the buoyant force acting upward on it, since no motion of the water occurs, and these are the only two forces at work. In other words, the buoyant force equals the weight of the water displaced, irrespective of shape.

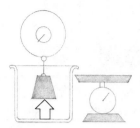

A dense ten-lb object displacing two lbs of water.

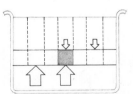

The buoyant force on a cube is the difference between the downward force on its top face and the larger upward force on its bottom face.

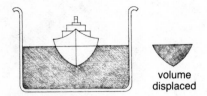

A floating object displaces its own weight of liquid.

If an object weighs more than the total amount of fluid it can displace, it sinks. A solid cube one foot on a side of gold, tin, aluminum, marble, concrete, or lead weighs more than the cubic foot of water displaced, and so each would sink, despite the 62.4-pound buoyant force it would experience.

If an object weighs less than the total amount of fluid it can displace, it will settle down until the buoyant force equals the weight and it floats partially submerged. A steel ship with a broad bottom encompasses a great deal of empty space (which gives it a low *average* density). However, it will have a large *displacement* and will float, cargo and all, provided that it weighs less than the maximum amount of water it can displace. Incidentally, the aircraft carrier U.S.S. *Nimitz* fully loaded has a record displacement of 95,100 tons.

Ships using canals are often charged passage by weight, which can be measured simply by recording the changes in water level (the volume displaced) when a vessel moves from one lock to the next. *The weight of water displaced by the submerged portion of an object equals the weight of the floating object.* A ship will often have depth markings painted on the hull that show how deep it's resting in the water and therefore how heavy its cargo is. The heavier the cargo, the more water the vessel must displace if it is to stay afloat, and the deeper down it settles.

Alligators and crocodiles swallow fair-sized rocks and store them in a muscular gizzard, where their food is ground. That tends to be helpful since they cannot chew, but whatever the reason, that dense load also gives them a low floating profile, which is handy when they go sneaking around the pond.

Of course the greater the density of the fluid, the greater the buoyant force. That's why it's easier to swim in the ocean ($\rho = 64$ lb/ft^3) than in a salt-free pool. A fresh egg, one in which little or no gas has developed, will sink in a glass of tap water, although adding a few tablespoons of salt will float it up to the surface.

Realize that you can displace more fluid than you actually have at hand. It is the *displaced* fluid that counts here. An object weighing a ton can be floated partially submerged in only a pound of water, provided it can displace a ton of the liquid. However paradoxical that may sound, it really isn't! Two close-fitting beakers illustrate the point nicely. The inner beaker floats surrounded by a fairly thin layer of water that has an actual volume far less than the volume displaced. Note that the beaker displaces the same amount of water here as it would floating in a lake.

An object whose weight exactly equals that of the total amount of fluid it can displace neither sinks nor floats but happily hovers beneath the surface under a zero net force. Some modern fish accomplish that feat by storing gas in an airtight bladder so that they do not have to be always in motion in order not to sink. Having no gas sac, sharks must literally swim or sink. By contrast, divers wear lead weights because the buoyant force on their totally submerged

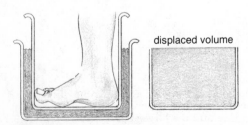

A way to displace more liquid than is actually present.

displaced volume

bodies is too great, and they would otherwise have to struggle to stay down. To rise, sink, or hover even without engines operating, submarines take on and discharge sea water from their ballast tanks. Dive! Dive! Dive!

Soon after Hieron II died, the Romans sent a powerful fleet led by General Marcellus against Syracuse. Thus began an incredible three-year siege pitting the might of Rome against the cunning of Archimedes, for the old man was then completely in charge of the city's defenses. His marvelous machines flung Roman ships onto the rocks, catapulted great barrages of stones against the advancing troops, and even (as legend has it) set the fleet ablaze with concentrated sunlight reflected from curved mirrors. Failing to storm the city, Marcellus resigned himself to strangling it by slow blockade, to starving it into submission. When the end was imminent, he ordered that Archimedes, for whom he had great admiration, not be harmed in any way during the inevitable sack and slaughter. A Roman soldier, fresh from the killing, came upon the solemn scholar completely engrossed in a mathematical vision he had traced in the dry sand. Haughtily the soldier commanded the great philosopher to come along at once, but the old man would not be moved until he had set the figure right. Before the impatient Roman returned to the fray, the fool slew the prophet, and we mourn the moment still.

The nuclear-powered U.S.S. *Los Angeles*.

EXPERIMENTS

1. You can weigh your own hand, at least approximately, with the following procedure. It assumes that your average density is close to that of fresh water, and since you surely float with most of your body submerged, that's a good enough approximation. Place a bowl partially filled with water on a platform scale. Now just read off the change in weight as you submerge your hand.

2. Make a stack of liquids of different density, floating one on top of the other in a tall narrow glass. Liqueurs are fun if you have them (grenadine 1.3 g/cm³, white crème de cacao 1.12 g/cm³, cognac 0.95 g/cm³, green crème de menthe 1.13 g/cm³). Or use water, carbon tetrachloride 1.6 g/cm³, trichloroethylene 1.4 g/cm³, benzene 0.88 g/cm³, ethyl or methyl alcohol 0.79 g/cm³. After you have made several layers, drop in some solids, such as cork, paraffin, and balsa wood. (Keep those organic compounds off your skin.)

3. Sprinkle some pepper on the surface of a glass of water. Touch the center of the pattern with your finger. Now try it again, but first rub some soap on your fingertip. What happened to the pepper and why?

4. Sprinkle some water onto a sheet of waxed paper. The droplets will stand rather tall, sagging under gravity but trying to be spherical. Now spray a very light mist of some household detergent over the drops. What happens and why?

5. Olive oil has a density of about 0.9 g/cm³. You can mix up a roughly 50% solution of alcohol (0.8 g/cm³) and water that will match it in density. Now gently introduce a drop of oil (from an eye dropper) into the solution. If all goes well, the drop will form a lovely sphere that hovers in the liquid and does not mix with it. This will probably take some trial-and-error fiddling with the alcohol and water. It may be convenient to remove the oil glob with an eyedropper in order to fine-tune the density of the bath.

REVIEW

Section 13.1

1. One can imagine the liquid as a transition state from the tumult of the _____ to the relative calm of the _____.

2. What are the two room-temperature elemental liquids?

3. Are the atoms or molecules of a liquid much farther apart than they were in the solid state?

4. What does *viscosity* measure?

5. Why do liquids assume the shapes of their containers?

6. Discuss the meaning of the term *surface tension*.

7. Is it true that the surface layer of a droplet will generally assume the least possible area?

Section 13.2

1. What is a *solution*?

2. Are all alloys solutions?

3. What is household ammonia?

4. Seltzer (club soda) is a _____ of _____ in _____.

5. What is the usual effect on the freezing and boiling points of a liquid when something is dissolved in it?

6. Define the term *colloid*.

7. Give an example of a *suspension*.

8. What is a *sol*?

9. Define the word *gel*.

10. When is a process said to be *endothermic*?

Section 13.3

1. What is the weight of two cubic feet of water?

2. Define the word *pressure*.

3. What are the common units of pressure in the United States?

4. How many N/m² equal 1 lb/in²?

5. The force exerted by a fluid at rest on a solid surface is always _____ to the surface.

6. The pressure within a fluid is given by the product of _____ and _____.

7. What does the existence of spherical bubbles in a glass of seltzer illustrate?

8. Why is the pressure in a liquid independent of direction?

9. State Pascal's principle.

10. What does Pascal's syringe illustrate?

11. Define the term *hydrostatics*.

12. Is the surface of a lake flat?

13. State Archimedes' buoyancy principle.

14. What gives rise to the buoyant force?

15. What is meant by the *displacement* of a ship?

16. When will an object float partially submerged?

17. When will an object sink to the bottom of a liquid?

18. Under what conditions will an object hover at any level below the surface of a liquid?

19. Why are nails usually pointed?

QUESTIONS

Answers to starred questions are given at the end of the book.

Section 13.1

* 1. Inasmuch as a little soap decreases the attraction between water molecules, will it be easier or harder to "float" a razor blade in soapy water than in pure water? Can you guess why soapy water is an effective cleanser?

2. Like soap, rubbing alcohol or acetone will reduce the mutual attraction of water molecules, thereby lessening the surface tension. Float two matchsticks or toothpicks parallel to each other about an inch apart in the middle of a pan of water. What will happen if you gently introduce a drop of water between the sticks? How about a drop of alcohol? Explain.

3. Concoct a bubble-making solution by adding water in the ratio of about 1 to 1 to any liquid dishwashing detergent. Throw in some glycerin as well if you have any (available at drugstores), but it's not crucial. Make a wire ring about two or three inches in diameter, and tie a small circle of thread across it. Dip the whole business into the soap, and then puncture the film within the sagging thread-loop. What will happen to the loop? Explain.

* 4. Suppose that two identical spherical droplets of a liquid came together as one. What would happen to the total amount of surface area? Imagine that the droplets were rubber balloons. What would be the difference in the total amount of energy stored in the stretched surfaces before and after coalescing? How would the temperature of the composite glob differ from that of the droplets?

5. Explain how a droplet of water can hang suspended from the mouth of a faucet.

6. What happens to the surface tension of a liquid as it is heated? Oil floating on water is usually pulled out into a thin sheet by the surface tension. What happens to oil floating on cool soup as the liquid is heated up?

* 7. Some dishwasher detergents are designed to reduce water spotting. What sort of properties might they possess?

Section 13.2

1. Pure benzene freezes at 5.5°C. What can you say about a sample of benzene that freezes at 4°C?

* 2. Provide another example of a suspension not listed in the text. *Hint:* Just think of "liquids" that settle.

3. What kind of colloid is cream?

4. List a few examples of a liquid dispersed within a gas.

5. List a few examples of a solid dispersed within a gas.

* 6. Which state is dispersed in which other state in Ivory soap and marshmallows?

7. List a few examples of a gas dispersed within a liquid.

8. Solutions are usually fairly transparent (to most colors of light) but colloids are fairly opaque. How well does this apply to the substances considered in the last six questions? Is blood a solution or a colloid? How about maple syrup?

9. Liquid mercury dissolved in a metal (e.g., gold) forms a solution known as an *amalgam*. Can you name another amalgam?

10. Why do people salt icy roads and sidewalks?

* 11. Most liquid salad dressings are emulsions. Why do you have to "shake before using"? What happens to the KE you inject into the system?

12. Should an opened bottle of soda be refrigerated after being recapped, or can it be stored as well on a shelf somewhere at room temperature? Explain. Should soda bottles ever be stored in hot places?

Section 13.3

1. You are in an elevator with a tank of water and two goldfish. A pressure gauge in the gravel on the bottom of the tank reads pressure in excess of atmospheric. (a) What will it read when removed from the tank? (b) Suppose it read ½ psi in place, under the water with the elevator car at rest. Will it increase, decrease, or stay the same when the car accelerates upward? (c) Will it change if the car accelerates downward? How? (d) What will it read if the car drops at 32 ft/s²?

* 2. Does the pressure at some fixed point below the water in a pool increase when a helicopter with pontoons lands on the surface?

3. Explain how a seltzer bottle or an aerosol can works.

4. Picture yourself in a rowboat floating in a swimming pool. On board you have an iron anchor, a banana, and a chicken. (a) What happens to the level of the water in the pool if the chicken gets up and flies away? (b) What happens to the level if you eat the banana? (c) What happens to the level if you toss the anchor overboard?

* 5. Suppose that we have five flat bathroom scales, each weighing five pounds, and we stack them one on top of another, placing a 10-lb watermelon on the uppermost one. What will each scale read?

6. Two walls of the same length and height are to be built to hold back two equally deep and wide bodies of water. One is a ten-mile-long lake; the other is a 10-foot-wide canal. Which dam would you make stronger and why?

7. A metal container completely sealed except for a threaded hole on top is filled almost totally with water and placed on a scale, which then reads 20 lbs. Will the scale reading change if you, standing next to the container, insert a rod down in the liquid, being careful not to let it touch the container? Suppose that the rod is exceedingly light, essentially weightless. Does that alter your answer? Now what would happen if you screwed the weightless rod down to the cover, with the same portion of it submerged as before so that you could let go? Would the scale reading change from 20 lbs?

* 8. Put an ice cube in a glass and fill it to the very brim with water. Will it overflow as the ice melts?

9. Why does a nickel float in mercury but not in milk?

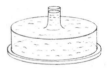

10. The narrow-necked vessel in the figure is filled with water and placed on a circular platform. Assume that the weight of the container is negligible. Does the pressure on the inside bottom of the vessel equal the pressure exerted on the platform?

11. A block of wood is floating in a pot of water on a table in an elevator. Is the height at which it floats in the water affected when the elevator accelerates downward at a rate less than g?

12. A cube of wood is pressed down against the flat bottom of a tank so that, although surrounded on five sides, it has no water beneath it. Will it experience a buoyant force? What is your advice to submarine commanders about clay regions on the sea bottom?

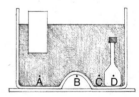

* 13. At which of points A, B, C, and D in the figure *must* the pressures be equal?

14. Will ice cubes float in alcohol? In gasoline? (See table on p. 283.)

15. Pascal set about to burst a sturdily made cask by fitting it with a very tall, narrow vertical tube opened on top and sealed into the lid of the barrel. The tube and cask were then filled with water (see figure). Explain what he was doing. Do you think he succeeded?

* 16. Will the bubbles in a glass of beer rise to the top of the liquid while the glass is in free fall? Why do bubbles generally rise?

17. Explain how a life jacket works. Will taking a deep breath affect your floating ability?

18. Is the pressure inside a submarine always equal to the pressure outside?

19. Is it possible to rupture a strong keg that is filled to the very brim by hammering a small solid peg into an opening in its lid? Explain.

20. The accompanying photo shows a glass tube open on both ends standing in a cup of cleaning fluid, trichloroethylene (of density 1.44 g/cm^3). Water with a few drops of black ink in it for contrast was poured into the tube. Describe and explain what you see.

MATHEMATICAL PROBLEMS

Answers to starred problems are given at the end of the book.

Section 13.3

1. Igor the Magnificent balances upside down on his 0.25-square-inch index fingertip. If he weighs 150 lb, what is the pressure on that wondrous digit?

2. What is the pressure exerted on the floor by a four-legged table with a mass of 10 kg on which sits 10 kg of physics books? Each leg has a square contact area with the floor of 2 cm on a side.

* 3. Assuming that salt water has a density of 64.4 lb/ft³, calculate the pressure at the bottom of the deepest ocean trenches, which surprisingly all reach about the same depth of 35,000 feet below the surface.

4. The density of sea water is 64.4 lb/ft³, and that of ice is 58 lb/ft³. What portion of a floating iceberg sticks out above the water? What fraction of an ice cube floats below the surface in a glass of pure water? (Check this out experimentally.)

* 5. Calculate the inward crushing force of a 1 m² hatch on the side of a submarine 100 m below the surface of the sea, where the water density is 1.03×10^3 kg/m³.

6. A 500-lb polar bear steps out onto an ice sheet one foot thick, and the ice sinks just beneath the surface. How big is the sheet? The density of ice is 58 lb/ft³, and that of sea water is 64.4 lb/ft³. How much does the ice weigh?

* 7. How much weight can be loaded onto a pine raft 1 ft by 10 ft by 20 ft for each inch it settles into the water? Assume the water is fresh and the wood has a density of 25 lb/ft³. How deep will it float with no load at all?

8. A 1.5-kg solid "gold" tray was purchased from a man in a raincoat on the corner of 14th and Main at a once-in-a-lifetime bargain price. When lowered into a filled fish tank, it displaced 100 cm³ of water. Should you be suspicious? The density of gold is 1.93×10^4 kg/m³.

9. A block of concrete (with density of 140 lb/ft³) sitting on a platform at the Al Capone Memorial Library in Chicago weighs 500 lb. How much did it weigh submerged while it was being hauled out of the river (fresh water)?

10. A water storage tank is 120 ft high. Water is pumped up and stored for continuous use. What is the street-level water pressure?

* 11. A hydraulic lift consists of two pistons acting on a common fluid. If the areas of the two are 10 in² and 500 in², which end do you put under a 2000-lb car to lift it? How much force must you exert on your end? If the car is to be raised five ft, how far must the other piston be lowered?

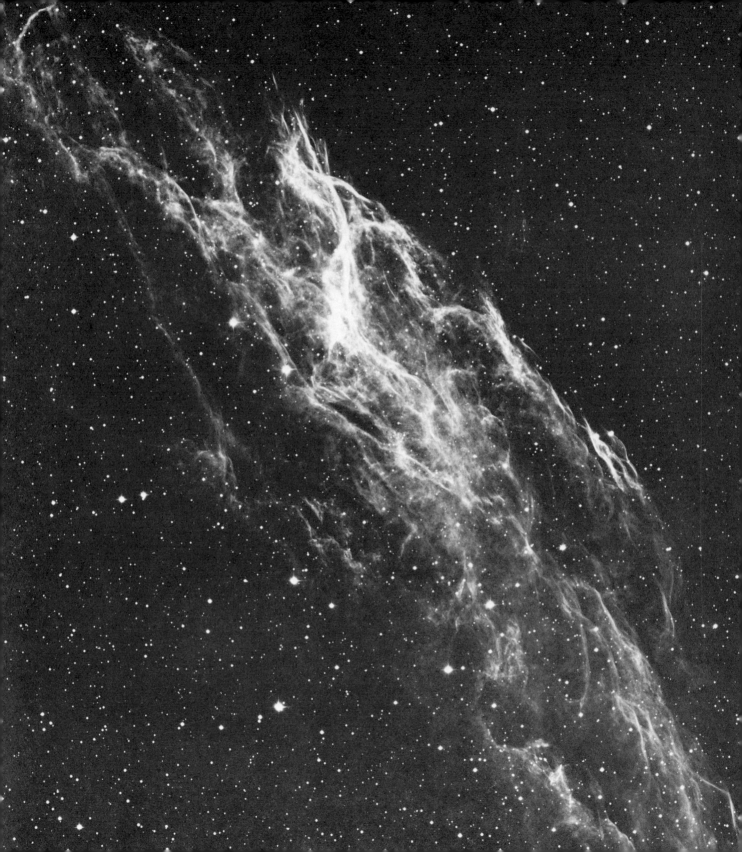

Gas and **plasma** constitute the last two states to be considered—the tenuous, weakly bound, chaotic bustling of widely spaced atoms and ions. The first section of this chapter introduces the dynamic imagery of the gaseous state, a ceaseless frenzy of high-speed colliding atoms. **Atmospheric pressure,** arising beneath the great weight of the surrounding sea of air, is the next focus of attention. The third section deals with some of the familiar fluid aspects of gases, that is, with exerted forces and **buoyancy. Boyle's law,** the relationship between the pressure and volume of a gas at constant temperature, is the theme of the fourth section. **A plasma is a gas that to some degree is charged or ionized; it's a collection of ions, electrons, and neutral particles that act in concert.** The final section of the chapter examines this tremendously important state.

CHAPTER **14**
Gases and Plasmas

Gases are freedom stuff—the almost uninhibited movers whose atoms (or molecules) possess enough kinetic energy to overcome the waiting web of binding force, to sail past one another without becoming ensnared, without condensing to liquid.

Most gases are transparent, invisible wisps like helium, nitrogen, argon, and fluorine, though a few are colorful enough: Chlorine is a pale yellow-green, iodine vapor is violet, and bromine gas is an orange-brown. Radon is radioactive and deadly, oxygen is the breath of life, nitrous oxide is amusing, and carbon monoxide is lethal. Hydrogen explodes, "natural" gas burns, and carbon dioxide extinguishes. The stench of hydrogen sulfide is rotten-egg-vile, ozone has a strange metallic sweetness, and water vapor has no scent at all. The smells of every day—of food and sweat and soil—reveal the vaporous traces of things, the wafting molecules of exceedingly faint gaseous clouds.

Though remarkably diverse, *gases are all compressible, all can be mixed in any proportion, and all tend to assume the shape and volume of their containers.* Their properties allow gases to blend readily, concealing their individuality. Because gases are far more subtle than solids or liquids, it was not until the seventeenth century that anyone even realized that there were different kinds of airlike substances. That revelation came to the Belgian mystic and alchemist Johann Baptista van Helmont (who had a long and difficult time with the Inquisition, primarily for mixing his physics with their religion). Somewhere around 1620, the great conjurer transmuted the Greek word for *chaos* into Flemish. What resulted was the word image *gas*, and a bustling chaos it is.

On facing page, the Veil Nebula in the constellation Cygnus is a great expanding cloud of gas, the glowing remnants of an ancient supernova.

313

14.1   RANDOM FURY

The "calm" invisible gaseous sea that envelops us all is in truth a random fury, a raging torrent of mayhem in miniature. The atmosphere, a mixture composed mostly of nitrogen (78%) and oxygen (21%), is a nonuniform layer of gas thinning out to almost nothing at a height of about 25 miles though, to be sure, it extends tenuously far beyond that. Activated by the warmth of the sun, its molecules are in endless motion and constant collision. At this very instant, each square inch of your face (and indeed, of every other exposed surface at sea level) is being pelted by two million million million million molecules every second! They collide, ricochet, sail off, smash into other molecules, and rebound over and again.

Under ordinary conditions of temperature and pressure, there are 6×10^{23} molecules in each 22.4 cubic liters of *any* gas (page 222); that's roughly 3×10^{19} per cubic centimeter, or 30,000,000,-000,000,000 molecules careening around within a tiny cube a mere one millimeter on a side. The distances between molecules, though constantly changing, seem to be typically about 10 times the size of the molecules themselves—roughly 3×10^{-9}m. Compared to the separations in the solid and liquid states, that average distance is quite large. Except for a very slight cohesion due to van der Waals force and even less to gravity, these far-flung molecules behave quite independently. At its greatest, the density of air near the planet's surface is only approximately 1/800 that of water.

Breezes aside, an air molecule on average travels at an incredible speed of about 450 m/s, or just about 1000 mi/hr. Typically it will get no farther than eight millionths of a centimeter before slamming into some other molecule, losing or gaining speed, and sailing off in a new direction until it collides again—going nowhere in a hurry. This aimless blasting goes on at a rate of about six thousand million collisions per second for each and every molecule in a gas under usual room conditions. Increasing the number of molecules present (i.e., the density) and/or their speeds (i.e., the temperature) would certainly result in a corresponding increase in that collision rate, and vice versa.

A jostling mass of high-speed, crashing molecules will spread out quickly within a container until they encounter and rebound from the walls. *They will rattle around everywhere within and fill the vessel more or less uniformly.* A gas puffs up and swells of its own internal violence. The room you are in right now is *filled* by the roughly 10^{27} to 10^{28} molecules of air rushing around within it. That's why we never have to worry about the unlikely possibility of walking into a truly *empty* region. Clearly the word "filled" would become far less meaningful if there were only one or two molecules flying about in the room.

Gravity, which is a long-range force, can exert a considerable influence in restraining the expansion of a gas, just as a material container can. Our atmosphere is confined to the planet by the mutual gravitational attraction between Earth and each and every air molecule. It's a delicate balance of molecular speed (i.e., tempera-

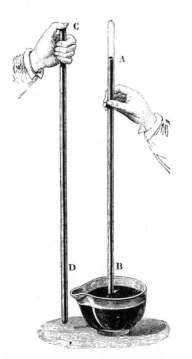

The atmosphere pressing down on the open mercury surface can support a 30-inch column.

ture) and strength of attraction (i.e., masses and distances) that keeps the film of air from escaping into the void, leaving us breathless. A drastic drop in temperature would make the atmosphere contract; a rise would make it begin to blow away.

Gravitational forces acting between the atoms within a great cloud of interstellar gas will cause it to gradually contract. As the density increases, the collision rate increases, gravitational potential energy is converted into kinetic energy, the temperature rises, and if all goes well, the ball of gas blinks on as a star (p. 369). Again, gravity influences gases, though here the setting is far more glorious than the ordinary laboratory one, where it's rarely a significant agent.

That familiar incantation, "gases expand *indefinitely,* uniformly, filling *all* the available space," really does not mean "indefinitely" or "all." It's much too grandiose to be taken literally. Still, as with all such textbook simplifications, it is often true enough. Because of Earth's gravity, the same amount of gas that would "fill" a 50-foot high vessel would not "fill" (in any ordinary sense) one 50 miles high, even if they both had the same volume.

14.2 A SEA OF AIR

Galileo knew that air had weight; he had determined that fact firsthand in a very straightforward way. He simply weighed a glass bulb sealed under room conditions and then opened it, forced in more air under pressure, and sealed and weighed it again. The maestro also knew (1638) that even the finest *suction* pump could draw water up a tube only to a height of about 34 feet and not "a hair's breadth" higher. Yet it remained for his student Evangelista Torricelli (1643) to bring the two seemingly unrelated ideas together as cause and effect.

About 14 times denser than water, mercury was far more convenient to work with since it would presumably rise only around 30 inches instead of 34 feet. Ingeniously, Torricelli sealed a long glass tube at one end, filled it to the very brim with mercury, and closing the opening with his finger, upended the tube into a bowl containing more mercury. When he released the submerged end, some of the heavy liquid in the pipe poured out, and the shiny column fell to a height of just about 30 inches (76 cm) above the mercury level in the bowl. Inasmuch as no air had entered the tube, the space at its top above the column was literally empty.

"On the surface of the liquid which is in the bowl," wrote Torricelli by way of explanation, "there rests the weight of a height of fifty miles of air." This air pressure, acting as if on an invisible ring-shaped piston, forces the open liquid surface downward. Pressure is in turn transmitted through the fluid and acts with an upward force on the narrow mercury column as if that cylinder of metal extending above the main liquid level were a piston in Pascal's hydrostatic setup. After being inverted and released, the column settles at a height such that the pressure it exerts equals the pressure beneath and supporting it; the total force is then zero, and any mo-

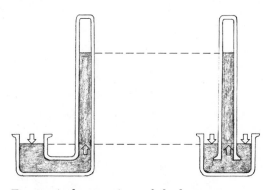

Two equivalent versions of the barometer.

A column of water 34 feet high produces the same pressure as a column of mercury 30 inches high.

water

mercury

tion of the column ceases. It may be helpful just to imagine that the pressure at the air piston equals the pressure at the column piston. Everywhere at the *level* of the open surface (both inside and outside the liquid) the pressure is equal. The column of mercury simply takes the place of a much taller though equally weighty column of air. The height of the mercury is therefore a direct measure of **atmospheric pressure.** Torricelli had constructed the first *mercury barometer*.

"We live immersed at the bottom of a sea of elemental air," he remarked—*a fluid sea within which pressures are generated precisely as they are beneath a liquid.* Roughly 6000 million million tons of air press down on the Earth, giving rise to a sea-level pressure *in all directions* of roughly 14.7 pounds per square inch (or about 100,000 N/m²). That is equivalent to a barometer reading of 29.9 inches of mercury* though pressure ordinarily varies over a range of 31 to 29 inches of mercury. By the way, rising air pressure (a higher barometer reading) usually portends fair weather, and falling pressure often heralds a storm.

The medicine dropper and its culinary cousin the meat baster are suction devices on the family tree of the barometer. With the rubber bulb collapsed, the tube end is immersed in liquid. When the bulb is released, it springs back to size, creating a *partial vacuum* within. Air pressure acting on the open surface then forces liquid up the tube. In the barometer, only the weight of the mercury column creates pressure in the tube, which counterbalances atmospheric pressure; there is no air above the thin shaft of liquid. In the medicine dropper there *is* air, which pushes down on the column, effectively adding to its weight. As a result, a much shorter column is forced up into the tube. Still, who needs a 34-foot eye dropper? The drinking straw and the pipette are the same machine, with the mouth replacing the rubber bulb. A slight drop in mouth pressure to about 97 percent atmospheric will satisfactorily operate the device, and of course, a pressure in excess of atmospheric drives the liquid column down below the open level and makes bubbles.

After almost 2000 years of philosophical speculation about the unattainable void, the "Torricellian vacuum," as it came to be called, was viewed as a miraculous accomplishment. With the top of the glass pipe made bulbous, fair-sized regions could be evacuated. Torricelli went on to demonstrate that the volume above the column was indeed empty. He cleverly did so by floating a few inches of water atop the mercury in the open bowl after the device was set up. When the tube was raised just out of the mercury bath, the heavy liquid metal ran down and out the tube as water rushed in, *completely filling the entire bulb.* All this had a tremendous effect on the intellectual habit of the era, which still embraced Aristotle's witless

---

* It's left as a problem to show that 29.9 in. of mercury (Hg) exerts a pressure of 14.7 lb/in² by way of $p = d \times \rho$.

notion that *Nature abhors a vacuum*. In fact, the universe is mostly emptiness, and if it made any sense at all to attribute passions to "her" (it doesn't), then one could say that *Nature adores a vacuum*.

Nowadays it's fairly easy to pump a chamber down to a thousandth of a millionth of the pressure in Torricelli's "vacuum," which was in fact faintly "filled" with mercury vapor. By contemporary standards, his was more a symbolic than an actual void, and yet, despite all our technological craft, we have never created anything that qualifies as a *perfect* vacuum. Perfection, after all, is something that does not quite exist when you go to measure it, though we like to believe that it should. Perhaps "witless" was too strong a word. In any event, Torricelli's experiment supposedly helped put an end to the stranglehold of Aristotelian physics. Strangely enough, 300 years later, there stood Miss Buckley, my sixth grade science teacher, pontificating on how Nature abhors a vacuum—the myth lives on.

As soon as Pascal heard accounts of the Italian experiments, he set out to test his notion that air pressure should decrease with increasing altitude. Not being physically up to it himself, he sent his young brother-in-law to climb the 3200-foot Mont Puy-de-Dôme carrying a mercury barometer in hand. To their utter delight, the column actually dropped three inches during the ascent. This was the final blow to Aristotle's doctrine of the *horror vacui* (horror of the void).*

At just about the same time (1646), Otto von Guericke was elected burgermeister of Magdeburg in Saxony, in appreciation for his help in rebuilding that city after its destruction during the Thirty Years' War. However busy with the business of government, he still managed to dabble in science, and within four years he constructed the world's first vacuum pump. This extraordinary man, a bold thinker and lavish spender, then put on a theatrical performance that surpassed even Galileo's showmanship when he demonstrated the telescope on the Campanile in Venice.

Von Guericke constructed a pair of matching hollow bronze hemispheres about two feet in diameter. These he placed against a soft gasket ring and then evacuated them through a valve with his pump, until the outside air pressure forced them together as a single sphere. On May 8, 1654, at Regensburg with Emperor Ferdinand III and the Reichstag looking on, he orchestrated a gigantic tug-of-war with horses against hemispheres. Only after sixteen powerful animals, four pairs on each side, had been hitched on was it finally possible to overcome the pressure of the atmosphere and yank the hemispheres apart!

Just as pressure at a point within a liquid is equal in all directions (and the force exerted acts perpendicular to any rigid surface),

14.3 AS A FLUID

The original Magdeburg hemispheres and the air pump used to evacuate them. Each segment had a broad lead lip. A gasket was compressed between the two lips to make the seal.

---

* Well, not quite. Descartes, who knew Pascal and had debated the point with him, never accepted the vacuum. Years later he even remarked that it seemed his young friend Blaise had a void in his head.

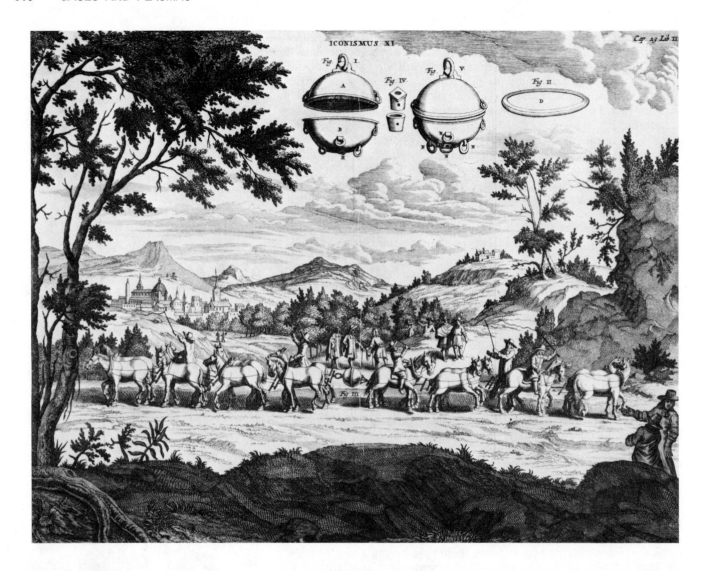

Von Guericke's version of the Magdeburg sphere demonstration from his *Experimenta Nova,* published in 1672.

so is pressure within a gas. We don't notice this rather considerable force because we are filled with and surrounded by air pushing in every which way. But if the air is drawn out of a container (as with a vacuum-sealed jar), that vessel had best be a strong one, or it will certainly be crushed by the 14.7 psi acting everywhere on its surface. Kids are always smashing paper cups simply by sucking the air out of them and wearing the crumpled remains as a muzzle. We can easily have the sea of air collapse something more formidable, even without a pump. Put a little water in an open gallon can, and gently heat it. The escaping steam will drive out all the air. Then cap it, cool it, and watch it collapse as the vapor condenses.

An ordinary window (3 ft by 4 ft) experiences a total force from the bombarding air molecules outside equal to 36 in $\times$ 48 in $\times$ 14.7 lb/in^2, or 25,401 pounds. This is an inward thrust, but it's usually balanced by an equal and opposite force pushing outward. If the outside pressure suddenly drops, as it might during a tornado, a house can literally explode. We have all felt the effects of a slight air pressure difference on the eardrum in rapidly moving elevators and planes.

Like any fluid, air exerts a buoyant force. That force is equal to the weight of the volume of air displaced by the object. Because air weighs only about 0.08 pound per cubic foot, the effect generally goes unnoticed. A person having a volume of roughly two cubic feet experiences a buoyant force of 2 ft^3 $\times$ 0.08 lb/ft^3, or only 0.16 lb (that is, 2.5 ounces). Helium has 1/7 the density of air, and it doesn't take much of it in a kid's balloon to displace a volume of air weighing more than the total of the balloon, the helium, and a bit of string. Yet 10,000 cubic feet of helium is needed to support only 700 pounds of load.

It has been known for centuries that warm air in a room rises, that it seems to float to the ceiling. Of course, the density of a given quantity of any gas will increase as its temperature (and therefore its volume) decreases until it liquifies. Hot air literally floats because it is less dense than cold air; each cubic foot of it is lighter.

A metal can crushed by having the pressure within it drop below atmospheric.

A partially filled hot-air balloon.

Filled and flying.

On a lovely afternoon in November of 1783, two young Frenchmen, Jean-François Pilâtre de Rozier and the Marquis d'Arlandes, climbed aboard a 74-foot blue balloon bedecked with golden zodiacal signs, portraits of Apollo, and delicate floral wreaths. That marvelous hot-air aerostat, built by the brothers Montgolfier, had been brought to Paris for the epochal flight. The mooring lines were cast off, and the ship slowly rose into the air. The two fliers busied themselves stoking the heater fire with straw and dousing small blazes that flared up around the open neck of the huge linen bag. For 25 minutes they soared over the rooftops in the first free flight in history with humans aloft. Hundreds of thousands saw them that afternoon, including the American envoy, B. Franklin.

Ten days later, the Montgolfiers' chief rival, physicist Jacques Charles, ascended with a companion in a hydrogen-filled balloon (that gas, the lightest of all, had recently been discovered by Cavendish). Story has it that after safely landing, Charles's companion unceremoniously disembarked, and as he did so, the craft lurched into the air to make Charles the first solo balloonist despite himself. Some 150 years later, the *Hindenburg* disaster (May 6, 1937) ended the use of explosive hydrogen in airships.

Nowadays a typical hot-air balloon is kept inflated by a small on-board propane burner. Hot air is not very buoyant; the bag has to be quite large and the gas must be kept warm, or else down it comes. The Montgolfiers' blue behemoth bulged with 2200 cubic meters of hot air and smoke.

14.4 BOYLE'S LAW

The Honourable Robert Boyle in England read about von Guericke's experiments in 1657 and immediately set out to match the remarkable feat. His newly built laboratory at Oxford was well suited for the task, and the work, which was conducted by Boyle's ingenious assistant, Robert Hooke, moved along quickly. By 1659 they had produced a vacuum pump that far surpassed von Guericke's. It was called—to Hooke's chagrin—the "Boyle engine"; penniless lab assistants are easily forgotten at christenings.

With the new pump they soon established that neither combustion nor respiration (at least for flies, bees, birds, and mice) can be maintained in the absence of air. They showed that an alarm clock ringing inside a vacuum chamber was quite inaudible (underscoring the belief that sound was transmitted as a vibration of the air). They verified Galileo's notion of weight-independent free fall by dropping a metal ball and a feather in an evacuated tube. They even put a Torricellian barometer in a chamber, pumped out the air, and watched as the mercury dropped from 30 to 0 inches.

Though others before had certainly performed experiments, Boyle's almost total devotion of life and fortune to research in the laboratory was something of a novelty at the time. Mid-seventeenth-century science more often reflected Spinoza's view that reason was

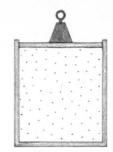

superior to experimentation, a view Boyle rejected despite his friendship with the Dutch philosopher.

Boyle's early work with vacuum naturally enough led to a study of gases. That research culminated in 1662 with his determination of a simple interdependence between the pressure and volume of any gas. Unlike solids or liquids, which are essentially incompressible, gases are as a rule easily reduced in volume, easily collapsed. Water is about 1600 times denser than steam at the same temperature (100°C), so steam contains a lot more empty space and can therefore be compressed. Although Richard Towneley and Henry Power were the first to actually discover the relationship (they had published it a year earlier), and Edme Mariotte independently arrived at it some time later, it's most often known as **Boyle's law** (in spite of his own acknowledgment to Towneley).

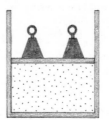

When the pressure on a gas doubles, the volume is halved; when it triples, the volume is reduced to one third.

The setup Boyle used in the research was remarkably simple yet elegant. He took a glass tube about eight feet long, bent it into the shape of a J, and sealed the short end. He then poured mercury into the open top, trapping some air, usually under pressure, within the smaller leg. When the device was tipped "so that air might freely pass from one leg into the other," the pressures were equalized and the mercury levels on both sides made the same. After the instrument was turned upright, air confined within the short leg was sealed in at atmospheric pressure, and the apparatus was then ready to use. As more mercury was subsequently added, its level rose unequally within the two tubular sections until the volume of gas compressed in the smaller segment was halved and the open column had risen just about 30 inches above the other. It was evident from that 30-inch difference in height that a pressure equivalent to 14.7 psi was being exerted by the additional mercury. The pressure in the trapped gas was then *twice* atmospheric,* the volume had been halved and the pressure doubled. At three times atmospheric pressure, the volume was further reduced to 1/3, and so on. In other words, keeping temperature constant, *the volume (V) of a given quantity of any gas varies inversely with the pressure (p)*. This is equivalent to saying that *pressure times volume is constant*, since doubling one halves the other, and vice versa. Symbolically, the relationship is simply

pV = constant.

To illustrate the idea, consider a bubble rising in a liquid (or a balloon floating higher in air). The external fluid pressure drops with altitude, and so the bubble's internal counterpressure must also lessen to match it. Boyle's law then maintains that since pV is to be constant, the bubble must expand in proportion. A fish that relies on a gas bladder to maintain buoyancy cannot quickly rise to the surface

---

* Newton's third law, which would lead to the conclusion that the pressure exerted on the confined gas equaled the pressure exerted by that gas, had not yet been enunciated. Even so, Boyle maintained that, as with a spring, the harder the gas was pushed on, the more it pushed back.

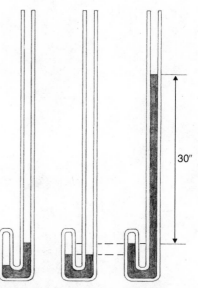

30″

The device used by Boyle to determine the law bearing his name.

An off-color cartoon by James Gilray (1802) mocking the "New Discoveries in PNEUMATICKS!" being made at the Royal Institution. The central figure, at the moment administering nitrous oxide, appears to be Thomas Young, professor of natural philosophy and chemistry. Next to him, bellows in hand, is Humphry Davy, who extensively experimented with and was not beyond enjoying "nitrous." Rumford, bemedaled, stands at the far right. The exhilarating effect of N_2O_2 on the libido caused many a titter during those highly fashionable lecture-demonstrations.

from the depths of the sea without causing that sac to blow up to lethal proportions.

To take another example, imagine someone sitting down on an air pillow. At first the cushion collapses, decreasing in volume until the rising internal pressure can support the person's weight, when the motion ceases. When the load is removed, the pressure drops, and the volume increases to its original value. Similarly, when a great quantity of gas is to be compressed and stored in a small container, a strong steel tank is the likely vessel since the pressure will rise proportionately.

Despite the fact that most prominent seventeenth-century thinkers embraced some form or other of atomism, nothing really substantial was done with that insight. The conception of a gas as an invisible swarm of bouncing, colliding atoms (the kinetic theory) was first

applied analytically in 1738 by Daniel Bernoulli, who actually succeeded in deriving Boyle's law from that imagery.

A molecule ricocheting off a chamber wall changes momentum in the process and (recall Newton's second and third laws) exerts a force on the wall. The staccato bombardment of countless millions of atoms blurs into what seems a constant force and pressure. Compressing a gas into half the volume doubles the *density*, giving the same number of molecules half as much room to fly around in. Accordingly, they will collide with the walls of the container that much more frequently. Doubling the number of collisions per second, in turn, doubles the pressure.

As is so often true, the seeming simplicity of Boyle's law is actually somewhat misleading; in a sense it is forced on a not altogether compliant Nature. The law is actually an approximation, an idealization, which applies exceedingly well over a broad range of temperature and pressure but does fail, nonetheless, under certain conditions. It works best (though Boyle did not know it) when the intermolecular forces exert a negligible effect, as they do when the molecules are far apart. Thus it's most in error whenever the molecules are relatively near to one another—at very high pressures and at temperatures close to liquefaction. A make-believe stuff that obeys the law exactly is known as an **ideal gas**—a useful theoretical fiction corresponding to noninteracting molecules. Insofar as all real gases more or less resemble the ideal gas, Boyle's law becomes a reliable gauge of the way Nature will generally behave.

Perhaps the most important aspect of Boyle's work on gases was that it began the long chain of events, the logical progression, that led step-by-step to Dalton and modern atomic theory.

14.5 PLASMA

When he took a job at the research laboratory of the General Electric Company in Schenectady in 1909, Irving Langmuir fully intended to spend only his summer vacation there and then return to teaching at Stevens Institute. Within days, though, he had become happily involved with a group trying to extend the life of filaments in incandescent lamps. Although the project was eminently practical, Langmuir's concerns were purely scientific. The gas-filled light bulb, one of his early important inventions, was for him merely a means of furthering fundamental research on the nature of atoms. Fifteen years later Langmuir was still at G.E., still probing into the behavior of heated gases (particularly atomic hydrogen). And it was around then, near the end of the 1920s, after having observed some rather extraordinary properties of charged gases, that he coined the word *plasma* to describe such systems.

In point of fact, ionized gas had been the object of laboratory scrutiny since the early days of discharge tubes (p. 236) in the 1850s. Prophetically, William Crookes, during a lecture at the Royal Institution in 1879, underscored the great significance of that subtle stuff within the glowing tubes by dubbing it the "Fourth State of Matter."

Aurora borealis.

The ancients had their four *elements* (Section 9.1), earth, water, air, and fire, which were not elements at all in any modern sense but better corresponded to our notion of four *states* of matter—solid, liquid, gas, and plasma.

Heat, the alchemist's wand, transforms one state into the next: solid to liquid to gas and finally to plasma. At temperatures around 4500°C, the solid state all but vanishes, melts away into liquid. Steel turns fluid at a mere 1500°C, whereas diamond holds rigid until more than twice that temperature. Somewhere before 6000°C (the temperature of the surface of the sun) the last remaining liquids bubble off into vapor and all is gaseous—iron, concrete, glass, everything vaporized. Depending on the binding forces, molecules split apart into their constituent atoms over a wide range of elevated temperatures. For example, by about 3000°C only about one quarter of any quantity of water present—as vapor, of course—ruptures into oxygen and hydrogen. Within the violence of 10,000°C, blasting collisions shatter most molecules into separate atoms, and many of those become ionized as well, losing one or more electrons in the tumult. Only raging tenuous swarms of atoms, ions, and electrons can survive at the surface temperatures of the hottest blue-white stars (roughly 50,000°C). And even in the furious blaze at 100,000°C, where all matter is generally ionized to some extent, neutral atoms do appear (as a result of spontaneous recombination of ions and electrons) if only to shatter again. *This sort of mixture of atoms and fragments of atoms (of ion cores and electrons), this whirlwind of conducting gas, which on average is electrically neutrol, constitutes a* **plasma.** Usually plasmas are quite hot (though they need not always be), so hot that below a million degrees or so a plasma is said to be "cold."

The transition from the gaseous to the plasma state is not abrupt; ions exist in the air in relatively small amounts, even at room temperature. The characteristic that is the hallmark of the plasma state is its collective behavior, arising from the interaction of and with its charges. Thus a plasma acting as a cohesive whole can be manipulated, through the application of electric and magnetic fields. Plasma beams can be pushed, heated, squeezed into shape, and even confined along specified paths.

Although plasma is not particularly abundant here on Earth, probably 99.9 percent of all the matter in the universe at large exists in the fourth state. The countless billions of stars ($\sim 10^{22}$), including our sun, are blazing plasma balls raging at millions of degrees. Interstellar and interplanetary space are both hung with delicate curtains of plasma, and great streamers of it pour from the sun (solar wind) to bathe the planets. Our immediate environment is too dense and too cool for the fourth state to survive very long naturally. Still, Earth's ionosphere is a faint, cold plasma that is significantly ionized beyond 60 km, not by thermal collisions but by the sun's ultraviolet radiation. Plasma streaming down along this planet's magnetic field glows as the aurora borealis, or northern lights. Immense lightning bolts and humble little flashes that crackle from a nylon shirt are both the fourth state sparking.

Alkali metals ionize readily, and a tiny bit of sodium, often inadvertently present in flames, is easily recognized by its bright yellow emanation (burn a bit of salt and see). Ordinary flames (matches, burning paper, etc.) are thereby weakly ionized, conduct electricity, and qualify—if only marginally—as the fourth state. The fire-plasma link has a certain ancient charm (earth/solid, water/ liquid, air/gas, fire/plasma).

Synthetic plasmas are everywhere, from gaudy neon signs and fluorescent lamps to blue-white mercury and yellow sodium vapor street lights. There are plasma rocket engines, plasma electrical generators (magnetohydrodynamic or MHD devices), and plasma lasers. Yet beyond all of these, the tremendous contemporary concern with the fourth state arises from the multibillion-dollar international race to harness the power of the H-bomb. At the very center of that effort, that drive to control the almost inexhaustible reservoir of thermonuclear energy (Section 16.2), is the blazing hydrogen plasma of the stars.

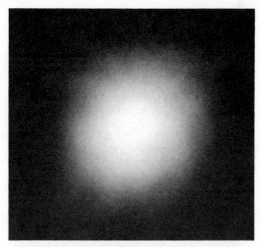

The star Betelgeuse, a blazing giant plasma ball. This is the first image of the face of a star (other than the sun) ever seen.

EXPERIMENTS

1. You can redo Galileo's experiment showing that air has weight. Simply weigh an "empty" basketball, and then pump it up and reweigh it. If you don't have a scale, balance two blown-up balloons on either end of a yardstick. When the arrangement is hanging horizontally, pop one of the balloons.

2. Lower an inverted empty glass straight down into a vessel of water. As the glass goes deeper, water will rise within it, compressing the air (you'll need plenty of depth to see this). Tilt the glass, and let all the air come bubbling up. Now raise the filled, submerged, still inverted glass so that much of it is *above* the surface. What happens? What is the limit on the size of the glass you can use before the water pours out?

3. Punch a hole in the cover of a jar, and insert a small funnel. Put some clay or gum around the funnel to seal it to the lid. Now pour some water into the funnel. Explain what you observe.

4. Put a piece of cardboard on top of a filled glass of water, and holding the cardboard in place, turn the glass upside down. Now take your hand away. Why does the cardboard stay in place? What would happen if you put a few coins in the glass along with the water? Does the glass have to be filled? How much air in the glass initially can you get away with? What happens if you move the glass downward and come to a sudden stop? (Do it over a sink!) Investigate the surface tension effects; i.e., see if a wet card will stick to the glass all by itself.

REVIEW

Section 14.1

1. Can gases mix in any proportion?

2. Solids and liquids are essentially incompressible. Are gases?

3. When the air is very, very still, are its molecules at rest?

4. How many gas molecules are there in one cubic millimeter under ordinary conditions?

5. What's a typical speed for an air molecule?

6. How many collisions does your ordinary, everyday, room-temperature air molecule undergo in a second?

7. Is gravity ever a significant influence on gases?

Section 14.2

1. What is the limiting height to which a "suction" pump can draw water?

2. How many more times denser than water is mercury?

3. In what sense are we immersed in a fluid sea of air?

4. Does Nature abhor a vacuum?

5. Was Torricelli's a hard (i.e., a good) vacuum?

6. In what year was the *perfect* vacuum first attained?

7. How do you blow bubbles with a straw in a glass of milk?

8. How did Torricelli show that the region above the mercury in the tube of his barometer was actually empty?

Section 14.3

1. What were the Magdeburg hemispheres?

2. How might you get the atmosphere to crush a tin can without using a pump?

3. Does the force exerted by a gas at rest filling a tank always act perpendicular to the walls?

4. The weight density of air at sea level is _____ pounds per cubic foot.

5. Estimate your body's buoyant force in sea-level air.

6. Why are hot-air balloons so large?

7. Would you prefer ascending in a hydrogen- or helium-filled balloon? Why?

8. How does a hot-air balloon work?

Section 14.4

1. The fascinating Robert Hooke is here again, this time with Boyle. What was he involved with?

2. List several of Boyle's vacuum experiments.

3. What was Spinoza's view of the relative merits of reason versus experiment?

4. A principle distinction between solids and liquids on one hand and gases on the other is that the latter are _____.

5. State Boyle's law.

6. What happens to the pressure in an air mattress when you lie on it?

7. Is the pressure exerted by a gas on the chamber in which it is confined actually constant when a typical gauge reads a constant value? Explain.

8. What happens to the pressure in a vessel when the number of collisions with the walls occurring each second is doubled?

9. Is Boyle's law exactly true for all gases under all conditions?

10. What is an *ideal gas*?

Section 14.5

1. What is a *plasma*?

2. When is a plasma said to be cold?

3. What is the hallmark of the plasma state?

4. Are plasmas common in the universe?

5. How would you classify the ionosphere?

6. What ionizes the gases in the ionosphere?

7. To which state of matter does a lightning bolt belong?

8. To which state of matter does a flame belong?

9. What is the aurora borealis?

QUESTIONS

Answers to starred questions are given at the end of the book.

Section 14.1

1. What is meant when it is said that a gas *fills* its container?

2. What are some of the limitations on the familiar statement "Gases expand indefinitely, uniformly, filling *all* the available space"?

* 3. What would happen to the Earth's atmosphere if the planet puffed up a bit, increasing in diameter but not in mass?

4. Describe what a gas would look like to an observer 10^{-10} m tall.

5. If gas molecules are flying around at 1000 miles per hour, why does it take several seconds to smell a newly opened bottle of perfume just a few feet away?

Section 14.2

1. Discuss the operation of a mercury barometer (incidentally, the word "barometer" was coined by Boyle). Must the tube be longer than 20 inches? What would happen to the reading, if anything, if someone floated a few coins on the open mercury surface? If a water spider tiptoed across it without breaking the surface?

* 2. Would Torricelli's barometer reading have been affected if a finger had been stuck into the open cup of mercury? Explain. If one had just poured in some more mercury?

3. Suppose that Torricelli's barometer was snugly fitted with a washer-shaped weightless disk that could ride up and down on the open mercury surface like a piston. Would the pressure reading change in any way? What if a bird landed on the disk?

4. A stopper is wedged into a hole with an area of one square inch in the side of a vacuum chamber. After the chamber is evacuated, the stopper is pulled out. How much force will it take to yank it free?

5. If there is a force of 14.7 lb acting on every square inch of your body, why don't you feel "put upon"?

* 6. Suppose that you lie down on your back and someone puts a flat board 10 inches on a side on your belly and loads it with weights until the pressure it exerts on you is 14.7 lb/in². How much weight would you be holding up? Would you feel it? If so, why it but not the 14.7 lb/in² of the atmosphere?

7. Why do you think the atmosphere becomes less dense with increasing altitude?

8. The cabin pressure in airplanes is generally main-

tained at levels somewhat lower than atmospheric. Would the level of mercury in an on-board barometer read more or less than 30 inches? Why are the planes pressurized in the first place? Why don't they pump the cabins up to atmospheric pressure? Why don't airliners have big picture windows?

* 9. Do we literally suck soda up a straw, or does the atmosphere push it up?

10. Suppose that we sealed a straw into an airtight bottle filled to the brim with water. Could you "suck" the liquid up the straw? The whole idea of "sucking" fluids is misleading, as this example underscores. Try it with a bottle half full. Punch another hole in the lid. What happens now?

11. The figure shows a rubber syringe bulb attached to some glass tubing. Describe what you see and explain what has taken place. What can you say about the densities of the two liquids?

12. Water vapor at the same pressure as air has a lower density; its molecules are much lighter than those of O_2 and N_2. How might this contribute to the notion that a falling barometer predicts bad weather?

* 13. Stick a straw into a glass of water, cover the end tightly with your index finger, and raise the straw out of the liquid. Why does the water stay within the straw?

Section 14.3

* 1. When a balloon ascends, it picks up KE and gravitational PE. Where does that energy come from? Did it take work to fill the balloon to start with?

2. Suppose that you had a helium-filled balloon floating inside the family car while out for a Sunday drive, and the vehicle came to a sudden stop. Which way would you be thrown? How about the balloon? Try it sometime.

3. Why does a rubber balloon filled with air sink to the ground rather than float?

4. The accompanying figure shows a barometer that weighs 1 N being supported by a spring balance. If the mercury in the tube weighs 10 N, what will the scale read?

5. Most people have played with darts tipped with rubber suction cups or have stuck a plumber's force cup (alias "a plunger") to a wall. How do these gadgets work? By the way, NASA is developing suction-cup shoes for use in space ships. Would they work if someone wanted to walk outside on the hull of the craft?

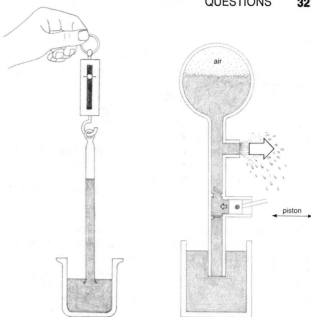

* 6. Liquids and gases are called fluids because they behave similarly in many respects. Still, there are many differences. What is the most striking difference?

7. The accompanying diagram shows a classic water pump, the kind once used on fire engines. The big brass dome on the engine was the air chamber. Discuss how such a pump works. The air chamber contains air compressed at the start of the cycle. Its function is to produce a sustained outflow of water rather than one in spurts. How does it function?

* 8. In order to siphon water from the container on the left to the lower one on the right (see figure), you first fill a tube with liquid, pinch it off, and lower it below level L_1. When you unpinch the tube, the liquid flows until the container is emptied or levels L_1 and L_2 become equal. How does this process work? *Hint:* Compare the pressure in the siphon at L_1 and L_2. Is there any limit to the height the bend can be above L_1?

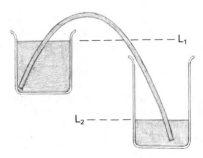

Section 14.4

1. Boyle put a mercury barometer in a vacuum chamber and pumped out all the air. What happened to the barometer reading? Why?

2. How was Boyle's work of interest to Newton while the latter was formulating his laws of motion?

3. A sample of gas at a constant temperature is doubled in volume. What happens to its pressure?

* 4. One kilogram of nitrous oxide (laughing gas) occupies a chamber at a temperature of 20°C and a pressure of 1 *atmosphere*. If the temperature is held constant, what will happen to the pressure if another kilogram of the gas is pumped in?

5. If a chamber containing carbon dioxide is compressed to half its volume while its temperature is held constant, what will happen to the pressure? Is the average speed of the molecules affected? *Hint:* Reread the section on temperature.

6. Boyle's law is $pV =$ constant. What are the units of that constant? Prove that it is equivalent to work.

* 7. Given that a gas goes from an initial pressure p_i and volume V_i to some final pressure p_f and volume V_f, show that Boyle's law can be equivalently written as

$$p_i V_i = p_f V_f.$$

8. The accompanying figure shows an underwater living chamber with an open bottom port so that people can enter and leave with ease. Why doesn't water rush in? How does the pressure inside compare with that outside in the water? Describe what would happen if the chamber developed a hole somewhere in the top. Note that as the chamber pressure drops, the gas volume decreases. Does this somehow violate Boyle's law?

MATHEMATICAL PROBLEMS

Answers to starred problems are given at the end of the book.

Section 14.2

1. Show that 29.9 inches of Hg exerts a pressure of 14.7 lb/in².

* 2. A gas tank sitting in the corner of a laboratory has an internal pressure five times that of the atmosphere. How much outward force will be exerted on each square inch of its walls?

3. How deep must we go under pure water before the pressure due to the liquid reaches 14.7 lb/in²?

* 4. The common pressure unit used in meteorology is the *millibar*, which is equal to 100 N/m². Show that the average sea-level pressure of 29.92 inches of Hg equals 1013 millibars. Mercury has a density of 1.36×10^4 kg/m³.

Section 14.5

1. Discuss the main differences between the gaseous and plasma states.

2. Is a beam of electrons a plasma?

* 3. As we go down each column in the periodic table toward the heavier atoms, the ionization energy (the energy it takes to remove the outermost electron) decreases. Can you figure out why? Why do you think cesium is the easiest element to ionize and one of the most commonly used for plasma research?

4. Describe what happens to the atoms and molecules within a block of ice as it is heated from −10°C to 100,000°C. Discuss the changes as they occur from one state to the next.

5. A pressure of one *atmosphere* (1 atm) is defined as the average sea-level pressure, equal to 1.013×10^5 N/m². Show that 1 atm = 76.00 cm Hg. Incidentally, just to confuse matters, the *pascal* is defined as 1 N/m². At 10 miles up, air pressure drops to about 2 lb/in². How many pascals is that? How many atmospheres?

Section 14.3

* 1. When the level of Pascal's mercury barometer fell three inches on the way up Mont Puy-de-Dôme, there was a corresponding pressure drop from 14.7 lb/in² to what?

2. Suppose that von Guericke had used half cubes instead of hemispheres. What force would each team of horses have

had to exert in order to pull apart a two-foot Magdeburg cube?

3. A thin-walled jar has a mass of 10 grams and contains 9×10^{-2} grams of hydrogen at atmospheric pressure. What would be the difference in the weight of the filled jar as measured in air and in vacuum? Hydrogen has a density at 1 atm of 9×10^{-5} g/cm³.

* 4. Calculate the weight of helium and of hydrogen needed to provide enough buoyance to lift a 1000-lb load ($\rho_H = 0.0056$ lb/ft³; $\rho_{He} = 0.0111$ lb/ft³).

5. The pressure in a car's tires is typically 25 lb/in² beyond atmospheric. How much area on each tire is in contact with the road if the car weighs 2000 lb?

Section 14.4

* 1. Suppose that a little fish on the bottom of a freshwater lake let loose a 1-cubic-millimeter bubble of gas. If the bubble was 10 cubic millimeters in volume when it burst at the surface at atmospheric pressure, how deep was the fish?

2. The pressure inside a hypodermic syringe is initially 1.01×10^5 N/m² (1 atm). What is the pressure when the gas is compressed to a third its original volume with no accompanying temperature change?

3. A 100-cubic-centimeter bottle of gas has a pressure of 152 cm of Hg. What volume will the gas occupy if fed into an expandable chamber at atmospheric pressure and the same temperature?

A liquid natural gas supertanker.

PART IV
Matter-Energy

Until now we have treated the concepts of matter and energy as if they were totally independent aspects of the physical world—often overlapping but always separate. We know, however, that that piecemeal perception is simplistic. Matter and energy are the two interchangeable manifestations of a single entity, of matter-energy, of essence of universe.

The next two chapters will evolve that interrelationship, first in general terms and then as it applies to the crucial processes of fission and fusion.

The open-spiral galaxy Triangulum—thousands of millions of stars in a cosmic pinwheel of matter and energy.

Matter and energy come together as a single entity in this chapter, much of which is centered about Einstein's revelation that $E = mc^2$; that **energy and mass are convertible, one into the other.** The first section explores that remarkable unity, ending in the reframed **law of conservation of mass-energy.** Armed with this potent new understanding, we return in the second section to a study of certain aspects of the atomic nucleus. To the two already familiar fundamental forces of Nature, gravity and the electromagnetic force, are added two more, the **strong** and the **weak nuclear interactions.** The interchange of mass and energy is basic to an understanding of how the component particles of a nucleus remain locked to one another, and **binding energy** is an important measure of this togetherness.

The chapter ends with a discussion of **antimatter,** the incredible mirror-image stuff that annihilates matter, converting it and itself into a blast of energy—substance vanishing into energy and vice versa. Nowadays it's commonplace to create matter directly out of the whirlwind of energy!

15.1 EE EQUALS EM CEE SQUARED

Even our brief glance at the special theory of relativity revealed that it was bulging with incredible ideas: time slowing down, distances shrinking, mass varying with speed. And yet we have still to study perhaps its most remarkable conclusion—certainly the result that Einstein himself thought was "the most important."

Earlier, in our not entirely successful attempts to define mass, we concluded that a fruitful approach would be to envision it as a measure of the resistance of a body to acceleration (Section 4.3). Later we learned that as a material object moves, its mass actually increases. In other words, the faster it travels, the more it resists any further increase in speed (Section 6.4). Thus, up near the speed of light, c, which the object can approach though never attain, even a tremendous force applied for a long time will only minutely increase the motion. The object's mass will become considerably greater, but its speed will hardly increase at all. Now this raises an interesting point. A force acting on the object does work; in effect, it pumps energy into the system, which here strangely manifests itself primarily as an increase in mass. Can there be some fundamental relationship between mass and energy? The answer is yes, and if there is a symbol of our age, a central insight of the twentieth century, this relationship is it.

The key formula, the **mass-energy equation,** is a logical consequence of relativity, which can be derived rigorously in a number of different ways from the postulates of the theory—all of them unhap-

Nature and Nature's laws lay hid in night.
God said, *Let Newton be!* and all was light.
ALEXANDER POPE

It did not last: the Devil howling Ho!
Let Einstein be! restored the status quo.
SIR JOHN C. SQUIRE

pily somewhat too complicated for our purposes. It is ironic that Einstein's own first derivation, which appeared in his 1905 paper "Does the Inertia of a Body Depend upon Its Energy Content?" was flawed; he came to the correct conclusion only after making a faulty assumption. In any event, we will leave it for one of the questions at the end of the chapter to show that if we take the relativistic expression for mass (p. 151) and apply the low-speed approximation of Mathematical Problem 6.4-4, we get a curious equation:

$$m_M c^2 = m_S c^2 + \tfrac{1}{2} m_S v^2.$$

On the far right we immediately recognize the classical expression for kinetic energy, but the two other terms, which are similar to each other, are new. Still they were not altogether unfamiliar in form to the physicists of 1905.* The term on the left deals with m_M, the mass of an object moving with respect to the observer, the *mass-in-motion*. The term on the right depends on m_S, the mass of a stationary object, the *rest-mass*.

Amazingly, this equation has a perfectly simple and yet profound interpretation, which by now has been extensively confirmed experimentally, namely,

total energy = rest-energy + kinetic energy.

In other words, *mass-in-motion times c^2 is taken to be total energy, and rest-mass times c^2 is thought of as rest-energy*. Note that KE involves mass multiplied by speed squared, as do the other two terms, so it's reasonable to expect them to correspond to some sort of energy as well (each has joules as its unit). The constant quantity $c^2 = 9 \times 10^{16}$ m²/s² is generally understood as simply a mathematical conversion factor. It translates a numerical amount of mass into a corresponding numerical amount of energy, much as multiplying your own mass by the constant g converts that number into a different number, which is your weight (Section 5.2) at the Earth's surface. Modern physics merely maintains that 9×10^{16} joules is equivalent to a kilogram. If we had known that a few centuries ago, we would not now have joules, BTUs, calories, and kilowatt-hours to fuss with; the kilogram alone would do for all forms of mass-energy.

Keep in mind that the presence of c does not mean that anything being considered is necessarily moving at that speed or at any speed at all (just as the presence of g in the weight-mass equation does not mean you are necessarily accelerating).

---

* For more than twenty years by that time, beginning with J. J. Thomson, theoretical work had been going on in an effort to relate mass to electromagnetism, and factors like these containing c^2 were common in those arguments. Moreover, the general idea that the mass of an electron would increase with speed was predicted and confirmed experimentally (1901) even before relativity. A lot of loose pieces to the puzzle were floating around by 1905.

The nuclear-powered aircraft carrier U.S.S. *Enterprise* with its crew on deck in whites spelling out $E = mc^2$.

Saying it again, but this time symbolically: The total energy, E, of an object equals its mass-in-motion (and it has become customary to drop any subscripts and just write m_M as m) multiplied by the speed of light squared:

$$E = mc^2.$$

What Einstein concluded was that **mass and energy are equivalent.** The two seemingly distinct conceptions are actually manifestations of one single entity: **mass-energy.** *Mass* is equal to and convertible into and from *energy*. The two flow back and forth as if we were pouring water from a measuring cup labeled "energy" and graduated in joules into another labeled "mass" and graduated in kilograms. *Every unit of energy, every joule, poured into a system—KE, gravitational PE, heat, whatever—increases the mass-in-motion of the system* by an amount equal to

$$\frac{E}{c^2} = \frac{1 \text{ joule}}{(3 \times 10^8 \text{ m/s})^2} = 1.1 \times 10^{-17} \text{ kg.}$$

Einstein, who conceived the mass-energy equation from considerations of electromagnetic energy, immediately generalized the conclusion to all energy. That bold step went beyond the derivation at hand, but he was relying on the vision that all forms of energy were equivalent. That conclusion has now been tested and splendidly confirmed in numerous cases with several different forms of energy, particularly on the nuclear level—not for all forms but for enough to be convincing.

The relativistic equation for mass tells us that as long as an object is moving, its mass-in-motion is greater than its rest-mass. As long as it is moving, the object's total energy is greater than its rest energy, the difference being the KE. All right, then, what happens when the object is at rest, when KE = 0? Does it mean that the total energy, which now equals the rest-energy, is constant? Is the rest-mass fixed and unchangeable? No, not at all.

The rest-mass of a single motionless particle by virtue of its very existence may be regarded as congealed energy, rest-energy. Beyond that, a body composed of several particles possessing internal energy (thermal, potential, whatever) when taken as a whole, has some additional rest-mass due to that internal energy. The rest-mass of a system of particles (such as an object composed of atoms) may well exceed the sum of the rest-masses of the individual particles. A hot apple pie sitting on a table next to an otherwise identical cool one has zero KE. Yet the hot pie has more rest-energy and therefore more rest-mass than the cool pie. It will be harder to accelerate and will weigh more on a scale—granted, only a tiny bit more, but more nonetheless! In other words, the atoms of the hot pie on the average have a higher KE and therefore more mass. The pie they constitute is motionless, but it must reflect this increase in the mass of its parts. Thus the rest-mass of the pie as a whole must increase with temperature. The hot pie presumably has no more substance, or matter, than

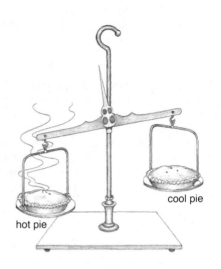

hot pie

cool pie

its cold twin, certainly no more atoms; the greater mass is due purely to its greater energy content.

A flashlight emitting a beam of electromagnetic energy becomes less massive in the process, just as a plant absorbing that light gains mass. A spring must weigh more after elastic PE is stored within it, after it has been compressed, than before. Rest-energy is increased, and so rest-mass must also increase; the two are one. A clock weighs more wound than unwound, a rubber band is heavier stretched than unstretched. These are the unambiguous conclusions of the theory. No one has ever measured the minuscule variations in mass involved with a spring or a flashlight or a rubber band, but as we'll soon see, there is ample confirmation elsewhere.

In the final analysis, all the familiar occurrences that liberate energy—from burning marshmallows to exploding dynamite—all transform minute quantities of rest-mass into energy. Ultimately that is *the* source of the reaction energy. All the dust and smoke, bits and fragments of a bomb blown up—every speck of its matter taken together weighs ever so slightly less than the bomb itself; the heat and light, the energy lost to the surroundings, is rest-mass lost from the bomb. And this is as true for energy produced chemically as it is for energy liberated from atomic nuclei in the process of radioactive decay.

Conservation of Mass-Energy

In Chapter 7 we talked about the conservation of mechanical energy, and in Chapter 8 we generalized that concept to include heat as well. The result was the first law of thermodynamics. Now it's time to do some more generalizing. Instead of two separate conservation laws, one for rest-mass and another for energy, there really is only one all-encompassing fundamental law:

> *The total energy of an isolated system always remains constant although any portion of it can be converted from one form to another, including rest-energy.*

In effect, we can neatly take care of our concerns about mass by simply including rest-energy within the general prescription. All right, then, how do we go about verifying this, perhaps the most remarkable principle in all of physics?

One thing that should be clear from the statement $E = mc^2$ is that c^2 is a very large number, so the energy stored in the form of mass is exceedingly concentrated. One single gram of matter is equivalent to 9×10^{13} joules, or enough heat to raise 200,000,000 kg (4.4×10^8 lb) of water from 0°C to 100°C. That is a lot of energy; it corresponds to the peak output of Boulder Dam operating for 19 hours, namely, 25 million kilowatt-hours. By comparison, the ordinary amounts of energy we are accustomed to dealing with in everyday life are puny. An exploding kilogram of TNT quickly liberates about five million joules, which is certainly formidable. Still, it represents a rest-mass loss of only about 6×10^{-11} kg—far too small to be detected in any practical experiment.

When a lump of coal burns, carbon and oxygen combine to form carbon dioxide, giving off light and heat in the process. Separately the C atom and O_2 molecule have potential energy, which is liberated as they come together and their electrons rearrange themselves. Alternatively, we can say that the C and O_2 have more rest-mass when they are apart than does the resulting CO_2 molecule. That's a very significant point: The mass of the whole is less than the sum of the masses of its parts. The chemical energy released in combustion of all common fuels is roughly 10 million joules for each kilogram burned. This corresponds to a conversion of mass into energy at a rate of only 1.1×10^{-10} kg transformed for each kilogram of fuel consumed—a quantity that would be quite impractical to measure. And that is precisely why the principle of conservation of mass in chemical processes, even though it was erroneous, was never seriously challenged in the past.

Confirmation of the mass-energy equation cannot be done, even at present, by chemical means. Einstein recognized that, and at the end of his epochal paper he suggested that the test of the theory be pursued in nuclear physics with radioactive materials. Indeed, the most favorable conditions for accurate determinations exist among the lighter nuclei, in which the changes in mass are proportionately larger and can be measured with great precision.

The first accurate verification of the principle was finally made in 1932. The idea was simple enough, but 27 years would elapse after the enunciation of the principle before the appropriate technique was available. If a nucleus could be bombarded and disintegrated in such a way that all the mass and energy before and after could be accounted for, then a straightforward tally would settle the question. J. D. Cockcroft and E. T. S. Walton, working at Rutherford's Cavendish Laboratory, fashioned a target of the lightest of all solids at ordinary temperatures, the alkali metal lithium ($_3^7$Li). Using a new particle accelerator of their own design, they fired a beam of low-speed protons, bare hydrogen nuclei ($_1^1$H), at the lithium. Flying out of the bombarded target came a stream of high-speed alpha particles, stripped helium nuclei ($_2^4$He). In effect, a lithium nucleus (three protons and four neutrons) captured an incoming proton, for an instant becoming a highly unstable nucleus composed of four protons and four neutrons. This immediately decomposed into two alpha particles, which were hurled out at tremendous speeds. The KE of the proton projectile before the collision was found to be vastly smaller than and quite negligible in comparison with the total KE of the two alpha particles. Measurements showed that the total kinetic energy was a rather hefty 27.2×10^{-13} joules. Moreover, the masses of the participants before

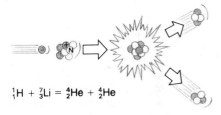

$_1^1H + _3^7Li = _2^4He + _2^4He$

Lithium bombarded by protons.

and after were also different, and that was the crucial point. Two alpha particles actually have a mass that is *less* by 3.07×10^{-29} kg than the combined mass of a lithium nucleus plus a proton. That vanished mass must have been converted into the KE of the alpha particles. Marvelously, 3.07×10^{-29} kg multiplied by c^2 equals 27.6×10^{-13} joules. For an experiment of that complexity, the agreement finally obtained between the two values was superb—all one could hope for.

It's interesting to note that in their original paper Cockcroft and Walton were much more concerned with producing artificial nuclear transmutation (for which they received the 1951 Nobel Prize) than with establishing the validity of the mass-energy relationship. *That* they appear to have taken almost for granted. Nonetheless, that experiment and countless others performed since, including ones using incident beams of photons, have all confirmed the mass-energy equivalence: $E = mc^2$.

15.2 NUCLEAR MATTERS

Within months of Chadwick's discovery of the neutron in 1932, Werner Heisenberg proposed his now familiar theory picturing the atomic nucleus composed exclusively of neutrons and protons. He further suggested that these two specks of matter were merely the charged and uncharged versions of the same basic particle. A great deal has been added to the theory since then, but its underlying conceptions remain intact. Nowadays, to underscore the inherent sameness of neutrons and protons, both are often simply referred to as **nucleons.**

The Four Forces

No sooner had the proton-neutron model of the nucleus been enunciated than there arose the obvious question: What held the thing together? A cluster of positively charged particles must repel one another with an electrical force—*like charges repel;* there was no doubt about that. Indeed, a simple calculation of the repulsion between two protons separated by a distance that puts them just about in contact within a nucleus yields a value for the force of around 50 newtons (about 11 lb). That may not seem much, but it really is enormous for the tiny masses of the protons. Evidently there is another kind of force operating within the nucleus, a cohesion of tremendous strength that can override the electrical repulsion—either that or the whole intricate construction is a fiction.

The **strong nuclear interaction,** as it unimaginatively came to be known, continues to be the subject of extensive research. Although our understanding of it even now is incomplete, we do know a bit about it. For one thing, the strong interaction exists between any two or more nucleons: protons (p $\leftrightarrow$ p), neutrons (n $\leftrightarrow$ n), or both (p $\leftrightarrow$ n). The evidence for an n $\leftrightarrow$ n attraction is inferential, but p $\leftrightarrow$ n and p $\leftrightarrow$ p interactions can be measured indirectly. Measurement is accomplished by shooting beams of neutrons or protons at a

The first particle accelerator (now in the Science Museum, London). Built by Cockcroft and Walton, it cost the then substantial sum of £1000, about $5000. (Modern machines go for hundreds of millions of dollars.) There was room for only one in the little lead-covered wooden booth in which the experiments were performed.

target consisting mostly of hydrogen (whose nuclei are protons) and analyzing the scatter. These experiments (performed with high-voltage generators or cyclotrons) show that the strong attractive force is extremely powerful within a minute range of only about 3×10^{-15} meters, whereas beyond that it decreases very rapidly with distance. This is a short-range force that acts effectively just over distances equivalent to a few nucleon diameters (a single proton has a radius of about 0.000,000,000,000,001 or 10^{-15} m). That much is reasonable because, after all, there are no pea-sized nuclei, whereas if the force extended far enough there apparently could be. The radius of even a large nucleus like Uranium-238 is only about 7.4×10^{-15} m, indicating that the force holding it together must be very limited in the extent of its influence; it simply cannot draw together very many more nucleons than that.

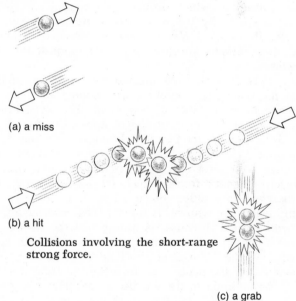

(a) a miss

(b) a hit

Collisions involving the short-range strong force.

(c) a grab

A rough comparison of the size of a typical atom with that of a proton, a gold nucleus, and a uranium nucleus.

(a)   (b) (c) (d)

particle diameters

(a) atoms $\approx 10^{-10}$m

(b) proton $\approx 2 \times 10^{-15}$m

(c) gold nucleus
    $\approx 10 \times 10^{-15}$m

(d) uranium nucleus
    $\approx 15 \times 10^{-15}$m

Incidentally, we are not altogether unfamiliar with other short-range effects, even though they are not fundamental in nature and not nearly so localized. For example, the interactions that bind a molecule into a distinct unit are fairly short of reach. Though basically arising from the electric forces between charges in the constituent atoms, the situation is such that the resultant interaction extends only over molecular dimensions. Beyond that, competing electrical influences cancel each other to the point where practically no force is perceptible from the outside. Atoms have to come very near one another before they can draw together.

A free neutron out on its own is unstable, decaying with a half-life of roughly 17 minutes into a proton, an electron, and a neutrino (actually an antineutrino). Compared with the lifetime of other unstable particles, this is remarkably long, and it suggests that the governing force is very weak. Moreover, if we use electrons as projectiles in experiments, it becomes evident that they are quite unaffected by the strong interaction. They simply don't "feel" it, and that is true of neutrinos as well.

Neutrons within the nuclei of several elements also undergo radioactive decay by beta emission—by the hurling out of an electron-neutrino pair and the leaving behind of the proton stuck in place by the strong force. It was in order to understand beta decay that Enrico Fermi introduced the concept of the **weak nuclear force** in his brilliant paper of 1933. In that paper, one of his major contributions, he developed the earlier suggestion of Pauli's (p. 172) that an as-yet unseen neutral particle was being emitted in the process. While answering a question one day, Fermi almost jokingly referred to that elusive speck as a **neutrino** (which means "little neutral one" in Italian), and the name stuck. By the way, that monumental paper was sent to the scientific journal *Nature*, which proceeded to reject it. With some disappointment, Fermi subsequently submitted it for publication elsewhere.

Fermi's work, then, marked the "discovery" of the fourth and last of the known fundamental interactions* in Nature—strong, electromagnetic, gravitational, and weak. The weak nuclear force is also short-range, so much so that it appears to function most effectively only *within* specific particles. Besides being approximately ten million million (10^{13}) times fainter than the strong interaction, its minuscule range is at most only a hundredth as great. Unlike the strong and electromagnetic forces, which act only on certain material particles, the weak interaction affects all of them, though minutely. Under its influence, the neutron is transformed on decaying. We use the word "transformed" because we believe that the electron-neutrino pair are never actually within the neutron (as the oak is never within the acorn or the flame within the lighter). Instead they are created on the spot out of the neutron at the instant of transmutation.

In addition to the handful of particles, the principal players we have mentioned thus far, there are actually many more subnuclear entities, approximately 300 in all—a great bewildering zoo of minimatter, some massless, some with ten times the mass of the proton. Most of these are fast-vanishing specks liberated in the incredible violence of nuclear collisions, blasting collisions orchestrated by the great sprawling machines (the synchrotrons and linear accelerators) of contemporary high-energy physics. The whole lot are known as **elementary particles**—a truly unsuitable name that implies far more than we have the right or reason to claim. All are usually categorized according to their response to the nuclear forces. Since by far most

---

* Several experiments during the 1970s have given strong support to the Weinberg-Salam gauge theory, which relates the weak and electromagnetic interactions. This gives hope to those who dream, as Einstein did, of understanding the four forces within the context of a single unified theory.

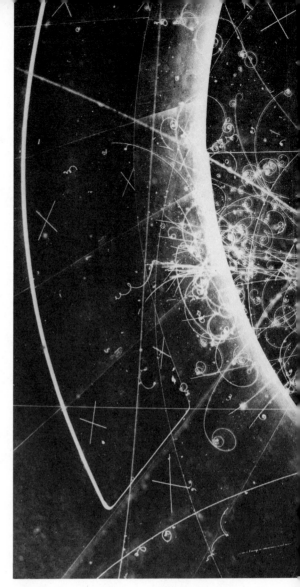

The tracks of subatomic particles in a bubble chamber.

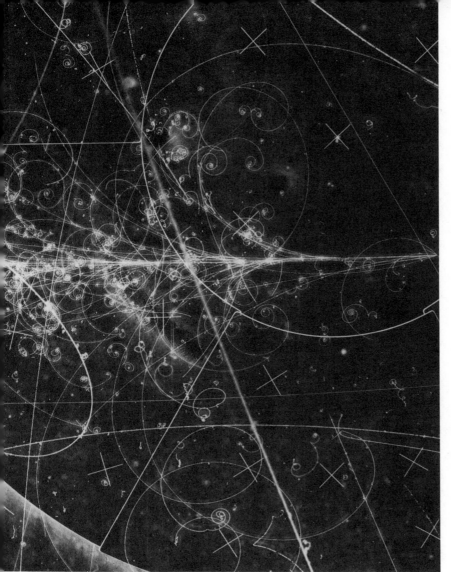

In descending order of strength, the four basic interactions governing the entire physical universe are:

1. Strong Nuclear: This is the powerful short-range binding force, operative between neutrons and protons, that controls the makeup of atomic nuclei.

2. Electromagnetic: Exerted between all charged particles, this long-range interaction affects the structure of nuclei, atoms, and molecules. Though about 100 times weaker than the strong force, it is the influence that binds bulk matter into its various forms.

3. Weak Nuclear: This force acts between all material particles but with such a short range that it is effective principally in beta decay. It is a hundred thousand million (10^{11}) times more feeble than the electromagnetic interaction.

4. Gravitation: This longer-range influence existing between all particles with mass is by far the weakest, being 10^{-40} times the strength of the strong force. Yet it rules on a grand scale. Holding the planets, stars, and galaxies together, it is the cosmic binding force.

of the subnuclear particles "feel" the strong interaction, they are called **hadrons** (from the Greek for large or strong). Most of these are both relatively massive and highly unstable. Only eight others out of the entire menagerie have long been known to be immune to the strong force and so are noticeably affected by the weak interaction. These few—and we count among them the electron and the neutrino—are accordingly referred to as **leptons** (from the Greek for small or weak).*

---

* For some time the lepton grouping has been thought to consist of four particles and their four corresponding antiparticles—eight in all. The four particles are the electron and its neutrino, and the muon (discovered around 40 years ago in cosmic-ray showers) and its neutrino. Now, however, there is some evidence for the existence of a fifth heavy lepton named tau and its mirror twin the antitau.

Now the smallest Particles of Matter may cohere by the strongest Attractions, and compose bigger Particles. . . .

There are therefore Agents in Nature able to make the Particles of Bodies stick together by very strong Attractions. And it is the Business of Experimental Philosophy to find them out.

I. NEWTON

Opticks

So, Nat'ralists observe, a Flea
Hath smaller Fleas that on them prey,
And these have smaller Fleas to bite 'em,
And so proceed *ad infinitum.*

JONATHAN SWIFT

On Poetry: A Rhapsody

This area of research is one of the frontiers, the high-tension points of modern physics, where theory is immature, creative imagination abounds, patterns of data are still forming fast, and the whole study is charged with the excitement of impending revelation. Particle physics now appears to be at the self-same stage that chemistry was a century ago, struggling with the growing collection of elements. All the participant particles have interesting names; there are proper lists of properties and reactions they undergo and tables that reveal lovely patterns of behavior; and everyone knows that no one really knows why any of it is happening.

Not one of the leptons has as yet revealed either a measurable dimension or any sign of an internal structure. For now, they all retain their pedigrees as elementary particles. This can no longer be said of the hadrons, however. There is very strong experimental evidence indicating that hadrons (including protons and neutrons) are actually complex forms.

How will we finally determine that our "elementary" particles are not really structured wads of finer dust, composed of still smaller specks? At the moment—not surprisingly—there is no good answer to that one.

Binding Energy

When a neutron is fired at a proton, each will "see" the other, will interact, only if they approach very closely. In a head-on encounter under the tremendous influence of the strong force, the two may violently rush together, emitting a burst of electromagnetic energy in the form of a gamma-ray photon. The end product of this microcataclysm is a **deuteron** (a proton and neutron stuck together), the nucleus of the heavy isotope of hydrogen called deuterium (p. 249). The mass of the deuteron is less than the sum of the masses of the separate neutron and proton by an amount precisely equal (by way of $E = mc^2$) to the emitted gamma-ray energy, namely, 3.5×10^{-13} joules. In coming so tightly together, the particles throw off excess energy and form a stable nucleus. This energy is known as the **binding energy** because if they are ever to be ripped apart, 3.5×10^{-13} joules will have to be pumped into the deuteron to make up that difference in rest-mass. Again, this is an enormous amount of energy on an atomic scale. By comparison, it takes a mere 2×10^{-18} joules, one hundred thousand times less energy, to pull the electron out of a deuterium atom, and even that is ten times more energy than is released when the "burning" of a carbon atom forms CO_2.

If we fire a proton at another proton, the two don't come together as one stable entity. If a composite object exists at all—there is evidence suggesting that the "diproton" has been created—it lasts for less than 10^{-18} seconds. Usually stable atomic nuclei have at least as many neutrons as protons (hydrogen's single proton and ^3_2He are the exceptions). And that, too, is understandable. You see, protons, although they certainly attract one another via the strong interaction,

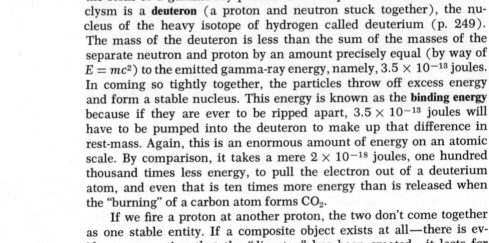

(a)

(b)

gamma
ray

The collision of a neutron and a proton
to form a deuteron.

also repel one another electrically. Since neutrons are uncharged, they only attract other nucleons and thus serve as a kind of nuclear glue.

Most of the first twenty elements have about equal numbers of protons and neutrons, but as the nuclei increase in size (in number of nucleons), the balance shifts to a preponderance of neutrons over protons. The nuclear attractive force, which has a very short range, becomes less effective in holding the larger clusters together against the electrical repulsion of the protons. Near a proton, that repulsive force is much weaker than is the strong interaction. Nevertheless, it diminishes much more gradually with distance. When there's an appreciable separation between two protons, the strong force vanishes, leaving only the long-range repulsion. In other words, whereas the strong force essentially acts attractively only between neighboring nucleons, the electrical force acts repulsively between even the most widely spaced pairs of protons. For nuclei with large numbers of protons, this means that proportionately more neutrons will have to be present if the cluster is to be stable and to resist flying apart.

A beam of deuterons sailing out of a 60-inch cyclotron at up to 28,000 miles per second.

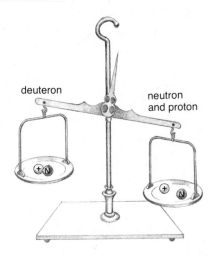

deuteron

neutron and proton

Up to the eighty-third element, all except technetium and promethium have stable (as well as unstable) isotopes. Elements with more than 83 protons do not have stable nuclei; despite their disproportionately large number of neutrons, all undergo spontaneous radioactive decay. They shed their overload of nucleons by hurling out alpha particles, dropping downward in weight, ultimately transmuting to some stable element.

The coming together of nucleons under the grip of the strong force is accompanied by a conversion of mass to energy—the resulting nucleus is always less massive than the sum of its individual proton and neutron parts. That difference in mass is equivalent to the total nuclear binding energy for a particular atom. Dividing that energy by the number of nucleons yields an average measure of how strongly each proton or neutron is fixed to the composite structure, namely, the *binding energy per nucleon*. It's like dropping a bunch of identical rocks into a well and then asking how much energy it will take to remove each one of them individually. Every kind of atom has a characteristic nuclear "well" with its own depth.

A plot of the binding energy per nucleon for the elements from hydrogen to uranium reveals a tremendously significant pattern. Hydrogen, with one proton, naturally has no nuclear binding energy at all; from there the curve rises to a spike for helium, falling back for lithium only to gradually rise again. Except for that spike, the graph is fairly smooth. Apparently the binding energy of the helium nucleus is exceptionally large. The two neutrons and two protons form a tight stable system, one that came to be known as an alpha particle before anyone realized that it was helium minus its two orbital electrons. This stability accounts for the fact that alpha particles are a common fragment in nuclear disintegrations.

The curve reaches its maximum with iron, which is therefore the most stable of all the nuclei, the one that is bound most tightly together. Each of its nucleons has in a sense been "squeezed" the most has lost the most mass, and requires an input of the most energy in order to be pulled out. From there on, the graph slowly drops all the way to uranium. In other words, as more neutrons and protons are added (beyond iron), the nucleus gets larger; the short-range strong force becomes less effective at drawing it all tightly together; and on the average, each nucleon retains a greater portion of the mass it had when totally free.

The fact that the curve falls off for elements above and below iron suggests (with the clarity of vision that comes of hindsight) a rather remarkable possibility. If we could begin with nuclei on either low side of the graph and alter their structures in such a way that they would move upward along the curve toward iron, a very large amount of energy could be liberated. Stated somewhat differently, if two light nuclei (say, of hydrogen) could be joined—fused—the resulting nucleus would be heavier, would reside higher up the curve, and would have a greater binding energy per nucleon. Each of its nucleons would be more tightly squeezed and individually less massive than it was

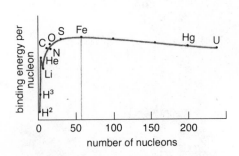

The higher the curve, the more tightly each nucleon is held in the particular nucleus.

prior to the new union. The difference, the mass lost, would be converted to energy by way of $E = mc^2$. If you haven't already guessed, the process is called *fusion*.

Alternatively, a large nucleus that split into two smaller fragments would again lose mass. The binding energy per nucleon for each fragment of nucleus would be greater than it was for the original prior to the breakup. Individual nucleons would be squeezed more tightly within each of the two smaller nuclei than they were in the large unshattered one. They would retain less mass than they had before the split. The difference would once more appear as energy in a process known as *fission*. It is no accident that the key elements of the nuclear age, hydrogen and uranium, are the conspicuous endpoints of the binding energy curve.

The Joliot-Curies—Irène, the elder daughter of Pierre and Marie Curie, and her husband, Jean Frédéric*—were rather disappointed when Chadwick announced his discovery of the neutron in 1932. They had certainly produced these new particles in their own laboratory many times before, but like almost everyone else, they had thought the particles were gamma rays, despite contrary theoretical indications.

Paris was then one of the great centers of research in radioactivity. After all, the grand Madame herself had studied the phenomenon there and was still working at the Radium Institute. Unhappily, she would not live to see her daughter's greatest triumph.† That spectacular discovery came out of an examination of the effects of alpha-particle bombardment on several light elements. In particular, the Joliot-Curies transmuted aluminum into phosphorus. The transformation of elements was already commonplace by then, but this time there was a surprising difference. The by-product was an isotope of phosphorus never before seen, and it was radioactive. The phosphorus went on emitting particles even when the alpha bombardment ceased—a fact they realized only after several months of work. They had **artificially induced radioactivity,** and within a year they had created a whole group of *radioisotopes*. The stage was set for our present-day medical, biological, and industrial use of these materials. Even then the implications were enormous.

In Chapter 1 we talked about the ideas of *making fact* and *being ready for discovery*. Certainly the Joliot-Curies had the neutron in their hands, but they weren't quite ready to "see" it. An even more ironic example of this nearsightedness had played itself out in California.

The Berkeley team, under Ernest Orlando Lawrence, the inventor of the cyclotron atom smasher, had been creating artificial radioisotopes without realizing it for years before 1934. In fact, because

Induced Radioactivity

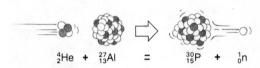

$$^4_2He + ^{27}_{13}Al = ^{30}_{15}P + ^1_0n$$

The transmutation of aluminum into phosphorus.

---

* To carry on the Curie name, Joliot added it to his own after their marriage.

† Like her mother, Irène also died of leukemia, probably brought on by long exposure to radiation.

Paul Adrien Maurice Dirac.

their Geiger counter annoyingly "misbehaved" (continuing to clatter away slightly even after the cyclotron was turned off), they ingeniously wired it to the same switch so that both would dutifully shut down together. On the afternoon of February 20, 1934, Lawrence dashed into the laboratory waving a copy of the Joliot-Curie paper. As he translated it aloud, they rewired the counter, positioned a carbon target within the deuteron beam, blasted it for a few minutes, and then turned off *only* the cyclotron. "Click . . . click . . . click went the Geiger counter," recalled M. S. Livingston, "it was a sound that no one who was there would ever forget." In that brief moment, they had created and recognized for the first time in history the telltale pulse of the radioactive isotope Nitrogen-13.

15.3 ANTIMATTER

One fantastic notion remains to be explored, one more surprise in a surprising universe of matter and energy. The notion is **antimatter,** and the vision it conjures is a kind of mirror image of ordinary stuff—matter beyond the looking glass.

In 1928 things were still simple, ominously simple: Atoms were composed only of electrons and protons; the neutron had not yet been discovered; and there were no other particles in the zoo except for lumps of energy, the photons. That was the year Paul A. M. Dirac at Cambridge published "The Relativistic Theory of the Electron," for which he would share the 1933 Nobel Prize with Schrödinger. It was Dirac's intent to reformulate quantum mechanics so that it was consistent with special relativity (i.e., independent of the motion of the observer). The result was an elegant masterpiece that beautifully accounted for some sticky theoretical matters that had been quite troublesome up to that point. But that splendid triumph did not come entirely free; the theory, however powerful, seemed ever so slightly flawed by one small aspect, one enigmatic mathematical quirk.

Special relativity abounds with square roots; they seem to underscore the geometrical character of space-time. That feature of the theory creates no particular difficulty in applications to the large-scale world, but it certainly did mischief in Dirac's quantum mechanical work. You see, a square root has two solutions, one positive and one negative; for example, the square root of 4 is either $+2$ or -2. Dirac had come up with a square root, too, but it was a square root of the electron's total energy, which therefore allowed both positive and negative values. But the energy of a free electron is a positive quantity; it has no physically meaningful negative value. Clearly, it made no sense to keep the negative solution; its implications were just too bizarre to be real. And so, as was quite customary in such cases, Dirac simply put aside the ugly twin as a mathematical aberration and went on with the work. But it soon became clear that this work was unlike the classical situation, in that the negative-energy portion could not be ignored, certainly not without seriously weakening the entire theory. Now, there was a dilemma—a richly potent theory embarrassed by what seemed a tiny bit of inexplicable nonsense!

It took Dirac until 1930 to come up with a bold, ingenious solution. It was either reinterpret or abandon, and he chose the former with a characteristic flair that was all his own. What followed was Dirac's fantastic *theory of holes*. All right, if we can't chuck the negative-energy states and have to suppose that they actually exist, what then? As we've seen, quantum theory has electrons jumping from one energy level to another, dropping downward on their own like stones skipping down a hill. So it seems that all the normal electrons in the universe may sooner or later radiate energy and descend to the lowest possible energy levels, namely, the negative ones. Since electrons do remain perfectly normal in this world, and we do not observe them all gradually tumbling into the oblivion of the negative Wonderland, something must be preventing that cascade. Rising to the challenge, Dirac proposed that all the negative states were already filled by electrons. Wonderland had no vacancies, and ordinary electrons simply couldn't drop in. The situation was like a large hotel with half its floors below ground and half above. With most of the upper rooms empty, there could be lots of traffic as guests shifted around, but below ground level all the rooms were occupied. No one new could move into a subroom before its occupant moved up and out.

The plot literally thickened. Instead of being empty, vacuum was now to be imagined filled with an invisible infinitude of electrons sailing around with negative energies. If one of them was somehow removed, pulled out and up to the ordinary positive world, it would leave behind an empty hole, a bubble in the negative sea of

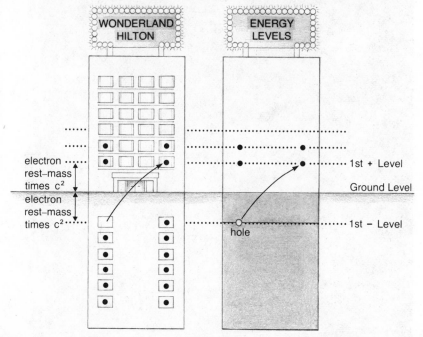

The creation of an electron-positron pair in the imagery of Dirac.

electrons (an empty basement room in the Wonderland Hilton). Dirac proposed that this hole, or absence of negative charge in a negative background, would appear to us as a positive charge. At first he thought that the hole was actually a proton, the only other known particle, but J. Robert Oppenheimer quickly showed that that was untenable. The hole had to behave as if it were the mirror image of an electron. "A hole, if there is one, would be a new kind of particle, unknown to experimental physics, having the same mass and opposite charge to the electron." So wrote Dirac in 1931. "We may call such a particle an anti-electron."

That incredible flight of creative imagination might have remained little more than a bewildering fantasy if not for some cosmic-ray work taking place on the other side of the world at Caltech. Carl Anderson, working with Millikan, was studying these particles (predominantly protons, possibly emitted by supernovas), which stream

A beam of antiprotons hitting a nucleus in the liquid hydrogen bath of a bubble chamber created a neutral particle. Leaving no track, that chargeless speck soon burst into an electron-positron pair. Having little mass and being charged, these were strongly influenced by a magnetic field, which bent their trajectories. The larger of the two oppositely winding spirals belongs to the electron; the smaller, tighter one is the positron's trail.

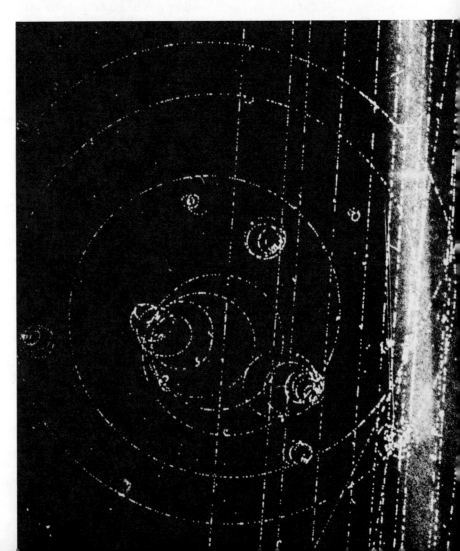

in on the Earth from outer space. Their paths were being observed in a cloud chamber, a device that produces a visible trail of tiny little droplets of condensed vapor. These minute beads of liquid form on the ions produced in the wake of the incoming charged particle and mark out its path like footprints in the snow. It was customary to position the chamber between the poles of a strong magnet. The tracks of the positive and negative particles shooting into the chamber would then be bent in opposite directions and could easily be distinguished. Among the thousands of photographs taken, Anderson noticed one that seemed rather curious. It clearly showed two oppositely curved tracks, one of which was certainly made by an electron; the other suggested an antielectron, but that conclusion was still "very radical at the time." Others had seen these positive tracks before over the preceding half dozen years but had either ignored them or dismissed them as "dirt." After all, there was nothing else they could be; nothing else fitted the accepted understanding. But Dirac's unusual paper changed that understanding, and the facts spread out within the cloud chamber changed accordingly. Now Anderson was ready for the new vision; he revised his experiment in order to better study that remote possibility. Amazingly, by the summer of 1932 he had clear evidence of the existence of the antielectron! It was Anderson who christened it the **positron.**

Today it is commonplace to create electron-positron pairs—to have matter materialize out of energy. A blast of pure electromagnetic energy, a gamma ray with zero rest-mass, can be made to disappear, and in its place an electron and a positron pop into reality. This ultimate alchemy occurs provided that the gamma-ray photon carries an energy equal to at least the total rest-energy of the two particles (by way of $E = mc^2$). Dirac predicted exactly that amount as the energy necessary to raise an electron up out of the negative sea into the real world, at the same time leaving behind a hole (a positron).

Just as this pair creation is possible, so is the inverse process of annihilation, and it has been equally well observed. A positron and *any* electron it approaches can come together and totally obliterate each other, vanishing in a puff of gamma rays (as if the electron had dropped back into Dirac's negative sea and disappeared into the hole). Nonetheless, so long as it can be isolated from ordinary matter, the positron is a stable addition to the particle zoo.

The contemporary theoretical picture has now been modified to the point where the negative-sea interpretation is no longer a necessity, and we need not break our heads with that powerful make-believe that gave birth to antimatter. Yet it's fascinating to keep a clear eye on the origins of scientific "understanding." A theory, however strange, that can predict something like antimatter seems to have some marvelous sympathy with the universe. Despite that, remember that a confirmed prediction does not prove the theory true.

The proton is 1840 times more massive than the electron. Its creation, which would require that much more energy, could not

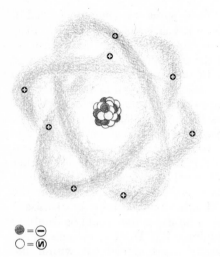

● = ⊖
○ = Ⓝ

Antimatter: a positron cloud surrounding a nucleus of antiprotons and antineutrons.

seriously be attempted until a new generation of high-energy particle accelerators became available. In the early 1950s something of a competition began between the mighty new machines, the *cosmotron* at Brookhaven (Long Island, N.Y.) and the *bevatron* at Berkeley (California). In the fall of 1955 the West Coast carried the day, announcing the discovery of the **antiproton.** This negative speck, having the same mass as an ordinary proton, possessed the telltale trait, annihilating itself and the first proton it embraced. Despite that self-destructive bent, it is possible on rare occasions for the proton-antiproton encounter to be less catastrophic, resulting only in the cancellation of the two equal and opposite charges. The neutral remnants of such a trade-off are a neutron and an **antineutron,** first discovered in 1956.*

Nowadays antimatter is a familiar stuff (or is it antistuff?) in the laboratory; there are even artificial radioactive isotopes, like ^{22}Na, that emit positrons on decaying and so serve as convenient sources. Furthermore, it is possible to produce beams of positrons and antiprotons on command, using accelerators. There's no scarcity of antimatter, even if it doesn't survive very long. Antinucleons have actually been joined together to make such compound structures as antideuterons and antialpha particles. Researchers have made positronium, a hydrogenlike atom composed of a positron and an electron bound together in a somewhat stable form. And an exotic atom called antiprotonic hydrogen, a proton and antiproton orbiting each other, was created in 1978.

Protons, electrons, neutrons, neutrinos—indeed, all subnuclear particles—have their antimatter twins, their mirror-image annihilators. So it's not difficult to imagine antiatoms (with negative nuclei surrounded by clouds of positrons), antimolecules, antiamino acids, antiproteins, and perhaps even antiphysicists and antilovers and anti-babes-in-arms no less real than us—but certainly *not* around here, not in our matter-crowded world, where they would vanish all too quickly. Still, it is possible—though not likely—that there are antistars in our Milky Way. It's far more probable that there are whole antigalaxies isolated in the dark void. Perhaps half the universe is antimatter, created with matter out of a blaze of gamma rays when everything was energy back at the very beginning. Perhaps—

---

* The similarity of the neutron and proton is further underscored by the fact that protons can be annihilated with antineutrons, just as antiprotons can be annihilated with neutrons. In general, nucleons don't just vanish into a burst of photons. Instead they leave behind several different unstable lighter particles known as pions.

REVIEW

Section 15.1

1. The total energy of an object is the sum of its kinetic energy and its _____ energy.

2. What is the quantity formed by the product of mass-in-motion and the speed of light in vacuum squared?

3. Define *rest-energy*.

4. What is the numerical value of c^2 in SI units?

5. How many joules are equivalent to 1 kilogram?

6. When is the mass-in-motion of an object greater than its rest-mass?

7. What sort of energy does an object have simply because of its existence?

8. A hot cup of coffee at rest has more _____ mass than it does after it's cooled off.

9. State the law of conservation of mass-energy. How does it differ from the first law of thermodynamics?

10. What is the significance of the fact that c^2 is a large number?

11. Is mass conserved in chemical processes?

Section 15.2

1. Define what is meant by a *nucleon*.

2. What is the range of the strong force?

3. Are free neutrons stable?

4. What is beta decay? Which force governs it?

5. List the fundamental forces that act on all material particles.

6. Do electrons and neutrinos reside within the nucleus?

7. List the fundamental forces in descending order of strength.

8. Why do we not experience the effects of the strong force in everyday life as we do with gravity?

9. State the distinguishing characteristic of the *hadrons*.

10. What is the major common property shared by the *leptons*?

11. Define the term *binding energy*.

12. What is a *deuteron*?

13. Compare the amount of energy it takes to ionize a deuterium atom with that needed to decompose its nucleus.

14. A nucleus is always less _____ than the sum of its nucleon parts.

15. Which is the most stable of all complex nuclei?

16. Explain briefly the processes of *fusion* and *fission*.

17. Who discovered the phenomenon of *induced radioactivity*, and what exactly is it?

Section 15.3

1. What role did Dirac play in the discovery of *antimatter*?

2. A *cloud chamber* is a device that produces visible _____ of charged subatomic particles.

3. Why was Anderson's cloud chamber placed between the poles of a magnet?

4. A *positron* is an _____.

5. To what extent have we succeeded in making complex antimatter nuclei?

6. What is *positronium*?

QUESTIONS

Answers to starred questions are given at the end of the book.

Section 15.1

1. Does the work done on a body always show up as an increase in its mass-in-motion?

* 2. When does the total energy of a body exceed its rest-energy? Its kinetic energy? Can the total energy of an object ever equal its kinetic energy? What if the object is a neutrino?

3. Discuss the role of the c^2 term in the mass-energy equation. If c was much smaller than 300,000,000 m/s, would the equivalence of mass and energy be more or less noticeable? Explain.

4. Imagine that you have a bunch of ants at rest inside an insulated box sitting on a supersensitive scale. What, if anything, happens to the weight of the system (box plus ants) if all the ants start running around? Does the rest-energy of the system change? Suppose that you violently shake the box and reweigh it. Why will the weight have increased—or will it?

* 5. Does $E = mc^2$ even when a body is moving at a speed v rather than c? Explain.

6. Suppose that you pumped 1 joule of heat into a cup of tea. By how much would its rest-mass increase?

7. Which weighs more, 6×10^{23} molecules of H_2O in the form of water or the same number of molecules as ice? Or are they equal in weight?

* 8. When an oxygen atom joins with two hydrogen atoms to make a water molecule, do they gain or lose rest-mass (or neither) in the process? Explain.

9. Go through the reaction produced by Cockcroft and Walton and account for all the neutrons and protons before and after the collision.

Section 15.2

1. Describe all of the characteristics you can of the strong nuclear interaction.

2. What are all the forces you know of thus far that act between two neutrons? Between a neutron and a proton? Between two protons?

3. List the forces you know of thus far that are at work between a proton and an electron. Between a neutron and an electron.

* 4. Within a distance of a nucleon diameter or so, gravity is by far the most feeble of forces. Yet it is the dominant influence acting on your body at this very moment. Explain.

5. Is it easier or harder to rip a nucleus apart if it has a large binding energy per nucleon?

6. Why is it reasonable to expect atomic nuclei to possess as many neutrons as protons—if not more?

* 7. Under what circumstances can the strong force between two protons actually be smaller than their acting electrical interaction.

8. Is there any reason to suspect that a helium nucleus (alpha particle) is likely to be a fragment thrown off by an unstable nuclear system? Can the same be said of lithium (Li)?

* 9. On average, is a proton in the nucleus of a uranium atom more or less massive than or equally as massive as a lithium proton? A proton in an iron nucleus?

10. The Berkeley team first created Nitrogen-13 by bombarding Carbon-12 with a deuteron. Write an equation for the process, accounting for all the protons and neutrons involved.

Section 15.3

1. Describe, as best you can, Dirac's image of antimatter.

* 2. If m_e is the rest-mass of an electron, how much energy does it take to create an electron-positron pair?

3. List the properties you can expect the antiproton to possess.

* 4. The radioisotope Chromium-49 decays by the emission of a positron ($_{+1}^{0}e$). Write out the appropriate equation, accounting for all the nucleons, assuming that a proton decays into a neutron and the positron.

5. The proton-proton chain begins with two protons combining to form a deuteron but only after one of the protons decays into a neutron and a positron. Write out the governing equation.

MATHEMATICAL PROBLEMS

Answers to starred problems are given at the end of the book.

Section 15.1

1. Beginning with the relativistic expression for the mass m_M (p. 151) and using the approximation of Mathematical Problem 6.4-4, arrive at the energy equation on p. 332.

* 2. Imagine that as a result of throwing a ball horizontally you impart 100 joules of kinetic energy to it. What is the corresponding increase in its rest-mass? In its mass-in-motion?

* 3. Suppose that 100 kilocalories of heat enter a pot sitting on a stove. Calculate its change in total energy; in rest-mass and mass-in-motion; in kinetic energy and rest-energy.

4. How much energy would be liberated if 10 grams of matter was totally converted to energy? Using the fact that

Argonne National Laboratory's 12′ (diameter) hydrogen bubble chamber.

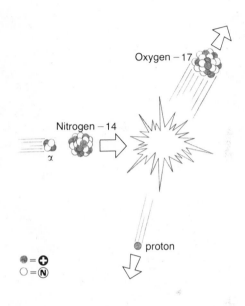

1000 tons of TNT is equivalent to 10^9 Cal (or 4.2×10^{12} joules), determine the released energy in kilotons.

Section 15.2

* 1. Rutherford (Section 10.3) became the first "real" alchemist when he transmuted Nitrogen-14 nuclei (2.325×10^{-26} kg) into Oxygen-17 nuclei (2.822×10^{-26} kg) by bombarding the former with alpha particles (6.645×10^{-27}

kg). The process ends in the production of oxygen and a stream of protons (1.673×10^{-27} kg).

a) Write an equation for the reaction accounting for all the nucleons involved.

b) In a measurement of the kinetic energy of the particles before impact (the alphas) and after (the oxygen and proton), the KE was found to have *decreased* about 2×10^{-13} joules. Where could that energy have gone?

c) Show that your conclusion that the total mass of the particles after the collision is *greater* than before is in agreement with the mass-energy equation.

2. Starting with a Carbon-12 nucleus, which has a mass of 1.992132×10^{-26} kg, calculate how much mass would be gained if it was entirely disassembled into 6 protons (each 1.67265×10^{-27} kg) and 6 neutrons (each 1.67495×10^{-27} kg). What, then, is the *average* binding energy per nucleon in joules? Would energy have to be pumped in, or would it be liberated during this remodeling?

3. With the previous problem in mind, calculate how much energy it would take to remove one neutron from the Carbon-12, that is, to end up with $^{11}_{6}C$ (1.827973×10^{-26} kg) and $^{1}_{0}n$ separately. Compare your result with the average binding energy per nucleon.

$$\alpha = \left(\frac{\hbar^2}{ec}\right)$$

The conversion of mass into energy is a twentieth-century nuclear gesture carrying the most profound implications for all of humankind. The first section of this chapter deals with **fission, the splitting of large nuclei into two roughly equal-sized fragments.** This rupture of the atomic core is accompanied by a loss in mass and a corresponding release of considerable amounts of energy. When the shattering of one nucleus triggers the shattering of others, the process is called a **chain reaction,** and it allows for the generation of a growing torrent of energy. The **nuclear reactor** is a machine that is capable of producing energy slowly for a long period of time via a **controlled** fission chain reaction. By contrast, the so-called **atomic bomb** is a machine that is capable of producing energy rapidly for a very short period of time via an **uncontrolled** fission chain reaction.

The second section of this chapter treats the process of **fusion, the joining of two light nuclei to form a single heavier nucleus.** Again there is a loss of mass (separately the two nuclei are more massive than when fused together), and again there is a release of energy. This is the primary energy-producing phenomenon of the universe; this is **star-fire.** At the moment, we can release that whirlwind in horrendous H-bomb spasms, but the gradual, controlled use of fusion power, with all its glorious promise, is still off in the future.

Enrico Fermi received his Ph.D. at Pisa just a few months before Mussolini, Il Duce, came to power in Italy. By 1927, when he was only 26, Fermi was already Professor of Theoretical Physics at the University of Rome—and already nicknamed "the Pope" because of his advocacy of the "new faith," the quantum theory.

Excited by the discoveries in nuclear physics, particularly those of the Joliot-Curies, Fermi shifted his almost inexhaustible energies from the purely theoretical work that had occupied him so fruitfully to a new adventure in the laboratory, an adventure whose outcome even he could never have imagined. Fermi's novel idea (and it even seemed "silly" to some at the time because the available sources were so weak) was to use neutrons as projectiles in order to create new radioisotopes. Because these neutral bullets experience no electrical repulsion, he reasoned that they would easily enter the nucleus. Remarkably, within weeks he published his first positive results. System-

16.1 FISSION

On facing page, Enrico Fermi.

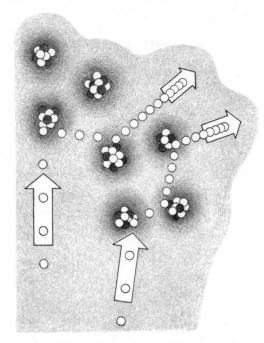

● = ⊕
○ = Ⓝ

The slowing down of high-speed neutrons via collisions with fairly light nuclei.

atically, the young Italian genius bombarded every element he could get his hands on, all the way up to uranium.

Along the way he realized that neutrons that had been slowed down in their passage through certain materials were even more potent at instigating transmutations than were ordinary fast neutrons. That fateful insight had come about almost by accident. Several young associates of "the Pope" had noticed that the level of induced radioactivity strangely depended on the surroundings—the kind of table the bombarded specimen was resting on, the type of materials nearby, that sort of thing. Remarkably, inserting a sheet of lead between source and target also produced a slight increase in activity. Almost immediately Fermi realized what was happening and suggested interposing a block of paraffin. As if by "black magic," the Geiger counter began clicking wildly—100 times more rapidly! Neutrons colliding with the light hydrogen atoms in the paraffin were losing kinetic energy and slowing down before escaping. These slow neutrons had a much better chance of being captured by the target nuclei and inducing transmutations. If that explanation was correct, large quantities of water should have the same effect. Down they rushed to the fountain in the garden behind the laboratory and there among the goldfish confirmed the amazing discovery.

Uranium bombarded by a stream of these slow neutrons exhibited some rather uncharacteristic behavior in that it subsequently emitted beta rays. The natural conclusion was that a neutron was swallowed up by the nucleus, which, finding itself unstable, emitted an electron. That left the nucleus with one more positive charge, a proton, than it had before and in turn placed it one box higher in the periodic table. Presumably, uranium had thus been transformed into a new superheavy element. That was the explanation Fermi announced in 1934. Although that sequence of events does occasionally take place, the predominant effect that had actually been observed was something even more spectacular, but it would be years before anyone knew it.

The idea of creating *transuranic* elements, i.e., elements possessing more than 92 protons, fired the imaginations of scientists around the world, and several soon began to repeat the work for themselves. Shortly after Fermi's announcement, Ida Noddack, a German chemist who with her husband had discovered the element rhenium, published an alternative scenario that seemed at the time nothing less than bizarre. She maintained that the incoming neutron had actually ruptured the uranium nucleus "into several big fragments which are really isotopes of already known elements." Her daring challenge to the opinion of one of the world's leading nuclear physicists met with instant neglect. Since no one took her prophetic insight seriously, she persuaded her husband to approach his old friend Professor Otto Hahn at the Kaiser Wilhelm Institute in Berlin. Hahn, probably the world's foremost radiochemist, dismissed the proposal with a simple "It's impossible" and an admonition not to continue voicing such wild ideas in public. Without knowing it, Hahn had thus taken one giant stride backward at a crucial moment, and that proved to be a fortuitous error. Until then, the pace of discoveries that would lead to fission and ultimately to the atomic bomb was ominously in step with the rise of Nazi Germany. But of course, no one was even vaguely aware of that then; no one had any idea of being a player in one of the greatest dramas of history.

In time, an erroneous suggestion was made that Fermi's uranium experiment had produced an isotope of protactinium rather than a transuranic element. Since Hahn and Lise Meitner were the discoverers of that element (1917) and had the most expertise in dealing with it, they were naturally drawn into the controversy. Joined by Frederic Strassmann, the junior member of the team who did most of the dirty work, they repeated the Italian experiments. The chemical analyses were exceedingly difficult, and even though the work continued for several years, their results remained inconclusive. But that was not their only worry; Dr. Meitner, the team's physicist, was an Austrian Jew, who managed to evade persecution in Berlin only on the mere technicality that she was not German. In the spring of 1938, the Nazis marched into Austria, and Fräulein Meitner became German. Despite Hahn's efforts and Max Planck's personal appeal* to Hitler, Dr. Meitner at the age of 60 was dismissed from the Institute, where she had worked for 30 years. Early that summer, without a visa but with the help of her colleagues, she fled to Holland and then to Stockholm.

Just about a week before Christmas in 1938, Hahn and Strassmann finally arrived at the realization that the substances created in the bombardment of uranium by neutrons were actually isotopes of the light element barium. That remarkable result seemed inescapably to suggest that the uranium nucleus had indeed split in two. How else could element 56 appear after they had blasted element 92? Excitedly they dashed off a letter to the editor of the prestigious journal *Naturwissenschaften*. Despite the meticulous care they had taken, Hahn was soon deeply troubled by the thought that they had somehow been misled, and he even wished he could withdraw the article, but it was too late for that. The only other person to whom a copy had been sent was their old friend-in-exile Lise Meitner, and they anxiously awaited her critique.

Meitner was visiting with friends in a small seaside resort in Sweden and had invited her nephew, a young physicist named Otto Frisch, to spend the Christmas holiday with her as he had done so many times before in Berlin. Frisch had also fled the Nazi race laws and was then working in Copenhagen in Niels Bohr's Institute. The morning after he arrived in Sweden, Frisch came down from his room to find Meitner engrossed, puzzling over the letter from Hahn. The two began to walk in the snow-covered woods, excitedly discussing the

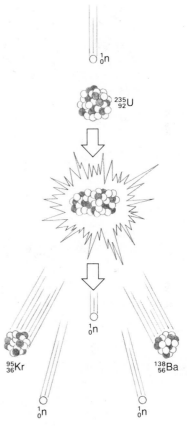

$^{1}_{0}n$

$^{235}_{92}U$

$^{1}_{0}n$

$^{95}_{36}Kr$

$^{138}_{56}Ba$

$^{1}_{0}n$

$^{1}_{0}n$

● = ⊕
○ = Ⓝ   The fission of U-235.

---

* In 1937 Planck was forced to resign his presidency of the society that bore his name, and in 1944 his son Erwin was executed for plotting to assassinate Hitler.

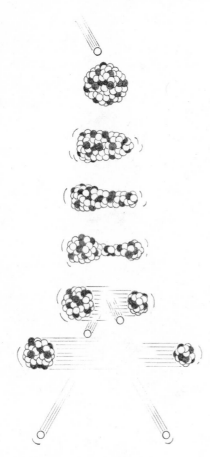

Nuclear fission.

startling revelations. How was it possible for a tiny slow neutron to shatter a massive nucleus? As fate would have it, Frisch was well versed in Bohr's new vision of the nucleus. The great man had pictured it not as a rigid solid but more like a liquid drop bound by a kind of surface tension (p. 294) arising from the mutual attraction of the nucleons.

They sat down on a tree trunk in the cold winter air and began making calculations on scraps of paper. Frisch determined that for a large nucleus like that of uranium, the electrical repulsion between protons would all but cancel the cohesive surface tension. It was possible that the uranium nucleus could be likened to an unstable liquid drop ready to elongate and snap in two with even the slightest nudge. As the nucleus deformed (into the shape of a peanut shell) the long-range electrical repulsion would overwhelm the short-range strong force, and the thing would jolt apart. But what about the energy that would have to become available since the two positively charged fragments would repel each other and instantly fly off at tremendous speeds? Remembering the values on the binding energy curve (p. 342), Meitner quickly found that the fragment nuclei would be less massive after the split by roughly 0.1 percent, an amount just about equal (via $E = mc^2$) to the needed energy. There it was, all neatly understood.

A few days later, Frisch was back in Copenhagen rushing to catch Bohr, who was about to sail to the United States to visit Einstein at Princeton. "Oh, what fools we have been!" Bohr blurted out, hands clasped to his head, "We ought to have seen that before." Hurriedly, Frisch and Meitner fashioned the paper announcing their conclusions over the telephone between Stockholm and Copenhagen. Borrowing a term from the biology of cell division, Frisch naméd the process nuclear **fission.**

Naturally occurring uranium is composed of three isotopes, U-234, U-235, and U-238, which differ only in their number of nuclear neutrons. By far the most abundant at 99.27 percent is U-238, followed by U-235 at only 0.72 percent, and a mere trace of U-234. On the basis of theoretical considerations, Bohr and his former student J. A. Wheeler at Princeton concluded that it was the rare U-235 that had undergone slow-neutron fission. In fact, it's the only naturally occurring element that can do that little trick. The common U-238 isotope can be made to split but only under bombardment by exceedingly energetic neutrons. That much was confirmed experimentally by April of 1940.

A slow neutron captured by a U-235 nucleus transforms the latter into highly unstable U-236. That in turn immediately and violently ruptures into two large pieces, which are not always the same in each such event. One of the common fragment pairs is Barium-138 (composed of 56 protons and 82 neutrons) and Krypton-95 (with 36 protons and 59 neutrons). Typically, about 0.1 percent of the rest-mass involved is transformed into energy. Thus for every kilogram of uranium split, one gram vanishes, liberating a fantastic 25,000,000 kilowatt-hours (p. 334), mostly as kinetic energy of the fission fragments.

The Chain Reaction

Few people in the 1930s realized that the world was already teetering on the the edge of a new and ominous age of nuclear energy. Most (including Einstein) shared Rutherford's view: "Anyone who expects a source of power from the transformation of these atoms," he quipped in September of 1933, "is talking moonshine." Leo Szilard, a short, stocky young Hungarian, was not the sort of person to let such a blanket, pontifical statement go unchallenged. Besides, he liked surprising people. Szilard had earned his doctorate at the University of Berlin and was on the faculty there, but with a remarkable ability to anticipate future events from present circumstances, he kept his suitcases always packed. It was clear to him that Hitler was no passing nightmare, and a man with his own Judaic credentials had best be wary. Within months of Hitler's becoming Chancellor in early 1933, Szilard left for England.

Several weeks after Rutherford's "moonshine" remark, Szilard conceived the notion of a **"chain reaction."** Prophetically he speculated that if a nucleus could be found that, when struck by a neutron, itself

The time will come when atomic energy will take the place of coal as a source of power. ...I hope that the human race will not discover how to use this energy until it has brains enough to use it properly.

SIR OLIVER LODGE
(1920)

The article in the right column reports Einstein's opinion (1934) that attempts at loosing the "energy of the atom" would be fruitless—an opinion he soon changed.

emitted *two* neutrons, then the process could continue building up on its own into a horrendous cascade. In an effort to control the use of his ideas, for he knew full well the military implications even then, he took out a patent on the work, assigning it to the secrecy of the British Admiralty. Sensing that Europe would soon be engulfed in war, Szilard accepted a post at Columbia University and arrived in New York in January 1939.

Another far more renowned nuclear physicist also arrived at Columbia that month. Enrico Fermi, whose mother's antecedents were Jews and whose wife Laura came from a prominent Jewish Italian family that had already experienced the tide of anti-Semitism, was now an immigrant. After receiving the 1938 Nobel Prize in Stockholm for his work with neutrons, Enrico with Laura and their two daughters sailed to the New World.

Only weeks after Fermi's arrival, Niels Bohr reached New York, bubbling over with the news of the discovery of fission. Fermi appreciated the possibility of a chain reaction using uranium almost from the beginning, but at the time there were many unknowns that stood in the way of establishing whether or not it was actually attainable. Only if each fission event emitted an average of at least two neutrons could a practical chain-reaction device be developed. In March, Szilard and Walter Zinn determined experimentally that on average fission was indeed accompanied by the emission of between two and three fast neutrons.

It was clear to Szilard very early that a nuclear bomb was a real possibility, and he valiantly tried to persuade his colleagues to keep their discoveries secret, to circulate information only among themselves. His attempts were abandoned after the French ultimately refused to withhold their work from open publication. When Hitler set an embargo on the export of all uranium from the rich Czechoslovakian mines, Szilard was convinced that the Nazis were well on the way to constructing a nuclear weapon. Accompanied by Eugene Wigner on one occasion and by Edward Teller on another, in August of 1939 Szilard went to Einstein with a letter to President Roosevelt on which the well-known and highly respected scientist put his prestigious signature. It advised "that extremely powerful bombs of a new type may thus be constructed." Although earlier efforts to awaken the U.S. government to the potential dangers of nuclear energy had fallen on deaf ears, this attempt was somewhat more successful. Even so, the "all-out" push to develop a bomb did not come until December 6, 1941. Ironically, that was just one day before Pearl Harbor was attacked by the Japanese.

A single fission event liberates about 3.2×10^{-11} joules, which is a great deal of energy in Atomland but nothing really much on the human scale of things. That was why the chain reaction—and in particular a chain reaction that fanned out and grew—was so crucial. If the first nucleus to split fired out two neutrons in the process, those two could split two other nuclei in the second generation, which then would emit a total of four neutrons and split four more nuclei in the

Albert Einstein
Old Grove Rd.
Nassau Point
Peconic, Long Island
August 2nd, 1939

F.D. Roosevelt,
President of the United States,
White House
Washington, D.C.

Sir:

Some recent work by E.Fermi and L. Szilard, which has been communicated to me in manuscript, leads me to expect that the element uranium may be turned into a new and important source of energy in the immediate future. Certain aspects of the situation which has arisen seem to call for watchfulness and, if necessary, quick action on the part of the Administration. I believe therefore that it is my duty to bring to your attention the following facts and recommendations:

In the course of the last four months it has been made probable - through the work of Joliot in France as well as Fermi and Szilard in America - that it may become possible to set up a nuclear chain reaction in a large mass of uranium,by which vast amounts of power and large quantities of new radium-like elements would be generated. Now it appears almost certain that this could be achieved in the immediate future.

This new phenomenon would also lead to the construction of bombs, and it is conceivable - though much less certain - that extremely powerful bombs of a new type may thus be constructed. A single bomb of this type, carried by boat and exploded in a port, might very well destroy the whole port together with some of the surrounding territory. However, such bombs might very well prove to be too heavy for transportation by air.

third generation, and so on. This all seems innocent enough, but as a branching cascade continues, it builds at an almost unbelievable rate. After only 76 generations an incredible 1.5×10^{23} nuclei have been shattered. Done quickly, say in about a millionth of a second, that produces a blast of energy equivalent to 20 thousand tons of TNT!

In reality, only about 85 percent of such collisions result in split nuclei. Not every shattered nucleus emits two neutrons, and some that are emitted escape or are otherwise lost. There was a good deal of experimentation to be done before it could be known with any certainty that the prime candidate, U-235, would produce a chain reaction. One overriding problem remained: Natural uranium contains 140 times more U-238 than U-235, and these nuclei absorb neutrons but don't usually split. A glob of pure natural uranium simply will not support a branching chain reaction. There were two possible solutions to the dilemma; both were tried and both worked. The obvious one was to separate out U-235 and use it alone, but that was exceedingly difficult to do because the two isotopes are chemically identical. In fact, only about a hundred-millionth of a gram of pure U-235 had been isolated up until that point, and no one then even knew how much would finally be needed for a bomb; the estimates ran from 2 to 200 pounds. This direct route led to one of the two types of atomic weapons used on Japan. The indirect route gave rise to the *nuclear reactor* (which established the feasibility of the chain reaction) and to a second type of A-bomb.

Nuclear Reactors

Fast neutrons emitted by a fissioning nucleus can be captured equally well by U-238 and U-235. But U-238 has much less of an appetite for slow neutrons than does U-235, and that difference was exploited in the design of the **reactor,** *the controlled chain-reaction machine.* If uranium was interspersed within a *moderator* (some light substance that slowed the neutrons down), it would have almost the same effect as removing the U-238 altogether: Slow neutrons simply wouldn't interact with it.

The most promising candidates for moderators were beryllium, heavy water (p. 249), and carbon. Beryllium was crossed off the list because there wasn't much of it around, and it wasn't very pure. Heavy water would have been perfect, but there were only a few quarts of it in the United States, and there wasn't time to prepare much more, certainly not the several tons of it needed. The choice fell to carbon in the form of purified graphite.

After the initial studies by Fermi, Zinn, Szilard, and others at Columbia, the project was moved to the University of Chicago. Pressed for space on campus (and particularly for rooms with ceilings more than 25 feet high) Fermi and his group set up shop in the squash courts under the abandoned football stands of Stagg Field. There on the floor they began stacking large solid graphite bricks, 40,000 of them. Alternate layers of the *pile* were composed of bored-out bricks into which were inserted chunks of uranium or briquettes of com-

-2-

The United States has only very poor ores of uranium in moderate quantities. There is some good ore in Canada and the former Czechoslovakia, while the most important source of uranium is Belgian Congo.

In view of this situation you may think it desirable to have some permanent contact maintained between the Administration and the group of physicists working on chain reactions in America. One possible way of achieving this might be for you to entrust with this task a person who has your confidence and who could perhaps serve in an inofficial capacity. His task might comprise the following:

a) to approach Government Departments, keep them informed of the further development, and put forward recommendations for Government action, giving particular attention to the problem of securing a supply of uranium ore for the United States;

b) to speed up the experimental work,which is at present being carried on within the limits of the budgets of University laboratories, by providing funds, if such funds be required, through his contacts with private persons who are willing to make contributions for this cause, and perhaps also by obtaining the co-operation of industrial laboratories which have the necessary equipment.

I understand that Germany has actually stopped the sale of uranium from the Czechoslovakian mines which she has taken over. That she should have taken such early action might perhaps be understood on the ground that the son of the German Under-Secretary of State, von Weizsäcker, is attached to the Kaiser-Wilhelm-Institut in Berlin where some of the American work on uranium is now being repeated.

Yours very truly,
A. Einstein
(Albert Einstein)

We are justified in reflecting that scientists who can construct and demolish elements at will may also be capable of causing nuclear transformations of an explosive character.
FRITZ HOUTERMANS
Berlin (1932)

The West Stands of Stagg Field, where the first self-sustained nuclear reactor was built.

pressed uranium oxide powder. They would have liked to use only pure metallic uranium, but there wasn't enough of it to be had at the time. In fact, when the program really got started at the end of 1941, there were only a few grams of metallic uranium in the entire country.

The idea was that fast neutrons spewing out of the fission reactions within any one small chunk of uranium would sail into the graphite and be slowed down inside it before reaching the next concentration of uranium. Several long cadmium control rods were built into the pile as it took form. Cadmium is a voracious absorber of neutrons; drawing the rods out or shoving them into the structure was equivalent to turning the flood of neutrons on or off. The reactor kept a leash on the chain reaction, allowing it to propagate but maintaining control so that it didn't branch out into a runaway cascade. The goal was to maintain a delicate balance, keeping the number of neutrons produced exactly equal to the number lost to the reaction by way of absorption and leakage out of the pile.

The crucial test of the whole conception came on the afternoon of December 2, 1942, with Fermi, slide rule in hand, directing every move. As a preliminary he had all but two of the cadmium rods withdrawn from the pile. "Zip out!" he ordered next, and Zip (the name of the independently operated security rod) was pulled from the dark, unshielded mountain of graphite and uranium. That left only the single control rod in place; it alone held the reactor in check. Everyone in the room was silent as Fermi ordered the last rod cautiously withdrawn, a few feet at a time. Neutron counters (their detectors hanging at the side of the pile) were already clicking away in a kind of nervous drone. The pen on the recorder continuously drew a graph of neutron production, which rose and then leveled off, going higher and higher each time the rod was pulled out a little, each time obediently stopping just where Fermi had predicted it would. Grinning now ever so slightly, he worked his slide rule for a moment and then confidently ordered, "Pull it out another foot." It was 3:20 when he announced, "This will do it. Now the pile will chain-react." Compliantly the graph rose again in a smooth curve, but this time it showed no sign of leveling off—the world's first self-sustained controlled nuclear chain reaction had begun.*

---

* There is evidence that a sustained chain reaction once took place naturally in a rich vein of water-saturated uranium ore in the Oklo deposit in West Africa. This presumably happened well over 1500 million years ago when the natural concentration of U-235 was as high as three percent.

A reproduction of a painting by Gary Sheahan of the Chicago *Tribune*. It depicts that electric moment at the squash court in the winter of '42 when CP-1 came alive.

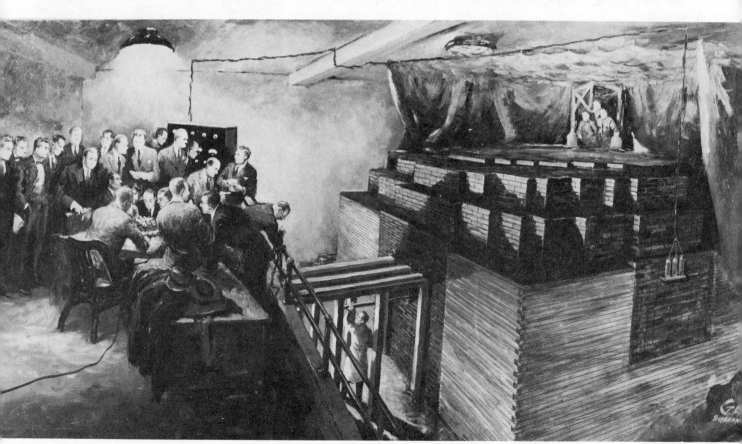

At 3:53 Fermi called out to Zinn "O.K., Zip in." As it slid into place, the counters abruptly slowed, and CP-1 (Chicago Pile One) was shut down. Wigner magically produced a bottle of Chianti he had been hiding behind his back, some paper cups were found, and everyone silently toasted the moment. News of the feat was cryptically relayed by long-distance phone: "The Italian navigator has landed in the New World."

Fermi's idea of creating transuranic elements was finally vindicated in 1940 at the University of California. When a $^{238}_{92}$U nucleus absorbs a neutron without undergoing fission, it is transformed into a new, still heavier uranium isotope, $^{239}_{92}$U. Highly unstable, that isotope breaks down in a matter of minutes. Emitting an electron, $^{239}_{92}$U decays into an element having 93 protons. This new substance was named neptunium (Np) after the planet Neptune, which orbits beyond Uranus, uranium's namesake. With a half-life of $2\frac{1}{4}$ days, neptunium in turn soon decays by beta emission into the long-lived element 94. For a while in 1941 "element 94" was referred to only by its code name, "copper." Sometime afterward it was christened plutonium (Pu) for the planet Pluto. All that nuclear alchemy became more than just an intellectual exercise after Louis Turner at Princeton predicted that plutonium would undergo slow-neutron fission even easier than $^{235}_{92}$U. The implications were momentous: $^{238}_{92}$U, relatively plentiful and comparatively worthless, could be converted into fissionable plutonium! And since it was an entirely different element, there would be no problem in chemically separating it from its parent uranium.

Out of those untested theoretical conclusions sprang the multi-million-dollar Plutonium Project. Three giant water-cooled production reactors in which to create plutonium were constructed in 1943 and 1944 on the Columbia River near Hanford, Washington. The Hanford

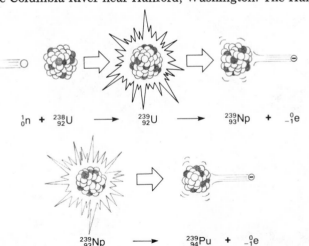

$$^{1}_{0}n \; + \; ^{238}_{92}U \; \longrightarrow \; ^{239}_{92}U \; \longrightarrow \; ^{239}_{93}Np \; + \; ^{0}_{-1}e$$

$$^{239}_{93}Np \; \longrightarrow \; ^{239}_{94}Pu \; + \; ^{0}_{-1}e$$

The production of Plutonium-239 from Uranium-238.

Uranium dioxide reactor fuel pellets being measured after fabrication. One such pellet can provide the energy of almost a ton of coal.

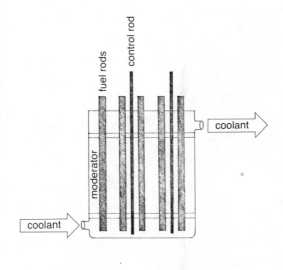

machines were basically similar to Fermi's pile, except that the uranium slugs were inserted into long cylindrical holes from which they could easily be removed. After "cooking" for a few months, the intensely radioactive "hot" slugs were taken out, and the $^{239}_{94}$Pu was extracted in remote-control facilities. Each reactor was designed for an average output of about half a pound of plutonium per day.

These machines were the forerunners of contemporary *power reactors*. Although they now come in many different designs (including minimodels for space satellites), the principal product of a typical power reactor is heat. That heat is then usually made to drive traditional turbines and generate electricity (about 10 percent of the present total U.S. output). Typically, a modern 1000 million watt reactor is charged with about 90,000 kilograms of uranium fuel, a third of which must be replenished each year. By comparison, a small portable reactor may use only 50 kg or so of U-235.

The Hanford reactors converted U-238 into Pu-239 under a deluge of neutrons emitted by the fissioning of U-235. Conditions were such that a bit more U-235 was always consumed than there was Pu-239 produced. It is possible, however, to surround the reactor core with a "blanket" of U-238 and actually end up creating more plutonium than the amount of Uranium-235 "burned" in the doing. The quantity of fissionable material in such a reactor continues to increase as the machine continues to generate power. Of course, that's at the expense of the U-238, which *is* diminished—this is no perpetual motion machine. Known as a **breeder reactor,** it can produce more fissionable fuel than it needs for operation.

The A-Bomb

During the war, while the scientists at Hanford were trying to manufacture Pu-239, others at a tremendous facility at Oak Ridge in the Tennessee Valley were at work struggling to separate U-235 from U-238. Not knowing which, if either, would be the more successful

Oppenheimer and Groves standing in front of the mangled remains of the vaporized 100-foot steel tower that held the first atomic bomb ever exploded. The desert sand was seared into a jade green glassy crust.

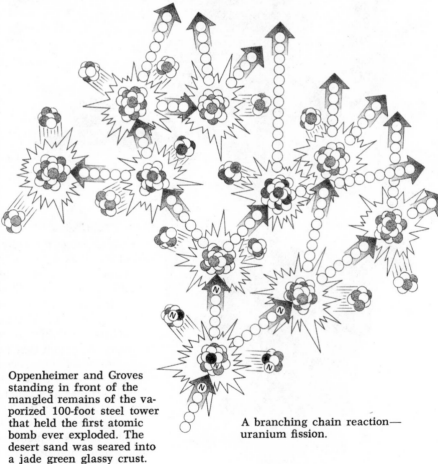

A branching chain reaction—uranium fission.

approach, they pursued both schemes equally frantically. By the spring of 1945, pounds of fissionable bomb stuff began trickling out of the processing plants, and the tension shifted to Los Alamos, New Mexico. There under the direction of J. Robert Oppenheimer ("Oppie" to almost everyone) a group of outstanding physicists had come to design the ultimate weapon. Many were already well-known figures—Bohr, Chadwick, Frisch, Fermi, and Bethe were among them. Others would become renowned later—Teller, Weisskopf, von Neumann, Alvarez, and Feynman. General Groves, who headed the entire bomb program (code-named the Manhattan Project), once remarked to his military staff, "At great expense we have gathered here the largest collection of crackpots ever seen."

Despite the momentous nature of the undertaking, not every moment was serious. Richard Feynman, now a Nobel laureate, was then a brash 25-year-old *enfant terrible* from Far Rockaway (Long Island), who loved to drive the security people mad. Annoyed when his mail was opened, he had his wife send him letters that were either written in code or torn into hundreds of pieces before they were posted. Once he went so far as to enter the main security area and leave a "Guess Who" message in the secret files.

The principle underlying an A-bomb is really remarkably simple. A small spherical chunk of fissionable material under bombardment by neutrons can support only a limited chain reaction; too many of the internally released neutrons will escape from the specimen, sailing out through its surface. As the radius of a sphere gets larger, its volume increases faster than does its surface area. That means that if we form a large solid spherical ball of U-235, more neutrons will be generated within it, and proportionately fewer will escape from its surface. There is therefore a **critical size,** at which the number of neutrons produced by way of fissioning nuclei exactly equals the number of neutrons lost. At that point the chain reaction is just initiated. Any increase in the amount of fissionable material beyond that will produce a branching, rapidly growing chain reaction. A solid sphere of U-235 about the size of a softball, struck by a stray neutron (and there are plenty of these flying around all the time) will roar into a furious, blasting, multimillion-degree plasma in a millionth of a second. That's all it takes, one chunk in excess of the **critical mass** (somewhat less than 5 kg for plutonium), and off it goes.

The first bomb to be designed was "Thin Man," a crude, brute-force beast, rugged but wasteful of its precious fuel, U-235. So simple was the design that it was used in war without even having previously been tested. Two pieces of U-235, each too small to ignite on its own, were rapidly brought together into a *supercritical* mass. Speed was absolutely essential. If the pieces approached each other too slowly, neutrons from one would affect the other as they came close. The reaction would start before the critical mass was reached: The pieces

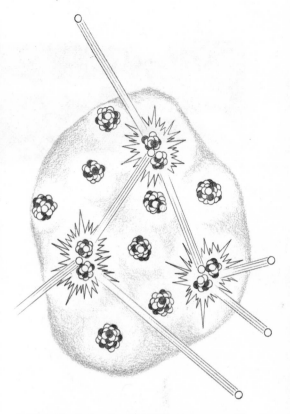

A subcritical chunk of U-235, one so small that many of the fission neutrons escape.

"Thin Man" was 28 inches in diameter and 120 inches long, and it weighed about 9000 pounds.

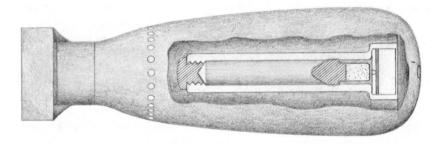

would heat up and melt, and the bomb would fizzle out without a full-scale explosion. Thin Man emerged as little more than a short cannon, approximately three feet long with about a three-inch bore. It simply fired a subcritical projectile of U-235 into another subcritical target chunk of the same stuff screwed down into the end of the muzzle.

Just when the Los Alamos team had settled on the gun-type bomb, they got a bit of a surprise. The test reactor was then beginning to produce its first samples of $^{239}_{94}$Pu, and a small problem soon arose. Appreciable amounts of $^{240}_{94}$Pu were being created out of and along with the desirable lighter isotope. This heavier isotope sometimes underwent spontaneous fission, splitting all by itself and firing out a continuous barrage of neutrons. Not much Pu-240 could be tolerated in bomb-grade plutonium, and furthermore its presence meant that the cannon approach (using two chunks of fissionable material) would probably be too slow for such a lively material. A completely different, much more rapid scheme to drive the Pu-239 into a compact

"Fat Man" was 60 inches in diameter and 128 inches long, and it weighed about 10,000 pounds.

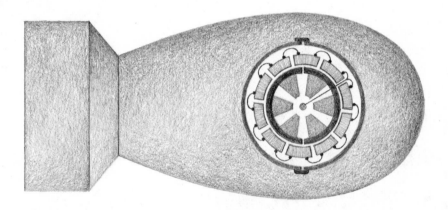

supercritical mass had to be devised. "Fat Man" was the alternative, a complex, technically elegant "nuke." This time the fissionable metal was fashioned into two pieces that together formed a thick, hollow, heavy sphere about the size of a cantaloupe. These were probably notched into segments—a little like chunks of chocolate in a candy bar—grooved so they would shatter neatly. Within the center of the plutonium was a small neutron source, which became activated as the bomb was triggered. The two hemispheres, slightly separated, were encased in a thin uranium shell, a "tamper," which served to reflect back wayward neutrons. Packed around the outside of this perpetually warm metal melon were 36 high-explosive charges. Tan in color and soapy to the touch, these were precisely shaped prisms—more than a ton of explosives. Multiple detonators, once ignited, started an inward-rushing blast wave designed to drive the hemispherical plutonium segments together, literally crushing them into a high-density super-critical mass.

ON THE ATOMIC BOMB

That is the biggest fool thing we have ever done. The bomb will never go off, and I speak as an expert in explosives.

ADM. WILLIAM LEAHY (1945)

Hiroshima.

AP INDICATES AIMING POINT

STATUTE MILE

Fat Man was far too complex not to be tested. Too much could go wrong. After five years and about a billion dollars, this five-ton horror machine was mounted on a tall steel tower in the remote desert of Alamogordo, New Mexico. At 5:29 A.M. on July 16, 1945, the Fat Man burst into a fierce, searing light. Gold, purple, violet—the colors swirled; an orange ball of flame rose silently into the still dark sky. Suddenly there was a tremendous blast of sound and a long, continuous roll of thunder.

Oppenheimer, standing in a distant blockhouse, his face lit by reflections from the sand, thought about Lord Krishna and the warrior Arjuna, about the giving of truth on the battlefield. Oppie had read the Hindu epic in the original Sanskrit, and at that breathless moment in the desert the lines came back to him:

> If the radiance of a thousand suns
> were to burst into the sky,
> that would be like
> the splendor of the Mighty One.

"There floated through my mind," he later recalled, "a line from the *Bhagavad-Gita* in which Krishna is trying to persuade the Prince that he should do his duty: 'I am become death, the shatterer of worlds.' I think we all had this feeling more or less."

On August 6, 1945, at 8:15½ in the morning Thin Man dropped from the belly of a B-29 over Hiroshima, Japan. Seconds later, 1500 feet above the ground, it burst into a fireball equivalent to somewhat less than 15 kilotons of TNT. Perhaps as many as 100,000 people died that moment and after. Three days later, on August 9, 1945, the Fat Man fell from 29,000 feet into the bright sunlight over Nagasaki; 74,000 people perished. "I am become death, the shatterer of worlds."

16.2 FUSION

Fritz Houtermans and Robert Atkinson were fellow students in 1927, undergraduates at Göttingen. While strolling together one day, they began to amuse themselves with the age-old problem of the seemingly inexhaustible source of the sun's energy. Einstein's $E = mc^2$ formula suggested that some sort of atomic transmutation involving a loss of mass might be the central phenomenon. In time they were able to show that in the interior of a star, where the density (100 times that of water) and temperature (15 to 20 million °C) are tremendous, nuclei of light elements like hydrogen could theoretically fuse together. Reactions of this sort, in which protons are captured and accumulated, might ultimately result in the creation and emission of alpha particles (helium nuclei). In effect, the sun would be transmuting hydrogen into helium while converting mass into energy.

Two years later Houtermans and Atkinson published their general ideas on star-fire in the austere German journal *Zeitschrift für Physik* —but only after they agreed to change the original title. "How Can

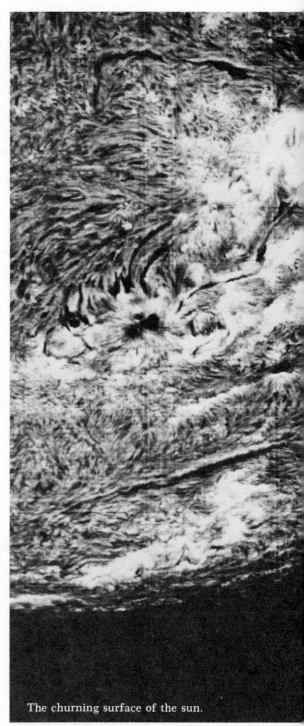

The churning surface of the sun.

One Cook a Helium Nucleus in a Potential Pot?" was too frivolous for the editor's tastes. Remember that these remarkable speculations came prior to the related discovery by Cockcroft and Walton (p. 335) that lithium could absorb a proton, split into two alpha particles, and give off energy.

A decade more passed before the work was developed in any detail. Then, one month prior to the announcement of the discovery of fission, Hans Bethe at Cornell revealed a sequence of nuclear events that could account for stellar energy production. Known as the **carbon-nitrogen cycle,** it began with the capture of a proton (^1_1H) by a Carbon-12 nucleus. After successive transmutations into several isotopes of nitrogen, carbon, and oxygen, accompanied by the capture of three more protons along the way, the scheme miraculously ended up with a Carbon-12 nucleus, a helium nucleus (^4_2He), and a lot of liberated energy. The carbon functioned as a sort of catalyst in the conversion ($4^1H \rightarrow {}^4He$), vital but unconsumed.

Another different process, which also had the effect of converting hydrogen into helium, was soon announced. Known as the **proton-proton chain,** this process began with two protons combining and transforming into a deuteron. Next that nucleus captured another proton, becoming ^3_2He. Two of these light helium isotopes fused together, finally liberating a total of one ^4_2He nucleus, two excess protons, and a great deal of energy.

Realize, of course, that all this is supposedly happening in the dense, seething belly of a blazing star—a plasma ball, where bare nuclei are flashing around at tremendous speeds, blasting into each other, sometimes rebounding, sometimes fusing. Indeed, only at temperatures of tens of millions of degrees will the particles, having formidable kinetic energy, be moving rapidly enough to drive in toward each other against their mutually repulsive electrical force. Only in high-speed collisions will they approach close enough to be caught by the strong force. Squeezed together into some new, more stable configuration of lower net mass, they liberate energy in a burst of **thermonuclear** fireworks.

Stars are born, live, and die; some disperse and others, latecomers, rise out of their ashes. A first-generation star probably began in the far distant past as primordial galactic hydrogen. By contrast, later generations contain, in addition to hydrogen, varying amounts of ancestral ash—elements created in earlier stars. Those twinkling points of light hung in the night sky are very individualistic, differing in size, age, mass, temperature, makeup, and behavior. But all are thermonuclear factories—generating energy out of matter, cooking up elements along the way.

A nascent star, an almost-star, is formed when a cloud of interstellar gas and dust gravitationally condenses, intensely heating up, becoming incandescent in the process. Composed mostly of virgin hydrogen (and hence rich in protons), it may reach a temperature and density (if the cloud is large enough) that triggers the onset of the proton-proton chain. The initiation of that thermonuclear reaction

Origin of the Milky Way by Tintoretto. The word "galaxy" derives from the Greek *gala,* which means milk.

Oh, these farm gardens with their lovely big red Provençal roses, and the vines and the fig trees! It is all a poem and the everlasting bright sunshine, too. . . . Just now we are having a glorious strong heat, with no wind—just what I want. There is a sun, a light that for want of a better word I can only call yellow, pale sulphur-yellow, pale golden yellow. How lovely yellow is!

VINCENT VAN GOGH

From a letter to Theo

marks the actual birth, the transition from collapsing gas cloud to self-sustained stardom. Puffed up by the internal release of energy, the star ceases to contract and stabilizes in size and output.

Our own star, the sun, is a dazzling yellow-white sphere, whose 2,200,000,000,000,000,000,000,000,000 tons of plasma swirl in an unimaginable inferno. Sol unleashes 3.8×10^{26} joules each second; each second it pours out into space the energy equivalent of three million million million gallons of gasoline. Every second roughly 660,000,000 tons of hydrogen are transmuted into helium ash through holocaustal alchemy. In the process, 4,600,000 tons (4.2×10^9 kg) of matter vanish, converted into a prodigious torrent of energy. And so it has raged for perhaps five thousand million years, fired up predominantly (90%) by the proton-proton chain.

At still higher temperatures, as deep within the Dog Star, Sirius, the carbon-nitrogen cycle prevails. But whether by "chain" or "cycle," hydrogen is fused into helium, and each He nucleus (664.5×10^{-29} kg) has less mass than the four separate protons ($4 \times 167.3 \times 10^{-29}$ kg) that went into forming it. The difference, 4.7×10^{-29} kg, corresponding by way of $E = mc^2$ to 4.3×10^{-12} joules, powers the stars. Most of the energy in the universe derives from this single fountainhead, and virtually all the energy on Earth is traceable back to the blazing sun.

Helium is only one product of the stellar furnace; all the known elements were probably created out of hydrogen in the stars.* For example, it has been demonstrated that under proper conditions, three helium nuclei can fuse, can "burn" to make ordinary carbon:

$$\tfrac{4}{2}\text{He} + \tfrac{4}{2}\text{He} + \tfrac{4}{2}\text{He} \rightarrow \tfrac{12}{6}\text{C}.$$

Helium, the ash left over from hydrogen burning, now becomes the next important fuel in the sequence. Stellar helium fusion of this sort is probably the principal source of carbon in the universe.

Hydrogen and then helium are the main star fuels. As a vigorous star ages, it goes through structural changes that alter its core temperature and allow it to burn heavier and heavier elements as the lighter ones are successively consumed. About 6000 million years hence, as the sun's hydrogen supply dwindles, it will first expand and then collapse, heating up as it does so. Helium accumulated in its core will then reach a high enough temperature to begin "burning."

---

* Recall W. Prout's (Section 10.4) hypothesis of 1815 that all the elements were built up from hydrogen. It was soon abandoned because the atomic weights simply weren't whole-number multiples of the weight of hydrogen. Not knowing about binding energy, neutrons, or protons, the scientists of the era made the wrong facts out of the right data and "proved" the theory wrong. But of course, in retrospect, rejecting Prout's suggestion was the only reasonable thing to do at the time, given their "understanding" of the universe. Note the difference between *rejecting* and *disproving*. As long as facts change, there's not much of substance that can be absolutely proved or disproved.

In time the helium will also be consumed, but its carbon ash, left behind in the core, will probably not ignite at the low prevailing temperatures, and our medium-sized sun will die. Larger, more massive, hotter-cored stars, having burned hydrogen and helium, may next shift to eating their accumulations of carbon, but our modest sun will not be able to do that. Thermonuclear "burning" of carbon, then oxygen, and then silicon occur in turn only in certain stars, those with tremendous temperatures in excess of 1000 million degrees.

While some stars are creating specific elements, others are consuming them. The latter process is believed responsible for the creation of still heavier nuclei. All the various elements up to iron are probably built in the stars through the fusion of charged light nuclei. Beyond that point the nuclei are too highly positive, exerting an electrical repulsion that restricts charged-particle reactions. Instead, the remaining heavy elements are slowly synthesized by way of neutron capture in second-generation stars that began their lives rich in metals, particularly iron. Later-generation stars—it seems the sun is one of these—are composed of all the stable elements. At birth they coalesced out of a rich galactic soup: hydrogen spiced with the debris of countless spent stars.

The finish finally comes to a large first-generation star when it is core-laden with iron. Stable iron, having the highest binding energy of all elements, can't be burned to liberate power in the fusion furnace. On the contrary, splitting or adding nucleons to it produces new, less tightly bound "plumper" atoms with more mass per particle. Iron is the dead end; altering it requires the *input* of energy. As a last gesture, a once mighty star, heavy with iron and low on fuel, blows apart as a catastrophic supernova. Its outer portion blazes into a horrendous explosion, unleasing a torrent of fast neutrons. During that mayhem, elements well beyond iron are quickly synthesized in a kind of rapid neutron feast. Erupting into interstellar space, this vast motherhood of matter spreads out to serve as star stuff and planet powder for another generation.

That is *our* personal history, atom by atom through the stellar foundry, for in the end *we* are star dust.

The Superbomb

The "Super," the hydrogen fusion bomb, was conceived way back in the summer of 1942, years before the atomic bomb was itself a reality. Edward Teller, who worked under Bethe at Los Alamos, was its chief advocate and principal conjurer. During the war and for several years after, the project made very little progress and generated even less enthusiasm. But there was an abrupt change in August of 1949, when the Russians exploded their first atomic weapon. In that atmosphere of shock and disbelief it wasn't hard for the proponents of the Super to win the day.

Thermonuclear fusion, of course, requires exceedingly high temperatures, much higher than could be provided by any sort of chemical detonator. However, an atomic bomb, generating a temperature in

excess of 100 million degrees, could easily serve as the trigger. Solar fusion is a leisurely business that perks away patiently at a fairly constant temperature in the stellar furnace. A bomb, on the other hand, would have to rush along rapidly while the atomic fire lasted. The lighter the fusing nuclei, the better, but ordinary hydrogen was out of the question because it burned too slowly. Bomb stuff would have to be the heavier hydrogen isotopes deuterium ($^2_1$H) and/or tritium ($^3_1$H). Several fusion reactions are possible:

$$^2_1\text{H} \quad + \quad ^2_1\text{H} \quad \rightarrow \quad ^3_2\text{He} \quad + \quad ^1_0\text{n} \quad + \quad 5.2 \times 10^{-13} \text{ joules.}$$
deuterium deuterium helium   neutron

$$^2_1\text{H} \quad + \quad ^2_1\text{H} \quad \rightarrow \quad ^3_1\text{H} \quad + \quad ^1_1\text{H} \quad + \quad 6.4 \times 10^{-13} \text{ joules.}$$
deuterium deuterium tritium   hydrogen

$$^2_1\text{H} \quad + \quad ^3_1\text{H} \quad \rightarrow \quad ^4_2\text{He} \quad + \quad ^1_0\text{n} \quad + \quad 28.2 \times 10^{-13} \text{ joules.}$$
deuterium   tritium   helium   neutron

The last of these, involving the fusion of deuterium and tritium, occurs at the lowest temperature and yields the most energy.

Although the energy liberated per fusion event is less than it is in fission, the masses involved are much smaller, so pound-for-pound, the two processes are roughly comparable, with fusion turning out to be several times more potent. In the A-bomb there are practical limitations on how much fissionable material can be safely packaged and still be efficiently assembled into a supercritical mass. There are no such restrictions on the H-bomb—the more fuel, the more horrific the blast. Deuterium, extracted from ordinary water, was cheap enough to use as the primary ingredient. A little tritium, manufactured in reactors (with no little difficulty), could serve as "kindling"; more of it would be created during the blast anyway. To obtain the needed densities, both components (which, like common hydrogen, are ordinarily gaseous) were liquified and maintained at an exceedingly low temperature.

The first thermonuclear "gadget," as they liked to call it, was installed on the islet of Elugelab, part of Eniwetok atoll in the Pacific. At dawn on November 1, 1952, Elugelab vanished in the whirlwind of a 10 megaton (released energy equivalent to 10,000,000 tons of TNT) thermonuclear fireball. However ferocious, this malevolent 65-ton liquid-hydrogen refrigerator was no practical weapon. Still, it had one main virtue; it utilized a configuration newly designed by Teller and the Polish mathematician Stanislaw Ulam that made it possible for a small A-bomb to trigger an arbitrarily large fusion reaction.

Another development in H-bomb technique appeared in both the United States and the Soviet Union at just about the same time. It wasn't actually a new insight (Hans Thirring had discussed the crucial idea in his book published in Vienna in 1946), but it simply hadn't been tried before. Instead of low-temperature liquified deuterium, why not use a solid compound, which would be dense and stable even without refrigeration? But the really clever stroke was to combine

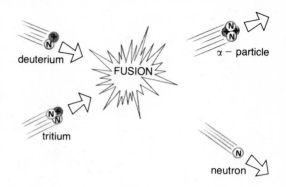

The fusion of a deuterium nucleus and a tritium nucleus to form an alpha particle.

deuterium with the relatively inexpensive light isotope of lithium, ^{6}Li, which captures a neutron and transforms into tritium:

$$^6_3\text{Li} + ^1_0\text{n} \rightarrow ^4_2\text{He} + ^3_1\text{H} + \text{energy}.$$

Lithium-6 deuteride (^{6}LiD) is an opalescent white solid composed of one atom of $^6_3$Li combined with a deuterium atom. Besides supplying the deuterium fuel, ^{6}LiD also provides the means of producing the needed tritium kindling. As soon as the A-bomb trigger fires, flooding the place with neutrons, the lithium is conveniently transformed into a whole supply of tritium right on the spot.

The last bizarre touch, which is optional but too often customary, is to surround the whole thing—A-bomb and lithium-6 deuteride—with a shell of U-238. The stupendous flux of very fast neutrons from the thermonuclear reaction will fission even cheap U-238. This blanket has no critical mass and so can contain as much U-238 as the delivery vehicle can carry. Known as a fission-fusion-fission device, this maniacal machine is the "dirty" H-bomb, dirty because the fission products descend to Earth as lethal radioactive dust—*fallout*—and the more fission, the more fallout.

In August 1953 the Russians fired off a relatively small, fairly crude fusion device, using lithium deuteride for the first time ever. It was no Super, but it was deliverable. Seven months later, the United States tested what was probably the first "practical" superbomb, code named Bravo. A lithium-deuteride fission-fusion-fission weapon, Bravo ushered in a new era of human folly with a 15 megaton roar.

Taming Star-Fire

Nuclear fusion has a double edge; it is both destroyer and provider. It is at once the raw unleashed power within the bomb and the potentially docile supplier of an almost limitless pool of energy. Taming star-fire is one of the great challenges of the twentieth century. The appeal is obvious: Every eight gallons of ordinary water contains approximately one gram (0.04 oz) of deuterium (about 3×10^{23} atoms), which can be readily separated out at a cost of only a few pennies. If these atoms are fused with one another they will liberate roughly 100,000 million joules, or the equivalent in energy of 24 tons of TNT or 770 gallons of gasoline. Spicing the brew with tritium and consuming the by-products as well could increase that output severalfold. Since there are about 10^{17} kilograms of deuterium in the oceans alone, we are talking conservatively about a supply capable of meeting all the world's energy needs, even if they are increased drastically, for well over 100 million years. As if that weren't enough, fusion has the additional advantage of being a safe, comparatively clean, almost pollution-free route. In other words, the potential environment problems seem manageable—far more so than with fission reactors.

The underlying goal that must be attained to achieve controlled fusion, regardless of method, is threefold: to maintain a sample at adequate *density* and high enough *temperature* for an appropriately long *duration*. Two distinct approaches are currently being pursued in

SEPTEMBER 12, 1933

Atom-Powered World Absurd, Scientists Told

Lord Rutherford Scoffs at Theory of Harnessing Energy in Laboratories

By The Associated Press

LEICESTER, England, Sept. 11.— Lord Rutherford, at whose Cambridge laboratories atoms have been bombarded and split into fragments, told an audience of scientists today that the idea of releasing tremendous power from within the atom was absurd.

He addressed the British Association for the Advancement of Science in the same hall where the late Lord Kelvin asserted twenty-six years ago that the atom was indestructible.

Describing the shattering of atoms by use of 5,000,000 volts of electricity, Lord Rutherford discounted hopes advanced by some scientists that profitable power could be thus extracted.

"The energy produced by the breaking down of the atom is a very poor kind of thing," he said. "Any one who expects a source of power from the transformation of these atoms is talking moonshine. . . . We hope in the next few years to get some idea of what these atoms are, how they are made and the way they are worked."

Lord Rutherford

the great worldwide fusion race. One approach is an attempt to establish a stable, contained plasma that could continuously churn away like a little star in a bottle. The other seeks to create minute hydrogen-bomblike explosions within a chamber that could absorb and carry off the liberated energy.

A high-temperature deuterium-tritium plasma can be confined by powerful magnets so that it stays suspended within a vacuum chamber without touching the walls. One of the most promising containment devices thus far is a bagel-shaped affair invented by the Soviets in the 1960s and known as a *tokamak*. When exceedingly large electrical currents are pumped through the circulating plasma, temperatures of tens of millions of degrees Celsius can be reached. Deuterons within a 10^8 K plasma flash along at 1500 miles per second. At about 350 million degrees, colliding nuclei will readily fuse, and the plasma will ignite into a minidonut star. Nearly 80 percent of the fusion energy released sails out as KE of the fast neutrons. Trapping these in an absorbing blanket converts that energy to usable heat. Unfortunately, no device to date has met all three criteria (density, temperature, and duration) at once. High hopes for success in the early 1980s are riding on the new giant Tokamak Fusion Test Reactor (TFTR) at Princeton.

The Soviet Tokamak-10 nuclear reactor on which a stable deuterium fusion was obtained in 1976.

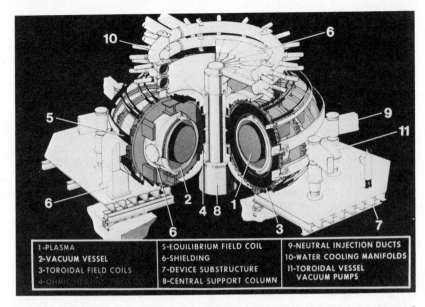

1-PLASMA	5-EQUILIBRIUM FIELD COIL	9-NEUTRAL INJECTION DUCTS
2-VACUUM VESSEL	**6-SHIELDING**	**10-WATER COOLING MANIFOLDS**
3-TOROIDAL FIELD COILS	**7-DEVICE SUBSTRUCTURE**	**11-TOROIDAL VESSEL**
4-OHMIC HEATING FIELD COIL	8-CENTRAL SUPPORT COLUMN	**VACUUM PUMPS**

Tokamak fusion test reactor.

The alternative approach consists of zapping a tiny spherical pellet of deuterium-tritium with an energetic beam, usually of either charged particles or photons. In one scheme the pellet, as small as 100 millionths of a meter in diameter, is blasted from all sides by a high-power laser as it drops into a multibeam crossfire. Its surface layer, instantly vaporized, rushes outward while the rest of the pellet implodes in reaction. Under tremendous pressure, in excess of a hundred thousand million atmospheres, the inner core compresses to a horrendous density of around 1000 g/cm^3. As it collapses, the bolt of laser energy heats the pellet to temperatures greater than 100 million degrees Celsius. At that point the thermonuclear burn begins, producing a blast of well over a million joules in less than 10^{-11} seconds.

Thus far, laser-induced fusion has been achieved on a very limited scale. We are still far from the "break-even" point, where the reactor generates at least as much energy as it uses. Ultimately, the fusion machine will drop and blast pellets with the ceaseless rhythm of a dripping faucet, producing many times as much energy as it consumes.

The thermonuclear burning of a fuel pellet. This approach is called *inertial confinement*. No externally applied force fields are needed to confine the plasma.

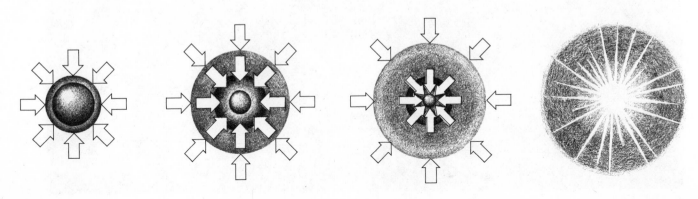

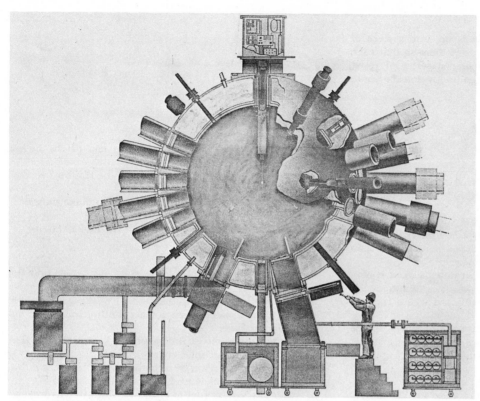

The target chamber of the Nova laser fusion system. This is a water-cooled vacuum bottle capable of withstanding the shock of micro-explosions equivalent to 50 lbs of high explosives.

There is another form of temptation even more fraught with danger. This is the disease of curiosity. . . . It is this which drives us on to try to discover the secrets of nature, those secrets which are beyond our understanding, which can avail us nothing, and which men should not wish to learn. . . . In this immense forest, full of pitfalls and perils, I have drawn myself back, and pulled myself away from these thorns. In the midst of all these things which float unceasingly around me in my everyday life, I am never surprised at any of them, and never captivated by my genuine desire to study them. . . . I no longer dream of the stars.

ST. AUGUSTINE

Confessions

Twinkle, twinkle, little star!
How I wonder what you are,
Up above the world so high,
Like a diamond in the sky!

ANN TAYLOR

Rhymes for the Nursery

EXPERIMENTS

1. Stick a few dozen matches, heads up, into a piece of clay on a plate or sheet of metal. Arrange them so that each match is near two others in a fan-shaped series of rows. *Being very careful,* light the first match and watch the chain reaction rapidly build.

REVIEW

Section 16.1

1. Fermi's novel projectiles were _____.
2. Why were slow neutrons of particular importance?
3. What experiment was performed in a goldfish pond?
4. What is a *transuranic* element?
5. Ida Noddack was the first to suggest that a uranium nucleus had been _____ by a neutron bullet.
6. Which element did Hahn and Strassmann isolate as a fragment from the uranium bombardment?
7. About _____ % of the uranium's mass is converted to energy on splitting.
8. What does the term *fission* mean, and where was the word borrowed from?
9. List the natural isotopes of uranium. Which is the most abundant? Which one undergoes slow-neutron fission?
10. Where does most of the liberated fission energy appear?
11. What is a *chain reaction*?
12. Describe the role played by Einstein in the development of the A-bomb (a role he later sorely regretted).
13. Why is Uranium-235 of special importance?
14. December 2, 1942, was a bitter cold day in the "Windy City." Gasoline rationing had just started, and auto traffic was way down; streetcars and the noisy elevated train were uncommonly packed. What else memorable happened that day in Chicago?
15. What was Chicago Pile 1?
16. What is a *moderator*?
17. For what purpose were cadmium rods stuck into CP-1?

18. Why is it ironic that Pluto is the god of the nether world, the god of death?
19. What was the primary function of the Hanford reactor facility?
20. Name the fissionable materials used to make *nukes*.
21. Describe the gun-type A-bomb.
22. Why was a new design needed for the Pu-239 bomb?

Section 16.2

1. The sun fuses the element _____ into the element _____.
2. Hans Bethe described the _____ _____ _____ as a mechanism for accounting for stellar energy production.
3. What is the name of the particular fusion process that is the primary source of solar energy?
4. Define what is meant by a *thermonuclear* process.
5. How is carbon made in the stars?
6. What is the next fuel to burn in a star after hydrogen?
7. Discuss the meaning of the term *supernova*.
8. What is the trigger for an H-bomb? Why is that necessary?
9. What is the function of Lithium-6 deuteride in the H-bomb?
10. What is a "dirty" H-bomb?
11. List the three criteria for controlled fusion.
12. What is a *tokamak*?

QUESTIONS

Answers to starred questions are given at the end of the book.

Section 16.1

1. Describe the process of uranium fission from the perspective of the *liquid drop model*. Why should a nucleus have a surface tension (Section 13.1)?
* 2. Write an equation for neutron-induced fission of U-235 into Barium-138 and Krypton-95, accounting for all the neutrons and protons.
3. Distinguish between the behavior of U-238 and U-235 as it relates to the design of a nuclear reactor (or "pile" as it used to be called).

4. During World War II, the only large producer of heavy water in the world was the Norsk Hydro plant at Vemork, Norway. Toward the end of 1942, a group of Anglo-Norwegian commandos carried out two fairly successful raids on that facility, which was then in Nazi hands. In the fall of 1943, after the plant had been rapidly repaired, American planes again struck at its power-generating station and the electrolysis works. At that point it should have been clear to everyone that we knew and that they knew that we knew. Explain all this. What was so special about heavy water?

5. Describe Fermi's reactor, explaining briefly how it worked. Did it utilize a branching chain reaction? Why do you think it was never allowed to operate in excess of a mere 200 watts?

* 6. Discuss the significance of the fact that Plutonium-239 can be produced by neutron bombardment of Uranium-238.

7. Explain the concept of *critical mass*. Would the critical mass increase, decrease, or stay the same if the specimen was made high in impurities? If it was in the shape of a cylinder rather than a sphere? If its density was increased under pressure?

8. It was Louis Slotin's job at Los Alamos, one for which he volunteered, to run experiments on fissionable materials to determine the actual critical mass of an assembly. He did so by bringing the two pieces close to each other and measuring the emitted neutron flux. One day soon after the war, while he was running a demonstration with six onlookers, his screwdriver apparently slipped, allowing the two pieces to remain too close. Realizing the enormity of what had happened, he lunged toward the apparatus and pulled the pieces apart bare-handed; nine days later Slotin was dead. Describe what you think was happening physically during the accident.

Section 16.2

1. Why are high temperatures necessary for fusion to occur? Distinguish between the roles played by the electrical force in fission and fusion.

2. What does it mean to say the stars are thermonuclear factories "cooking up elements"?

3. It seems that a star ought to collapse under the influence of gravity. What keeps it from doing just that all of its stable life?

* 4. Draw a sketch of each of the two deuteron-deuteron interactions, showing the distribution of nucleons before and after.

5. Why is fusion power so appealing?

6. Describe the fusion technique that uses a magnetically confined plasma. Why don't the walls of the chamber melt when the plasma is at hundreds of millions of degrees?

7. Describe the laser-induced fusion technique.

MATHEMATICAL PROBLEMS

Answers to starred problems are given at the end of the book.

Section 16.1

1. How many fission events *ideally* occur within the tenth generation of a branching chain reaction in which two neutrons are liberated per event?

* 2. Uranium-235 can capture a neutron and undergo fission in any one of a whole range of different ways, producing different fragments. One possible route results in Rubidium-90, Cesium-143, and neutrons. Write an equation accounting for all the nucleons.

3. The equation

$$^{235}_{92}U + ^{1}_{0}n \rightarrow ^{90}_{38}Sr + ^{136}_{54}Xe + 10^{1}_{0}n$$

describes a commonly occurring A-bomb–produced reaction that gives rise to the infamous Strontium-90 radioactive fallout. Using 1.675×10^{-27} kg, 3.9031×10^{-25} kg, 1.4930×10^{-25} kg, and 2.2568×10^{-25} kg as the masses of the neutron, uranium, strontium, and xenon, respectively, compute the energy given off in that fission reaction.

4. On the average, a typical Uranium-235 fission event liberates about 3.2×10^{-11} joules. Calculate how many tons of this fuel would be needed to supply the whole world's energy requirements for a year, about 2.4×10^{20} joules (1 kg $\rightarrow 1.1 \times 10^{-3}$ tons).

* 5. In order to show how the critical mass is reached by increasing the size of the spherical sample, calculate the ratio of surface area ($4 \pi r^2$) to volume ($4/3 \pi r^3$) for spheres with radii of 1, 2, and 4 centimeters. Note how the ratio decreases as the radius increases. With a density of 18.7 g/cm³, how large would a 5-kg U-235 sphere be?

Section 16.2

1. A gallon of gasoline corresponds to a stored energy of about 1.3×10^8 joules while 1000 tons of TNT is usually taken to be equivalent to about 4.2×10^{12} joules. Verify that the sun's energy output per second equals that of 3×10^{18} gallons of gasoline. How many 10 megaton H-bombs per second is that equivalent to?

* 2. Given the mass of a proton (1.673×10^{-27} kg) and the mass of an alpha particle (6.644×10^{-27} kg), compute the energy liberated in the process

$$4^1H \rightarrow {}^4He.$$

3. Compute the loss in mass in the tritium (5.007×10^{-27} kg)- deuterium (3.344×10^{-27} kg) fusion interaction, which results in an alpha particle (6.645×10^{-27} kg) and a neutron (1.675×10^{-27} kg). Show that that corresponds to an energy of about 28×10^{-13} joules.

PART V Electro-magnetism

Much of what we have studied of the material universe up until this point is associated with that primary property of matter, mass. The next five chapters, however, focus on the specific effects of an equally fundamental though somewhat more subtle characteristic, *charge*.

The whole complex of phenomena that are attributable to this single aspect of matter are drawn together under the one designation of *electromagnetism*. This includes the interactions of charges at rest, currents of charge, magnetism, and all forms of radiant energy from gamma rays to light.

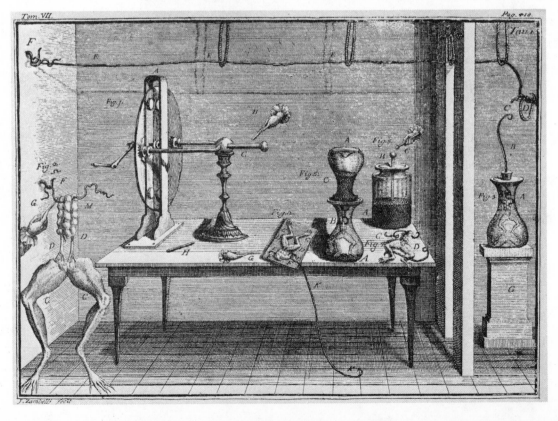

Matter exhibits its basic characteristics via interactions between material objects. The interaction that over centuries has come to be known as **electrical** arises from the fundamental property of matter we call **charge.** This chapter deals in particular with **electrostatics, the study of the interplay of charges at rest.** The first section is concerned with a general description of the behavior of charge: how it is transferred; the fact that there are two kinds of charge; how they respond to each other, etc.

The second section evolves the relationship that describes, in quantitative terms, the force between any two charges—**Coulomb's law.** As it turns out, **this formula varies directly as the product of the two charges and inversely with the square of the distance separating them.** The obvious similarity to the universal law of gravitation is both striking and inexplicable.

The last section develops the concept of the **electric force field,** as introduced in the mid-nineteenth century by Faraday. It ends with a discussion of **electric potential.** This quantity is more familiarly referred to as **voltage**—a practical notion we will continue to be concerned with, particularly in the next chapter.

CHAPTER 17
Electrostatics

Earlier we observed that material bodies change their motion in a way that can be understood if we suppose that they interact with each other even when not in direct contact. The logic leads to both the concept of a *gravity force* and the notion of *mass*. If there is a force, "something must be causing it," and so mass (in particular gravitational mass) becomes the physical entity giving rise to that influence. An apple falls from a tree—that we see. Gravity pulls it downward—that we believe. Its mass interacts with the mass of the Earth—that we provide as explanation.

When rubbed with a woolen cloth, a chunk of yellow amber or a plastic comb can pick up small wads of lint and hair—the ancient Greeks knew that (Section 10.1). It was clear centuries ago that some independent influence was at work, an influence that could even surpass the action of gravity. The upward leap of a tuft of lint demands the presence of a nongravitational force. We are led by our understanding of the laws of motion to both the concept of an *electrical force* and the notion of *charge*. If there is a force, "something must be causing it," and so charge becomes the physical entity giving rise to that influence.

Man is nothing else than his plan; he exists only to the extent that he fulfills himself: he is therefore nothing else than the ensemble of his acts, nothing else than his life.

JEAN-PAUL SARTRE
Existentialism and Human Emotions

On facing page, Galvani's experiments on animal electricity.

In a relatively modern painting (1903) by A. A. Hunt, William Gilbert demonstrates electrostatics to a remarkably young-looking Queen Elizabeth I.

Charge *is the embodiment of that which produces electrical force,* and we are now just beginning to figure out how it may manage to do that. Alas, when we are finished describing its behavior, we are finished defining it. Like mass, charge is a fundamental quantity that cannot be described in terms of simpler, more basic concepts. I can tell you what *salami* is but not what *charge* is—at least not conceptually. We know it more by what it does than by what it is. If you like, *it is what it does,* and that's that.

An electron is said to be charged because it exerts a specific, recognizable, nongravitational influence on other charges. For example, it repels other electrons, and it attracts protons. There is no discharging an electron, no way to peel away its charge and have it naked neutral. If the electron is truly a fundamental particle, as many suspect, then charge is an inseparable aspect of the thing as a whole.

It is rather puzzling, then, that the proton has a charge that is equal to though opposite from that of the electron. Recent experiments have indeed shown that the charges of the two are so nearly identical that if their ratio differs from 1 at all, it does so by a minuscule amount, less than 10^{-20}. That's fascinating, especially since the proton seems to have a complex structure. What is going on is still a mystery, but physics is happily full of mysteries. Thus far, charge has been found to exist only in whole-number multiples of a basic amount, namely, the electron's charge, q_e. Whether positive or negative, **charge is quantized**—it comes as one or more equal-sized lumps, never as fractional pieces. Thus q_e is the fundamental indivisible unit of charge. That much we know, even though at present there is no theoretical reason for it to be so. Realize, too, that the names "positive" and "negative" don't mean anything profound; they're the totally arbitrary bequest of old Ben Franklin, the great statesman, "electrician," and writer of racy prose. Perhaps "charge" and "anticharge" would make more sense, but we're stuck with + and −.

In addition to charged entities, there appear to be neutral things, like the neutron, devoid of any total charge. These bits of matter neither attract nor repel other matter electrically; they are quite indifferent to charge. Measurements cannot prove that the neutron is absolutely neutral. There will always be some ambiguity, some unavoidable experimental uncertainty. To date, however, laboratory work has established that if the neutron carries any charge at all, it's at least a hundred million million million times smaller than q_e. Similarly, experiments have confirmed the general belief that a quantum of light, a photon, is also chargeless (at least to within a minute error of a 10^{-16}th of q_e). We know that electrical charge is a property of matter (and antimatter), but we don't yet know enough to say, "An electron is negatively charged because...."

Unlike mass (which can be converted into energy) charge seems to be scrupulously conserved. *The total amount of charge (positive minus negative) within an isolated system is always constant, regardless of the reactions taking place.* That is the **law of conservation of charge.** Whether it is created or just shifted around, when positive charge appears in one region, somewhere else there will be an exactly equal quantity of negative charge. As far as we can tell, the total amount of charge in the universe is a constant—probably equal to zero, but obviously nobody knows for sure.

Unlike mass, which varies relativistically with speed, **charge is an invariant,** independent of its motion. That, too, is a curious and sharp distinction between these two fundamental aspects of matter.

Another remarkable difference is that the electrical force is both **attractive and repulsive,** whereas the gravitational force seems always

Electricity is not like red paint, a substance which can be put on to the electron and taken off again, it is merely a convenient name for certain physical laws.

BERTRAND RUSSELL (1923)

Philosopher

ELECTRICAL ACTION-REACTION
***BEFORE THE* PRINCIPIA**

It is commonly believed, that Amber attracts the little Bodies to itself; but the Action is indeed mutual, not more properly belonging to the Amber, than to the Bodies moved, by which it also itself is attracted; . . ."

MAGALOTTI (1665)

Florentine Academy

to be attractive. This inexplicable distinction severely limits the influence of the electrical force, which generally all but cancels in the region beyond neutral matter. The sun is a violent plasma ball of swirling positive and negative charge and yet, since there is as much of one as of the other, the star as a whole is neutral. Not far beyond it, certainly down here on Earth, the electrical pushes and pulls working against each other become practically impossible to detect. From here the sun appears quite chargeless. Although its electrical influence cannot even raise a tuft of lint, its gravitational influence holds the planet in orbit and raises the tides.

Gravity comes on as the more influential because it has no negative opposition, no repulsive aspect. Nonetheless, it is faintly possible that the interaction of matter and antimatter might produce just such a repulsion, despite the strong theoretical arguments against it. Settling that issue will have to wait awhile—the gravitational forces on the tiny specks of antimatter available until now have been too small to measure accurately.

Now that some of the subtleties have been examined, we can explore the more apparent behavior of charge *en masse*. This chapter deals specifically with **electrostatics,** *the study of charges at rest.*

17.1 A CHARGE ACCOUNT

We are surrounded, literally inundated, by charges that form a subtle electrostatic environment whose varied influences usually go quite unnoticed. Still, everyone has probably felt the tiny sparks that often punctuate a short shuffle across a carpet on a dry winter's day. Most of us have rubbed balloons on shirts to make them stick to walls or fought with a sweater fresh out of the dryer and clinging. From a bolt of blue-white lightning to that annoying layer of dust that's always sticking to the TV screen, the manifestations of charge are everywhere. Yet if the universe as a whole is neutral, why is there such a prevalence of charge, and what is the peculiar link to rubbing?

Atoms are surely neutral; they possess as much positive charge as negative charge, as many electrons as protons. But the outer electrons, the ones farthest from the nucleus, are the least strongly bound, and they are often very easily shed. The process whereby these electrons pass from the surface of one object into another is not entirely understood, though basically we have a good idea of what's happening. Different substances have different affinities for electrons. When two substances are put in contact, one of them may give up some of its loose electrons with little struggle while the other draws them into itself. Suppose, for example, that a sheet of plastic is pressed against a metal plate (assume that both substances were neutral to begin with). Simply coming into close contact results in the transfer of some electrons, leaving both materials charged when subsequently separated. In this instance the metal is the grabber; it ends up with an excess of electrons and becomes negative. The plastic, having lost electrons, now contains some number of immobile positive ions on its surface and, as a whole, has become positive.

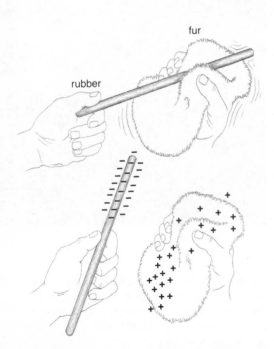

fur

rubber

Charging by contact. Electrons are transferred from fur to rubber.

In much the same way, if a hard rubber rod is stroked with a piece of fur, the rod will draw off electrons, becoming negatively charged, and the donor fur will assume an equal positive charge. The rubbing seems to do little more than increase the area brought into intimate contact. *Note that charge is not being produced but merely redistributed; electrons are being transferred.*

There are all degrees of grabbers, and a substance that can snatch electrons away from one material may well find itself serving as a donor when confronted by a more potent taker. Glass rubbed with asbestos will draw off electrons from that fibrous material and become negative, but if stroked with some persistence against silk or flannel, the glass will emerge positively charged, having lost electrons.

You may well ask at this point, "How does anyone know the polarity of the charge generated on the rod?" You can rub a plastic pen with a piece of wool, and it will pick up scraps of tissue paper, but that doesn't tell you whether the pen is electrified + or −. The most rudimentary and straightforward way to determine the kind of charge is to hang the various samples from threads like pendulums and observe which ones attract and which repel. Two pieces of glass rubbed with silk swing apart, repelling each other. When both are rubbed with fur, the two again repel. But if one is stroked with silk and the other with fur, then they come toward each other—they attract! The conclusions are simple. **There are two different kinds of charge.** Furthermore, **like charges repel and unlike charges attract.** Now, if we arbitrarily define glass-rubbed-on-silk as *positive,* anything that attracts it is *negative,* and we're ready for all comers.

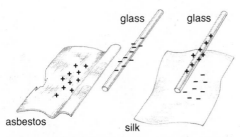

Electrons are transferred in contact from the asbestos to the glass and from the glass to the silk.

Asbestos
Fur (rabbit)*
Glass
Mica
Wool
Quartz
Fur (cat)
Lead
Silk
Human skin, aluminum
Cotton
Wood
Amber
Copper, brass
Rubber
Sulfur
Celluloid
India rubber

On contact between any two substances shown in the column, the one appearing above is supposed to become positively charged, the one listed anywhere below it negatively charged.

* I have my personal doubts about this one's position above all kinds of glass, but this is the customary listing. There are a lot of variables operating here: surface condition, impurities, temperature, etc.

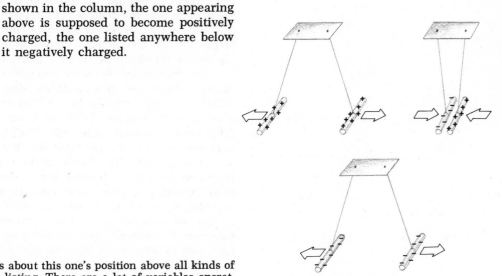

Attraction and repulsion of charges.

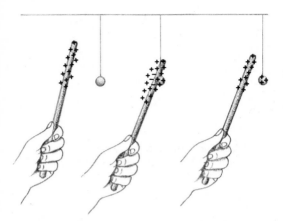

Charging a piece of pith positively by direct transfer of electrons from ball to rod.

Insulators and Conductors

Isolation of charge on a nonconductor.

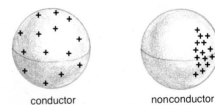

conductor   nonconductor

The distribution of charges placed on the surfaces of a conductor and a nonconductor.

Of course, if you happen to be concerned about the polarity of a charged elephant, hanging it from a thread might be a bit impractical. There's really no problem, though. Just suspend two identical, light, convenient neutral objects—two Ping-Pong balls, two balloons, or two pith balls (the physicist's traditional electrostatic dangler).* Now touch a rod of glass-rubbed-on-silk to one of the pith pair. The glass, which is positive and therefore deficient in electrons, will draw electrons from the pith, leaving it positive (just as if some mobile positive charge had actually been transferred from the rod to the ball). If you then back the electrified pachyderm into the second pith ball, it will take on some of the unknown elephant-charge. If the two electrified balls then attract, the elephant was negative; if they repel, it was positive.

Now any astute mahout might ask, "Is it not possible that one end of our elephant is charged and the other not?" The answer is "Yes, it is possible." If you rub the middle of a comb with a piece of wool, that region will have the ability to pick up tissue tufts, but the comb's ends won't; they remain neutral, even though the middle is electrified. This behavior is representative of a whole class of materials (such as wood, plastic, air, hair, cloth, glass, resin, paper, mica, salt, leather—and elephant), all of which are known variously as **insulators, nonconductors,** or **dielectrics.** In other words, *when an insulator receives a charge on its surface, it retains that charge, confining it within the localized region in which it was introduced.* By contrast, a **conductor** *is a substance that allows any charge introduced onto its surface to flow, freely redistributing itself.* Metals (e.g., copper, steel, aluminum, brass, gold, silver, tin, mercury, etc.) are among the best conductors. The fact that there were both good and bad conductors of electricity was not realized until 1729, when Stephen Gray conducted a series of brilliant experiments, despite the severe handicap of being a pensioner in a London poorhouse.

The distinction between conductor and nonconductor (and it isn't always unambiguous) arises from the presence within the material of mobile charges. Recall (Section 12.3) that metals contain a tremendous number of free electrons, roughly one per atom, which drift around almost unimpeded within the specimen. Any added electrons just join the unbound sea that moves among the unmoving ions. On the other hand, the atoms composing a nonconductor are far more possessive, holding on to their family of electrons much more tightly. Any added electrons are similarly anchored, somewhat immobilized by their interactions with the fixed atoms. Despite all of this, *no material is a perfect nonconductor;* all allow some redistribution of charge, some meandering. Thus human skin is a respectable nonconductor by comparison with copper, although it is a

---

* Pith is the pulpy, soft stuff in the center of certain plant stems. It was the Styrofoam of generations past, used in work with electricity for centuries and still to be found in physics storage rooms everywhere.

fairly decent conductor when compared with glass or plastic. Perhaps it would be better just to say that it is both a poor conductor and a poor insulator. Pure water is a fair insulator, but a pinch of some dissolved impurity like salt will provide enough ions to turn it into a good conductor (*ion* means *traveler*). Indeed, water vapor in the air carries away charge quite effectively; that's why if you get shocked after walking on a rug, it's likely to be in the winter when the heat is on indoors and the relative humidity is very low. Few things are more frustrating than trying to demonstrate electrostatics on a damp day, when the charge laboriously built up quickly leaks off to the surroundings.

Charges tend to bunch up on the pointed regions of a conductor.

Like all gases, air is a good insulator, particularly dry air, even though it contains about 300 ions per cubic centimeter. Nevertheless, if enough negative charge builds up on an object, electrons under the influence of their mutual repulsion can be propelled into the surrounding gas. The gas itself has some of its own electrons ripped off in the ensuing maelstrom, becoming ionized and creating a temporary conductive path along which the bulk of charge then flows. Collisions with the gas increase its temperature and raise some of its atoms into excited energy states, from which they descend, emitting light. The result is that familiar glowing trail known to all as a **spark**—a blast of charge from one body to another through a gas.

The fact that electrons can flow freely within a conductor is responsible for several very useful properties. First, we anticipate that any cluster of like charges introduced on a conductor will experience a mutual repulsion that will send them all scurrying. Constantly pushed apart, they will move until they can separate no farther and are as distant from one another as possible. Charges tossed onto a metal sphere will very quickly stream around until they are *uniformly* distributed and at rest on the outer surface. *No matter what the shape of the conductor is, excess charge will always reside on its outer surface.* The charges simply push each other to the very extremities of the object. If the object is nonspherical, the charge distribution will be nonuniform, bunching up somewhat in the remote regions. Each charge tries to get as far away from as many others as possible, even if that means getting closer to a few charges in the process. That's why charge tends to concentrate on a conductor's protrusions.

In 1786 Rev. Abraham Bennet introduced an improved device that, until the modern era of electronics, was the mainstay of electrostatic metering. The **gold-leaf electroscope** is basically two extremely thin metal leaves hanging parallel to each other from a conducting wire, all surrounded by a protective glass enclosure (see figure).*

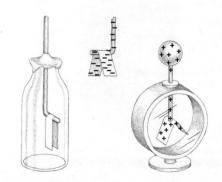

Electroscopes homemade and otherwise. No longer used in the laboratory, the electroscope is now primarily a teaching device.

---

* The electroscope is very easy to make. Use clay or wax to seal the support wire into a jar. Any wire will do as long as it's not insulated on the ends. Very thin gold leaf is best, but the light aluminum foil on chewing gum wrappers works fine. Strip it from the paper backing and clean off any wax residue. Fix the foil over the support wire and tack it in place with a little glue. Make sure the foil and wire are in good contact.

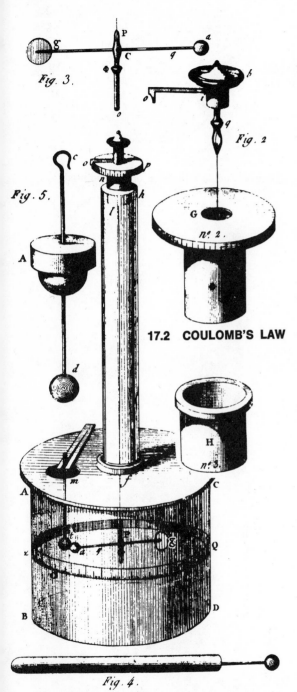

Fig. 3.

Fig. 2

Fig. 5.

17.2 COULOMB'S LAW

Fig. 4.

C. A. Coulomb's torsion balance (from *Histoire et Mémoires de L'Académie Royale des Sciences*, 1785).

Suppose that the top of the support wire is touched by something carrying a negative charge. Electrons will be transferred to the wire, and as they move apart under their mutual repulsion, some will enter both leaves. The leaves, now identically charged, will repel each other and spring apart at an angle proportional to the charge. There they'll stay, leaves parted, until the charge is deliberately removed or it ultimately leaks off.

The very earliest electroscopes at the beginning of the eighteenth century were primitive devices, usually consisting of no more than two hanging threads, sometimes carrying pith balls at their ends, often not. By the end of that century, angular scales marked off in degrees were installed, so that the separation between the leaves (or threads) could be read with some precision, and the device became a charge-meter, an *electrometer*. That instrument in its various forms was used to *detect the presence* of charge, to *establish the polarity*, and to *determine the relative size* or quantity of charge. Absolute measurements would have to wait awhile. At that time there was no established unit of charge, so one was unable to say "Ah, there are now exactly five '*shmiggles*' of charge on my balloon" and be sure everyone knew precisely what was meant.

By the time Charles Augustin de Coulomb published his definitive study (1785), it was already widely suspected that the electrical force acting between two charged objects varied inversely with the square of their separation (p. 235). Working independently, Coulomb was one of the inventors of a device known as a *torsion balance*, which could measure exceedingly small interactions.* Armed with that remarkably delicate instrument, he set out to determine the force law directly.

Basically the apparatus consisted of a light insulating rod hung horizontally from its middle on a long, very fine, vertical silver wire. On one end of the rod was a small pith ball; on the other, counterbalancing it, was a paper disk (a damping vane). Positioned behind the suspended pith ball was another one held rigidly in place so that initially both were in contact. The two were then simultaneously electrified by being touched with a charged rod. Immediately the suspended ball swung away from the fixed one, twisting the silver wire in the process and finally coming to rest some distance from where it had been at the start.

The key point was that Coulomb had already determined how much force it took to twist that wire through any angle. He had done so by hanging certain geometrically shaped masses on the wire

---

* Rev. John Michell also invented the torsion balance in order to measure gravity. He died soon after constructing the instrument, and it ultimately passed, unused, to Henry Cavendish. Modifying the device somewhat, Cavendish proceeded to carry out his now famous measurement of the average density of the Earth (p. 124) or equivalently the constant G. That was in 1798.

and studying the way they oscillated. It was a complicated mechanical problem but one well understood. In any event, when he was done, he knew that a given twist meant a certain force. Now, twisting the wire further by hand, he could successively drive the two charged balls closer and closer together against their mutual repulsion. While doing that, he measured their separations and determined the forces holding them at each location in turn. His results were quite clear: "The repulsive force between two small spheres charged with the same type of electricity is inversely proportional to the square of the distance between the centers of the two spheres." Alternatively, we can say that the electric force, F_E, depends on or is proportional to (symbolized by $\propto$) 1 over the distance (r) squared,

$$F_E \propto \frac{1}{r^2},$$

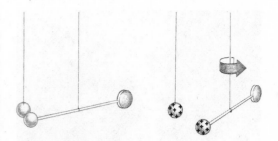

A detail of Coulomb's device.

just as is the gravitational force (p. 124). Modern experiments have confirmed that if the power to which r is raised differs at all from 2, it must deviate by *less* than about 3×10^{-16}. That's way out in the sixteenth decimal place, so 2 is a good number to use.

Coulomb went on to ingeniously determine how the force law varied with the magnitude of the two interacting charges, q and Q. First, he showed that if two identical conducting spheres, one electrified and the other not, were made to touch, the original charge would be shared equally. It would redistribute itself half on one and half on the other. That much he confirmed by subsequently measuring the force exerted by each of these on a third electrified sphere. In effect, by touching the original with a twin sphere, he had halved the charge it carried. What's more, he could go on to neatly reduce that charge to a quarter, an eighth, etc., of its initial value simply by repeating the touch-and-halve procedure with the two spheres.

That done, Coulomb next varied the charges on the two interacting spheres and measured the resulting variation in force between them. He found, for example, that halving the charge on one halved the force, whereas halving it on both quartered the force. Accordingly, Coulomb concluded that the force was proportional to the product qQ, and so

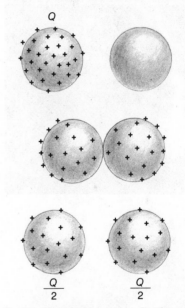

Coulomb's technique for dividing charge on identical conductors.

$$F_E \propto \frac{qQ}{r^2}.$$

The similarity with the gravity formula is obvious and striking though still not explained theoretically; they have the same form, but nobody knows why.

The right side of this relationship could be made identically equal to the left side (not just proportional to it) by simply defining the unit of charge appropriately. One unit of force, whatever it was (newtons, pounds, etc.), would result when one unit of charge (one something-or-other) resided on each sphere, separated by one corresponding unit of distance (meters, feet, whatever). That was done

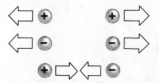

1752

in the early days (in a system that used *grams*, *centimeters*, and *dynes* rather than *kilograms*, *meters*, and *newtons*). Now, however, we can more accurately define the unit of charge from measurements of currents rather than electrostatically. When we do that, we have to introduce a constant (K) because there's no way the two sides of the expression are going to be exactly equal, independent of the units selected. Depending on what we pick as the unit of charge, the numbers for q and Q will differ in any particular situation, but so will the value of K. We decide on a unit of charge and then determine the appropriate constant, which will make the expression become an equality.

The unit itself could be anything convenient and reproducible. For example, the "shmiggle" of charge might be defined as the amount found on a medium-sized green banana after it has been rubbed 10 times with cat's fur—not that that is either convenient or reproducible. Instead we define the unit of charge via a precisely constructed current setup, and then Coulomb's law becomes

$$F_E = K\,\frac{qQ}{r^2}.$$

Presently the internationally accepted unit of charge is the *coulomb* (C), and the constant K then turns out to be very nearly equal to 9×10^9 newton-meters^2 / coulombs^2. In other words, two spheres, each carrying one coulomb of charge and separated from the other by one meter, will experience a force of nine thousand million (9×10^9) newtons. That's a tremendous force (about a million tons), and it arises because a coulomb is really a tremendous amount of charge to have just sitting around. It corresponds to more than six million million million electrons (each having a charge of -1.6×10^{-19} coulomb).

As an example of the law, let's calculate the repulsion between two electrons separated by a millimeter (10^{-3}m). In that case,

$$F_E = 9 \times 10^9\,\frac{(-1.6 \times 10^{-19})(-1.6 \times 10^{-19})}{(10^{-3})^2}$$

and $F_E = 2.3 \times 10^{-22}$ newton. By itself this tiny force isn't very impressive, but it becomes more interesting when compared with the gravitational interaction that exists between the particles. In this instance the electrical force is just about 10^{43} times larger than the corresponding gravitational force. The electrical interaction between two charges (electrons, protons, positrons, etc.) is immensely stronger inherently than their gravitational interaction.

By human standards the coulomb is definitely a great deal of charge. We're actually accustomed to taking our charge in far smaller doses, usually in the form of sparks that carry much less than a millionth of a coulomb. Storing even one coulomb is a formidable task, and yet the Earth appears to be carrying a horrendous negative charge of roughly 400,000 coulombs. In fact, in storm-free re-

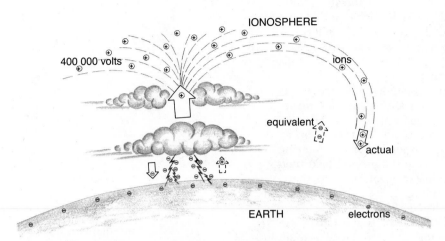

IONOSPHERE

400,000 volts

ions

equivalent

actual

EARTH   electrons

The flow of charge between ground and atmosphere. In a storm
region, positive ions are carried upward from a thunderhead into the
ionosphere, and electrons pour downward as lightning.

gions, the ground leaks about 1500 coulombs every second to the
atmosphere. That flow is returned by lightning bolts (up to 20 coulombs each), flashing almost constantly in the several thousand
thunderstorms that run continuously over the surface of the planet.

    This ability of the Earth to store vast quantities of charge makes
it an ideal dumping ground for excess electrons. Any conductor in
good contact with the Earth (like the water pipes in a building)
will carry off all the charge one could possibly want to get rid of.
That's exactly what is meant by the phrase "to **ground**" something.
Gasoline trucks trail conducting tails along the road in order to
dump the negative charge usually picked up by vehicles that ride
on insulated wheels. Those whiplike wires rising out of the blacktop
that tickle cars just before they reach a toll booth are there to ground
the buggy and save the coin collector from endless shocks.

Return now to the electroscope for a moment, and realize that it's
not actually necessary for a charged object to physically touch the
device in order for the leaves to respond to its presence. A negatively
electrified body located anywhere near the top of the electroscope
will certainly repel free electrons within the support wire. These will
be forced down and away into the foil strips, rendering the top
portion of the device positive and the bottom negative. The leaves
will then jump apart and stay that way as long as the charged object is nearby maintaining the imbalance. The closer it comes, the
greater the force, the more electrons enter the leaves, and the farther
apart they spread. When it's removed, the displaced electrons will
immediately flow back to where they were originally; the leaves will

Inducing a charge on an electroscope.

Electrostatic Induction

hang vertically; and the electroscope, which was neutral in total throughout, will revert to its normally unsegregated charge distribution. Instead of being transferred to the electroscope, the charge is said to have been **induced.** In exactly the same way, bringing a positively charged object into the vicinity of the top of the electroscope will induce a positive charge on the foil strips.

Suppose that the electroscope is now grounded (a conducting wire is attached to it and to the faucet on a sink), and we repeat the game above, bringing a negatively charged rod nearby. The mobile electrons in the top of the device can get a lot farther away from the repulsive intruder by flowing into ground than by entering the leaves, and that's where they go. Removing the ground lead then isolates the scope, rendering it shy a clutch of electrons and therefore positively charged with leaves standing apart. Again charge was induced—the scope was never touched by the rod.

Electrons within a dielectric, such as a tuft of tissue paper, are far less mobile than those of a conductor and ordinarily remain atom-bound. When a negatively charged rod is brought near such a nonconductor, it repels the electron clouds surrounding the atoms and in effect distorts them. Each nucleus, now slightly exposed, represents the positive end of the elongated atom, and the electron cloud bunched up on the opposite side constitutes the negative end. The result as a whole is that one side of the dielectric is electrified positively, the opposite side negatively. A charge has been *induced*, and it shows up on the surfaces of the specimen.

This is just what happens when you negatively charge a comb with wool or your own hair and then pick up bits of Styrofoam, dust, lint, cellophane, or tissue. The region of the tissue closest to the comb takes on a positive charge; the region farther away becomes negative. Since by Coulomb's law the force drops off with distance squared, the mutual attraction of comb and tissue exceeds their slightly weaker repulsion because of the greater distance, and the two objects experience a net tug compelling them toward each other. That's why an electrified balloon sticks to a nonconducting wall. It induces an opposite surface charge, and the two cling together until the excess electrons leak off the balloon.

Occasionally a highly charged object will pick up a bit of light nonconducting material in the usual way, but after a little while the scrap of lint or tissue will literally leap off, jumping free. That happens because some charge has actually been transferred to the scrap while it's in contact with the electrified surface. No longer neutral, the tissue, carrying the same charge as the object, is repelled and flies off.

Here we are again, this time rod in hand wondrously lifting lint and pulling pith and not even touching either. Rub the "magic" glass wand, wave it near an electroscope, and watch the leaves flutter in rhythm as if some invisible puppeteer were pulling imperceptible

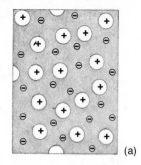

(a)

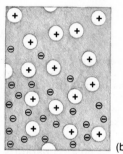

This is the way a charge is induced on an otherwise neutral conductor. The electron "sea" simply shifts, effectively leaving the conductor oppositely charged on its two ends.

(b)

ions (+)   electrons ⊖

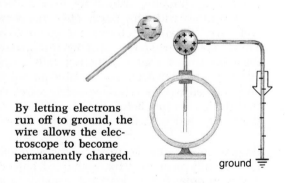

By letting electrons run off to ground, the wire allows the electroscope to become permanently charged.

ground

17.3 THE ELECTRIC FORCE FIELD

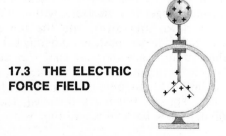

A polarized atom.

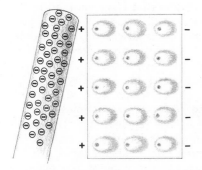

An electrically polarized dielectric.

A charged comb picking up bits of tissue paper.

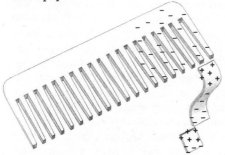

strings. Action-at-a-distance once again, the fundamental puzzle of moving without touching (Section 5.2). That remains to this day a mystery, no matter how we try to cloak it in mathematical notation. We are still struggling to learn the means by which one speck of matter reaches across the void to influence another—still struggling and not nearly content even now.

Today the prevalent theoretical imagery is that of an electromagnetic field (recall the gravity field) permeating space around any and all charge. Nowadays we take our fields to be quantized—to be, in some sense, granular—and the energy lumps, or quanta, of the electromagnetic field are photons. These "subtle particles" are envisioned as continuously emitted and absorbed by charge, which itself is the source of the field. The back-and-forth exchange of photons occurring between two charges is the suspected mechanism for transmitting the Coulomb force. This esoteric formulation, whether "true" or not, provides a splendid, if not totally satisfying, understanding of electromagnetic phenomena. It is to appreciate the central creation in all this—the electromagnetic field—that we go back now more than 100 years to the man who gave it form, the consummate experimentalist, Michael Faraday.

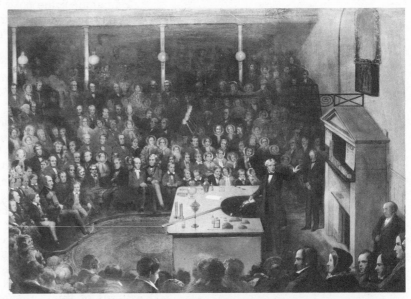

Faraday delivering one of his famous Christmas lectures (December 1855) at the Royal Institution. In the front row on the left is the Prince Consort between his two princely charges. The charming hall is much the same today as it was then, and the Institution still offers lectures in that very room.

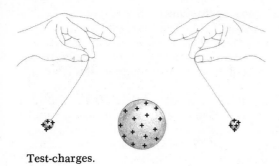

Test-charges.

I can hardly imagine anyone who knows the agreement between observation and calculation, based on action at a distance, to hesitate an instant between this simple and precise action on the one hand and anything so vague and varying as lines of force on the other.

SIR GEORGE B. AIRY (1801–1892)
British astronomer and mathematician

Mapping the force field surrounding a charge.

One of ten children of a blacksmith in London, Michael was apprenticed to a bookbinder in his early teens. Fortunately exposed to the printed word by a generous employer, he read a good deal during those years even though he had little formal education. In 1812 at the age of 21, Faraday was given several free tickets to attend the popular evening lecture series on chemistry delivered by the famous Sir Humphry Davy at the Royal Institution. Faraday had been attending scientific lectures for a few years already, but there was something special about these. Perhaps it was the fashionable audiences, or perhaps it was Davy, a handsome, poised, brilliant showman. The young man was so dazzled by the experience and so eager "to escape from trade . . . into the service of Science" that he sent Davy an illustrated leather-bound copy of his meticulous notes, accompanied by a request for a job as an assistant. A while later Davy fired his lab man for brawling and, remembering Faraday's flattering gesture, offered him the job of bottle washer. The bright young bookbinder accepted and quickly rose from lackey to protégé to rival, a breathless pilgrimage in prowess that Davy came to resent openly. In the course of his life's work, Faraday was at once a pioneer in cryogenics (low-temperature physics), the discoverer of benzene, and with Davy a founder of the study of electrochemistry. Although any one of these accomplishments would have brought him renown, his greatest fame came from his discoveries in electricity.

Imagine some positively charged object, which we now confront, pith ball in hand. If our tiny dangling detector (sometimes called a **test-charge**) is positively electrified, it will be repelled from the object under study, no matter where that object is located. Of course, the farther away, the weaker the repulsion, but repulsion there will be, everywhere in the surrounding space. At every point the detector will reveal the *size* and *direction* of the electric force, and we could map both in a diagram, drawing little arrows proportional in length to the magnitude of the repulsion. What is thereby visualizable is the field of force as if emerging from the charged object, filling all space beyond it.

Faraday was the first to introduce this sort of visual representation of the **electric force field.** His scheme, in which he used what he called **lines of force,** is an equivalent and somewhat more convenient alternative to our field of arrows. Again the force experienced by a *positive* test-charge is along these continuous lines, but this time the strength of the force field is indicated graphically by the density of lines. The more lines drawn in a region—that is, the denser the concentration of lines—the greater the field they represent; the farther apart the lines, the weaker the field. To Faraday these intricate patterns of lines, streamers of force, in time came to stand for an invisible physical reality. The field pervading space became an entity that reached from puller to pulled. An electrified object sent out its field into space, and the pith ball immersed within that web interacted with that field in proportion to its own charge. Charge-field-

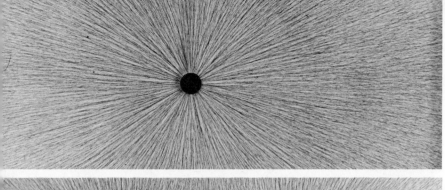

Fine rayon fibers suspended in oil tend to align themselves when in the vicinity of a charged object. The fiber patterns can be thought of as revealing the electric field distribution. A single charge. Two like charges. Two opposite charges.

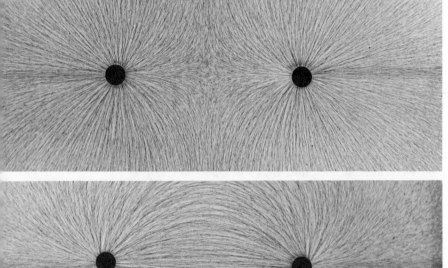

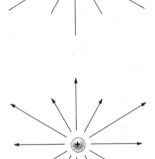

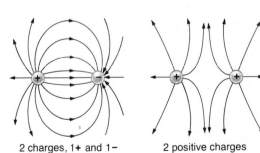

2 charges, 1+ and 1−   2 positive charges

Electric field surrounding various arrangements of charge. Where the lines are more closely packed, the field is stronger.

charge—that was Faraday's answer to the magic of action-at-a-distance. The field was not only an illustrative device; it was a reality—though no less mysterious.

Faraday's was a prophetic pictorial approach born of remarkable intuition and yet devoid of mathematical authority—after all, he was no mathematician. The formal structure would crystallize only later in the hands of his incomparable disciple, James Clerk Maxwell, who would impart to field theory a mathematical elegance and rigor (Section 20.1). Even so, as long as the charges are at rest and the lines of force are unchanging, the two conceptions of actions-at-a-distance and field are essentially equivalent. Not surprisingly, Faraday's picture of lines of force had few early adherents. The real power of the vision doesn't become evident until currents of charge have to be dealt with.

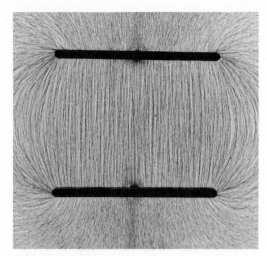

The electric field surrounding two charged parallel plates.

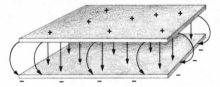

The *strength of the field* $(\mathcal{E})$ at any point in space is defined as *the force that would be experienced by a test-charge, divided by that charge*. In other words,

$$\text{electric field strength} = \frac{\text{force}}{\text{charge}},$$

or

$$\mathcal{E} = \frac{F_{\mathrm{E}}}{q},$$

where the units are newtons/coulomb. The direction of the field is the direction of the force acting on a positive charge. Although Coulomb's law makes it clear that the force on the probe, the pith ball, certainly depends on the size of the test-charge it carries, dividing by q removes this dependence. The field strength is determined by whatever charge distribution exists and, reasonably enough, is independent of the test-charge on the probe, whether it is a pith ball or a single proton. That's exactly why the notion of an $\mathcal{E}$-field is used rather than the familiar one of force. Force depends on the probe, and electric field does not; it is *force per unit charge*.

Imagine that we have two large, parallel, horizontal metal plates separated by some distance, one above the other. Now suppose that the two are equally electrified, the upper one positively and the lower one negatively. Charge would distribute itself nearly uniformly on both conductors, bunching up slightly only near the edges. A positive test-charge located anywhere in the gap would be repelled from the top plate and would at the same time be attracted toward the bottom plate—down it would go. The electric field would then be *uniform* and directed straight down (except for some fringing at the very edges, where the field lines curve). *Each $\mathcal{E}$-field line can be imagined as emanating from a positive charge and terminating on a negative charge.*

This was the arrangement Millikan used (p. 238) to suspend electrified oil droplets between two plates whose charge could be varied. Negatively charged oil globs, experiencing a force oppositely directed to the $\mathcal{E}$-field, were pushed upward against gravity. Similarly, charged plates provided the means whereby J. J. Thomson deflected streams of electrons, which were made to traverse the uniform field (p. 237). To this day, charged parallel plates control the motion of the beams in cathode ray tubes used in display systems like oscilloscopes.

The mobility of free electrons within a conductor makes for particularly simple $\mathcal{E}$-fields in the electrostatic case. Indeed, what could be simpler than the fact that **within a solid conductor (that is, anywhere beneath its surface) the electric field is zero?** Suppose that some excess charge, say a million electrons, is introduced into the interior of a metal object. An internal electric field would certainly exist, but only momentarily. The sea of free electrons responding to the field would rush about immediately, redistributing itself so as to cancel any force. In other words, any and all electrons experiencing a force would continue to travel among and interact with their fellows until they could ultimately move no more. The net force on each would then be zero, and therefore the $\mathcal{E}$-field would have vanished. At that point an amount of charge equal to the original excess, in this instance one million electrons, would be distributed over the conductor's surface. In effect, the charge, repelled as far as it could be, would shift around on the surface until the internal field canceled as a natural consequence of the Coulomb interaction.

Much the same thing would happen if a conductor was placed in the $\mathcal{E}$-field of some other object. Free charges would shift around on the conductor's surface, setting up a field that would cancel the applied field within the body. Again the $\mathcal{E}$-field inside the conductor would be zero, and this would be equally true if the conductor happened to be hollow. That's an important point; it means the electric field is unlike gravity in that *we can shield against it simply by surrounding the region to be isolated with a closed conductor*.

Electronic tubes and transistors are often encased in little metal cans. For example, many of the wires in your hi-fi set are surrounded by braided copper sheathing to keep out stray electrical fields. If you wrap a portable radio in aluminum foil or put it in a closed or nearly closed pot, the $\mathcal{E}$-field signals will not be able to penetrate the conducting barrier and reach the antenna. That's why you can't listen to the car radio while driving inside a long metal-encased tunnel and why the signal weakens appreciably even on a steel bridge.

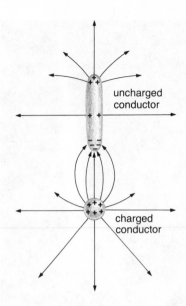

uncharged conductor

charged conductor

Below left, an electric field does not penetrate a closed conductor. Here the unaligned fibers within the metal ring are evidence of the absence of a field in that region. Below, a conductor and a nonconductor in a uniform electric field.

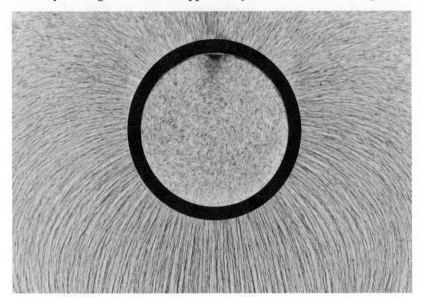

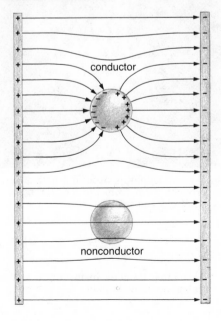

conductor

nonconductor

Electric
Potential

Braided metal shielding around a pair of
wires protects them from the interference
of external electric fields.

Work must be done in moving a charge
against an electric field.

Not long ago the CIA had the windows of the U.S. Embassy in
Moscow covered with wire mesh to keep out the intense electric fields
(microwaves) being beamed at the building by their counterparts
in the KGB. The Americans had long been eagerly scanning much
of Moscow's microwave communications, using instruments inside
the Embassy, and now the Russians were trying to jam the eaves-
dropping. The placement of conducting mesh over the windows
helped to isolate the staff from the flood of radiation, which already
left several of them feeling ill. These goings-on are reminiscent of
a dramatic experiment Faraday performed, in which he constructed
a small room within a room and covered the inner enclosure with
tinfoil. This *Faraday cage*, as it has come to be known, clearly demonstrated the shielding phenomenon. No matter how highly charged
the cage became, he could detect no field at all while he was inside.
Metal automobiles and airplanes are reasonably good cages and
therefore do a fair job of protecting their inhabitants from things
like lightning.

It takes a force to raise a mass a few feet up in the Earth's gravity
field. We do *work* (equal to force times distance) on the object
against the force field, and up it goes from one level to another,
from one value of gravitational potential energy to another. Doing
work against the pull of the field increases the gravitational PE. The
same can be said for the electrical PE. Pushing a positively charged
pith ball against an $\mathcal{E}$-field requires the exertion of force and a corresponding expenditure of work. Consequently, it is accompanied by
an increase in electrical PE. Similarly, an electron released in the
vicinity of a positively electrified object will be drawn toward it,
work will be done *by* the field on the electron, its electrical PE will
decrease as its speed, and its kinetic energy will increase. The electron will fall to the object, just as a rock falls to the Earth. In fact,
that's just the way we accelerate high-speed beams of electrons in
all sorts of devices, including the TV picture tube.

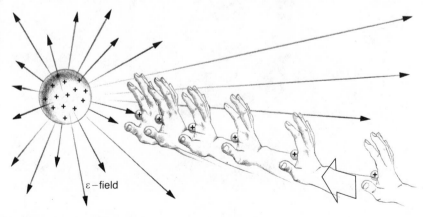

ε – field

A piano held up in midair side-by-side with a pea has more potential energy with respect to the floor than its little green companion because it has more mass. Similarly, a positively charged pith ball held away from a quantity of negative electricity experiences a greater attractive force and has more electrical PE than does a single proton, simply because it possesses more charge. All that is well and good. It tells us a lot about piano and pea, pith and proton, but it doesn't say much specifically about the fields. *Potential energy is a property of the object in the field.*

In the study of electricity, where the physical notion of field originated, it was natural for the idea of potential energy to be augmented with something independent of the test-charge. What we want is some related concept that depends only on the point-by-point location within the field—something that describes the field's link with energy. An immediate possibility suggests itself: Take the electrical PE of a test-charge at all points within the field, and divide each of those values by the charge itself. That gives you a map of *potential energy per unit charge* at all points, which would be the same for any test-charge used and would depend only on the field. Accordingly, the **electric potential*** is defined as follows:

$$\text{electric potential} = \frac{\text{electrical PE}}{\text{charge}}.$$

The units, of course, are joules per coulomb. In honor of Alessandro Volta, 1 joule/coulomb is taken to equal 1 volt (V), and electric potential is often referred to as **voltage.**

A positive particle with a charge equal to that of an electron ($q_e = 1.6 \times 10^{-19}$ coulomb), which moves from one point in the field to another, dropping, say, 1 volt in potential, will thereby decrease its electrical PE by $1 \times 1.6 \times 10^{-19}$, or 1.6×10^{-19} joules. Energy is conserved, and any decrease in the particle's potential energy will appear as an increase in its kinetic energy. This particular amount of energy, picked up as a charge of $+q_e$ falls through 1 volt, is spoken of as an **electron volt,** abbreviated eV. This convenient unit is used predominantly in atomic physics, where lots of things are happening on a very small energy scale. For example, the energy needed to tear the orbital electron free from a hydrogen atom (i.e., the ionization energy) equals 13.6 eV. It takes 4.5 eV to split apart the two hydrogen atoms in an H_2 molecule. A photon of light emitted by any kind of atom has an energy of roughly 1.6 to 3.2 eV.

Just as a rock will fall from a place where its gravitational PE is higher to one where it's lower, an electric charge will spontaneously descend from a higher to a lower potential as if it were dropping down a mountainside. An electron in a TV picture tube is typi-

---

* The idea of potential, which had been around in the abstract, was first applied to electricity by the English mathematician George Green, a contemporary of Faraday.

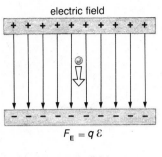

electric field

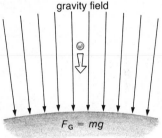

$$F_E = q\,\mathcal{E}$$

gravity field

$$F_G = mg$$

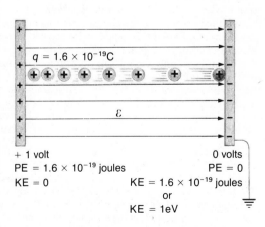

A proton falling through a potential difference of 1 volt.

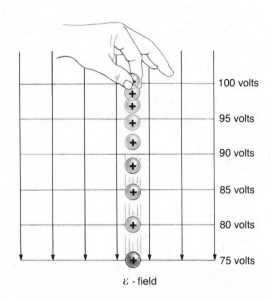

ℰ - field

After falling through a potential difference of 25 volts, the charge has lost an amount of potential energy (25 q) equal to its gain in kinetic energy.

cally accelerated toward the front face through a potential difference of some 20,000 volts and arrives at a tremendous speed with a kinetic energy of 20 keV (20 kiloelectron volts). Air breaks down, ionizing and supporting sparks when the voltage across it is about 30,000 volts per centimeter. A little spark ⅛ inch long, the kind that comes from rug shuffling, corresponds to a difference in potential of almost 10,000 volts.

Since potential energy is a relative quantity, inasmuch as its zero level can be placed anywhere, *potential* is also relative. It has become customary for us Earth-chauvinists, being practical people, to take *ground* as the zero of potential and in most cases simply to measure voltage with respect to it. The potential of an object may be *above* (+) or *below* (−) that of ground. Thus, at some instant the voltage of a power line strung overhead may be either plus or minus 100,000 volts with respect to ground. If it is +100,000 V, it requires the input of 100,000 joules of work for every coulomb of positive charge carried from ground up to that wire. If it is −100,000 V, the 100,000 joules will be delivered as the coulomb of positive charge descends in potential.

The primary goal in the next chapter is to discuss how we can raise charge to some desired electric potential so that on command it can come pouring down to a lower potential and in the process deliver energy—turn a motor, heat a toaster, make light or sound, whatever.

EXPERIMENTS

In all the experiments that follow, dampness is the primary detractor. Therefore it may be difficult to perform them on days when the humidity is high. The invisible film of water sometimes coating equipment can make it very difficult to retain charge. Warm the material being charged, work in a dry place, and keep things clean by occasionally wiping with alcohol. Tubing, particularly clear plastic, seems to make the best "magic wand." One of my treasures is a long cylindrical plastic container that was once the packaging for a set of windshield wipers. It builds up charge remarkably well. Try a lot of different things, as both the rubber and the rubbed. A plastic pen or comb will charge nicely. Glass requires more effort but works well. A filled balloon is very easily electrified.

1. Charge a wand of plastic by rubbing it against fur or cloth; almost any kind will do. Use it to pick up feathers, small scraps of paper, plastic food wrap, thread, etc. Will it lift bits of aluminum foil? Run a slow trickle of water from a faucet, and without touching it with the wand, bend the stream. Is it attracted or repelled by the wand? Explain.

2. The brilliant sixteenth-century scientist William Gilbert devised a simple instrument he called a *versorium* with which to detect electrification. Simply balance a length of

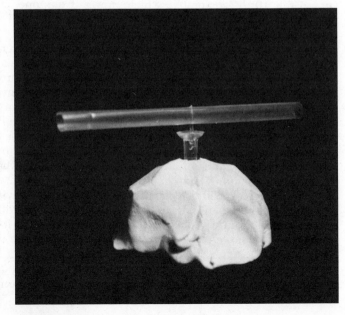

any material "lightly pivoted on a needle." A drinking straw or even a pencil stuck on a tack set in clay will work fine, as long as it is free to rotate. Fold a narrow strip of paper and balance it, or hang a needle from its middle on a thread. Now rub something, thereby presumably electrifying it, and use the versorium to see if you've succeeded—that is, wave the object near the detector (as you would a magnet before a compass), and watch it rotate.

3. Construct the leaf electroscope described on p. 387. By charging it with a known polarity (+ when touched by glass-rubbed-on-silk), you can determine the sign of the charge on anything else. If you don't have silk, a plastic comb passed through dry hair or stroked with wool will be charged negatively. If the leaves move farther apart when approached by the electrified object, both carry the same charge. Some books claim that glass rubbed on rabbit's fur (often found lining leather gloves) gets charged negatively; other writers suggest using cat's fur. Try though I may, I've always ended up with + glass, even when assisted by Willow, the family cat. See what you come up with. The only thing I have ever found that leaves glass negative is asbestos (available in the form of paperlike sheets at some hardware stores, but watch out for the dust because it's cancer-producing when inhaled). There are lots of different kinds of glass, so I haven't given up yet.

4. Position both the versorium and the electroscope in front of a TV set (black and white works better than color), and then turn the box on. What happens? Shutting off the electron beam should be easily detected by either device. Can you feel the charge building up on the screen via the hair on the back of your hand?

5. Take your best "magic wand" and a fluorescent light bulb into a dark closet. Rub the bulb with fur or wool—or anything else convenient, like a polyester or nylon shirt—and the bulb will glow. In fact, I had my son Jamey hold a lamp, and without telling him what to expect, I charged the rod nearby in the dark, and the bulb—untouched—glowed brightly in his hand. Stroke the bulb with the wand, or just shuffle around on a rug in the dark while holding it, and it will give out light as long as you keep moving.

For a different but equally fascinating effect, get a *fresh*, unused fluorescent lamp. If it contains, as some lamps do, a drop of mercury, you can hear it crash on impact whenever the tube is upended. Hold the bulb vertically, invert it, and watch the sparks. Shake it violently while looking at the ends.

6. In the dark, rapidly pull off a length of ordinary transparent plastic tape. Sparks will blaze brilliantly at the point of contact with the roll.

7. Take a mirror and some rock candy (or Wint-O-Green Lifesavers) into a dark room. Snap the candy by hand (or bite it in two in front of the mirror), and it will spark at each break.

REVIEW

Section 17.1

1. Charge is the embodiment of that which produces electrical _____ just as _____ is the entity giving rise to gravitational force.

2. Can an electron be discharged as a prune can be pitted?

3. Suppose that you measured the charge on a speck of lint and found it to be 55.5 q_e. Would that be amazing? Explain.

4. What, if any, is the significance associated with using the names *positive* and *negative* in referring to the two types of charge?

5. State which of the following are charged: photons, protons, electrons, neutrons, neutrinos, antiprotons, positrons, antineutrons.

6. State the *law of conservation of charge*.

7. What does it mean to say *charge is invariant*?

8. What is electrostatics?

9. In what respect is an atom neutral?

10. Does rubbing a rod with fur create charge? Explain.

11. Define what is meant by the term *insulator*.

12. Define what is meant by the term *conductor*.

13. Excess charge deposited anywhere on or within a conductor will always appear on its _____.

14. Is the distribution of charge on a conductor always uniform, regardless of its shape?

15. Why do the leaves of an electroscope jump apart when the device is touched by a charged object?

Section 17.2

1. What sort of device did Coulomb use to determine the electrical force law?

2. To what accuracy do we know that Coulomb's is an inverse *square* law?

3. How did Coulomb reduce the charge on his sphere to $\frac{1}{16}$ its original value?

4. What is the name of the SI unit of charge?

5. The proton carries a positive charge of _____ coulombs.

6. Is the Earth charged? If so, with which polarity?

7. What kind of charge streams downward from the ionosphere to Earth on a clear day?

8. Is lightning positive or negative electricity?

9. Why do some cars have straplike bands dragging around behind them?

10. What do we mean when we say something is *grounded*?

11. Define the term *induced charge*.

12. Describe how charge can be induced on a nonconductor even though it has no free electrons.

13. How does a charged balloon stick to a nonconducting wall?

14. In 1676 Newton sent a note to the Royal Society describing an experiment in which he charged a glass disk and held it horizontally above "some little fragments of paper." These he observed "sometimes leaping up to the glass, and resting there a-while; then leaping down and again resting. . . . Sometimes they skip in a bow from one part of the glass to another." Although he certainly could not have known what was happening, you surely can. Please explain.

Section 17.3

1. Contemporary theory maintains that the Coulomb force between charges arises from their ongoing exchange of _____. Who first gave the field concept a new sense of reality?

2. Who first gave the field concept a new sense of reality?

3. A tiny quantity of charge used to examine the force produced by some electrified body is called a _____ _____.

4. What are *lines of force*?

5. A small negative charge in the vicinity of an electrified

body will experience a force (opposite to/in the same direction as) the body's $\mathcal{E}$-field.

6. Define *electric field strength*.

7. What are the units of the $\mathcal{E}$-field?

8. Why is the quantity $\mathcal{E}$ a more appropriate concept to use for the field than F?

9. How might you create a uniform electric field in some small region of space?

10. Is the $\mathcal{E}$-field of a tiny positive charge uniform everywhere in the surrounding space?

11. The static electric field within a conductor is _____.

12. Can a region of space be shielded from the electric field? Explain.

13. Is it fairly safe to be inside a car during a thunderstorm?

14. Work must be done on a positive charge if it is to be moved in an $\mathcal{E}$-field in the _____ direction to that of the lines of force.

15. Define the quantity known as the *electric potential*.

16. What is a *volt*?

17. What is a *Faraday cage*?

18. In what sense is potential a relative quantity?

19. It's common to take _____ as the zero of potential.

20. How much energy will it take to move 1 coulomb of positive charge through a potential difference of $+1$ volt?

21. Which of the two oppositely charged parallel plates considered in this chapter has the higher potential—the positive or the negative one?

QUESTIONS

Answers to starred questions are given at the end of the book.

Section 17.1

* 1. If there were a third kind of charge, how might it interact with other charges so as to be recognizable?

2. Why can't measurements prove that the neutron is absolutely uncharged?

3. Make a list of several ways in which charge behaves quite differently from mass.

4. State the charge that each of the following will take on when rubbed with wool: glass, sulfur, amber.

* 5. One of my students once rubbed the middle of a thick glass wand with admirable vigor and yet completely failed to pick up any of the tissue fragments when he approached them with the end of the rod. Explain his oversight.

6. Describe what a spark is, discussing the role played by the gaseous medium. If a blast of charge traversed a vacuum, would we see the usual flash or hear the familiar clatter?

7. Why is a metal a good conductor of both heat and electricity?

8. While living at Charterhouse Inn (a poorhouse in London), Stephen Gray attached a lead ball to the end of a glass tube with a three-foot length of moistened parcel string. Under the ball he placed some fragments of brass leaf (very thin foil). When he rubbed the rod, the leaf was attracted up to the ball. Explain what was happening. What conclusions might he have come to? By the way, he later redid the experiment using 80 feet of string, and "the Tube being rubbed, the Ball attracted the Leaf-Brass." How is that similar to what's happening to the young woman in the engraving (from 1748) on the right?

* 9. Another way to test our charged elephant (p. 386), instead of backing it up into the second pith ball, would be to use what was once called a *proof plane* (a conductor on an insulated handle, e.g., a penny waxed to a glass rod). How might you perform the test?

10. Why are long strips of plastic tape freshly pulled from

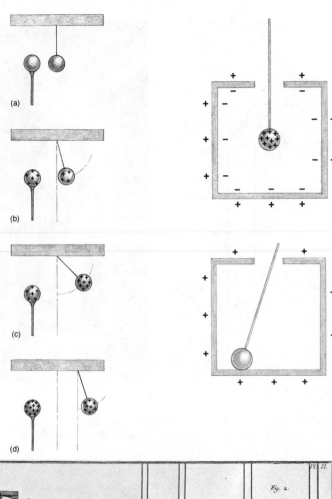

the roll such a nuisance to handle, usually crumpling and sticking to everything nearby?

11. If you rub a strip of nylon cloth on a piece of paper, the cloth will become positively charged. Will it attract or repel a balloon rubbed on wool or cotton?

Section 17.2

1. How does the force between two small charged spheres change if their separation is doubled?

* 2. Suppose that we have three identical metal spheres and that one of them is given a charge, Q. If all three are brought together and then separated, what charge will appear on each of them?

3. The accompanying diagram shows several different arrangements of pith balls. Discuss what's happening in each case.

* 4. How will the force between two small electrified objects vary if the charge on each is doubled and their separation is doubled, too?

5. Why are people that work in hospital operating rooms very concerned about static electricity?

6. If you're stretched out in a bathtub full of water, are you likely to be grounded?

7. Given a glass rod, a piece of silk cloth, and an electroscope, how would you go about determining the kind of charge carried by some object without in any way altering it?

* 8. The accompanying figure shows a nearly closed neutral conductor into which was placed a positively electrified metal sphere. Describe the distribution of charge, and explain how it came about. Incidentally, this setup is called a Faraday ice pail.

9. Suppose that the charged sphere in the previous question is now made to touch the inner wall of the pail. Describe the resulting charge distribution.

10. Will a charged balloon stick to a conducting surface? Explain. Try it!

11. The *lightning rod,* an invention of Ben Franklin, is simply a tall conducting shaft fixed to the highest point of a structure and connected by a heavy metal cable to another conducting rod thrust deep into the ground. How does such a device protect a house from lightning? This major practical discovery had profound implications. After all, many of the tall church steeples in Europe had been rebuilt several times after being hit by lightning rather frequently over the centuries.

12. Suppose that someone with very fine, dry, straight hair is well charged by some sort of electrostatic generator. Why does that person's hair stand on end straight out?

* 13. The accompanying diagram shows one variation of the most widely used electrostatic generator, a device that car-

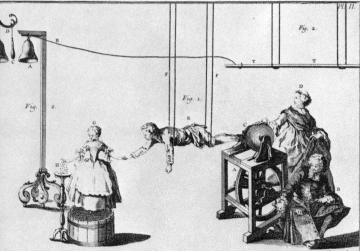

ries the name of its inventor, Robert J. Van de Graaff (1902–1967). The two pulleys are covered with different materials so that when they are rubbed by the motor-driven belt, the belt acquires a negative charge from the bottom disk and a positive charge from the top one. See if you can figure out how the thing works. Why is negative charge shown only on the outside of the top metal ball? How might the ball be charged positively? If the base was not grounded would it become charged?

14. Two neutral conducting spheres in contact are brought near a negatively charged rod with one sphere behind the other. The spheres are then separated, and the rod is removed. Describe the charge distribution on each. Finish the diagram by putting in any missing charge.

15. Suppose that you approach an elephant in succession with each of two oppositely charged pith balls, in order to determine whether or not the elephant is charged. The first ball is attracted to the elephant, the second repelled. What does that tell you? Which ball provides the more unambiguous evidence? Expain your answer.

Section 17.3

1. The diagram (p. 395) of the field associated with two equal charges (one plus, the other minus) shows a higher density of lines in the region between the charges. What does that mean, and why should it be so?

2. From the figure on p. 395 what would you guess was the value of the electric field at a point midway between and on the line connecting the two equal positive charges? Explain why that is a reasonable conclusion.

* 3. Is it correct to say that a positive test-charge released from rest will move along the electric field lines in space, provided that those lines are straight? Will it do so if the lines are curved? *Hint:* Remember Newton's first law.

* 4. *Electric field lines are always perpendicular to the surface of a conductor,* regardless of its shape, so long as equilibrium has been reached and there is no motion of charge. Explain.

5. In most practical situations we are not moving around batches of positive ions. What is actually going on when we say that we have placed a positive charge on an object?

6. Suppose you were testing someone who presumably had ESP. Why might a Faraday cage be useful?

* 7. Prove that the units of the $\mathcal{E}$-field could be taken as volts/meter.

8. A proton drops in potential by 1 volt. How many joules of KE will it pick up in the process? How many eV of KE will it acquire?

9. Suppose that we have two unequally charged conductors and we connected them together for a moment with a copper wire. Describe what, if anything, will happen. What can you say about the final potentials of the two objects?

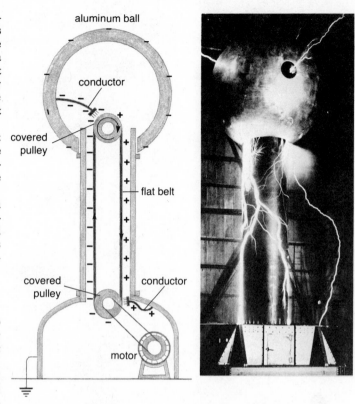

aluminum ball

conductor

covered pulley

flat belt

covered pulley

conductor

motor

A huge Van de Graaff generator housed within a blimp hangar. It was built in the 1930s at South Dartmouth, Massachusetts.

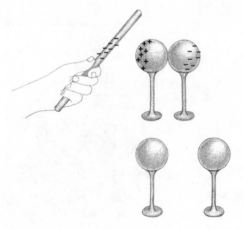

10. During the time Newton was president of the Royal Society, its curator of experiments was one Francis Hauksbee. The following is a demonstration he performed, which seems to be a very early hint, a faint suggestion, of what Faraday would someday evolve into *lines of force*. Onto a semicircular wire frame, Hauksbee fastened "several pieces of Woollen Thread . . . so as to hang down at pretty nearly equal distances." At the center he placed a glass "Tube," which was subsequently charged by rubbing. The threads immediately swung about so that they pointed directly at the center of the glass (see figure). Explain what was happening to the threads and why this suggests lines of force.

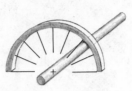

* 11. Explain why two electric lines of force can never cross each other.

12. Is it true that field lines point in the direction of decreasing potential? Explain.

13. When an electron undergoes a drop in potential, does its potential energy increase or decrease? For example, if it is carried from the positive to the negative plate in the figure on p. 399. How is this different from what happens to a proton?

14. Is it true that an electric line of force cannot begin and end on the same conductor?

MATHEMATICAL PROBLEMS

Answers to starred problems are given at the end of the book.

Section 17.2

1. Compute the gravitational attraction between two electrons ($m_e = 9.1 \times 10^{-31}$ kg) separated by 1 mm, and compare that to the electrical repulsion.

2. Calculate the ratio of the electrical to the gravitational force, F_E/F_G, as it exists between a proton and an electron.

3. By how much does the mass of an object change if it acquires a negative charge of 1 microcoulomb (10^{-6} C)? ($m_e = 9.1 \times 10^{-31}$ kg; $q_e = 1.6 \times 10^{-19}$ C).

* 4. A proton and an electron are separated by a distance equal to the radius of a hydrogen atom (0.5×10^{-10} m). Determine the attractive electrical force betwen them. ($q_e = 1.6 \times 10^{-19}$ C.) Explain the significance of the sign.

Section 17.3

1. Imagine a small positively charged sphere with N lines of force emanating straight out uniformly from it. Show that the density of lines decreases as $1/r^2$. *Hint:* Surround the charge with imaginary concentric spheres, and think of the number of lines per unit area coming through each.

* 2. A $+10$ microcoulomb (10^{-6} C) test-charge at some point in space beyond a charged sphere experiences an attractive force of 40×10^{-6} newtons. Determine the value of $\mathcal{E}$ at that point.

3. A small positively charged object falls from rest in a uniform downward $\mathcal{E}$-field. Write an equation giving its speed, v, after it has fallen a time t in terms of its mass m and charge q.

4. Calculate the size and determine the direction of an $\mathcal{E}$-field if an electron ($m_e = 9.1 \times 10^{-31}$ kg) placed in it is to experience a force that will exactly cancel its weight at the Earth's surface.

* 5. Show that the difference in potential ($\mathcal{V}$) between two charged parallel conducting plates separated by a distance Δd is given by $\mathcal{V} = \mathcal{E} \times \Delta d$. *Hint:* The work done in moving a test-charge across the gap is $F_E \times \Delta d$.

* 6. Two large charged conducting parallel plates separated by 1 millimeter (10^{-3} m) encompass a uniform $\mathcal{E}$-field of 4×10^5 N/C. Determine the difference in potential between the plates.

7. An electron ($q_e = -1.6 \times 10^{-19}$ C) moves from a point where the potential is 100 volts to a point where it is 50 volts. Calculate its change in electrical PE in units of joules and in eV.

Charge can be set moving like a stream of water, and this **electric current**—this flow of charge—carries energy and can deliver power. The transmission of electrical energy is one of the primary notions of this chapter, which is concerned quite generally with charge in motion.

Considered first is the challenge of how to create and sustain a current. **Capacitors,** which can store and dump considerable amounts of charge, were the earliest devices that could provide formidable currents, though only in short bursts. It was not until the beginning of the nineteenth century, when the **battery** was invented and served as a practical source of continuous current, that electricity began to become the bloodstream of technology.

In most common applications **electricity is a flow of electrons** confined within the boundaries of a conductor. The second section of the chapter deals with the relationship between the **voltage across** and the corresponding **current through** some physical element, a conductor—a piece of wire or a chunk of carbon. Ordinary conductors more or less impede the passage of current through themselves and so are said to have **resistance.** The important equation relating these parameters, known as **Ohm's law,** can be used to anticipate the behavior of a circuit consisting of various components. **Power** is the next theme to be considered, power expressed in the language of electricity, in terms of current, voltage, and resistance.

The third section deals with a comparison of the utility of **alternating** versus **direct current** (leaving the actual means of producing AC for later). The chapter terminates with an electrifying discussion of the adverse effects of passing electricity through people.

ANDRÉ MARIE AMPÈRE
(Mathematicien et Physicien),
Membre de l'Académie des Sciences,
Professeur au Collège royal de France et à l'Ecole polytechnique, des sociétés d'Edimbourg, de Cambridge, de Genève, Helvétique, etc.
Né à lyon (Dep¹ du Rhône) le 20 Janvier 1775.

If there is a life's blood in this technology of ours, it is most certainly electricity—electricity surging along wire veins, delivering energy and doing work, trickling through a metal nervous system just as it trickles through our own. *The flow of charge is called* **current,** whether we're talking about electrons meandering down a wire, ions floating through a solution, or protons sailing in an accelerator beam. *Charge in motion is electric current.* Specifically, the amount of charge (Δq)

On facing page, a beam of electrons (60 kiloamps, 3 MeV) deflected by both air scattering and a magnetic field. This is as much an electrical current as any that ever negotiated a toaster.

But when I took the frog into a closed room, laid it on an iron plate and began to press the hook that was fixed in its spinal cord against the plate, lo and behold, the same contractions and the same kicks!

GALVANI

flowing past in a given time interval (Δt) is defined as the current:

$$\text{current} = \frac{\text{charge}}{\text{time}},$$

or using I as its symbol,

$$I = \frac{\Delta q}{\Delta t}.$$

The unit of current is the *ampere* (A) or simply the amp, where 1 amp = 1 coulomb/second. In other words, the equivalent of 6.25×10^{18} electrons flowing by each second is one amp (1 A).

18.1   SOURCES

The Abbé Nollet's electrical experiments (1764).

In the earliest days currents were produced in the form of sparks. Electrostatic generators—rubbing machines—could develop prodigious voltages but only tiny trickles of charge, *microamps* (millionths of an amp) at best. Not until the mid-eighteenth century could charge be stored (in capacitors), built up, and then dumped, thereby delivering appreciable amounts of electrical energy for the first time in history.

That delightfully vigorous era was a time of puttering; of shocking everything from turkeys to one another; of trial and error, with the amateur as welcome as the philosopher. Little theory existed; certainly when compared with the Newtonian masterpiece, electrical notions were crude indeed. After all, many a professional viewed the business as hardly more than an amusement of no apparent consequence. The short-lived, powerful surge of charge, the transient current was the hallmark of the times. To this day, pretty much the same techniques are used wherever tremendous jolts of electricity are called for.

The development of chemical means for providing continuous currents marked the next major advance. That came at the very start of the nineteenth century with the invention of the battery, and it meant that electrical energy could be delivered and utilized in an uninterrupted stream. Practically, that was tremendously significant. One could then tap electrical energy just as one had tapped thermal energy with the steam engine ever since the beginning of the Industrial Revolution.

Capacitors   Once charge could be generated electrostatically in moderate amounts, the next logical step was to devise a scheme for storing it, for building up the amount that could be dumped in a single blast. This feat was accomplished, rather spectacularly though quite serendipitously, in 1745 by Georg von Kleist and again independently a few months later by Peter van Musschenbroek. Both of these European gentlemen, like so many others, were experimenting with electricity and happened to insert the conductor being charged at the moment into a hand-held glass jar. Reasonably enough, they were probably attempting to collect "electric fluid" in a bottle. Von Kleist electrified a nail inside a small vial, and when he attempted to remove it with his other hand, he received a stunning shock. Van Musschenbroek, for his part, dangled a brass wire (attached to a gun barrel that was being charged) into a flask "partly filled with water." All at once his body convulsed "as if it had been struck by lightning; . . . I thought it was all up with me," he later recounted after catching his breath.

Each of them had unknowingly constructed a device in which a conductor (the wire) was separated from another grounded conductor (the sweaty hand) by an insulator (the glass bottle). Not having taken any precautions to isolate themselves, they were grounded, hand and all. So dramatic were the blasts of charge, so startling the effect by comparison with the previous capability, that the work was widely hailed and duplicated long before anyone understood how dangerous it was or even what was happening.

A conducting object cannot continue indefinitely to receive charge from a source. Sooner or later the accumulated charge will exert a repulsive Coulomb force so great that no additional charge can be impelled by the source to the object's surface. In other words, the conductor will reach the same potential as the source, and there will be no further transfer of charge.

The new arrangement of conductor-insulator-conductor worked to substantially forestall that cutoff. In effect, the charge put on one conductor, the central wire, induced an equal and opposite charge on the other conductor, the moist hand. That induced charge, having the opposite polarity, acted to reduce the wire's repulsive force, which was in turn responsible for limiting the additional arrival of charge. The result was a considerable increase in the charge stored on the wire before it reached its cutoff point. Stated slightly differently, much more charge was accumulated before the device reached the potential of the source.

The *Leyden jar*, as it came to be called, was an instant success, an invention that had come at a moment when all the world was eager. It was quickly improved, most effectively by simply coating the inside and outside with tinfoil, thereby doing away with the need to clutch it in a damp palm. Everyone even remotely interested in electricity had a Leyden jar or could at least see one in the many traveling shows that delighted audiences in Europe and the Colonies. Sparks were flashing everywhere. The Abbé Nollet had 180 of Louis XV's fearless guards join hands in a circle, or *circuit*. With the first man holding the outer terminal of a charged Leyden jar, the last victim gleefully touched the center wire and shocked the whole assembly as charge drained through all of them. Even the austere Carthusian monks cloistered in Paris played the circuit game.

That renowned tamer of lightning, Ben Franklin, had his Leyden jars, to be sure, but he went even further. Though not the inventor, Franklin was among the first to use a new configuration consisting of two flat metal plates separated by a pane of window glass. That simple arrangement allowed for a dramatic increase in the size of the

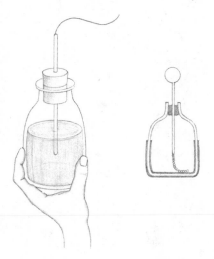

A Leyden jar.

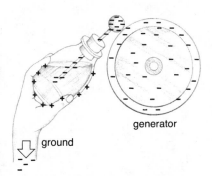

Charging a Leyden jar with the help of a generator and a grounded experimenter.

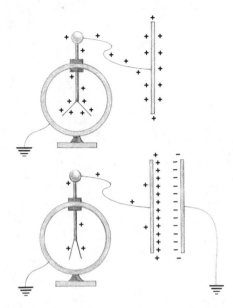

The operation of the capacitor as a charge storer.

conductor-insulator-conductor sandwich. Any such charge-storing device was later dubbed a *condenser* by Volta, but that term, though popular for quite a while, has now been almost totally replaced by the word **capacitor**. Thus Franklin's flattened Leyden jar was a *parallel-plate capacitor*.

It was Volta, at the end of the eighteenth century, who first carried out quantitative measurements of the potentials of large charged objects. The easiest way to do so was simply to attach the test object via a conductor to an electroscope. The leaves would spring apart in proportion to the charge impressed on them, which was in turn proportional to the potential of the object. Thus if we attach a positively charged metal plate to an electroscope (see figure), the angle the leaves make will be proportional to the potential.

We can simplistically think of potential as related to the net "push" or "pressure" exerted on the charge by their own Coulomb interactions or, if you like, as the "height" of the electrical mountain they're about to tumble down. Any one of these analogies is useful insofar as it aids in an intuitive appreciation of the process, but none of them should be taken literally. Accordingly, a positive plate at a high potential can "push" a lot of positive charge down into the leaves of the electroscope (or as actually happens, draw a lot of electrons out of the leaves, an action that is quite the equivalent).

Now suppose a second identical grounded plate is brought near the first electrified one, thereby forming a capacitor. The leaves of the electroscope will gradually descend as the plates approach each other. In effect, the negative charge induced on the new plate will draw off positive charge from the leaves and hold it in place on the near side of the plus plate. Introducing the second conductor lowers the potential of the first, which can then be restored to its original value by adding more charge. The *capacity to store charge* is therefore considerably increased by the presence of that second plate. **Capacitance** is the measure of that storage capability and ultimately depends only on the materials, the size, and the geometry of the device. Capacitance is defined as *the ratio of the amount of charge on either plate (since they are equal) to the potential difference between the plates:*

$$\text{capacitance} = \frac{\text{charge}}{\text{potential difference}}.$$

To honor Faraday the unit of capacitance is the *farad*, equal to 1 coulomb/volt.

Inasmuch as the early interest in such devices was purely for storage purposes, it's not surprising that Franklin was inspired to wire almost a dozen parallel-plate capacitors together to produce gigantic blasts of rushing charge. Borrowing appropriately from the jargon of the artilleryman, he called his group of blazing capacitors an "electric battery," just as a group of cannon was known as a battery. However, that phrase would also come to change its meaning somewhat in time.

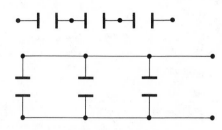

Series and parallel arrangements of capacitors.

When wiring together the individual capacitors, Franklin used two different basic schemes, which we now call **series** and **parallel.** In the former, one and only one plate of each capacitor was attached to an adjacent capacitor. In the latter, one plate from each pair was attached to the same wire, and all the remaining plates were connected to another wire. The second arrangement is equivalent to having just two large segmented plates forming one composite capacitor. The total capacitance in the parallel arrangement is equal to the sum of the individual capacitances. Great racks of interconnected Leyden jars became a common sight throughout the Western world.

With the advent of modern electronics, capacitors have come to serve a great many functions in addition to storing charge. These days they can be found in radios, air conditioners, calculators, computers, car ignitions (where they're still called condensers), watches, TVs, submarines, and space ships. In fact, whenever you twirl a knob to change the station on a radio, you are physically moving the plates of a variable capacitor, altering the capacitance and thereby retuning the pickup circuit.

The Battery

It had been widely known since ancient times that there were certain sea creatures that could deliver stunning blasts of some mysterious force. The Torpedo ray, the infamous "putter to sleep" (which could develop a potential of up to 220 volts), became the focus of considerable attention when people realized that its strange power produced the same physiological effects as did a discharge from the newly invented Leyden jar. By the mid-1700s, the electrical nature of that phenomenon was confirmed, and the study of "animal electricity" was well under way. The first wrenching shocks delivered by the Leyden jar set researchers off almost immediately to probe the electrical excitation of muscles in animals. The frog, whose muscular legs were a popular delicacy, became a logical martyr to the cause. In time dissected frogs' legs would be twitching, convulsing on command, from one end of Europe to the other. Even Faraday kept a froggery in the basement of the Royal Institution.

Luigi Galvani, a noted Italian anatomist, was 43 when he began his studies of animal electricity at the Institute of Sciences at Bologna. In 1780 he made the first of two accidental discoveries that would arouse a storm of controversy in the scientific community and inevitably change the course of history. Having just dissected a frog, he happened to place it on a table not far from an operating electrostatic generator. An assistant poking at the frog with a scalpel was surprised to see that it jerked violently when the blade gently touched a crural nerve. Another technician, who was cranking the nearby generator, noticed that the spasms occurred only when that machine sparked.* Galvani instantly recognized that this startling behavior

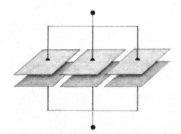

Several capacitors in parallel are equivalent to one large capacitor whose plates simply happen to be segmented. The connecting wires, of course, tie the plates together.

---

* What they were observing was electromagnetic induction (Section 19.4), but they were unaware of that. It was as if the nerve-muscle system were a radio set with the scalpel as antenna picking up signals from the sparking generator.

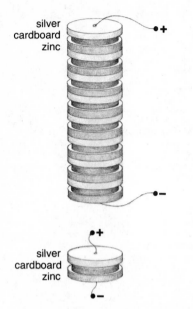

silver
cardboard
zinc

•+

•−

silver
cardboard
zinc

•+

•−

The voltaic pile, in which the elementary
cell was a zinc-cardboard-silver sandwich.
Actually, Volta wrongly thought that it
was the zinc-silver contact that provided
the effect.

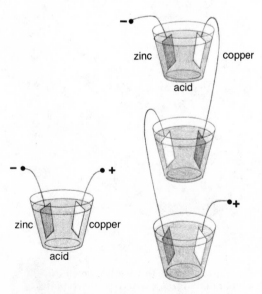

−

zinc   copper

acid

−   +

zinc   copper

acid

+

+

A single voltaic wet cell and three wired
in series to form a battery.

was of great significance, and he determined to arrive at its cause.
After a thorough investigation in the laboratory with what he called
"artificial electricity," he set out to see if "natural" atmospheric electricity would work as well. A long wire was attached to a dissected
frog via a nerve, and another wire from the feet was grounded down
a well. Again the muscles twitched, this time in rhythm to lightning
flashes.

Then came the second fortuitous discovery, the more important
one. Each little green beastie was "humanely" terminated by being
impaled, through the spinal cord, on a sharp bronze hook. Thus prepared, the frogs waited their turns hanging on an iron trellis in the
garden. To his utter surprise, Galvani noticed a rather weird spectacle one cloudless day. The disembodied legs would spastically convulse
every now and again for no apparent reason. From these and other
observations, Galvani finally (and wrongly) concluded that he had
discovered "in the animal itself" a natural source of electricity; he
believed that the metal hook touching the trellis simply closed the
path, enabling the "nerve-fluid" to flow to the muscles.*

The publication of these remarkable results astonished scientists
everywhere and prompted a vigorous flurry of work in the field. Alessandro Giuseppe Antonio Anastasio Volta, already renowned as an
"electrician," soon became Galvani's major critic. It seemed to Volta
that the source of the electricity was not the animal itself but the
two different metals brought into contact, the hook and the trellis.
The frog was no more than an extraneous current meter.

Recognizing that his own tongue was a highly sensitive and
quite convenient muscle, Volta became his own guinea pig (or in
this case, frog). He cleverly put different pairs of metals on his tongue
and brought them into contact. Volta expected to feel a contractive
spasm, but instead it produced a flood of strong tastes by stimulating
the sensory nerves.† He realized that the taste sensation lasted as
long as the two metals were in contact, implying the incredible idea
that in that simple arrangement "the flow of electricity from one
place to another is continuing without interruption."

Soon thereafter Volta produced two revolutionary devices that
were the forerunners of the modern-day electrical battery. In both he
replaced the hook, frog, and trellis with something more convenient.
The *voltaic pile* was a stack of small disks of silver, zinc, and brine-
soaked cardboard layered in order—zinc-cardboard-silver, zinc-cardboard-silver, and so on. Each zinc-cardboard-silver sandwich formed

* Although mistaken in this bimetallic business, Galvani was right in
principle. We now know that the nerve-muscle control system is electrical
and that animals, including ourselves, do generate electrical currents.

† Volta was not the first to perform this little experiment, nor did he
ever actually understand it. There is in fact such a thing as *contact potential* between different metals; Volta was right about that, even though it
was not the operative mechanism here. Volta used tin and silver or gold,
but aluminum foil and a piece of sterling silver will work quite well. Try it.

a unit (much like the wet tongue experiment), which was then repeated about twenty times to build up the voltage. The result was a device that, when the top and bottom plates were connected with a wire, produced an appreciable continuous current.

The other variation on the theme was the *voltaic wet cell*. It consisted of a drinking glass filled with brine or a dilute acid (sulfuric works nicely) into which were immersed two metal strips, one of zinc, the other of copper. Again, when these plates became charged, they could supply a considerable continuous current. By putting several of these cells in a series, Volta created the world's first **electric battery** (in the modern sense of the term). That was in 1800, and within months Nicholson and Carlisle in England had already used a voltaic battery to decompose water into hydrogen and oxygen.

After a command performance in Paris before the Emperor Napoleon in 1801, Volta was decorated with the Legion of Honor and made a count. He had won the struggle with Galvani and in the process began a whole new era—the nineteenth century was to be the Age of Electricity.

Volta demonstrating his "pile" to Napoleon in a painting by a contemporary, A. E. Fragonard. Some years later, Volta gave Faraday a battery of this sort. You can see two such "piles" on the latter's table in the painting on p. 393.

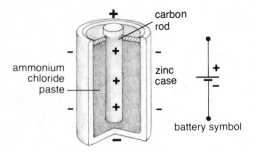

The basic dry cell.

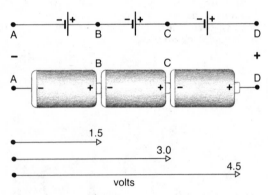

Cells in series.

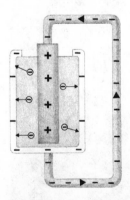

The flow of charge through a dry cell.

Within just a few years Humphry Davy began his work in electrochemistry (Section 9.2), an effort that underscored the relationship between electricity, elemental matter, and chemistry. Indeed, it made the point that chemistry was physics. Always in competition, the French and English held it a matter of national pride to possess the world's largest battery. Benefiting from the contest, though always aloof from such things, Davy used a giant 2000-cell monster in the Royal Institution to power his newly invented carbon arc lamp.

By 1838 the Great Western Railway already had a voltaic pile powering a telegraph link. The granddaddy of the cheap "flashlight battery," the *carbon-zinc cell*, was conceived by Bunsen in 1842. In 1859 Planté had given the world the lead-plate-dipped-in-sulfuric-acid storage device that we have all come to know and hate as the automobile battery.

Almost any two different solid conductors immersed in a variety of conducting solutions will function more or less as a battery. Energy stored in the interatomic bonds as chemical potential energy will be converted into electrical PE as the solution and one or both of the plates become involved in chemical reactions. Charges will separate, and one of the two conductors will become *positive* with respect to the other. The *negative* plate or terminal, on which there is a continuous excess of electrons, will have the lower potential.

A given kind of cell will generate a voltage difference that is determined by its chemical makeup, independent of size. Thus an ordinary flashlight "battery," a *dry cell*, maintains a potential difference of 1.5 volts. A *mercury cell*, one of those little button-sized camera, watch, and hearing-aid "batteries," generates 1.4 volts. The old rechargeable *lead storage cell* of auto fame is a 2-volt creature. The *nickel-cadmium cell*, making up expensive rechargeable electronic calculator batteries, has a voltage of 1.2 volts.

To boost the potential difference provided, cells are customarily connected in series. Franklin did as much with his Leyden jars, and Volta's pile was a series array of individual zinc-cardboard-silver cells. The crucial point is that *the voltage across the series-connected battery is the sum of the voltages across each constituent cell.* Using the diagram, go in imagination from the negative terminal at point A across the first cell to point B, and the voltage rises 1.5 volts. Point B can be thought of as either the + terminal of the first cell or the − terminal of the second—they're connected and therefore at the same potential. Consequently, point C is 1.5 volts above B and 3 volts above A, and D is 4.5 volts above A. This sort of series stacking of cells is exactly what you are doing when you load D cells, top to bottom, into a flashlight or a portable radio. Moreover, it's the way the cells are wired within your car battery to yield 6 or 12 volts. That's why there are often six separate little capped openings into which I forget to put water.

When the cells are arranged in series, the same amount of current enters and leaves each cell, and that's the current provided by the battery as a whole—the voltages add, and the current is unaltered. By contrast, cells in parallel form a battery whose voltage is equal to the voltage of each component cell but whose current capacity is the sum of the individual current outputs. The situation matches that of the capacitors: Putting the cells in parallel essentially makes them into one large cell (with segmented plates) having the voltage of each individual cell but the current capacity of the total of all of them.

Reasonably enough, the larger an individual cell is, the more current it can usually supply. Nowadays manufacturers provide (though not often right on the battery) a crude measure of the current output capability in the form of an *amp-hour* rating. A little 1.5 volt AA cell, the common "penlight" finger-sized "battery," is rated at about 0.6 amp-hour. Its big brother, the ever popular standard flashlight D cell, delivers 3.0 amp-hours. Presumably one can draw a steady 3 amps for 1 hour or 1 amp for 3 hours or 0.3 amp for 10 hours from such a D cell before discharging it totally. It will never be able to put out 300 amps for 0.01 hour no matter how it tries, but we get the manufacturer's point.

These are rather hefty currents, but of course, the voltages are generally quite modest as compared with the tens of thousands of volts produced by "your everyday" electrostatic generator. Then again, those rubbing machines were and are rather frugal with current, usually dealing in microamps (millionths of an amp). A giant modern million-volt van de Graaff generator might put out only around a milliamp (a thousandth of an amp). Compare that to a heavy-duty 12-volt auto battery with an amp-hour rating of near 100. It could provide tremendous currents—and even 10 amps is a great deal—but only at a meager 12 volts. It's a little like having the whole Atlantic Ocean behind a dam one foot deep; there's lots of water stored but little pressure.

There is a great diversity in conductivity, the ability of materials to conduct electricity, and Georg Simon Ohm set himself the task of finding some little unity in that chaotic variety. His experiments had to be on a modest scale; Ohm was a high school teacher in Cologne and never far from poverty. He had been inspired to the particular challenge by the recently published writings in thermodynamics of J. B. J. Fourier. That superb physicist and mathematician established that the rate of flow of heat along a conducting rod was proportional to the temperature difference between its ends. Ohm rightly wondered whether the rate of flow of charge along a conducting rod was likewise proportional to the voltage difference between its ends.

At first he tested wires of different materials—gold, copper, brass, etc. He placed them, one at a time, across the terminals of a voltaic battery and measured the current passing through each one with a kind of magnetic torsion balance (p. 450). What he found was that the current the battery could force through a specimen depended on

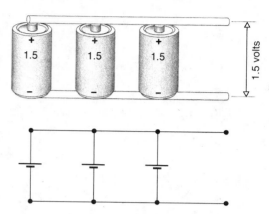

Three 1.5-volt cells in parallel.

18.2 CURRENT AND VOLTAGE

ON THE VOLTAIC PILE

It may be conjectured that we have carried the power of the instrument to the utmost extent of which it admits; and it does not appear that we are at present in the way of making any important additions to our knowledge of its effects, or of obtaining any new light upon the theory of the action.

JOHN BOSTOCK

Account of the History and Present State of Galvanism (1818)

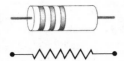

Although resistors come in many forms, the most common is the little striped cylinder.

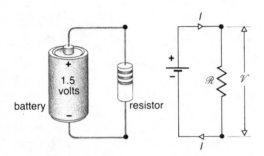

Current (*I*) is generally pictured in the direction of flow of positive charge, even though we now know that what is really flowing is a stream of negative electrons. That convention has been around since B. Franklin introduced it, and it persists, however awkward.

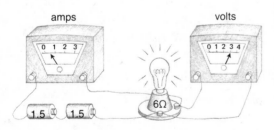

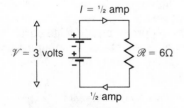

A circuit consisting of two 1.5-volt cells and a lamp, all in series.

the material used, its cross-sectional area, and its length. The battery was the supplier of electricity, with a given "push" corresponding to the potential difference or voltage between its terminals. How much electricity it could actually propel through the specimen depended on the latter's **resistance** to the flow of charge. That was a term Ohm introduced and quantified. The greater the resistance (symbolized diagrammatically as ‑‑ᴧᴧᴧ‑‑), the less current would pass through it.

In 1826 Ohm published a formula relating the constant resistance ($\mathcal{R}$) of a particular specimen to the voltage ($\mathcal{V}$) impressed across it and the corresponding current (I) flowing through it, namely,

$$\text{current} = \frac{\text{voltage difference}}{\text{resistance}},$$

or

$$I = \frac{\mathcal{V}}{\mathcal{R}}.$$

Ohm's law, as this relationship has come to be called, deals with a rather limited, rather special set of circumstances, and yet it is of very great practical concern. It applies to conductors, notably the common metals and several nonmetallic conductors as well. Happily I is proportional to $\mathcal{V}$, or stated another way, R is independent of both I and $\mathcal{V}$ for all the materials we are likely to use—copper, aluminum, etc. That's certainly not true, however, if the specimen is an ionized gas; then I and $\mathcal{V}$ have a complex interrelationship. Even so, *the resistance of any conductor (a simple length of copper wire or a rod of carbon) is not really a constant; instead, it varies with temperature.* Ohm's law ($\mathcal{V} = I\mathcal{R}$) is an important, useful relationship, but it's a restricted, practical statement, nothing like the grand fundamental pronouncements of Coulomb's law and the universal law of gravity.

To honor Ohm—though recognition for his work was depressingly late in coming—the unit of resistance was named the *ohm* and abbreviated by the Greek letter omega, Ω. Ohm's law serves, in fact, to define resistance as $\mathcal{R} = \mathcal{V}/I$, and so 1 ohm = 1 volt/ampere. These days we have *ammeters*, which measure current flowing in a circuit, and *voltmeters*, which measure voltages across any element in the circuit. Accordingly, suppose that two D cells connected in series supply electricity to a little bulb, just as they would in a flashlight. If the voltage drop across the bulb is found to be 3 volts and the current through it 0.5 amp then its resistance, $\mathcal{V}/I$, is 3 volts/0.5 amp, or 6 ohms. Similarly, if a hot plate draws 10 amps when plugged into a 120-volt wall outlet, its resistance is 12Ω.

All sorts of electrical devices, including the very wires that connect them, have resistance. Even so, there are specific devices called **resistors,** *circuit elements whose primary function is to introduce a certain known resistance,* ranging from fractions of an ohm to megohms (millions of ohms). Used in almost all electronic devices from radios to computers, they serve to regulate the flow of current. You may know resistors as those little brown cylinders whose values are color coded in several painted rings. The most common variety is composed simply of a carbon rod of precise diameter to which two wire leads are affixed.

Note that, as with those monks holding hands with the Leyden jar, current flows around a continuous or *closed* circuit. In effect, the electrons leave the negative terminal of the battery, meander through the light bulbs or what have you, around the circuit, and back to the positive terminal. *Current doesn't get used up and it doesn't bunch up.*

Here "meander" is the right word. Electrons make progress only very slowly along a wire, typically at speeds of about 1 millimeter per second. One might well ask: How, then, can the telegraph and telephone manage to transmit electrical signals along their clutter of wires at close to the speed of light? The answer is quite simple: The electron we start pushing in New York is not the one that tickles the telegraph receiver in San Francisco. That starting electron will take sixteen minutes to move the first meter of the journey; it may not even be out the door before the message is ended! The situation is a little like having a long tube filled with greased marbles. You push one in at this end, and almost immediately one pops out at the other end. Do it in Morse code, and you've got a telegraph.

Electrons are propelled along a wire by an internal electric field established as a result of charges built up on the source, the battery. Ordinarily a free electron in the wire experiencing the field will accelerate for a short while, picking up speed before it undergoes a collision that slows it down and diverts it from its course. Again it accelerates, and again it smashes into some impediment, such as an impurity atom, a defect in the crystal, or a grain boundary between the many tiny crystals making up the wire. All the while thermal vibrations are jostling the atoms, and the wandering electrons are kicked around every which way as they gradually progress along the wire.

If there was no electric field constantly urging them along, the electrons would go careening around at random, going nowhere in particular. It's the electric field that can be thought of as moving down the wire at near the speed of light, carrying the signal, setting the electrons in motion before it.

All this banging around within the conductor is what actually gives rise to resistance. Interestingly enough, there are many materials, usually ones that are ordinarily only fair conductors, that undergo a sudden remarkable transition at very low temperatures (lower than about 25 K), *losing all traces of resistance.* These **super-conductors** are increasingly finding their way into applications where their unsurpassed current-carrying ability more than justifies the considerable expense and annoyance of maintaining such low temperatures. Currents that have been set in motion have thereafter

APPROXIMATE RESISTANCES OF SEVERAL MATERIALS

Material	Resistance* (in ohms)
Silver	1.6
Copper	1.8
Gold	2.4
Aluminum	2.8
Tungsten	5.6
Iron	10
Lead	21
Mercury	96
Carbon	3500
Salt water	20,000,000
Germanium	50,000,000
Distilled water	5×10^{11}
Glass	10^{20}

* Of a sample 1 square millimeter in cross section and 100 meters long at room temperature.

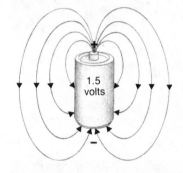

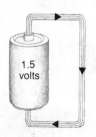

The electric field set up by a battery gets confined within an attached conductor, causing a current therein.

continued to circulate on their own within a closed superconducting circuit (without any driving source) for up to a year, showing not even the slightest sign of diminution. Indeed, it has been estimated that such a *supercurrent* would persist for not less than 100,000 years. The theory suggests that electrons in the superconductor interact with the crystals through which they travel in a way that causes them to pair up. Roughly speaking, once electrons are paired, they tend to keep each other from scattering off obstacles as they move. No scattering, no resistance.

Power   Imagine that we have a piece of "magic" wire with no resistance at all, connecting the terminals of a battery. Free electrons would sail from the negative terminal, picking up speed as they went around to the positive terminal. Their initial potential energy, provided by the battery, would soon be converted entirely into kinetic energy. As they came crashing back into the battery, that KE would be passed along to the molecules of the terminal as heat. Each electron would lose an amount of PE equal to its charge times the voltage difference ($\mathcal{V}$) through which it passed—that was the definition of voltage in the first place. Of course, it wouldn't be long before this "short circuit" would run the battery down completely.

In a real wire with a substantial resistance, the electrons would give up whatever energy they had progressively in collisions with the material. The wire would then heat up, and the electrons would re-enter the battery with little or no gain in KE. There would be a conversion of chemical PE to electrical PE within the battery and a subsequent transformation of that electrical PE momentarily to KE and then to thermal energy in the wire. The electron is in effect the recyclable medium transmitting energy from battery to wire.

Of course, all this happens not with one but with many electrons, so if a total charge of (Δq) coulombs sails the circuit, an amount of energy equal to (Δq)$\mathcal{V}$ will be transformed. Inasmuch as a current $I = \Delta q / \Delta t$ is circulating, the process is ongoing, carrying charge and converting energy continuously. Thus the amount of charge (Δq) making a complete journey around the circuit in any given amount of time (Δt) is $I(\Delta t)$, and in the process it transforms an amount of energy, $I(\Delta t)\mathcal{V}$. During the time (Δt), this is the extent of the work done on the electrons by the battery, and it's also a measure of the thermal energy produced. Dividing $I(\Delta t)\mathcal{V}$ by (Δt) yields the more convenient quantity, the energy-per-unit-time or **power,** P, expressed in joules per second (Section 7.3). In other words,

power = current × voltage,

or

$$P = I\mathcal{V},$$

where the units of power are watts, current is in amps, and voltage is in volts.

If a resistor is placed between the terminals of a battery, a current (I) will flow through it, and a voltage drop ($\mathcal{V}$) will exist across

An electrical energy meter.

it. For example, if the current is 2 amps and the voltage is 9 volts, 18 watts of power will be dissipated in the form of heat by the resistor. That's the only thing resistors do with the energy they draw. Note that Ohm's law ($\mathcal{V} = I\mathcal{R}$) can be used to arrive at an alternative expression for *the power dissipated via electrical resistance*, namely, $I(I\mathcal{R})$, or

$$P = I^2\mathcal{R}.$$

Joule was the first to recognize this current-squared dependence experimentally, and the process is often called **Joule heating.** The last time you toasted a slice of bread or lit a cigarette with a car lighter or fried an egg in an electric skillet or dried your hair with a blower or electrocuted a frankfurter, you were utilizing Joule heat—pumping electrons through a resistor.

In general, when a current passes through some gadget and drops in potential in the process, an amount of power ($I\mathcal{V}$) will be delivered although it need not appear as heat. A 100-watt light bulb plugged into a 120-volt socket will, in accord with Ohm's law, draw 0.83 amp and indeed consume 100 watts of electrical power. Unfortunately, only about 3 percent of that will appear as light; the rest, an incredible 97 percent, will be wasted as heat. An incandescent bulb is just an enclosure containing a thin tungsten spring (a filament), which becomes white hot ($\approx$2800 K) via Joule heating. Door bells, vacuum cleaners, toy trains, TVs, dishwashers—all try hard to produce something other than heat for their input of energy.

The central idea in all this is that electricity has become the principal pipeline of power in our everyday lives. Electrons are as much the working substance of our age as steam was the working substance, the energy carrier, in the early days of the Industrial Revolution. Electricity, energized by batteries and power plants, silently flows into home and factory, delivering its bounty of power to do work, on command, night or day.

Batteries were the only practical source of electric current in the early 1800s. However rudimentary, they nonetheless served to power a few blazing carbon arc lamps, run the growing railway telegraph system, and give rise to the whole electroplating industry that would soon be busily providing the new middle class with the affordable illusion of wealth in silver-plated forks and spoons.

Charge flowing from a battery proceeds in one direction. Electrons leave the negative terminal, slowly swing through the circuit, and ultimately return to the positive terminal. That unidirectional flow between terminals of a fixed polarity is known as **direct current,** or DC for short. The first several decades of the nineteenth century were exclusively a time of DC. Consequently, when the primitive electric generator was born in the early 1830s (Section 19.4), it was seen as a useless curiosity. After all, those hand-turned devices produced an **alternating current,** or AC, in which electrons moved first forward, then backward, oscillating essentially in place at some given number

ELECTRICAL POWER CONSUMPTION

Appliance	Typical wattage
Clothes dryer	5200
Hairblower	1300
Dishwasher	1200
Iron	1100
Toaster	1100
Refrigerator (big, double-door)	800
Vacuum cleaner	600
Washing machine	550
Blender	400
Fan	200
TV (black and white; tubes)	190
Humidifier (small)	40
Pocket calculator	8
Clock	4

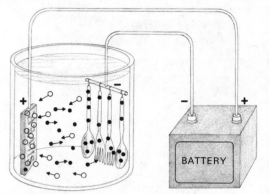

(Ag NO$_3$) silver nitrate solution

o NO$_3^-$ ions

• Ag$^+$ ions   The process of silver plating.

18.3 AC-DC

A circus poster from 1879.

of cycles per second. Instead of maintaining a fixed polarity, each terminal of the AC generator, though always opposite to the other, alternated between + and − over and over again, like a flip-flopping battery.

Joule heat arises with AC, just as it does with DC, as a result of the collisions of free electrons with crystal imperfections, impurity atoms, etc., within the conductor carrying the current. In AC, the electrons that lose their KE through collisions are *oscillating;* in DC, the electrons are migrating *one way.* The effect is the same in both: Electrical energy, initially in the form of organized motion, is transformed into heat.

When AC generators first appeared, their alternating nature was seen as a definite drawback. That seeming difficulty was overcome in about two years, however, when a modification in the generator converted it into a DC power machine known as a **dynamo** (from the Greek *dynamikos,* meaning powerful). One could crank the wheel by hand or with a steam engine, and out came electrical current. There were immediate uses for that tireless source in the electroplating industry, and the generator quickly made the transition from toy to workhorse.

By the 1870s the mechanical generation of electrical power, both AC and DC, was more or less a practical reality. In those days an electrified factory, store, lighthouse, or large ship would have its own steam-driven generator. That was usually a DC device, in part out of habit, in part because the DC motor was already well perfected and, of course, because DC could be used to recharge storage batteries, which were still commonplace. Carbon arc lamps, which were far too bright for domestic use, could operate on either AC or DC, but the former was often preferred. When the practical incandescent lamp was introduced by Edison in 1880, however, each installation came with its own 110-volt generator, and it was DC.

In order ultimately to supply the small user, the dozen-light-bulb consumer, Edison constructed the first permanent central power plant in the world—a DC facility known as the Pearl Street Station in New York City. The arteries of electrical power began to spread out, city by city across the country and the world. By the end of the 1880s the AC motor was at last an operational reality, and alternating current started to look even more appealing, particularly to shrewd visionaries like George Westinghouse. Initially, whenever AC was produced in quantity, it was generally converted at substations to 110 volt DC, which was sent on to the customers. Slowly, as the market grew, there developed a titanic struggle for control of the industry, a struggle between the hustlers of *high-voltage AC* (primarily Westinghouse and General Electric, the J. P. Morgan combine to which Edison would ultimately sell out) and those of *low-voltage DC* (led by Edison).

When central generating stations came into being and especially when they were constructed near remote energy sources like Niagara Falls, the electrical power they produced had to be transmitted over

long distances. That would become the crucial determining factor in
"the great American AC-DC dilemma." You see, the very wires carrying electricity have some resistance, and that poses a major problem.
A medium-sized city might easily require around 10 million watts of
power. Now if that amount is to be provided at a modest 100 volts or
so, then 100,000 amps will have to be supplied ($P = I\mathcal{V}$). There's the
difficulty—transporting power tens or even hundreds of miles. The
Joule heating of the transmission wires ($I^2\mathcal{R}$), the power lost as current moves along them, varies as I^2, not just as I. Even a moderate
resistance will waste tremendous amounts of power at such high
current levels.

Shy of superconducting transmission lines, which are not quite
practical even now and were unthought of then, there was no eco-
nomically feasible way out but to lower the current. Clearly, if the
voltage was increased to 100,000 volts, the same power could be
delivered rather efficiently by 100 amps! Since there were very simple
ways to raise and lower the voltage of AC (via transformers) but no
comparable means for DC, the contest ultimately went to the high-
voltage AC people. The Niagara Falls project was a turning point that
established the practicality of AC and set the world oscillating.

Usually power is generated at around 10,000 volts and boosted
to several hundred thousand volts for transmission to the region of
consumption. Substations outside each area then step it back down
to a local distribution voltage of several thousand volts, carried overhead by those endless ugly wires. Thereafter it's again reduced, usually by small pole-mounted transformers, each of which feeds a
cluster of buildings. *Line voltage* in the United States is ordinarily
rated at 110, 115, or 120 volts and can be anywhere in that range,
actually varying somewhat from one moment to the next. To conserve power in inadequately supplied areas during heavy use periods,
the producers will occasionally drop the voltage 5 or 10 percent for
a while; the result is the increasingly familiar "brownout."

As shown in the figure, AC voltage builds, peaks, drops off to
zero, reverses polarity, peaks again negatively, and comes back to
zero, completing one cycle. Sixty such complete alternations will occur each second, and we say the voltage has a *frequency* of 60 *cycles
per second*, or in the more modern jargon, 60 *hertz* (60 Hz). There's
nothing particularly sacred about this frequency; indeed, most of
Europe and Asia use 50 Hz (220 volts) AC. Except for synchronous
motor-driven gadgets (clocks, record players, etc.), most devices are
fairly indifferent to frequency. At the time decisions were being
made, the generated frequencies had to be only high enough to keep
light bulbs from flickering, but 60 Hz was convenient for electric
clocks as well.

It's a common misconception that an alternating current somehow flows out of an outlet into and through an electrical appliance.
In fact, nothing but energy flows into a toaster. The free electrons
already within the appliance are set into oscillation by the electric
field produced at the generator. We can imagine an electrical dis-

Austrian-born Nikola Tesla was one of the
leading figures in the development of alternating current. The spectacular display
crackling around Tesla was produced by
high-frequency (20,000 Hz) high-voltage
AC. The handwritten inscription reads "To
my illustrious friend Sir William Crookes
of whom I always think and whose kind
letters I never answer! June 17, 1901."

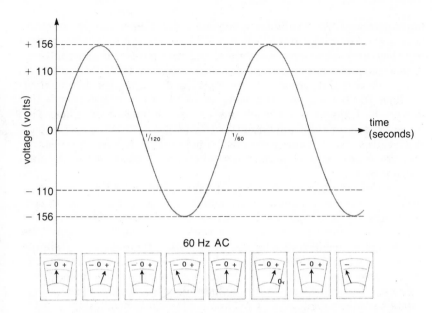

60 Hz AC

A 60-Hz AC voltage with an average value of 110 V and a peak of 156 V.

turbance as a moving interface behind which electrons have already been set in motion and in front of which they are as yet unaffected. It's that disturbance that propagates at near the speed of light for both AC and DC. Electrons within the toaster are pushed and pulled back and forth, as if they were all connected by little springs and the outlet terminal somehow took hold and rattled the first one. The "rattle" corresponds to the energy, and it moves down the line until all the free electrons are oscillating together at 60 Hz.

Note that a line voltage of, say, 110 volts refers not to the peak value but to the *effective* value that would deliver the same power if it was DC. In other words, an AC voltage with a peak value of 156 volts will provide power at the same average rate as would a source of 110-volt DC.

Typically, several wall outlets and light sockets are connected in parallel on a single circuit. Each appliance plugged into one of these outlets experiences an average of 110 volts across its terminals and draws a commensurate current determined by its own resistance. An 1100-watt (11 Ω) toaster will draw 10 amps, and a 100-watt (121 Ω) bulb 0.91 amp. If both are on simultaneously, the sum of 10.91 amps will come surging through the kitchen circuit from the power lines. The relatively thin wires within the walls cannot safely carry an unlimited amount of current. They will heat up and even melt unless protected from overloads. If five or six toasters were plugged in at once on the same circuit, the feed wires coming up from the basement would be called on to carry 50 or 60 amps, greatly exceeding their safe capability. Accordingly, a fuse, typically rated at 15 or 20 amps, is placed in series with the feed wire so that all the current in the circuit passes through it. When that current reaches the fuse rating, the fuse melts, not the wires. Every installation, office, factory, home—whatever—is divided up into a number of circuits, each separately fused and all in parallel. Consequently, each circuit draws electricity directly from the power lines, and each is isolated from the others. When the kitchen fuse blows, the lights stay on in the bedroom.

Usually there are two lead wires (and therefore two little plug holes in each outlet), the "hot" one oscillating from about +156 volts to −156 volts with respect to the other, which is "neutral," or grounded. When you switch off the lights or blow a fuse, you are simply causing a break in the hot lead; the ground wire, which is literally connected to the ground via a water pipe, should always be continuous.

About 1000 people a year are accidently electrocuted in the United States alone. Included are the foolhardy who balance plugged-in electrical appliances on bathroom shelves within reach of the tub, as well as the poor souls who invariably try to retrieve kites from high-voltage lines, usually with long sticks and often in or after a rain.

When a substantial amount of charge flows through muscles, it causes wrenching spasms. If those muscles are in the hand, the effect may be no more than tiny burns and a deep ache. If they are cardiac muscles, the heart can go into a lethal condition of ventricular fibrillation (irregular movement). The idea is to avoid letting currents flow through the body in general and certainly to keep them away from the heart! Electricians and TV repairmen will often position one arm well away from the circuit; they risk blasting the fingers they're working with, but they avoid creating a hand-to-hand path across the chest. It takes only about 100 mA (100 milliamps or 0.1 amp) flowing through the body for a second or more to be fatal. The vibrating cramps from 10 mA will be remembered for quite a while, even though that dose isn't lethal. Indeed, at above 15 mA or so, one

Shocking

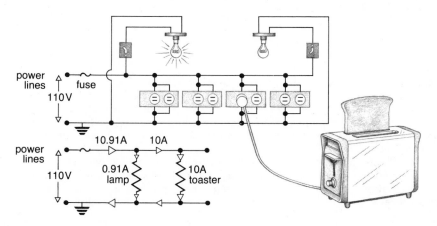

A simple house circuit.

An elderly Thomas Alva Edison posing in front of some special lightning-simulating equipment. The man with the cigar—who in fact was rarely without one—is Charles Proteus Steinmetz, already the foremost scientist in the new field of AC electricity.

loses voluntary muscular control—rather awkward if you're holding on to a "hot" wire and can't let go.

The resistance of the body is determined to a very large degree by the contact resistance with the outer layer of skin. The wet human stuff within each of us is rich in ions and a fairly good conductor. Therefore, depending on the condition of your skin and the area of contact, your resistance (hand-to-hand or head-to-foot) may vary from perhaps a thousand to several hundred thousand ohms, with about 20,000 Ω as not too atypical. Thus, from Ohm's law, 120 volts applied between one hand and the other will ordinarily push a current of 120 volts/20,000 ohms, or 0.006 amps, through the body—not lethal but not pleasant either.

More often than not, the victim of a serious domestic electrical accident is unknowingly grounded (touching some damp plumbing, standing on a wet basement floor, taking a bath, etc.). All that's necessary then is to come in contact with something at 120 volts, and if the current path crosses the chest on the way to ground—that means trouble!

As a rule, be exceedingly careful with 120-volt AC, and don't even consider puttering around with 240-volt AC. Although batteries have plenty of current capability, they usually have such low voltages that they are perfectly safe to play with—provided you don't put a few dozen in series. There are special commercial batteries with outputs of several hundred volts, and those can be lethal.

On August 6, 1890, technology took one of its ghastly little detours in the name of humanity—this time in the death chamber of Auburn Prison in New York. The use of the "electric chair" has become highly controversial today, but at the turn of the century, when it was fresh in the mind, Edison often raised its grotesque specter as the symbol of AC horror, the ever present danger of alternating current. A bit of macabre humor popular at the time even suggested that being electrocuted was the equivalent of getting "Westinghoused."

EXPERIMENTS

1. This is more an experience than an experiment. Go outside and trace the overhead power lines until you locate the cylindrical pole–transformer feeding your house. If you live in a big city apartment building or dorm room, forget this exercise because the feed cables are probably buried. Everyone else should be able to locate the three wires leading to the "entrance head" on the building. Three wires essentially provide two 120-volt circuits, which can be combined to supply 240 volts. Follow the wires down to the power meter (usually outside small buildings or in the basement).

Observe the rotating metal disk, whose speed is directly proportional to the power being consumed, and the little dials that read off the energy used in kilowatt-hours. Next locate the main panel containing either the fuses or breakers. Out of it come the various branch circuits that supply the building.

2. You can crudely examine the conductivity of various materials with any one of the circuits in the figure by using a battery wired to a bulb. Be careful to match the voltage ratings of the components—3-volt flashlight bulbs will burn

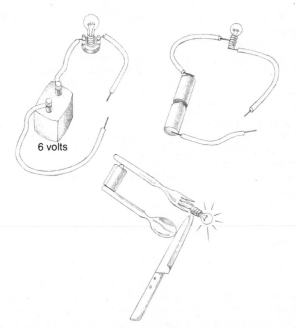

6 volts

out by the dozen if you're using a 6-V battery and are not exceedingly careful. Put whatever you wish to test across the open terminals, thereby closing the circuit. If the object is a good conductor, the lamp will light brightly. Try various

metals, aluminum foil, plastic, paper, glass, etc. Remove the "lead" from a pencil and test it. Insert the two wires into a glass of water, and then add some salt.

3. An electrical battery is very easy to construct, although without a meter to measure current and voltage output, it's not so easy to tell that you have actually succeeded. These days everyone has an electrical speaker somewhere about the house, and that device will serve our purposes nicely. Make a wet cell like Volta's, using a strip of copper (some wire or a penny) and a strip of zinc (the casing from a D cell or a piece of galvanized steel), and immerse them in salt water. Such a gadget will put out about 0.6 volts (around 0.7 milliamp). Adding some vinegar will raise its voltage to around 0.9 V. Attach the cell to two wires. Fix one of them to a terminal of your speaker, and tap the other lead on the remaining terminal. The speaker will click at each touch. Try different combinations of metal plates: copper and aluminum (folded up foil), zinc and aluminum—whatever is handy. I have a 0.7 volt (0.5 milliamp) lemon battery that I made by sticking a carbon rod (even a pencil lead will work) into a lemon and then surrounding it with about 20 large-headed galvanized nails. When I replaced the carbon with a paper clip, the voltage dropped to 0.2 volts, but it still managed to make the speaker click faintly. Instead of a lemon try a grapefruit or even a glob of sauerkraut. Moisten a piece of paper with lemon juice or vinegar, and place it between two metals—a penny and an old silver quarter work well.

REVIEW

Section 18.1

1. What constitutes an electrical *current*?
2. The ratio of the amount of _____ flowing past a point per unit _____ is defined as electrical current.
3. What are the units of current?
4. One *ampere* equals 1 _____ per _____.
5. What is the meaning of the word *microamp*?
6. What is a *parallel-plate capacitor*?
7. Define *capacitance*.
8. One farad equals 1 _____ per _____.
9. When are any two or more circuit elements (capacitors, resistors, etc.) in *series*?
10. When are any two or more circuit elements (capacitors, batteries, etc.) in *parallel*?
11. Which little green creature gave a "leg up" to the invention of the battery?
12. Briefly describe a *voltaic pile*.
13. Without actually knowing what he was doing, Volta arranged the cells in his "pile" in _____.
14. What does the *voltaic wet cell* consist of?
15. Nicholson and Carlisle used an early voltaic battery to

_____. By the way, these two gentlemen are now widely believed to have intended to "rip Volta off," as we might say in the contemporary jargon. In any event, after learning of Volta's invention, they built and used a pile without making any mention of him whatever.

17. What is a *dry cell*?
18. Distinguish between a *cell* and a *battery*.
19. A *lead storage cell* maintains a voltage of _____ across its terminals.
20. The accompanying figure shows an ordinary flashlight.

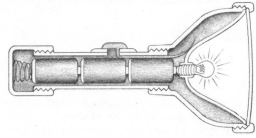

Explain how it works. Are the D cells in series or parallel?

21. Usually the larger a cell is, the more _____ it can supply.

22. Define the *amp-hour* rating of a battery.

23. Some people would never say "a flow of current," insisting that the expression is redundant. What do you think?

Section 18.2

1. State Ohm's law.

2. Define *resistance*.

3. What are the units of resistance?

4. Is the resistance of a piece of metal always constant?

5. Electrons constituting a current drift along a wire at a speed of about _____.

6. We can think of electrons in motion constituting a current, which is propelled along a wire by a(n) _____.

7. Does the fact that electrons collide with crystal imperfections in the medium through which they move show up as anything observable?

8. What is it that can be imagined traveling along a wire at near the speed of light?

9. What is a *superconductor*?

10. In a circuit carrying a current, I, what does the quantity $I(\Delta t)$ correspond to?

11. Write an expression for electrical *power*.

12. Discuss the meaning of the phrase *Joule heat*.

13. State an expression for the power dissipated by a resistor.

14. Write an expression for I in terms of $\mathcal{V}$ and $\mathcal{R}$. For P in terms of $\mathcal{R}$ and $\mathcal{V}$.

Section 18.3

1. *Direct current* corresponds to a _____ flow of charge.

2. Describe what is meant by AC.

3. In a conductor the organized motion of free electrons in response to a potential difference is converted into _____ motion via collisions. The result is a transformation of electrical PE into _____.

4. What is a *dynamo*?

5. Who introduced the first practical domestic light bulb in the 1880s, thus ultimately giving rise to the power indus-

try? There are still many Victorian houses around with the gaslight fixtures that were converted to electricity by simply wiring through the old gas pipes. In fact, many of the early fixtures were for both gas and electricity because you couldn't rely on the power supplier for uninterrupted service.

6. Describe "the great American AC-DC dilemma."

7. Why was high-voltage AC more attractive to the power industry than low-voltage DC? By the way, we are now seeing a growing interest in transmitting power over long distances via very-high-voltage DC. This was made possible by advances in rectifiers, which convert AC to DC at prodigious voltages of 200,000 V and more.

8. In the United States AC power is generally supplied at a frequency of _____ hertz. Incidentally, airplanes often use 400-Hz electrical systems.

9. The line voltage in Nepal is 50 Hz, _____ volts.

10. Does AC current literally flow out of the wall outlet into your TV set?

11. Distinguish between *peak* and *effective* values of AC voltage.

12. What is the purpose of *fuses* or *circuit breakers* (the more modern resettable equivalent)?

13. The "hot" lead in domestic wiring is usually black and carries a voltage with respect to the white ground lead that oscillates from _____ to _____ volts.

14. What is meant by the symbolic representation "1 mA"?

15. Most people can feel 1 mA, and currents above _____ will block voluntary muscle control. Beyond 30 mA, breathing becomes impeded, and by 70 mA or so, each breath is exceedingly labored. Currents of from roughly _____ up to 200 mA will cause lethal heart fibrillation. In excess of 200 mA, the heart simply stops beating altogether, and breathing ceases. As long as no major internal damage occurs, the heart can be restarted with prompt first aid. Correcting fibrillation, however, demands more drastic medical intervention. Thus it's actually better to get blasted with a bit more than 200 mA than with a bit less.

16. Someone touching a source of more than 200 volts will experience electrical currents capable of piercing the skin, making tiny burns that gradually lower the overall contact resistance. Why is that not a welcome development to a person unable to let go of the Devil's tail?

QUESTIONS

Answers to starred questions are given at the end of the book.

Section 18.1

* 1. What is the current in amps if +100 coulombs of charge flows past a point every five seconds?

2. Suppose that 10 coulombs of positive charge and 10 coulombs of negative charge flow in the same direction past an observer every second. What is the net electrical current?

What would it be if the two streams of + and − charge were traveling in opposite directions, as they do in a plasma?

3. Explain why more charge can be stored when there are two opposing plates than when there is one. In other words, how does a Leyden jar work? Was it necessary that van Musschenbroek be grounded for the jar to become fully charged? Why?

4. It was commonplace in the 1700s to electrify great hulking conductors, such as cannons and even people (remember, van Musschenbroek was charging up a gun barrel when he got blasted). Discuss why large objects were often used. Talk about potential.

* 5. Imagine that you come upon a cylindrical circuit element with a wire jutting out of each end. Now suppose that you break it open and find two sheets of metal foil separated by a strip of paper wound up into a cylinder, with each plate attached to one of the wires. What function might such a device serve? What would you call it?

6. Two identical electroscopes with their metal cases grounded are resting on a table. One is in contact with a metal ring snatched from a carousel. The other touches a large bronze bust of Galvani's pet frog. The ring carries a charge of +1 microcoulomb whereas the bust is charged with +1000 microcoulombs, and yet the leaves make the same angle on both. Explain.

7. What would happen to the amount of charge that could be stored by a parallel-plate capacitor if the voltage impressed across it was doubled?

8. What is to be gained by putting capacitors together in parallel, inasmuch as the voltage across each is the same? What is the advantage of arranging capacitors in series?

* 9. How would the capacitance of a parallel-plate device vary with changes in the area of the plates?

10. Describe Volta's tin-silver-tongue battery. Why don't you taste anything until the two metals touch? Volta thought that the current of "electric fluid" in his pile arose from a "contact force" between the two metals. It was not until about 25 years later that the chemical action taking place was finally recognized (by C. A. Becquerel, A. De La Rive, G. F. Parrot, and others) as the operative mechanism! Ironically, neither Volta nor Galvani was totally right or totally wrong.

11. The dry cell was first introduced in pretty much its modern form by Leclanche in 1868. Write down everything of significance you know about the device. Ask your parents or any other ancients if they remember how leaky D cells were when they were kids.

12. Suppose that you scaled up a D cell to five times the size of an ordinary flashlight cell. What would happen to output current and voltage capabilities?

13. Make a clear distinction between current and voltage.

* 14. There are several large batteries in common use composed of dry cells. The 6-volt lantern battery is an example. Discuss how it might be constructed. Design a high-current 1.5-volt battery with an amp-hour rating in excess of 10.

Section 18.2

* 1. Ohm remarked that the laws of electricity "are so similar to those given for the propagation of heat . . . that even if there existed no other reasons, we might with perfect justice

draw the conclusions that there exists an intimate connection between these natural phenomena." To what similarity was he alluding? For conductors he was closer to the truth than he could have imagined. Why?

2. The accompanying figure shows the voltage across a tungsten filament plotted against the corresponding current flowing through it. Discuss the filament's resistance at different currents.

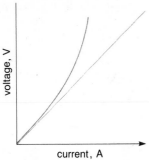

3. The accompanying figure shows water flowing around a circuit at a rate that keeps the levels constant. Draw the corresponding electrical circuit, and describe each component in the analogy.

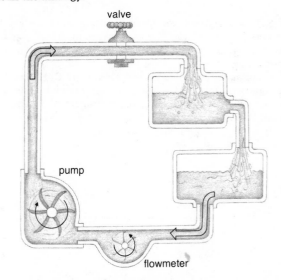

4. Given the conditions in the previous problem, present a watery argument for the position that current is never used up in its meanderings through a circuit. What does the pump actually "pump" *into* the system? If the system in the figure is isolated thermally from the rest of the universe, what will happen to the circuit? What would happen to the corresponding electrical circuit? No wonder nineteenth-century scien-

tists liked the notion of "electric fluid." Have you ever heard contemporary electricians mutter things like "turn on the juice"?

5. With the liquid analogy in mind and considering friction, explain how the flow of a fluid would be affected by passing through a long narrow-bore tube (see figure). How would electrical resistance vary with the length and diameter of a wire? Now that we know that fluids are ultimately granular (as is electricity), the analogy is a rather good one.

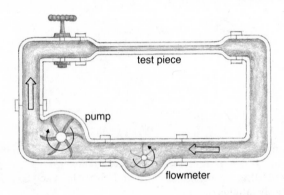

* 6. Suppose that you double the voltage across a resistor. What will happen to the current traversing it? If you keep the voltage constant and halve the resistance, how will the current vary?

7. Imagine that you have two strings of Christmas tree bulbs, one wired in series, the other in parallel. Determine which is which in the accompanying diagram. If one bulb blew out in each string, what would happen and why?

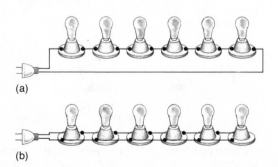

(a)

(b)

8. What voltage would you measure across a battery made of two D cells attached + to +?

9. Which bulb in the figure burns brighter (i.e., through

which does more current flow), the one with the single D cell or the one with the double D cell? Explain your answer.

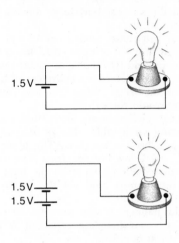

* 10. Which bulbs in the figure burn brightest (i.e., through which does more current flow), the single, double, or triple arrangement? Explain your answer.

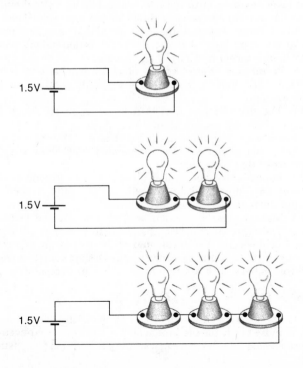

11. Determine whether the resistors in each circuit in the figure are in series or in parallel.

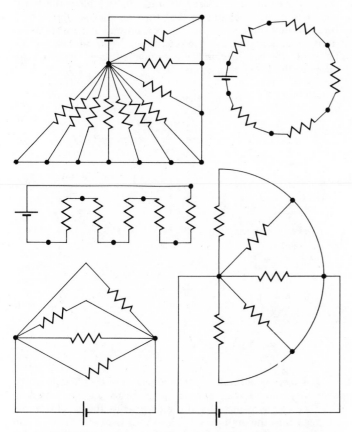

12. The diagram shows several 10-Ω resistors connected to a 1.5-V battery. What is the voltage across the various resistors? Explain.

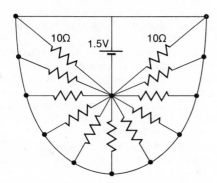

13. Two identical resistors are placed in parallel across a battery. What is the voltage across each? What can you say about the current through each one? What can you say about the current passing through each resistor in the previous question?

***14.** The accompanying figure shows three light bulbs with resistors of 5 Ω, 10 Ω, and 2 Ω across a 10-volt battery. What is the voltage across each resistor? Use Ohm's law to determine which will draw the most and the least current. Which will dissipate the most and which the least power?

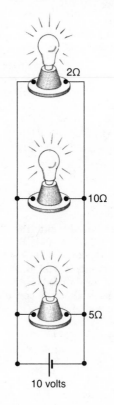

10 volts

Section 18.3

1. In the transmission of electrical energy, what is the significance of the fact that the power dissipated by a resistive element varies with I^2?

***2.** How many times per second does AC voltage momentarily become zero?

3. Why would 110-volt AC feel far more unpleasant than 110-volt DC?

4. Why does heavy-duty equipment (such as electric lawn mowers and air conditioners) need its own extra-thick-wired extension cords?

5. An electroscope (once calibrated) can be used as a *high-voltage* voltmeter. Thus if a DC source of, say, 120 volts was connected across the central wire and the housing, the leaves would spring apart. What would happen if we attached the "hot" terminal of an AC outlet to the scope's central wire and grounded its metal casing? *Hint:* Remember that the leaves have inertia.

6. The figure shows a metal can attached to a microammeter, which in turn is grounded. Describe what you would observe if a negatively charged rod is rhythmically inserted and removed from the can.

microammeter

* 7. A light switch essentially cuts the "hot" lead, opening and closing it at will. With the switch in the "off" position (see figure), is there a potential difference across (a) the bulb, (b) the two terminals A and B of the switch? (c) If the bulb was removed or burned out, would that change matters? Would you get a shock if you touched A and B at one time (d) with the switch open, (e) with it closed?

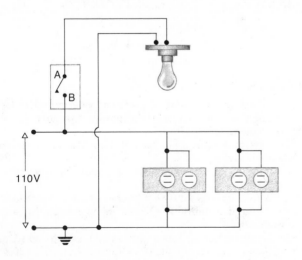

110V

8. The accompanying diagram shows a hodgepodge electrical system for a car. List which of the switches, A, B, C, D, E and F, must be closed in order to (a) blow the horn, (b) turn on the headlights, (c) turn on the tail lights, (d) turn on only the parking lights, (e) activate the inside dome light. When do the side markers go on? *Note that the whole body of the car, including the engine block, is generally taken as the common ground.*

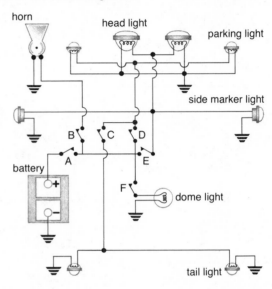

9. I own a portable electric typewriter that has a sharp decorative metal trim around the edge of the carrying-case lid. One day someone closed the case while the machine was plugged into the wall outlet. The sharp trim cut through the insulation on the power cord, there was a bright flash, a bit of wire melted into little beads, and the room lights went out. Describe what happened in detail.

* 10. High-voltage cables are generally uninsulated, and a bird sitting on one of them may have its feathers puffed up. Why? Why aren't our little feathered friends electrocuted on their high-tension perches? Could a chicken safely walk on the third rail of an electrified railroad? What would happen if it stepped off with one foot?

11. When lightning hits the ground, it often spreads out radially to some degree as it penetrates downward. Thus it is possible for a large four-legged animal standing with its tail end closer to the bolt than its front end to be electrocuted. Explain. Considering ground currents, what is the best posture for a human out in a thunderstorm—lying down, squatting, or standing up?

12. The modern wiring scheme uses an extra third lead to ground the metal casing of an appliance. The idea is to pro-

tect against situations in which the insulation on the "hot" lead becomes frayed, whereupon it can short out to the casing of the device. Describe the dangers of having a washing machine cabinet at 115 volts. Make a diagram showing how the third wire could save someone's life. What would happen to the fuse in the circuit?

* 13. Generally, because batteries have low voltage, they are not at all likely to represent an electrical hazard. Still, I once saw someone accidentally drop a wrench across the termi-nals of a car battery. There was a tremendous sparking flash, and molten beads of metal flew all over the place. Explain what happened. Why did the wrench melt only where it touched the battery?

14. Under what circumstances is it possible to plug several appliances into the AC wall outlets and yet not get 110 volts across them, even though 110 volts is being supplied at the fuse box? Have you ever experienced lights dimming for a moment or the TV picture shrinking on the sides?

MATHEMATICAL PROBLEMS

Answers to starred problems are given at the end of the book.

Section 18.1

* 1. In the course of a one-minute survey, 6000 coulombs of charge in the form of a beam of electrons pass an observer's nose. How much current does that correspond to?

2. The photo on p. 406 shows an electron beam representing a current of 60,000 amps. How many electrons flow out of the device each second? ($q_e = 1.6 \times 10^{-19}$ C)

* 3. How much charge can be stored on the plates of a 1-microfarad capacitor if the voltage across it is 100 volts.

4. Suppose that an isolated sphere carries a charge of 1 C and is 1000 V above ground. What is its capacitance?

5. The Torpedo, a giant saltwater ray, is covered with cells known as electroplaques, each of which develops a potential difference of about 0.15 volts. Several thousand rows (each made up of a series-connected array of cells) are then connected in parallel to build up the needed current. How many cells would you guess form each row? Why do freshwater electric fish in general develop higher voltages?

Section 18.2

* 1. Show that 1 ohm = 1 kg-m²/C²s.

2. If electrical contacts are placed on the scalp, time-varying differences in potential will be observed. These can be recorded by an electroencephalograph. Usually voltage differences of around 0.5 millivolts will appear across resistances of about 10,000 Ω. What size currents are involved?

* 3. A wooden stick in contact with a 100,000-volt Van de Graaff generator carries a current of 2 microamps down to ground. Calculate the ruler's resistance.

4. Suppose that someone falling out of a tree grabs an overhead power line. The wire has a resistance of 60 micro-ohms per meter and is carrying a current of 1000 amps. With hands a meter apart, will the unfortunate soul get much of a shock?

5. Calculate the price of electrical energy in dollars per kilowatt-hour as supplied by a 35-cent D cell. Compare that with less than 10¢ per kWh for AC on tap.

6. Suppose that you drew 2 amps from a 1.5-volt dry cell constantly for two hours. How much energy would that correspond to in joules and in watt-hours?

7. A bulb rated at 100 watts for 110 volts is connected to a source of 220-volt AC. If its resistance is unchanged, how much current does it draw, and how much power will it dissipate?

* 8. Determine the resistance of a lamp that dissipates 30 watts when operated at 6 volts.

Section 18.3

* 1. Suppose that you are cooking breakfast in a 1400-watt frying pan while drying your hair with a 1300-watt blower. Will that pop a 20 A fuse if the line voltage is 115 volts?

2. What are the resistances of a 100 W bulb and a 10 W bulb, both operating at 110 volts? Compare them.

3. For AC the peak or maximum voltage $\mathcal{V}_{max}$ is related to the effective voltage $\mathcal{V}$ by way of

$$\mathcal{V}_{max} = \sqrt{2}\mathcal{V}.$$

Compute the peak voltages corresponding to effective values of 110, 115, and 120 volts.

* 4. Two 100 W bulbs are plugged into a 120 V AC line. Calculate the current drawn by each. Now replace that arrangement with an equivalent single bulb that dissipates the same power and draws the same total current. What is its resistance? *The "equivalent resistance" of two or more resistors in parallel is less than any one of them individually.*

5. The diagram shows a three-way light bulb. Explain how it works. With both filaments on, what is its total resistance (assume 110 volts)?

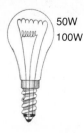

50W
100W

M Faraday

Likeness from a recent Photograph from life

Naturally enough, the underlying theme that runs through everything in this chapter is **magnetism.** To start, the basic ideas concerning **permanent magnets** (attraction, repulsion, poles, etc.) are examined. Earth's own magnetism was a primary focus of early research. After all, for well over two thousand years the only "things magnetical" that were available to the experimenter were natural permanent magnets: pieces of iron and the Earth's pervading influence. The second section of this chapter introduces our contemporary understanding of what, on an atomic level, gives rise to the different kinds of magnetic behavior manifested by all forms of matter (e.g., **diamagnetism, paramagnetism,** and **ferromagnetism).**

As the first half of the chapter deals principally with permanent magnetism, the second half is devoted to **electromagnetics,** magnetism arising from electrical currents and vice versa. Oersted's wonderfully simple discovery—**a current generates magnetism**—would revolutionize the entire subject and profoundly change the whole course of human history from that moment on. **A charge moving across a magnetic field experiences a force.** That's another closely related insight that stands as the key concept governing the operation of both the electric **motor** and the **generator.** The last section of the chapter treats the process by which **currents can be produced through the application of magnetic fields.** This is the scheme humming away somewhere behind all those silent wall outlets, the primordial idea at the very heart of our twentieth-century technology. The steam engine all but vanished from the scene, replaced, for the most part, by the electric motor. Yet even though we tend to forget it, at the very beginning of the modern power chain there is usually a steam turbine turning the generator that makes the current that activates the motor that drives the gadget that does the work "in the house that Jack built."

Our contemporary understanding of magnetism is an intricacy of whirling electrons, of fields and photons and currents, and even so, it is still most assuredly incomplete. Magnetism began to be an area of human concern sometime in the distant past, when the ancients

A lodestone with a few steel nails clinging to it.

INTRODUCTION: MAGNETICAL

Galileo's demonstration of the strength of one of his lodestones. It could grab onto and support the heavy iron coffin-shaped weight beneath it. That morbid gesture was just Galileo's way of responding to the legend of Mahomed's suspended coffin.

I have seen Samothracian iron rings even leap up and at the same time iron filings move in a frenzy inside brass bowls, when this Magnesian mineral was placed beneath.

LUCRETIUS (First century B.C.)

first discovered the strange power of the lodestone, the ability to draw together its own fragments and to cling to iron. Exactly when that awareness came, however, no one now knows.*

Chinese legend has it that Emperor Hwang-ti (ca. 2600 B.C.) was once guided in battle through a dense fog by a small pivoted figure mounted on his war chariot that always pointed "south," a lodestone embedded in its outstretched arm. That wondrous capability of the compass to align itself with the north-south axis of the "universe" must have been utterly awe-inspiring then. It's no less marvelous even now, when we know it's only a local effect of the Earth's own invisible magnetic field.

The **lodestone,** or loadstone (from the Anglo Saxon *laedan,* to guide or lead), is a dark, pitted, rocky mineral, a crystalline oxide of iron (Fe_3O_4) known as *magnetite.* Permanently and strongly magnetized pieces of that particularly rich iron ore occur naturally in many parts of the world.

For all the centuries from Greek antiquity through Rome and the Middle Ages, through the Renaissance and the Enlightenment, up to the 1800s, the only practical source of magnetism, apart from the Earth itself, was the dark lodestone. Behind every experiment, every compass needle, every charlatan's trick was a chunk of "the stone." Indeed, the very word *magnet* was probably derived from the ancient Greek colony of Magnesia, where lodestones were plentiful 600 years before Christ.

After a long development that never got much past the rudiments, the study of magnetism experienced a surge of activity and innovation that began in the nineteenth century. It followed on the heels of Hans Christian Oersted's startling discovery that electrical currents give rise to magnetic forces (Section 19.3). That was a triumphant turning point. For literally thousands of years, electricity and magnetism were taken to be two distinct powers in Nature, two unrelated primitive forces, and now they were somehow connected, somehow one! Guided by people like Ampère, Faraday, Henry, and Maxwell, electromagnetism became the unified imagery. *Charges generate electric fields. Charges in motion, currents, generate magnetic fields.* The two fields, the two kinds of forces are simply different manifestations of a single phenomenon—*electromagnetism.* All magnetic effects are ultimately traceable to charges in motion, and we'll find our explanation of the mysterious power of the lodestone in terms of the no less mysterious power of its spinning electrons, the primordial current (Section 19.1).

Hundreds of different kinds of robust permanent magnets have been created to date, and the electromagnets that now exist range from the humble door chime ringer to superconducting giants that

* The oldest magnetic artifacts to date are a collection of 4000-year-old stone statues found in Guatemala. It was discovered in early 1979 that several of these great roly-poly figures had magnetized navels. Apparently they were deliberately carved around natural magnetic deposits by sculptors who had already mastered the lodestone.

control the most modern particle accelerators. All the motors and power generators, all the obedient electrical movers are things magnetical—descendants of "the dark stone."

Perhaps the earliest of the ancient scholars to study the attractive forces of Nature was Thales from Miletus, a Greek colony in Asia Minor on the coast of the Aegean Sea not far from Magnesia (in what is now Turkey). It was the sixth century B.C., the time of Pythagoras, Confucius, Buddha, Zoroaster, and the biblical Babylonian Nebuchadnezzar. Thales, discoverer of the wonders of rubbed amber (Section 10.1), was also the earliest to teach about the lodestone's power to attract iron. By then iron and steel were in fairly common use and magnetite was a well-known ore. That was the first step, the first simple observation of a nongravitational interaction—*a magnet attracts iron*—although at the time there was nothing simple about it.

Almost two hundred years later, the renowned sage Socrates dangled a long chain of many separate soft iron rings clinging one to the other from beneath a lodestone. That's a game all of us have no doubt played with a good magnet and a box of paper clips or nails. Socrates' student Plato (who was in turn Aristotle's teacher) quotes the master as remarking, "This stone not only attracts iron rings but also imparts to them a similar power whereby they attract other rings." Just as a charged comb can induce charge on a scrap of paper, a magnet can magnetize a nearby piece of iron. The process, called **magnetic induction,** is the means by which the second paper clip sticks to the first and so on down the "chain"—each clip becoming a magnet.

There are many different variations on the basic "chunk" of iron, and they all behave somewhat differently. Cast iron is a hard, brittle material rich in carbon (from 2% to about 7%) whereas pure iron is actually rather soft. Between the two, in a sense, is *steel*, the alloy of iron and less than 2 percent carbon (with dashes of things like nickel, manganese, chromium, etc., added as desired). Soft iron retains its induced magnetized condition only so long as the inducer (the stone) is kept nearby. When it is removed, the specimen quickly demagnetizes, just as charged paper quickly depolarizes. The same is true of low-carbon soft steel, the stuff of paper clips and common nails. The clips and nails become magnets, to be sure, but merely temporary magnets.

In contrast, a piece of hardened steel, once magnetized, retains much of its power, and we speak of it as a **permanent magnet** although that's a slight exaggeration. Lucretius the Roman atomist came close to that realization. He, too, made chains of magnetized finger rings, in particular the ones sold as souvenirs by the priests on the island of Samothrace. These rings may have been made of hard steel because he commented, "It also happens that at times iron is repelled by this stone." The key word here is *repelled*, and that certainly could have happened if the rings had been permanently magnetized. As we will see in this chapter, magnets, depend-

19.1 MAGNETICS

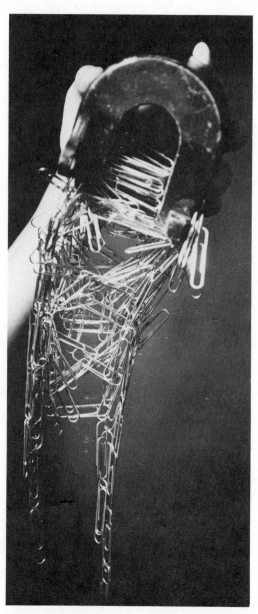

My son Jamey, a magnet, and a box of paper clips.

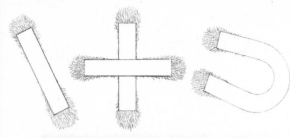

The poles of a few differently shaped magnets.

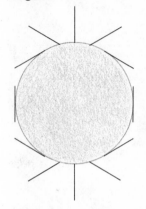

Several small steel needles attracted to the surface of a spherical magnet.

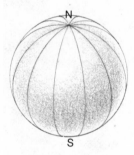

ing on their relative orientation, can either attract or repel each other just as charges can.

The Chinese were well versed in the art of making permanent magnets by the beginning of the second century A.D. and very likely had learned to do so long before. A manuscript of that period suggests stroking an iron rod or needle from end to end along a lodestone, repeatedly and *always in the same direction.* Try it with a magnet and a scissor or screwdriver made of hard tool steel.

By the Middle Ages the study of magnetism had progressed very little beyond those few humble observations of the Greeks and Romans. The only practical application was the mariner's compass, which was in very limited use in the West sometime at the end of the twelfth century.

Poles

The year 1269 has a special significance in the development of the physics of magnetism. On a hot summer's day in that year a French engineer, one Pierre de Maricourt—alias Petrus Peregrinus, Peter Peregrine, Peter the Pilgrim—sat down to write a long letter to a friend. Peter, whose assumed name suggests the credentials of a returned Crusader, was whiling away time in the trenches with an army under Charles of Anjou. They were laying sluggish siege to a city in southern Italy, and there was plenty of time for writing. As one of the very few medieval experimentalists, Peregrine was both an inspiration to his famous friend Roger Bacon and a forerunner in the use of the empirical approach to science as it would be practiced by Galileo centuries later. The young Crusader set about describing his extensive researches on magnetism, and the result was a rather remarkable discourse, especially considering that it was written 700 years ago.

Peregrine was probably the first to introduce the concept of the **magnetic pole.** He had a lodestone rounded into a sphere, and he found that a needle placed anywhere on its surface would doggedly align itself in a particular way. By drawing lines on the stone in the various directions assumed by the needle, he found that they all crossed at two opposing points, just as "all the meridian circles [lines of longitude] of the Earth meet in the two opposite poles of the world." Furthermore, he observed that if a small piece of the needle was broken off and placed in contact with the stone, it would stand straight upright at and only at the two poles. Peregrine was in effect probing the field of the magnet as Faraday would do six centuries later, when that vision had matured.

Most magnets have two poles, two places where the force field is clearly strongest, clearly concentrated. Just sprinkle iron filings over a magnet. The places where the iron dust is most attracted, where it clusters most, mark the poles. A straight bar magnet is the simplest two-pole configuration (sometimes called a *dipole*). It is possible for a magnet to have more than two poles. Indeed, some modern flexible magnets are made in long strips with hundreds of pairs of poles. Certainly two bar magnets could be welded into a four-pole cross, and a bar magnet could be cut halfway down lengthwise and bent into a three-pole Y.

Peregrine next put his magnet in a small wooden bowl set afloat in a large vessel of water. Around it spun, bowl and all, swinging into persistent alignment, the north-seeking or **north pole** always pointing northward, the south-seeking or **south pole** always pointing southward. This, of course, was a compass, much like the crude floating needle versions already in use, but our pilgrim was the first to recognize (or should I say "create") the significance of the poles in the scheme.

Armed with another magnet, whose poles had likewise been determined and marked, he approached his floating stone. When the north pole of one was brought near the south pole of the other, the little boat lunged toward the hand-held stone, but when either two north poles or two south poles were positioned near each other, the little boat fled. Peregrine had discovered the characteristic behavior of all magnets, which can be summarized in the statement that **like magnetic poles repel and unlike magnetic poles attract.** This he confirmed as well for bar magnets he had made of iron.

Naturally, Peregrine tried to isolate a single **monopole,** a chunk of magnet that was simply and only north polar or south polar. And what better way to do that than to split a magnet in two? Surprise! No matter how one breaks a magnet, the fragments are always bipolar. *The monopole cannot be isolated.* It's as if a magnet were composed of a succession of microscopic bar magnets with opposite poles touching and neutralizing each other everywhere but at the naked ends. When the magnet is split the appropriate poles appear as if the rupture exposed rather than created them.

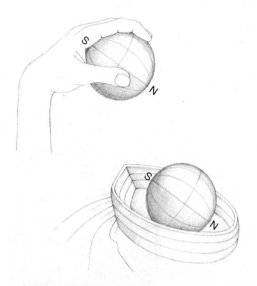

Peregrine's experiment with the forces between magnetic poles.

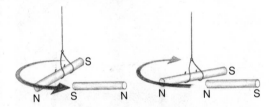

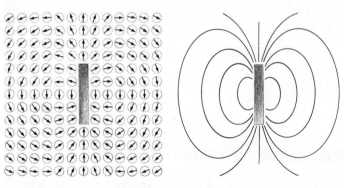

The force field around a bar magnet, as revealed by an array of small compasses.

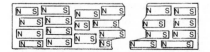

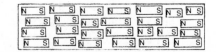

The fragments of a magnet always have two poles, as if the magnet were composed of tiny bipolar units, tiny magnets (dipoles).

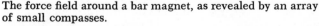

Peregrine's principal error befitted his times: He saw the compass needle pointing not to a spot on our little magnetized dirt ball but to the very pole of the heavens, to the pivot of the Aristotelian-Ptolemaic universe out on the celestial sphere.

The Earth's Magnetism

Iron filings lining up in the vicinity of a small bar magnet.

The Chinese of the eleventh century had long been aware of the fact that the compass needle does not align itself everywhere along a true north-south direction. Yet it was none other than the intrepid "Admiral of the Ocean Sea," Christopher Columbus, who on his first voyage in 1492 had the unhappy surprise of finding out firsthand just how much deviation there is. The difference between the compass reading and true north as determined by the stars, known as **declination,** gave Columbus a few bad moments, as attested to in his journal. It's actually far more common over the Earth for the compass needle to point in some direction other than true north. For example, a compass in Chicago does indeed align itself almost due north, but one in Los Angeles tilts about 15° to the east, and one in New York leans almost 15° to the west of north.

Another rather surprising observation was made about fifty years after Columbus's voyage by the Vicar Hartmann although it became common knowledge only when it was rediscovered by Robert Norman, an English navigator and compass maker. Norman's book of 1581, one of the first on Earth's magnetism, revealed that a compass needle free to swing in *any* direction actually pointed *downward* as well as north. The north-seeking end of the compass in his London shop *dipped* 71 degrees 50 minutes below the horizon. Of course, Norman could have concluded that its tail end was jutting up to the heavens, but this was the Renaissance, not the Middle Ages. With his feet firmly planted on the ground, he decided that the needle was thrust downward toward a point far below the Earth's surface somewhere to the north.

Actually this **magnetic dip,** as it's called, is more or less zero near the equator and increases as one moves north. Below the equator the south-seeking pole of the compass points increasingly downward toward the ground as one moves further south.

Now all this should be somewhat reminiscent of Peregrine's needle standing on end from a spherical lodestone. Indeed, it was not long before the similarity was explored with superb skill by Sir William Gilbert (p. 382), a tremendous figure in English science. Gilbert was a renowned medical practitioner, a devoted experimentalist, and perhaps the earliest important exponent of Copernicanism in the realm. In 1601 he became court physician to Elizabeth I. The year before (the year Bruno was put to the stake), Sir William published his famous treatise *De Magnete* (*Concerning the Magnet*).

In it he likened the Earth to a large spherical lodestone, maintaining that the compass needle was drawn to the planet's poles, not the heavens. *It was simply a matter of one magnet pulling on another.* He pointed out that the force between magnetic poles in general grew weaker with the increase in their separation. That much is immediately evident to anyone who has held two strong magnets anywhere near each other. Indeed, Coulomb (Section 17.2) almost two centuries later found the force to vary *with the inverse square of the separation,* just as do the gravitational and electrical forces.

Gilbert probed the region surrounding a magnet, the region he called the "orbe of virtue," with a little compass. He concluded that "Rays of magnetick virtue spread out in every direction in an orbe." The statement calls up the ancient image of rays of light streaming from a source, and it seems almost identical in spirit to Faraday's *lines of force of the magnetic field.* One needs only to connect the little compass arrows with smooth arcs to transform the imagery from an "orbe of virtue" into a "magnetic field." Gilbert compared the compass indications (the field directions) in the region around a spherical magnet and found them to agree nicely with those of the planet, underscoring his main contention that "the Earth itself is a great magnet."

Forty-four years later, Descartes carried the mapping process begun by Peregrine and elaborated by Gilbert one step further. He sprinkled iron filings around the magnet, and they aligned themselves like minute compass needles to form continuous curved filaments, which even more potently suggest lines of force. Indeed, Faraday would be inspired by the same marvelous patterns a hundred years later.

As for geomagnetism, however, matters were not nearly as simple as Gilbert had thought. The Earth does behave as if it contained, as its major component, a relatively short bar magnet embedded at its center and tilted over at about 11.5° (along the 70° W meridian). But the field as a whole is a good deal more complex than that single *dipole.* For one thing, it changes direction somewhat. There's even evidence that it has reversed itself perhaps a dozen times or more in the past several million years and may have done so just 30,000 years ago. At present the terrestrial magnet is still in a weakening phase, having dropped in strength about five percent over the last century. It's possible that we are now experiencing that part of the cycle in which the field decreases to zero prior to reversing.

As a further complication, the *dip poles,* where the field points straight up and down, not only wander but are about 500 miles from the *geomagnetic poles,* the poles of Gilbert's magnet toward which a compass leads. Furthermore, neither the geomagnetic nor the dip poles are actually at the geographic poles, as fixed by the Earth's spin axis. The final "irony" is linguistic: The Earth's northern magnetic pole is actually a *south* pole, and the southern magnetic pole is—you guessed it—a *north* pole. That's why the north pole of a compass needle is attracted toward the geographical north. There's a *south* magnetic pole up there tugging on it! (Ah, "What's in a name?")

It is likely that the geomagnetic field arises from electric currents circulating in the molten iron outer core within the planet, but that *dynamo theory* is still highly speculative. We really don't know for sure what causes the Earth's magnetic field.

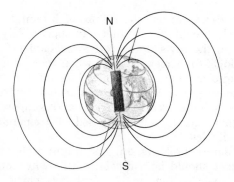

The Earth's magnetic influence resembles that of a tilted bar magnet.

Having made many and divers compasses . . . I found continuallie that after I had touched the yrons with the stone, that presentlie the north point thereof wouilde bend or decline downwards under the horizon in some quantitie.

ROBERT NORMAN

The Newe Attractive (1581)

As we saw earlier, the failure of attempts to isolate a monopole by naïvely breaking a magnet into successively smaller pieces is a good clue to the invisible operative mechanism. The logic is the same as the logic the Greeks applied when they imagined they could cut matter into smaller and smaller bits until they reached the atom, the ultimate grain. Well, one can also very fruitfully imagine an ultimate submicroscopic magnet and suppose that all ordinary-sized magnets are merely composed of tremendous numbers of aligned minidipoles.

Before pressing on, one might well ask whether the ultimate magnet should be a dipole at all—why not two separate entities, a south-pole particle and a north-pole particle, just as there are electrons and protons? In 1931 P. A. M. Dirac presented a lovely theoretical argument for just that—pairs of monopoles. Besides establishing a symmetry between electrical $(+, -)$ and magnetic (N, S) particles, the existence of monopoles would also provide a ready explanation of the quantization of charge into lumps equal to the electron's charge. These monopoles would be fascinating little devils, exerting an attraction between their opposite numbers almost 5000 times greater than the attraction that exists between the electron and proton.

The monopole would indeed be a prize if someone came upon it, but alas, nobody has found one, despite the greatest effort. Monopoles have been sought in deep-sea sediments and in moon rocks, in experiments with cosmic rays and giant particle accelerators, but still they remain elusive. Every once in a while (most recently in 1975) a great din of excitement whirls around a scientist who has announced that monopole leavings have at last been found. Unhappily all such sightings have been repudiated or at best gone unconfirmed. But of course that incredible shyness may spring from the fact that monopoles just do not exist and never have existed. So there we are—on the most profound and fundamental level, wonderfully tantalized and confounded once again!

As we will see, charges in motion produce magnetic force, and currents following in circular paths generate magnetic dipoles. Thus it now seems highly likely that *each atomic electron is itself the primal current, the ultimate submicroscopic magnet*. According to our contemporary understanding, electrons consistently behave in a way that suggests they are perpetually spinning, each about its own axis. This whirling charge is believed to constitute a current and therefore a submicroscopic magnetic dipole.

The orbitlike motion of an electron about the nucleus of an atom also constitutes a current and also produces a magnetic field but one far weaker than the spin field. Together these two electron mechanisms (spin and orbit) account for the magnetic behavior of all the various forms of matter, *for there most certainly is no such thing as a nonmagnetic material*.

19.2 A BIT OF THEORY

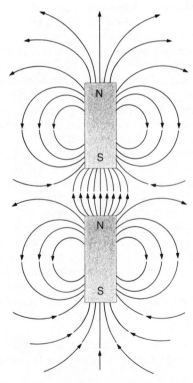

The field produced by two magnets with unlike poles facing each other.

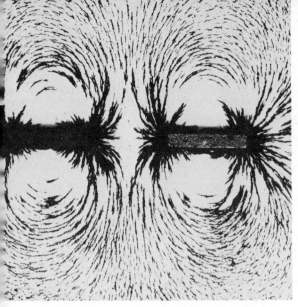

The field produced by two magnets with like poles facing each other.

Steel nails align themselves to reveal the field configuration surrounding the opposite poles of two extremely powerful superconducting magnets. The frost around the necks of the stainless steel Dewar flasks is a clue to the fact that they're operating at liquid helium temperature (around 4.2 K or about −452°F).

Many substances, such as water, glass, copper, lead, salt, sulfur, rubber, diamond, graphite, liquid nitrogen, mercury, bismuth, hydrogen, silver, quartz, etc., are very weakly magnetic in what may at first seem a startling way. They are actually *repelled* by the pole of a strong magnet, pushed out of the region where the field is strongest toward a region where it is weakest. Faraday was the first to observe this peculiar phenomenon in 1845, and it's still amusing to swing a small piece of glass dangling on a thread into a field 100,000 times stronger than the Earth's, only to have the glass pop right out and stay out at some gravity-defying angle.*

All these substances (which include many inorganic compounds and most organic compounds) are said to be **diamagnetic,** and even for the strongest of the lot, bismuth, the effect is relatively faint. Diamagnetism is believed to arise from the orbitlike motion of the electrons and so is present in all substances although it is apparent only when not swamped out by the stronger spin effects. Usually atomic electrons inexplicably arrange in pairs, spinning in opposite directions and effectively canceling each other's magnetic fields. When that happens almost completely, the underlying diamagnetic response shines through, despite its weakness.

More often, however, not all the spins are compensated for by opposites. Then an atom may have a resultant spin-magnetic field. In a substance made up of countless atoms, these submicroscopic dipoles en masse will produce only a feeble magnetic response because the ordinary thermal agitation of the atoms will tend to keep them disoriented. Such substances, called **paramagnetic,** include aluminum, oxygen, sodium, platinum, uranium, etc. All will be drawn in toward a magnetic pole (as some of their dipoles struggle into alignment) but only weakly.

The last of the three major classes of substances we will consider is known as **ferromagnetic.** They are what most lay people recognize as being "magnetic." The list, to which newly concocted materials are always being added, includes magnetite, a troop of alloys like steel and Alnico, and five elements—iron, cobalt, nickel, gadolinium, and dysprosium. These are materials that on the average have a greater number of uncanceled, unpaired, spin dipoles per atom. Just as important, *these dipoles enter into large-scale cooperative alignments.* In other words, the uncompensated spin dipoles of each atom interact strongly with the dipoles of adjacent atoms, locking together in a parallel orientation that tends to persist even at room temperatures.† Such substances are strongly attracted to the poles of a magnet and are therefore of the greatest interest.

---

* Humans are mostly water and should actually feel this sort of repulsion in a nonuniform magnetic field. But that would require a tremendously strong field, and very few people have ever experienced it.

† That powerful coupling is a quantum-mechanical electrostatic effect, first explained by Heisenberg in 1928. Still, why some elements have uncompensated spin dipoles in the first place is not completely known.

The magnetic-domain structure of a poly-crystalline sample of cobalt samarium.

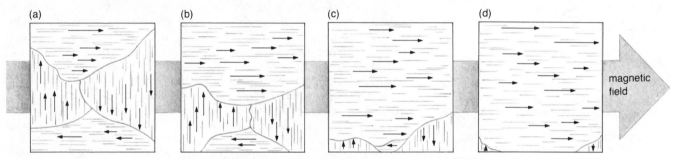

(a) (b) (c) (d)

magnetic field

The growth of domains aligned in the direction of an applied magnetic field.

Domains

The formidable magnetic power of ferromagnetic substances arises out of a preexisting internal order, an order that occurs naturally without any human intervention. Such materials are composed of very many microscopic **domains,** islands of marvelous order, throughout each of which tremendous numbers of spin dipoles are all aligned parallel to one another. Thus each domain is itself a tiny (roughly 5×10^{-5}m across) magnet, which can be viewed and photographed under a microscope although the individual dipoles, of course, are too small to see. In an unmagnetized specimen the orientations of these domains are random, and their fields cancel.

When a ferromagnetic sample is brought into the field of a magnet, its domain structure can be drastically affected in two different ways. The low-energy process (which occurs, if at all, even with fairly weak fields) is one in which *the many domains that happen to be already aligned with the applied field grow at the expense of the domains that are misaligned at the start.* This is just what occurs when soft iron is placed in a field and becomes magnetized by

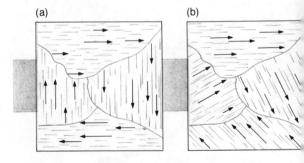

(a) (b)

The reorientation of domains, bringing about an alignment with the applied field.

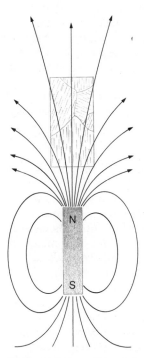

Magnetizing an iron rod.

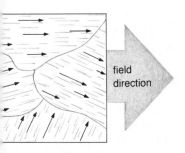

What led me more or less directly to the special theory of relativity was the conviction that the electromotive force acting on a body in motion in a magnetic field was nothing else but an electric field.

ALBERT EINSTEIN

induction. The aligned domains grow; the iron as a whole becomes a magnet and is drawn toward the inducer magnet. When the external field is removed, the domains more or less spring back to their original configuration, and the iron spontaneously demagnetizes. All this can be watched as it happens, so we are confident in our understanding of the microscopic mechanism.

The other process, which requires higher fields, results in the *irreversible reorientation of the domains*. All the electron dipoles coupled together within each domain can literally be rotated into alignment with the applied field—just like a compass needle. This is the mechanism that prevails in substances that become permanently magnetized. Domains in materials with irregular internal structures, such as steel, can not easily change shape. They rotate instead, and having once done so, they tend to stay put. When a steel knife blade (some stainless steel is nonmagnetic, so it won't work) is stroked with a magnet, the effect is simply to rotate the domains into that direction, to "comb" them into alignment.

We can grind up a ferromagnetic material into tiny powdery pieces even smaller than the size of a domain and suspend the stuff in some liquid plastic. Before it hardens, each and every grain (which consists of a single domain) can then be aligned by an applied field. What results is a flexible, castable, machinable, strong, permanent magnet. You'll find such magnets beneath the grommets on most modern refrigerator doors.

The same sort of structure exists in the most powerful permanent magnets, such as those of Alnico 5 (a blend of aluminum, nickel, iron, cobalt, and copper). When slowly cooled in a magnetic field, it forms a structure of tiny ferromagnetic grains distributed throughout a nonmagnetic alloy. Each grain consists entirely of one domain rotated into alignment and so permanently magnetized, frozen into place.

Anything that will reorient any of the domains will thereby diminish the specimen's "permanent" field. Banging on a magnet with a hammer will do just that by knocking regions out of alignment. Two magnets placed with like poles near each other will slowly repel domains and so weaken each other. And of course, if the temperature is made high enough, the vibrating atoms will jostle themselves out of alignment and disrupt the whole business. Pierre Curie was the first to realize that there was a limiting temperature (now called the **Curie point**) for each substance above which ferromagnetism vanishes. For iron it is 770°C, and as with most materials, this temperature is well below the melting point, which for iron is 1539°C. By contrast, gadolinium is ferromagnetic only below 16°C and dysprosium below a frigid −188°C.

Since the Earth's crust hits a temperature of about 800°C at depths of only 35 kilometers or so, it's rather unlikely that anyone will ever find the mother of all permanent magnets buried at the center of the planet.

19.3 ELECTROMAGNETICS

A charge, whether at rest or in motion, always has associated with it an electric force field. That's nothing more than saying that charges *always* interact with one another electrically. Yet when that same charge is moving, it also possesses a magnetic force field. The key point is that *magnetism arises out of motion, and motion is relative.* A charge possesses an electric field impressing forces on other charges all the time, but it may or may not seem to have a magnetic field, depending on the relative motion of the observer. A beam of electrons flashing past is a current and as such exerts both electric and magnetic forces. And yet if we run along with the flow, essentially freezing it, the magnetic force vanishes. It must, because there's no longer a current with respect to us, no longer a source of the magnetic field.

What comes out of Einstein's special theory is that magnetism is actually a facet of electricity, a relativistic aspect born of the effects of motion. The electrical interaction is enormously powerful, and so even a slight modification of it, arising from relative motion, can appear as a substantial force in its own right. Until this century that force was thought to be a totally separate affair—the magnetic interaction. Now electricity and magnetism are welded, unified into *electromagnetism* by one of the centralmost conceptions of all, *motion.*

Basic Interactions

Volta's invention of the battery (1800) began a whole new era of sustained, flowing electricity, an era that has continued and continues at this very moment. Even at the beginning, researchers attempted to find some relationship between electricity and magnetism, despite the fact that many held that static electricity and battery-delivered or "voltaic" electricity (called galvanism during that period) were totally different phenomena. That may seem a bit weird to us, but consider their vastly different origins: rubbed glass as opposed to brine-soaked cardboard.

Evidence that some kind of relationship existed between electrical and magnetic phenomena had been at hand for decades. For example, the 1735 volume of the *Philosophical Transactions of the Royal Society of London* carried a paper entitled "Of an Extraordinary Effect of Lightning in Communicating Magnetism." Lightning struck a tradesman's house, blasting apart a box full of knives and forks, hurling them "all over the room . . . but what was most remarkable" was that they were all strongly magnetized afterwards!

During the two decades following Volta's triumph, all sorts of experiments were tried in order to tie the two effects together. Some were baseless, like the one in which a voltaic battery was floated in a large vat of water to see if it would align itself north-south. The efforts of other workers were more on target, but the connections they came up with were too subtle, the links too ambiguous, or the work too inconsistent to make much of an effect, either on them or on the community of scientists. When the masterstroke of simplicity and clarity finally came, it came like a bolt from the blue on July

An 1840s daguerreotype of H. C. Oersted.

21, 1820, and within weeks the incredible news had spread through the world's scientific community.

Hans Christian Oersted, professor of physics at Copenhagen University, while delivering a lecture to some advanced students, by chance placed a wire leading to a voltaic pile *parallel* to and *above* a compass that happened to be on the table along with other paraphernalia. When the circuit was closed, a startled Oersted noticed the needle swing around perpendicular to the current-carrying wire as if gripped by a powerful magnet. Soon after he announced his finding, physicists all over Europe and America were duplicating the feat in a burst of activity. It was determined in rapid succession that the magnetic force experienced by a compass needle in the vicinity of a current-carrying straight wire acted in planes at right angles to the current along a series of concentric circles. Moreover, such an energized wire attracted iron filings and magnetized bars of iron and steel, just as an ordinary magnet did.

Nowadays we say that *a current-carrying wire generates a circular (or more accurately, a cylindrical)* **magnetic field** $(\mathcal{B})$ *in the space surrounding it, a field that is constant at any given perpendicular distance from the wire and gets weaker as that distance increases.* That's why Oersted's compass rotated until it was perpendicular to the wire. It aligned with the $\mathcal{B}$ field, which was everywhere at right angles to the current. *The direction of that field (by convention, the direction in which the north-seeking pole of a compass needle points) is the direction in which the fingers on the right hand curl when the thumb is pointing along the current (by convention, the direction of flow of positive charge).* Just that peculiar perpendicular character of the force was itself probably enough reason for it to have taken twenty years to detect. This was the first force ever discovered to act in a direction other than along a line connecting the interacting objects.

Oersted's demonstration of the magnetic effect of a current-carrying wire on a compass.

The circular magnetic field surrounding a current-carrying wire, as revealed in an iron-filing pattern.

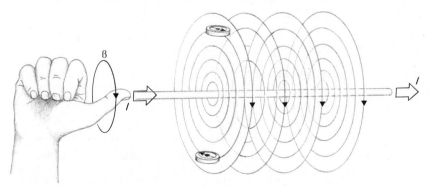

When the thumb on the right hand points in the direction of the current (I), the fingers curl in the direction of the circular magnetic field $(\mathcal{B})$.

TALKING ABOUT AMPÈRE
He further deduced from this analogy the consequence that the attractive and repulsive properties of magnets depend on electric currents which circulate about the molecules of iron and steel.

D. F. J. ARAGO (1820)
French physicist

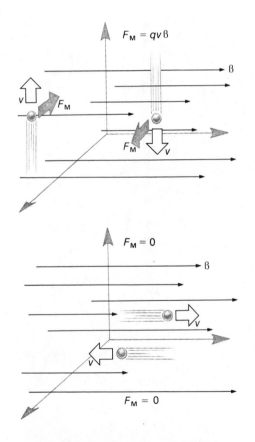

The relationship between speed and force for a positive charge moving either perpendicular or parallel to a field.

To establish more firmly that the electric current was not just somehow affecting the compass but was indeed a magnet in its own right, André Marie Ampère did away with the iron altogether. He passed currents through two parallel wires, one of which was suspended so that it could swing either away from or toward the other. *When the currents were in the same direction, the wires attracted each other; when they were oppositely directed, the force was repulsive and the wires separated. Since both wires were electrically neutral, the force could not have been electrostatic and so had to be magnetic.** By the way, the present-day unit of current, the *ampere*, is defined by exactly this interaction as measured in practice on an exceedingly delicate parallel-wire balance.

Since currents exert forces on magnets, as Oersted discovered, it follows from Newton's third law that magnets ought to exert forces on currents. Ampère's parallel-wire experiment (which, incidentally, he performed within one week of hearing of Oersted's discovery) is an example: There the magnetic field of one current exercised a force on the other, and vice versa. In actuality, magnetic forces act only on moving charges (the electrons traveling within the conductor), which in turn transmit the force to the wire as a whole via collisions. A beam of free-flying charges will certainly experience a magnetic force without the presence of any conductor.

The simplest case to deal with is the one in which the charge (q) moves (with a speed v) parallel to $\mathcal{B}$, whereupon the magnetic force is zero. As the angle between the direction of motion and $\mathcal{B}$ increases, the force F_M increases until it reaches its maximum value when the angle is 90°. With the motion perpendicular to the field, the force is given by

$$F_M = qv\mathcal{B}.$$

Note that when the charge is negative rather than positive, the force is negative and therefore oppositely directed.

The internationally accepted unit of magnetic field is called the *tesla* (abbreviated T) and as suggested by the equation

$$1 \text{ tesla} = \frac{1 \text{ newton}}{\text{coulomb (meter/second)}},$$

it is defined in terms of force over charge times speed. A pith ball carrying 1 coulomb of charge sailing along at 1 meter/second through and at right angles to a 1-tesla magnetic field will experience a force of 1 newton.

---

* Using the Lorentz-FitzGerald contraction (Section 6.4) of relativity, one can show that concentrations of charge will be "seen" by charges within each of the wires, and the resultant force between the conductors therefore appears to them as an ordinary electrical interaction. Only to someone standing outside at rest, where both wires are neutral, is the presence of a magnetic force a necessary hypothesis. Of course, no one in the early nineteenth century could possibly have known that.

The tesla is actually a rather hefty unit. For example, the Earth's field at sea level, which does vary from place to place, is roughly 0.00003 tesla. By comparison, an ordinary bar magnet musters 0.2 to 0.3 tesla and a good strong permanent magnet about 1 tesla. A very large laboratory electromagnet can hit up to 10 teslas routinely. But values much greater than that take a good bit of doing. Indeed, the highest steady field attained was 22.5 teslas, and the most powerful field ever achieved in the laboratory—only for an instant in an explosive device—was 1000 teslas.

Recall that when J. J. Thomson discovered the electron at the end of the century, he bent the cathode-ray beam with a magnetic field (Section 10.1). For that matter, a pair of current-carrying coils (called a *yoke*) wrapped around the neck of your TV tube generates a magnetic field that steers the electron beam as it "paints" the picture on the screen. Controlled-fusion devices (Section 16.2) like the tokamak, which confine plasma beams, do that bit of juggling with powerful magnetic fields. The great accelerators (the cyclotrons, bevatrons, and synchrotrons) that put out high-speed particle beams all bend the swarming charged specks into circular paths using $\mathcal{B}$-fields.

It's even possible to store *neutrons*, to have them revolve around a circular track confined by complex magnetic fields. Now there's another of those little mysteries—the neutron behaves like a magnetic dipole, but it has no *total* charge. It seems to be a complex spinning structure with a distribution of as much positive as negative charge—hardly the description of an *elementary* particle!

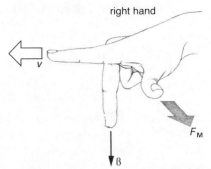

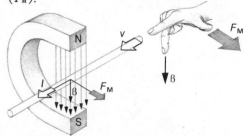

A positive charge moving at a speed (v) perpendicular to a magnetic field ($\mathcal{B}$) experiences a force perpendicular to both (F_M).

The force on a current-carrying wire in a magnetic field.

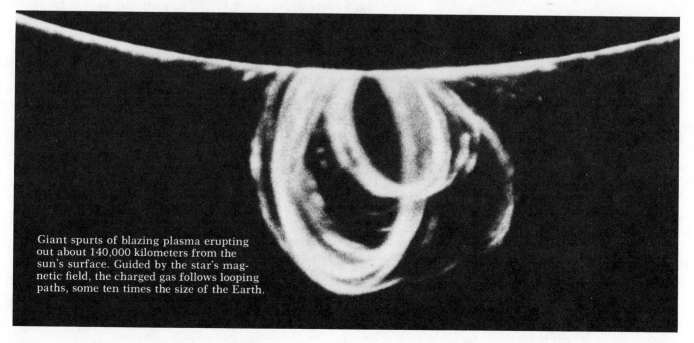

Giant spurts of blazing plasma erupting out about 140,000 kilometers from the sun's surface. Guided by the star's magnetic field, the charged gas follows looping paths, some ten times the size of the Earth.

Dipoles and Telegraph Poles

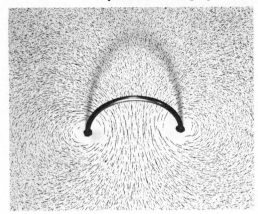

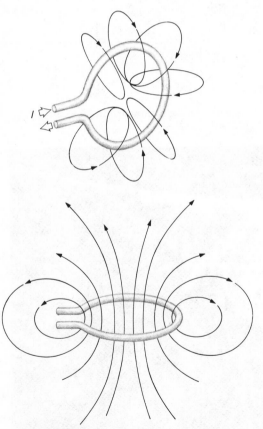

The magnetic field of a single current loop.

During that first rush of excitement Ampère had yet another lovely thought. Since a straight current-carrying wire was surrounded by concentric rings of magnetic force, bending the conductor into a circular loop should concentrate that force, increasing it within the loop. Using field imagery, we can picture the lines of force made less dense outside the turn of wire and compressed, crowded together within the region encompassed by the circular flow of charge. That field rises out of the circle of current just as the field of a bar magnet rises out of its north pole. In point of fact, the *current-circle constitutes a magnetic dipole*. Recognizing as much, Ampère rightly suggested that currents were the basic cause of all magnetism, including the Earth's.

If one loop intensified the magnetic force, then many parallel loops, many successive turns of wire should increase it even more. The resulting coil, or helix, when carrying a current has a relatively strong internal field, where the lines of force from each individual turn of wire reinforce one another.

Sometime around 1825, W. Sturgeon wrapped eighteen turns of bare copper wire around a varnished iron bar and sent a current through the coil; in so doing he created the first powerful **electromagnet.** The axial field set up by the current aligned the domains within the iron to produce a combined magnetic field of unprecedented strength. That is not quite the way anyone at the time would have explained it, but the enhancement due to the iron was obvious enough. The great unsung American genius Joseph Henry heard about the feat in 1829 and immediately set about to better it. Story has it that he tore apart one of his wife's petticoats so that he could insulate his wires with the silk stripping. By wrapping many turns of insulated wire in many layers around an iron core, he could enhance the field while keeping the current fairly low. In 1831 Henry produced a modest-sized device powered by an ordinary battery that could lift more than a ton of iron! When the current was interrupted, the field vanished, the core demagnetized, and the load was released.

That same year, in a third-floor classroom at Albany Academy in New York where he taught math and physics, Henry strung more than a mile of wire leading from a battery around the room to an electromagnet and back. Between its poles he positioned one end of a pivoted permanent bar magnet and near the other end placed a gong. Closing the circuit activated the coil, swung the bar magnet, and clanged the gong. That crude machine was the first electromagnetic telegraph. Henry believed in free access to the gifts of science and consequently never patented any of his inventions. Nor was he grudging in the extensive help he gave to a curious and determined fellow he happened to meet, a painter named Samuel Finley Breese Morse.

The electromagnet in a wide range of sizes and shapes is everywhere nowadays: banging chimes at the front door, switching cycles on the washing machine, clattering in telegraphs and telephones,

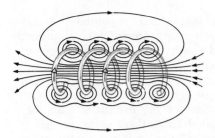

The magnetic field of a current-carrying coil.

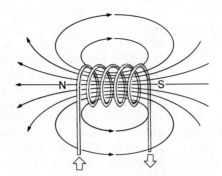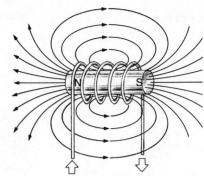

The field of a current-carrying coil with and without a ferromagnetic core.

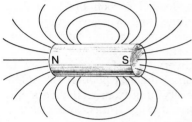

Comparison of the magnetic fields of a bar magnet and a current-carrying coil.

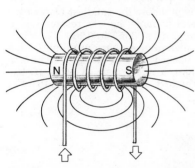

driving the speakers in your hi-fi, activating mechanical devices on distant space probes, and of course, ruling the world's junkyards with an iron will.

Suppose we place a little coil (an electromagnet), free to pivot, between two permanent magnetic poles (as in the figure). Depending on which way the current flows through it, the electromagnet will take on a certain polarity and swing either clockwise or counterclockwise, aligning itself with the fixed field. If we mount a pointer on the coil and attach a little unstretched spring to it to keep it at a fixed position when the current is zero, we have a **galvanometer**. That has been the basic electrical meter mechanism for ammeters and voltmeters and most other kinds of electrical meter (e.g., resistance, light, tuning, pressure, etc.) for the past hundred years or more although the electronic digital read-out will probably replace it in time.

A junkyard electromagnet.

A coil free to pivot in a magnetic field.

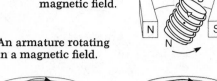

An armature rotating in a magnetic field.

A galvanometer mechanism.

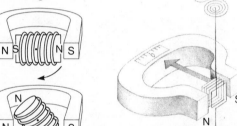

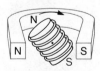

A simple electric motor.

With the first arrangement—no spring and no pointer—the pivoted electromagnet, or *armature,* will swing around into alignment with the fixed magnets; north and south poles will come together. Actually, the electromagnet will slightly overshoot the mark and swing back, oscillating for a moment or two about the equilibrium position. Note that if the polarity of the armature is inverted (by reversing the current through it) just at the instant it sails past alignment, it will be violently rotated from there clear around to where it started. If it's reversed again just as it passes the poles, around it will go for a second time. These current reversals are easily accomplished automatically whenever the armature rotates through 180° by feeding the current in through a split-ring arrangement known as a *commutator.* Soft spring-loaded carbon blocks, called *brushes,* usually press against the rings, and whenever they pass over a split from one half of the commutator to the other, the current reverses direction through the armature. Of course, this continuously whirling machine, this converter of electrical energy to kinetic energy, is a rudimentary version of the ever popular DC motor —the great mover that has powered everything from trolley cars and golf carts to auto starters, cocktail stirrers, toy trains, and portable vibrators.*

---

* W. H. McLellan constructed what is probably the smallest motor ever made. In the pursuit of that peculiar claim to fame, he was responding to a public challenge made by Richard Feynman (p. 365) in 1960. The challenge was to build an electric motor smaller than a quarter-millionth of a cubic inch—a dare Feynman personally backed with $1000. McLellan ultimately collected, but he had to bring along a microscope to prove it. His one-millionth of a horsepower device weighed only a half-millionth of a pound.

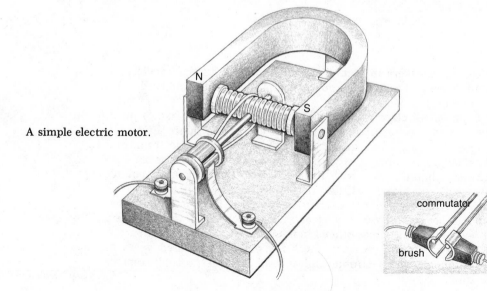

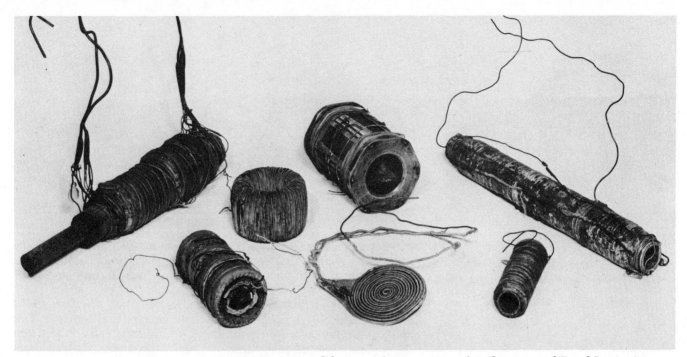

Several of Faraday's original coils preserved in his magnet laboratory (now a museum) in basement of Royal Institution.

"Convert magnetism into electricity"—that was the brief remark Faraday jotted in his notebook in 1822, the challenge he set himself with an easy confidence that made it seem so reasonable. If electricity could produce magnetism, surely magnetism could produce electricity. Not until August of 1831 did he do just that, but the results he got in that memorable experiment were really quite unanticipated.

Being familiar with Joseph Henry's electromagnets, Faraday decided to similarly concentrate the *magnetic force* within an iron core —even he was not yet talking about *fields* or *lines of force*. He wound two coils around opposite ends of a soft iron ring, one connected to a switch and a battery, the other to a galvanometer. The anticipated steady current was not observed, but the galvanometer needle certainly did jump every time the switch was either opened or closed, only to return immediately to zero. A changing current in one coil had induced a current in the other coil, if only a transient one. Faraday was the first person to report the successful *generation of electricity via magnetism*, a phenomenon called **electromagnetic induction.** And yet Henry, harried by his teaching duties, had actually made the discovery a bit earlier but didn't find the time to publish until a year later.

Faraday went on to explore the phenomenon, to try to figure out what was going on, in a series of clever, probing experiments. He first removed the iron ring, and the effect, though weakened, persisted. Then he removed the left side of the setup (the battery-

19.4 ELECTROMAGNETIC INDUCTION

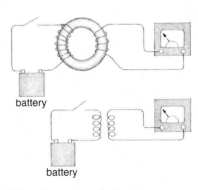

Electromagnetic induction.

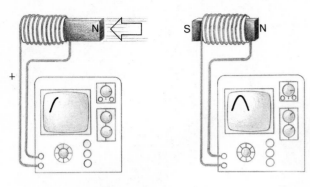
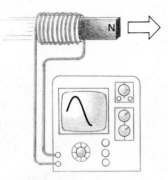
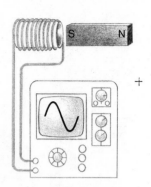

The moving magnet generates a voltage that varies along with the motion.

energized coil), replacing it with a permanent magnet. Whenever the magnet was either thrust into or removed from the remaining coil, the galvanometer indicated a pulse of current. He even waved a coil through the air—through the Earth's magnetism—thereby producing jolts of current with each moving gesture.

The common theme in all these situations was the presence of a conducting loop, a magnetic force field changing with respect to it and a resulting electric current. Faraday ultimately came to visualize the process in terms of his lines of force: *A change in the number of lines of magnetic force passing through a loop is accompanied by an electric force* (and therefore by a potential difference and a current). This revelation, stated in a more modern way—*a magnetic field varying in time produces an electric field*—is one of the crucial pieces to the puzzle of electromagnetic radiation (Section 20.1).

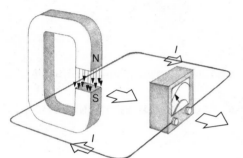

The current induced in a wire moving across a magnetic field.

Faraday's electromagnetic generator.

The Generator

Faraday's experiments on induction led to a remarkable device, the **generator.** When a wire loop moves through a fixed magnetic field, in a direction other than parallel to it, a current will flow in the loop. There are two relativistically equivalent ways to understand what's happening. You can move along with the loop, thereby freezing it motionless with respect to you. Then the number of lines of force threading the loop changes in time. To put it in more modern terms, the magnetic field changes; after all, it's moving by. That in turn produces an electric field, a resulting force on the charges within the conductor, and finally an induced current. Alternatively, you can remain at rest with the $\mathcal{B}$-field and will then see the wire moving with a speed v. If for simplicity the motion is perpendicular to the field, then each charge within the conductor is also moving at v and therefore "feels" a force $F_M = qv\mathcal{B}$ directed along the wire. That in turn causes the mobile charges to move down the wire, and there you are again with a current. In either interpretation a conductor moving in a nonparallel direction with respect to a magnetic field carries an induced current.

In 1831 Michael Faraday devised the world's first electromagnetic generator. It was a seemingly simple thing; a copper disk that could be hand-cranked to rotate between the poles of a permanent magnet. Charges in the metal, set in motion through the field as the disk turned, experienced a force acting radially along the disk. Pushed in that fashion, the mobile charges streamed through the copper, constituting a continuous current. Flowing between the rim and the axis, current persisted all the while the thing rotated. The current output was feeble and the device impractical, but its effect on the world was as far-reaching as any other single invention in all of history.

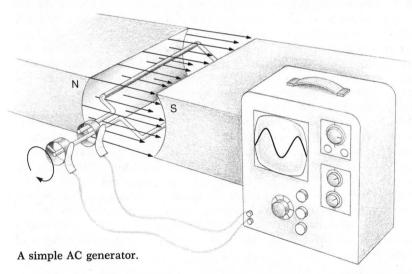

A simple AC generator.

One gets the impression that Faraday had some inkling of its significance from a story that is often told about it. It seems that the Prime Minister once asked Faraday if his research had any practical value. The great electrician is said to have replied prophetically "Why, sir, someday you will be able to tax it."

A simplified version of the modern AC generator is shown in the accompanying figure. A coil, usually of many turns, is rotated in the constant field of a magnet. Brushes rubbing against the two rings carry off the induced current. As the coil rotates, its two long parallel

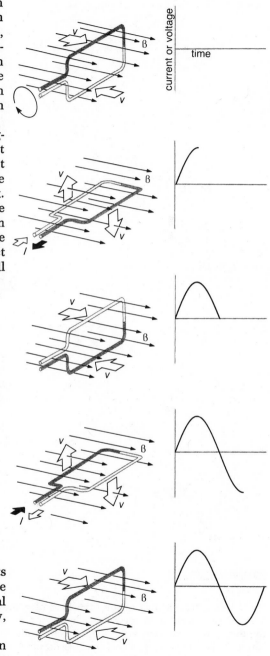

Various stages in the cycle of an AC generator.

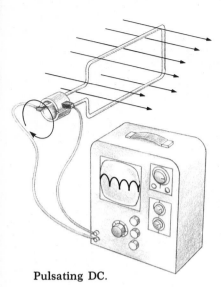

Pulsating DC.

sides will move through the field, charges within each will experience forces (F_M), and a current will flow. At any instant when the coil is vertical, the long wires are moving in a direction that is along the field. Then F_M is zero and the current is therefore zero, a condition that happens twice in each rotation or cycle. By contrast, whenever the loop is horizontal, the two wires are moving perpendicular to the B-field, F_M reaches its maximum value, and the induced current hits a peak. Just as the coil inverts itself with respect to the field once each cycle, the induced current reverses itself correspondingly. The result is the familiar gradual rising and falling of AC, first in one direction, then the other, first positive, then negative.

With a split-ring arrangement like that on the DC motor (p. 450), the current direction can be flipped at will. Thus the negative half of the AC signal can be reversed and made positive. The result is a bumpy direct current, which rises and falls but never goes negative, never reverses its direction. DC generators of this sort were called dynamo-electric machines in the latter part of the 1800s, when they were being developed as practical instruments.

It's curious that the similarity between the dynamo and the DC motor was not recognized until mid-century and not really appreciated until a remarkable accident occured in 1873. Not only do our simple versions of these two devices look alike; they are actually identical. The action of the dynamo is really just the reverse of the action of the DC motor. The dynamo converts mechanical energy into electrical energy; the armature is turned mechanically from the outside, and out comes a current. The motor converts electrical energy into mechanical energy; in comes a current from the outside, and around the armature turns.

There were many dynamos at the 1873 Vienna Exhibition. The large hall was filled with gadgets. One of the Gramme dynamo-electric machines driven by a steam engine was pouring forth electrical power when an unknown workman unwittingly connected the output leads from another dynamo to the energized circuit. In an instant the second device began to groan, to spark and whine and come alive, whirling around at great speed. Transformed by the inrush of current, the dynamo had become a motor. That impromptu mating of machines gave birth to the future as we have come to know it.

EXPERIMENTS

1. Using iron filings to map the field of a magnet is the kind of fun experiment everyone should try. You can make the filings by—you guessed it—filing some iron, the finer the better. If you haven't any soft iron, settle for soft steel, such as a wire coat hanger. Make about a teaspoonful, and put it in a salt shaker. Place a magnet or two on a piece of paper, and simply sprinkle the filings while gently tapping the table.

2. Put a toothpick through a small piece of Styrofoam or cork, and hang it on a wire frame, as shown in the figure. Now pierce the foam with a sewing needle, and balance the whole thing so that the needle is horizontal. Magnetize the needle by stroking it with a magnet, but don't shift its position in the Styrofoam. In addition to pointing north, your hanging compass will also roughly show the local dip in the field. Try waving a magnet near it. Make a floating compass

by sticking a magnetized needle through a piece of cork and sailing it in a cup of water. Check out your home's iron radiators; they should be magnetized north-south by the Earth's field and therefore should deflect the floating compass accordingly.

3. Oersted's experiment is very easy to duplicate (see figure). Just align a compass due north (the floating needle in the previous experiment will work fine). Now stretch a

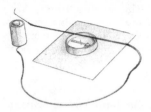

foot or so of any ordinary insulated wire from north to south directly over and about 1/4 inch above the needle. A 1.5-volt D cell attached to both ends will clearly deflect the needle. This is essentially a "short" across the battery, and you'll be drawing in excess of an amp, so just touch the lead to the terminal for an instant. If you want to leave it on for a while, either use a much longer wire or put a flashlight bulb in series with it. Either one will increase the resistance and lower the current. Orient the setup so that the wire's $\mathcal{B}$-field is opposite to the Earth's, and see if you can swing the compass around 180°.

4. You can make a very delicate device capable of responding to just a few milliamps using a compass (or a "studfinder") and some wire. Since it has no scale, it might better be termed a galvanoscope than a galvanometer. In any event, if you have a compass two inches or so in diameter, that will do nicely. If you have a smaller one, place it inside a little box. Now wrap 20 or 30 turns of wire around the compass (or box), spacing the turns of wire enough so that

you can still see the compass between them. A thin insulated wire, such as #28, works well. Try it out with a 1.5-volt D cell. Now you have an instrument that will respond to the milliamp currents put out by the homemade batteries of the previous chapter. I've used mine on a copper-zinc-vinegar 1-volt cell, and it works nicely.

Note: Read through the next two experiments before performing either, so that the coil you make will be appropriate for use in both.

5. To duplicate one of Faraday's induction experiments, you will need a strong bar magnet or a horseshoe magnet and an iron or steel rod. Find some hollow cylinder (I used a 1″ diameter test tube) on which to wind a coil. It should be wide enough for the bar magnet to fit into it. Now wrap at least 250 turns of insulated wire in several layers, making a coil 5 or 6 inches long. Attach the two leads from the coil to the galvanoscope you made in Experiment 4. If your magnet is strong enough, every time you thrust it into or yank it out of the coil, the compass needle will jerk. If that doesn't happen, increase the number of turns on the coil.

6. Let's make an electromagnet but one that fits into the coil of the previous experiment and thus serves a double purpose. Find something that will function as an iron core. I used an old ¼-inch-diameter iron hinge pin from a door. Several large nails taped together or a steel rod will do. You can wrap the coil right on the iron core, but it's best to make a cardboard or plastic sleeve (perhaps a drinking straw) and build the coil on it. That way the rod can be taken out conveniently. Put about 150 turns of insulated wire (#22 or heavier) on the sleeve. Make sure it's not too thick to fit within the coil of the previous setup. Attach the leads to a 1.5-volt battery. A D cell will do; just don't have it on for more than a few seconds at a time. See how many paper clips you can pick up. Try it without the core. Will it affect your compass? Now use this electromagnet in place of the bar magnet in Experiment 5.

REVIEW

Section 19.1

1. What is a lodestone?
2. Charges generate _____ fields. Charges in motion generate _____ fields.
3. Define the term *magnetic induction*.
4. Does soft iron retain its magnetized state once removed from the field that induced it?
5. Why do paper clips hanging from a magnet pick up other paper clips?
6. Explain how to magnetize a steel ruler.
7. What is a *magnetic pole*?

8. How might you locate the poles of a magnet using iron filings?
9. What is a *north pole* in the jargon of magnetism?
10. State Peregrine's interaction law, the one with the like poles and the unlike poles.
11. What do you get if you break a bar magnet in two?
12. Elaborate on the meaning of the term *magnetic dip*.
13. Is the terrestrial magnetic field constant? Has it always pointed in its present direction?
14. If the Earth's field were due to a bar magnet at its center, where would its north-seeking pole be?

Section 19.2

1. Discuss what is meant by a *monopole*.
2. Do monopoles exist?
3. The motion of atomic _____ constitutes the primal current, the ultimate magnetic _____.
4. Name something truly nonmagnetic.
5. List several common *diamagnetic* materials.
6. Diamagnetic substances are _____ by the pole of a magnet.
7. What is the cause of diamagnetism?
8. Describe the characteristics of a *paramagnetic* material.
9. Is water paramagnetic?
10. A _____ substance will be weakly drawn toward the pole of a magnet.
11. List the *ferromagnetic* elements.
12. Describe a magnetic *domain*.
13. In what two ways do domains change when immersed in a magnetic field?
14. Describe the general internal structure of powerful permanent magnets.
15. List several ways of demagnetizing a magnet.
16. What is the *Curie point*?
17. Why is it unlikely that the Earth actually contains a permanent magnet at its center?
18. Alcomax is similar to but an improvement over Alnico. What is it used for?

Section 19.3

1. A charge is always accompanied by a(n) _____ field.
2. After 1800 there was a widespread shift in emphasis. The great rubbing machines of electrostatics all but vanished, and the Leyden jars disappeared. What took their place?
3. Prior to 1800, was there any reason to suspect that electricity and magnetism were related somehow?
4. Describe the $\mathcal{B}$-field surrounding a straight current-carrying wire.
5. What was so surprising about the force field in Oersted's experiment?
6. Do magnets exert forces on electric currents?
7. In order for a charge to interact with a magnetic field, the charge must be _____ with respect to the field.
8. What is a *tesla*? Incidentally, the somewhat eccentric Nikola Tesla was the inventor of the AC induction motor.
9. If a beam of electrons and protons is fired into a uniform magnetic field that deflects the electrons upward in a vertical plane, in which direction will the protons be pushed?
10. Is the force on a charged particle moving through a $\mathcal{B}$-field always perpendicular to both the field and the direction of motion?

11. What was Ampère's profound and at the time very controversial suggestion as to the ultimate source of magnetism?
12. What is an *electromagnet*?
13. How did Henry (who later became science advisor to President Lincoln and was the great proponent of "ironclads") improve the electromagnet?
14. Who built the first electromagnetic telegraph?
15. What is a *galvanometer*?
16. A friend who says to you, "The brushes are sparking to the commutator," is talking about (a) a paint set in a thunderstorm, (b) a chimney sweep who goes to work by train, (c) a TV newsperson kissing a broom, or (d) none of the above.

Section 19.4

1. Define what is meant by the term *electromagnetic induction*.
2. What does a change in the number of lines of force passing through a loop produce?
3. What is the main function of an electrical *generator*?
4. Discuss the principle underlying the generator.
5. Was Faraday right about taxing electrical energy?
6. Was the very first generator (the copper disk device) AC or DC?
7. How are generators and motors related?
8. What is a *dynamo*?
9. What would happen to Faraday's copper disk generator if you connected the two output terminals to a powerful battery?
10. Where does the input energy come from for a bicycle generator?
11. Assume that the tuning fork in the figure is magnetized. Explain the signal displayed on the scope.

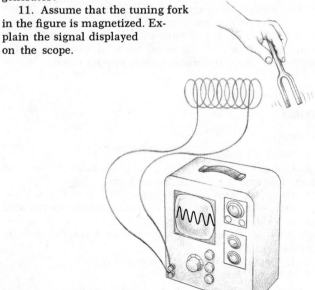

QUESTIONS

Answers to starred questions are given at the end of the book.

Section 19.1

1. Lucretius noticed that his rings, once magnetized, often repelled each other later on. Explain. A scrap of paper picked up by a charged rod may, a moment later, sail off under a repulsive interaction. Why doesn't that happen with paper clips picked up by a magnet?

2. Why is a chunk of soft iron attracted to either pole of a magnet? What does this tell you about the reliability of attraction as a test of whether an object is itself a magnet? Is repulsion a better test?

* 3. A classic puzzle involves two seemingly identical rods, one of soft iron and not magnetized, the other of steel and magnetized in the usual way, with poles at its ends. The problem is to determine which is the magnet, using nothing but the rods and neither breaking nor bending either one. How would you do it? If allowed to break one or both, how would you do it?

4. Is it possible to have a magnet in the shape of a disk like a penny with one face a north pole and the other a south pole? With the central region north and the edge south?

5. Suppose that you had a compass needle pivoted on a pin, which lined itself up north-south as usual. If you didn't know which was which, how could you use the compass to determine north from south?

6. I once found that the steel wastepaper baskets in the place where I was working (a space satellite factory in New Jersey) were all magnetized. Explain how that might happen naturally, and figure out the most likely polarity for the baskets. It's a breathtaking thought, but most of the garbage cans in the world are probably magnetized.

* 7. Suppose that you wake up in a strange land with only a bar magnet and a peanut butter and jelly sandwich. Balancing the magnet at its very center, you find that it sits almost perfectly horizontal, pointing more or less north-south. Couple that with the fact that a zebra is looking in the window at your sandwich. Roughly where are you?

8. What advantage might a horseshoe magnet have over a bar magnet as far as the field is concerned?

9. St. Augustine in *The City of God* (ca. 400 A.D.) tells "with astonishment" how he saw someone "take the said stone and hold it under a silver plate upon which he laid a piece of iron: and as he moved the stone underneath the plate, so did the iron move about, the plate not moving at all." Explain what was going on. Incidentally, this same scheme is used nowadays in several commercial devices for stirring chemicals. The piece of iron is coated with a tough plastic, and the base is motor-driven, but otherwise it's pretty much the same.

10. Given only a compass and an odd-shaped magnet, how would you find the magnet's poles?

11. I have a cloudy movie-memory of Tony Curtis, standing at the prow of a befogged Viking boat, dangling a lodestone on a string, and yelling "Odin, Odin!" At any rate, given a nondescript hunk of lodestone, how would you make a compass out of it?

12. An automobile compass, equipped with movable pieces of iron inside it, comes with instructions on how to adjust the thing when the car is pointed north-south. Why is all of that necessary? Will you have to do it again after you take your barbell set out of the trunk?

* 13. Will a floating compass be drawn to either of the Earth's poles, so that it moves bodily in addition to rotating into alignment? Will it move if you approach it with the pole of a large bar magnet?

Section 19.2

1. Suppose that there is a uniform magnetic field pointing horizontally from left to right somewhere out in space. Describe what would happen to a small bar magnet, a dipole, inserted in the field at some random orientation and floating freely.

* 2. Repeat the previous problem, but instead of a dipole, consider the mythical *monopole*. (By the way, that's a perfect name for a unicorn, and perhaps neither will ever be found.)

3. The accompanying figure shows a device for measuring the vertical force on a sample in a magnetic field. Describe the force (large, small, up, down) that would be exerted on each of the following: copper, aluminum, sodium chloride, a piece of apple, cobalt, a rusty nail.

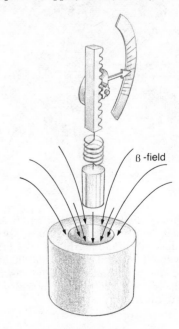

β -field

* 4. A very old technique for magnetizing an object is known as "double touch" because it requires two magnets. Suppose that the item to be magnetized is a steel bar. The process begins by placing a north pole of one magnet and a south pole of the other at the middle of the bar. Finish describing both the procedure and the final state of the bar.

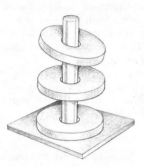

5. Two of the three disk-shaped objects in the figure are floating above the third, all kept loosely in place by a wooden post. Which, if any, are magnets, and where are their poles located?

6. Gilbert makes the following suggestion (in his book of 1600) concerning a "glowing mass of iron": "Let the smith be standing with his face to the north, his back to the south. Let him always, while he is striking the iron direct the same point of it toward the north and let him lay down that end toward the north (during the cooling process)." Explain what will happen to the iron.

7. State which of the following materials can be used to make a permanent room-temperature magnet: pure iron, aluminum, steel, gadolinium. By the way, it is possible to make an alloy of nonferromagnetic elements that is itself ferromagnetic. Manganese, copper, and silver produce such an alloy. On the other hand, Elinvar is a nonferromagnetic alloy of iron. Is there anything significant about the location of manganese in the periodic table?

8. Suppose that we immerse a magnet in a thick liquid, such as oil or glycerine, and mix into the brew millions of fine strands of nickel wire or iron filings. The iron would certainly reveal the field lines in three dimensions. Would the nickel? Explain.

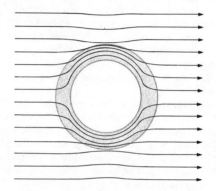

* 9. The figure shows a soft-iron ring in a magnetic field. Note how the lines of force crowd into the metal, a property Lord Kelvin called *permeability*. There are alloys (like supermalloy, a nickel-iron combination) that are roughly 10,000 times more permeable than iron and a million times more so than a vacuum. To what use might these be put? Diamagnetic materials are *less* permeable than vacuum. What does that mean as far as the field lines are concerned?

10. Is your wind-up wristwatch marked "antimagnetic"? What does that mean, and how would it be accomplished? *Hint:* The idea is not to let the balance wheel and hairspring become magnetized (see the previous problem).

11. Describe what you think will happen to a long glass tube filled with very fine iron filings when it is placed in a strong magnetic field, shaken, and then gently removed. What effect will shaking now have on it? Compare this with a solid bar of iron from the perspective of domains.

Section 19.3

1. A long wire is strung horizontally out to a tree, around it, and back. The ends are connected to a battery, and a current passes through the two long parallel segments. Describe the electromagnetic forces, if any, acting on the wire.

2. Is it possible to make a piece of copper pick up iron filings? Explain.

3. Describe the relationship between the directions of v, $\mathcal{B}$, and F_M for a charge moving perpendicularly through a magnetic field.

* 4. A beam of charged particles traveling in a horizontal plane passes between the two poles of a magnet, which are also in that plane. Given that the north pole is on the left and the south pole on the right of the beam, and that the beam is bent up out of the plane, are the particles positive or negative?

5. The American Henry Rowland in 1876 placed some static electric charge on a nonconducting disk, which he then rotated at high speed near a delicate compass. What do you think happened, and what did it prove?

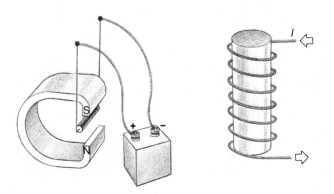

6. Which way will the movable bar swing in the accompanying figure? Explain.

* 7. The figure shows a coil wrapped around an iron bar. Given that the wire carries a current as shown, label the poles of the electromagnet.

8. Repeat the previous problem for the horseshoe electromagnet as shown.

9. An electric door bell is basically an automatically interrupted electromagnet. Describe how it works.

10. A charged particle moving perpendicularly within a uniform magnetic field and not appreciably losing energy will swing out a circular orbit in the plane at 90° to the field. Using what you know about F_M, explain this behavior.

* 11. In 1821 Faraday devised a primitive motor he called a *rotator*. Actually he produced two different versions of it in a single unit. One had a pivoted wire carrying a sizable current swinging around a vertical, fixed bar magnet immersed in mercury. The other had a pivoted magnet rotating around a fixed current-carrying wire, again in mercury. Explain how each worked.

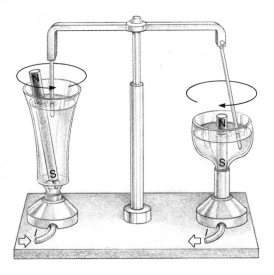

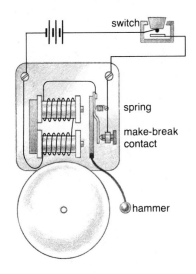

Section 19.4

1. What is the source of the energy that appears at the terminals of a generator as electrical energy?

* 2. It takes a certain amount of energy to crank a generator, even when it's "open circuited"—that is, even when it's not delivering power. Where does that energy go?

3. The wire segment in the figure is being moved with a speed *v* through a magnetic field. What is the direction of the induced current, if there is any?

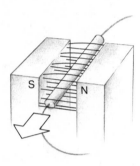

4. The wire in the accompanying drawing is pushed straight up through the B-field. What is the direction of the induced current, if there is any?

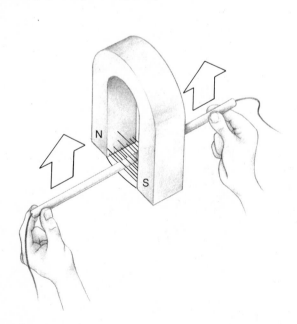

5. Suppose that the ring in a setup like Faraday's early induction experiment (p. 451) was made of steel. Field lines confined within it would certainly circulate around it. Does that mean it would become magnetized, even though it had no poles? What would happen if one removed a piece of it, opening the ring? Would the piece and/or the remainder have poles?

6. At what rate must the armature of the simple AC generator (with one set of poles) rotate in order to produce a 60 Hz output frequency?

* 7. Imagine a long hollow coil held vertically with its two lead wires attached to an ammeter. If a bar magnet is dropped through the coil without touching, will it hit the ground at the same speed as it would without the coil? *Hint:* Think in terms of energy.

8. An electric motor draws a lot more current while it's starting than while it's running. Why? Is there an analogous situation with a generator?

9. Imagine that a steam engine is attached to a DC generator by a bicycle-chain arrangement. The dynamo, in turn, is hooked up to a storage battery, which it is charging. Everything is turning and humming, when all of a sudden the chain breaks. However, the dynamo keeps right on whirling. Explain.

* 10. The simple DC generator (p. 454) puts out *pulsating* direct current, which can be smoothed considerably by adding more armature coils and segmenting the commutator.

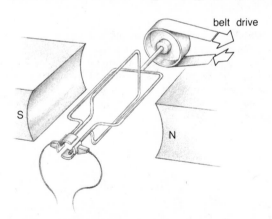

Describe the operation of the two-cell generator shown in the figure. Draw a picture of the output signal, which is the sum of the outputs from each separate coil. *Hint:* It no longer drops to zero anywhere in the cycle.

11. A device known as a rotating coil magnetometer measures magnetic fields. A small coil mounted at the end of a long rod is spun by a constant-speed motor. The coil leads are attached to a galvanometer calibrated in teslas. Explain how it works.

MATHEMATICAL PROBLEMS

Answers to starred problems are given at the end of the book.

Section 19.3

1. Another common unit for the magnetic field strength in addition to the tesla is the *gauss* (1 tesla $= 10^4$ gauss). The magnetic field in intergalactic space is probably around 10^{-6} gauss, as compared with the 8000-gauss field of a powerful samarium-cobalt permanent magnet. Express these quantities in teslas.

* 2. Imagine a wire of length L carrying a current that is moving perpendicular to a $\mathcal{B}$-field with a drift-speed v. If n is the number of mobile charges (each of q_e) per unit length, show that

$$F_M = nLq_e v\mathcal{B},$$

where F_M is the total force on the wire.

3. Using the results of the previous problem, prove that

$$F_M = IL\mathcal{B},$$

where I is the current.

4. If a current of 10 amps is carried by a wire with a 10-cm-long section at right angles to the 0.5-tesla field of a permanent magnet, what is the force on the wire? *Hint:* Use the result of the previous problem.

* 5. For a charged particle moving perpendicular to a magnetic field, F_M serves as a centripetal force (Section 5.2), where

$$F_c = \frac{mv^2}{R}.$$

Show that the particle will move in a circle of radius $R = mv/q\mathcal{B}$.

Section 19.4

1. The figure shows an AC generator with *two* sets of poles instead of just one. At what rate must it rotate in order to generate 60 Hz current? This result underscores one of the main reasons for using multiple sets of poles.

AC output

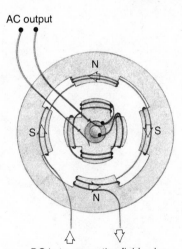

DC in to magnetize field poles

At last we come to the second form of cosmic essence, that which along with material "stuff" constitutes our universe—**radiant** or **electromagnetic energy.** The first section of this chapter treats the wavelike aspects displayed by radiant energy, beginning with Maxwell's wonderful theoretical prediction. It goes on to deal with wave phenomena in general, to distinguish between the two main classifications of **longitudinal** and **transverse** waves, and to examine the interrelationship of the concepts of **wavelength, frequency,** and **speed.**

The second section considers the various types or bands of radiant energy, which we rather arbitrarily distinguish according to wavelength (or frequency). This **spectrum** can be visualized as a progression of decreasing wavelengths, starting with what we have come to call **radio waves** and diminishing successively to **microwaves, infrared, light, ultraviolet, X-rays,** and finally to the shortest of all, **gamma rays.** Although they overlap somewhat, each such band has characteristics that are more or less distinctive, and yet all are intrinsically the same; all are electromagnetic energy.

The final ingredient in the cosmic scheme as we have come to know it is energy devoid of the usual aspect of the material—massless and chargeless—energy existing on its own independent of matter; energy unleashed and on the wing; pure energy, *radiant energy.* Radio, microwave, infrared, light, ultraviolet, X-ray, gamma ray—the long list of names beclouds a profound "unity in the variety." It's all one, all part of the same electromagnetic whirlwind.

Our present conception of light has evolved over millennia out of experiment and inspiration. Take a moment now to briefly trace the unfolding ideas from their start; to see that today's truth was not yesterday's and may not be tomorrow's; to see how subtly all the visions intertwine.

The ancients had many different notions concerning the basic nature of light and vision. Often confused and always rudimentary, most are of little concern here, though in them were sown the seeds of what would come. The atomists in the line of Democritus of Abdera were among those who believed that light was a hail of particles streaming out from visible bodies. Aristotle, who could not abide the idea of a vacuum, let alone one laced with beams of flying particles, understandably took the opposite position. To him all space was filled with aether, and "perception arises from a movement, produced by the body we perceive, in the interposed medium." Light was a kind of pulse in the aethereal sea.

CHAPTER 20
Radiant Energy

20.1 PHOTONS AND FLUTTERS

Light is not an emanation of particles but the transmission of a movement.
ROGER BACON (1220–1292)
Opus Majus

Thomas Young
(1773–1829)

LIGHT

Caloric is the name assigned by the new nomenclature to the element of heat. It is the most subtile of all bodies, and exists in nature in very great quantities. . . . Light is occasioned by the transmission of caloric, or the matter of heat, which travels through the air, in one second of time, about 170,000 miles. . . . It is heat projected. The identity is the same; and consists of infinitely small particles thrown off in every direction from a luminous body.

H. G. SPAFFORD

General Geography and Rudiments of Useful Knowledge (1809)

Leonardo da Vinci seems to have been the first (of whom we have record) to grasp the similarity between light, sound, and water, a similarity framed in the vision of waves. Sometime in the early 1500s he wrote about light rippling, spreading out from a body in "circles," filling the surrounding space "just as a stone thrown into water becomes the cause and center of various circles." That same appealing imagery would be called on throughout the centuries to follow.

During the mid-1600s, Descartes's philosophy prevailed. Light was seen as a pressure propagated in a dense particle-filled universe to be perceived as the blind man perceives the world through the vibrations of his cane. But that dominance of Cartesian thought was only a temporary situation, which was swept aside soon enough by the more powerful Newtonian world view.

Sir Isaac's arch rival, the ever present and extraordinary Robert Hooke, was an outspoken proponent of the **wave model,** as was the great Christian Huygens. To them, as to Leonardo, light was a motion, a ripple set up in the surrounding transparent medium. Newton objected to this wave picture, it seems, for two principal reasons: The seventeenth century had given rise to what Boyle termed the "corpuscular philosophy," the age of atomism, and in the spirit of the times, Newton favored a corpuscular picture of light. Moreover, and more important, since everyone "knew" that light traveled in straight lines casting sharp shadows, it seemed quite impossible for it to be a wave. Waves spread out as they expand, bending around objects; a stream of particles simply would not do so.

Newton failed to recognize that light does bend as a wave would—subtly, but certainly. He was aware of the observations that would later "prove" as much, but he interpreted them quite differently. Although he preferred to report on "the bare discovery" and not to propose hypotheses to explain it, he did indeed hypothesize, if only to satisfy certain of his readers. Ultimately, Newton suggested that light could be fruitfully envisioned as a stream of particles that set the surrounding aether into vibration, affecting in turn the behavior of the light itself—corpuscles of light accompanied by waves of aether.

The eighteenth century was an age commanded by the potent science of Sir Isaac. His doting disciples and devotees, ardent apostles and reverent followers forgot the master's doubt and in his name simplistically became convinced—light was to be a stream of particles. The Newtonian spell was not lifted easily. Not until the start of the nineteenth century, until the excesses of the Industrial Revolution gave fire to the Romantic rebellion (Section 7.1) was there intellectual room to move. It was then, in the first few decades of the 1800s, that the wave theory was reborn with a new mathematical rigor. Resurrected, principally and independently by Thomas Young and Augustin Fresnel, the wave conception slowly but surely gained wide acceptance. Their masterful analysis, backed by ingenious experiments, could explain all the knotty problems of the past. Light was again an elastic wave in the all-pervading aether.

By the end of the 1800s the work of Faraday, Kelvin, Maxwell, and Hertz had established that *light was actually electromagnetic in nature*. It was a wave of varying electric and magnetic fields flashing through the aether, not simply a mechanical trembling of that invisible gel.

In this century, special relativity put aside the aether, leaving light to waiver on its own through a web of fields. Aether vanished and the fields took on a complete, self-sustaining reality. Finally quantum theory brought the business full cycle (Section 11.2), demanding yet another variation in the vision. The energy carried by light was found to be not smoothly spread out across the wavefront but concentrated in lumps, in photons. The emission and absorption of light, its interaction with matter (i.e., the transfer of energy and momentum to a material object) takes place in a decidedly corpuscular way. *Light is a stream of particlelike concentrations of electromagnetic energy, which exist only in motion at the speed* c, *and which propagate (just as "particles" of matter do) in a wavelike fashion.* The wee beasties of the micro world don't play our game of waves *or* grains—of "either/or." Protons and photons each behave in what we have come to think of as both wavelike and particlelike ways. This, then, is the latest construct—surely not the last, but the latest.

It is the exalted business of theory to make prophecy, to predict occurrences beyond what is already known. Perhaps the most splendid prophecy in all the history of physics was rendered by a good-humored, witty, brilliant young man, one James Clerk Maxwell of Kirkcudbrightshire in the Lowlands of Scotland.

William Thomson (later to become Lord Kelvin, the Grand Old Man of nineteenth-century British science) in the 1840s was the first to begin to mathematize the notion of *lines of force*. Faraday himself was quite ill at ease with mathematics. Never having had any formal education in the discipline, he conceived of lines of force in a pictorial rather than a symbolic way. Thomson's lovely though disconnected insights were a stimulus to Maxwell, who pursued his own researches into electricity and magnetism within a month or two of graduating from Cambridge in 1854. About a year later, he published the first of several papers on the subject, entitled "On Faraday's Lines of Force," a paper whose mathematical sophistication was obvious at first glance.

In time Maxwell mathematically drew together almost everything that was known about electricity and magnetism into a set of four equations: Two of them, credited to Karl Gauss, related the way the electric and magnetic fields extend through space to the presence and distribution of their sources (charges and "monopoles"). The third equation was Faraday's *induction law*, which states that *a time-varying magnetic field is always accompanied by an electric field*. The last equation was *Ampère's law* describing the

PROPAGATION OF LIGHT BY THE ETHER

As the atoms of matter vibrate in the ether in which they are immersed, they communicate their vibration to it. The vibrations thus started in the ether are propagated through it in every direction in minute waves and with an inconceivable velocity.

J. A. GILLET AND W. J. ROLFE

Natural Philosophy for the Use of Schools and Academies (1882)

Maxwell's Waves

James Clerk Maxwell

A LETTER FROM MAXWELL TO THOMSON

Trin. Coll., *Feb.* 20, 1854.

Dear Thomson

Now that I have entered the unholy estate of bachelorhood I have begun to think of reading. This is very pleasant for some time among books of acknowledged merit wh one has not read but ought to. But we have a strong tendency to return to Physical Subjects and several of us here wish to attach Electricity.

Suppose a man to have a popular knowledge of electrical show experiments and a little antipathy to Murphy's Electricity, how ought he to proceed in reading & working so as to get a little insight into the subject wh may be of use in further reading?

If he wished to read Ampère Faraday &c how should they be arranged, and at what stage & in what order might he read your articles in the Cambridge Journal?

If you have in your mind any answer to the above questions, three of us here would be content to look upon an embodiment of it in writing as advice. . . .

SOURCES OF LIGHT WAVES

Just as sound waves are disturbances set up in the air by the vibrations of bodies of ordinary dimensions, so light waves are disturbances set up in the ether probably by the vibrations of the minute corpuscles, or *electrons,* of which the atoms of ordinary matter are supposed to be built up.

ROBERT A. MILLIKAN AND HENRY G. GALE

A First Course in Physics (1906)

magnetic field set up by a current. The last relation seemed to Maxwell not to represent the whole truth, and so he boldly amended it by adding on another term. In effect he argued that *if a varying B-field created an E-field, then a varying E-field should create a B-field* as well. Thus, in addition to the fact that currents produced magnetic fields, as Ampère had realized, changing electric fields would also give rise to them. Similarly, in addition to the fact that charge produced electric fields, changing magnetic fields (as Faraday pointed out) would also give rise to them. Ah, the symmetry was irresistible. This set of four all-embracing mathematical formulas is now known throughout the world as **Maxwell's equations.** They are to electromagnetic theory what Newton's laws are to mechanics.

On December 8, 1864, Maxwell delivered before the Royal Society one of the most pregnant papers ever written: "A Dynamical Theory of the Electromagnetic Field." Combining his four field equations, he arrived at a complicated expression that he immediately recognized as the well-known formula describing all forms of wave motion from vibrating drum heads to water ripples. In the places where the equation would ordinarily have a variable describing the medium (e.g., the displacement of a string or the pressure in sound), there sat instead $\mathcal{E}$ and $\mathcal{B}$. The theoretical implication was utterly amazing. Electric and magnetic fields serving as a medium could support waves. In what must have been a heart-pounding moment, Maxwell realized that the all-important stroke was there in front of him, waiting for him. The crucial term, which he knew had to correspond to the speed of the wave, was within easy reach. In his equation, that speed was represented by the square root of a cluster of electric and magnetic constants. Carefully he inserted their numerical values (which had been measured several years before), multiplied and divided appropriately, and then took the square root. Amazing! The predicted speed of propagation of these electromagnetic waves turned out to be just about 3×10^8 meters per second, almost exactly equal to the measured speed of light. "The agreement of the results," he calmly reported, "seems to shew that light . . . is an electromagnetic disturbance propagated through the field according to electromagnetic laws."

Maxwell never knew his final triumph; he did not live to see the confirming experiments of Heinrich Hertz. Maxwell died of cancer in 1879 at the age of 48, nine years before Hertz generated electromagnetic waves in the laboratory and showed them to be considerably longer but otherwise quite similar to light (Section 11.2). Curiously, Maxwell's grand synthesis of classical electricity and magnetism took place prior to the discovery of the electron (before anyone had any idea that charge was even granular) and while aether still survived. Yet the main conclusions more or less prevail to this day: *Light is certainly wavelike and it is certainly electromagnetic.*

Maxwell died a century before this present moment, when we casually fill the air with channels and chatter, when every room in every house is aflood with electromagnetic waves, with CBS and ABC and NBC and BBC. The planet is ablaze with electromagnetic radiant energy, a kind of poignant hymn to Maxwell and the field.

It is adequate for our purposes to simply define a physical **wave** as *a disturbance that, once begun, propagates on its own through a medium, moving in space and time, carrying energy.* Such a disturbance may be a single brief pulse—the sound of a gunshot rushing through the air, the shock wave from a passing SST, or the flash of light from a spark. Alternatively, it may be a more or less steady, sustained vibration, in which the repetitive nature of the process is far more obvious—the continuous ringing tone of a bell or the parade of water ripples spreading out beneath a rhythmically wiggled toe. In every case there is an underlying medium, a sustainer of the disturbance, something that is disturbed. It may be water, air, string, or spring; it may be the head of a drum or the whole Earth ringing like a gong from a quake; or it may even be the electromagnetic field or perhaps the gravitational field.

In General

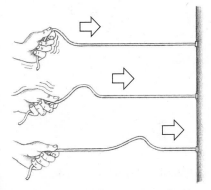

A wave pulse sailing along a taut rope.

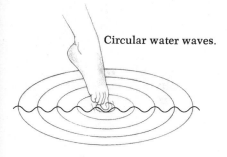

Circular water waves.

A wave pulse of light, just a few millimeters long, flying by at 300,000,000 m/s.

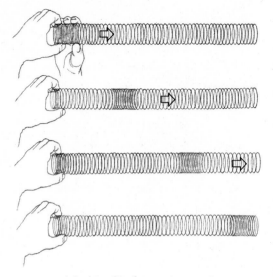

A longitudinal wave in a spring.

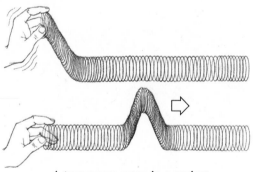

A transverse wave in a spring.

A sound wave.

We can easily distinguish two basic kinds of waves, **longitudinal** and **transverse.** Imagine a long loose-coil spring (like a "Slinky") held at its ends in such a way that the coils are fairly straight and horizontal. The spring itself is the medium that will carry the disturbance. For starters, push one end of the spring forward, compressing it, reducing its length for a moment. Then quickly return it to the original position. That region of compression will rapidly sail away, moving down the whole length of the spring without any further help from you. *This sort of wave, caused when the displacement of the medium is parallel to the direction of propagation, is called longitudinal.*

For example, sound waves are pressure fluctuations, compressions and rarefactions of a medium, usually air, traveling at roughly 750 miles per hour (331 m/s). In general, with sound waves, the molecules of any elastic medium are displaced parallel to the propagation direction, but they move only a little from their original positions. It's the disturbance that constitutes the wave, and it's the disturbance that does the traveling and carries the energy. The individual molecules pretty much stay put, just as the individual coils on the spring stay close to where they began, even as the wave flashes by. If one yells a short blasting "Beep," that pressure *pulse* of crowded molecules will push on the next layer of molecules and then the next and so on. In that way the disturbance will sail out as a wave, propagating on its own across the room, just as the pulse moved along the spring. Anyone downrange who happens to pick up a little of that wavefront receives energy that will set the ear machinery vibrating in step, producing the perception "Beep."

Alternatively, if the end of a long loose spring is displaced straight upward a bit and then back down, a hump will be generated—a pulse that will traverse the whole length of the spring. *This sort of wave, caused when the displacement of the medium is perpendicular to the direction of propagation, is called transverse.* A plucked guitar string oscillates as a transverse wave, as does the taut skin of a drum. Light is of this variety—a *transverse* electromagnetic wave-like phenomenon.

One of the most useful visual and mathematical representations is the **periodic wave**—an idealized disturbance that is perfectly regular, repeating over and over, oscillating endlessly. Since no such wave pattern ever actually exists, the imagery is really a fiction, but it can be well approximated in practice. The pure tone of a tuning fork comes near to doing that with sound, and so does the spectacular light from a laser.

The periodic wave reproduces itself again and again, such that the distance, λ (the **wavelength**), corresponds to the extent of the basic pattern being repeated. In the simplest case, we can imagine a constant stream of risings and fallings, of smooth peaks and troughs sailing past at some **speed,** v. The number of wavelengths that flash by per unit time is called the **frequency,** f, given in hertz.

In the case of sound, the deep base vibrations, the familiar hi-fi rumblings that go right through the walls, constitute the low-frequency (or low-*pitch*) end of the range, and the squeaky stuff, the treble, corresponds to the high-frequency (or high-*pitch*) end. A young ear (if not too battered by the pounding of overamplified rock) can at best hear from about 15 Hz to 20,000 Hz (women on the average do slightly better than men at both extremes).

A wave on a string consists of something literally going up and down, and so it's easy to appreciate why the wave is represented by a line wiggling up and down. But a sound wave is a longitudinal increasing and decreasing of pressure, and an electromagnetic wave is a transverse increasing and decreasing of field strengths. In either case, there's actually nothing to be seen vibrating up and down, because we're talking about oscillations in pressure or field strength. These oscillations can be more easily visualized (at least in principle) if we watch the indicator needle on an appropriate meter swing back and forth (plus and minus) as either a sound wave or an electromagnetic wave passes. The smoothly oscillating motion of the needle corresponds to the undulation of the wave, and that's what is represented pictorially in the figures.

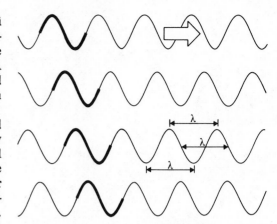

A periodic wave of wavelength λ, advancing to the right.

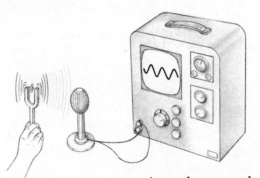

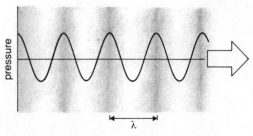

A sound wave graphed as a variation in pressure.

That quantities v, λ, and f are mutually interdependent is one of the basic characteristics of all sorts of vibrations, including light and sound. Suppose that a periodic wave of some kind is streaming past. The number of wavelengths that will sail by in a second is f, and each one is λ meters long. Thus f times λ is the overall length of wave that passes each second, but that quantity, corresponding as it does to distance over time, must be the speed of the wave. In other words,

speed = frequency × wavelength

or

$$v = f\,\lambda.$$

The rising and falling of the electric and magnetic fields of an electromagnetic wave.

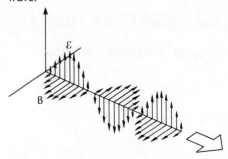

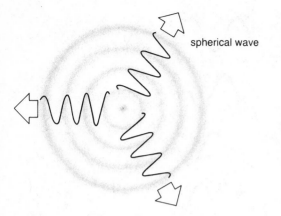

spherical wave

A spherical wave propagating out in all directions.

A plucked G string on a guitar vibrates at a frequency of 384 Hz. Since sound travels at a speed of about 331 m/s, our equation tells us that the wavelength of that tone must be a bit less than 1 meter (0.86 m, to be precise). Note that for any given kind of wave, *as the frequency increases, the wavelength decreases, and vice versa.*

Sound and light are, of course, three-dimensional phenomena, and their associated waves generally spread out into space instead of just running along essentially in one direction as did the disturbance confined to a spring. Imagine a pulsating tiny balloon rhythmically expanding and collapsing at the center of a pool of water. Ideally it would produce **spherical waves,** disturbances emanating from this point source uniformly in all directions. Similarly, a burning candle fills a room with light and can be seen at almost every angle. Though not actually spherical (no real waves are), these wavefronts resemble spheres, more so when one is far from the flame than when one is near it.

When light from any source impinges on an object, the bombarded surface atoms, responding to the incoming energy, momentarily "take in" the light and then almost immediately reradiate it in a process known as **reflection.** We can therefore think of every atom on the surface of this page (indeed, every atom on the surface of every object seen) as a point source of light. If the object is self-luminous (like the sun), the light is directly emitted; if not, the light is reflected. In either case, what we see is the combined effect of light streaming out in essentially spherical waves from each atom on the visible surface of an object.

In the extreme, when an observer is exceedingly distant from a point source, the ever expanding spherical waves tend to flatten out. As their radii become immense, the small portion of the disturbance entering an eye or a camera comes to resemble a flat sheet, a **plane wave.** Certainly a star is no point source of light, but when you're standing a thousand million million miles away looking at it with a telescope only a few feet in diameter, the segment of the incoming wavefront will be flat enough to be called planar. Indeed, when a camera is focused at infinity, it is set to receive plane waves originating from point sources interminably far off. Even though the mountains on the horizon being photographed are surely not infinitely distant, the waves of light coming from them are so flat that we cannot tell the difference.

20.2 ELECTROMAGNETIC RADIATION

All the forms of electromagnetic radiation propagate through vacuum at the same incredible speed, $c = 299,792,456.2 \pm 1.1$ m/s (which corresponds to traversing a distance of about one foot in a mere thousandth of a millionth of a second). Whether light, UV, or X-rays, each is a web of oscillating electric and magnetic fields inducing one another and expanding out into space, hand over hand. A fluctuating electric field creates a magnetic field perpendicular to itself,

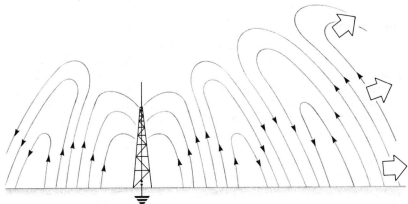

Electromagnetic waves from a transmitting tower.

surrounding and extending beyond it. Because that magnetic field is also varying, it creates a perpendicular electric field that moves out as well, and so on, one field generating the other indefinitely. This energy, rumbling through the electromagnetic field, is initially provided by electrical charge—accelerating charge.

As a rule, all the various forms of radiant energy begin with charges in nonuniform motion, and although it's not clear what exactly happens internally when an atom radiates, we do know with some certainty that light is emitted from the "orbital" electrons. Moreover, we know in general that free charges (those that are not bound within an atom) emit electromagnetic radiation when accelerated. And this relationship holds for charge sailing around in circles within a synchrotron, moving at a changing speed in a straight line, or simply oscillating back and forth—*if charge accelerates, it radiates.*

An accelerating charge essentially converts its kinetic energy into radiant energy. But remember that there is also another conversion mechanism: Matter itself can be annihilated, transformed directly into a burst of radiant energy (Section 15.1). In a way, electromagnetic radiation is rather like a ribbon of energy that begins and ends on matter and while in between is free to fly the universe unleashed and unseen.

Perhaps the simplest electromagnetic wave-producing mechanism to visualize is the oscillating dipole—two charges, one plus and the other minus, vibrating to and fro along a straight line (in much the way currents pulsate up and down a radio transmitting antenna). Electric field lines begin and end on charges, and they close on themselves only when the two charges overlap and effectively vanish. Moving charge constitutes a current (in this instance an oscillating current) accompanied by a magnetic field whose closed circular lines of force are in planes at right angles to the motion (p. 445). *The $\mathcal{E}$- and $\mathcal{B}$-fields are everywhere perpendicular to each other, as well as to the propagation direction.* Realize that the radiation pattern is three-dimensional, roughly resembling a series of almost spherical concentric tires expanding radially outward.

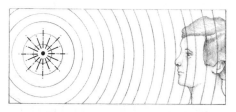

A distant point source emits spherical waves that increasingly resemble plane waves as their radii grow.

The pattern of electromagnetic waves radiated by an oscillating dipole. The loops are electric field lines. The dots and crosses depict the emergence and exit of the circles of magnetic field, which are in planes perpendicular to the page and surround the dipole.

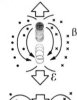

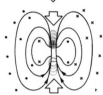

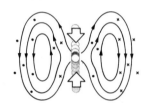

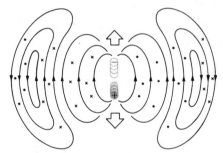

The Spectrum

The whole spread of radiant energy (which ranges in wavelength between zero and infinity) is referred to as the **electromagnetic spectrum.** It is usually subdivided into at least 7 more or less distinct regions. The motivation for subdividing according to wavelengths (or frequency) was more historical than physical, and so there is a good deal of overlapping. Needless to say, light was "discovered" first, then infrared (1880), ultraviolet (1881), radio waves (1888), X-rays (1895), gamma rays (1900), and soon afterward, microwaves.

Great, slowly rising and falling electromagnetic waves (more than 18 million miles long) have been detected impinging on the planet, streaming in from outer space. The grouping that includes these faint cosmic flutters at one extreme and extends continuously all the way down to wavelengths of about 0.3 meter is known as **radio waves.** This collection contains the usual transmission frequencies of TV, AM and FM radio, and shortwave, as well as the 60 Hz hum radiated from AC power lines. The range of FM is from about 88 to 108 million cycles per second (or megahertz, MHz), corresponding to wavelengths of about 3 meters. In contrast, AM waves are about 100 times lower in frequency and 100 times longer in wavelength.

Microwave antennas.

The energy of each quantum of radiation ($E = hf$) depends directly on its frequency, and since radio frequencies are comparatively low, the individual photon energies in this range are quite small. Thus in a flood of radio emission, the granular nature of the radiation is generally obscured, and one detects only a smooth wave-like transfer of energy. Analogously, as a water wave sweeps by, it's not at all apparent that we are watching the harmonious concert of countless individual molecules.

By the way, those little electronic calculators that are now so commonplace transmit radio waves—not very powerfully and therefore not very far, but far enough to be picked up. Just place one near an AM radio, turn the station dial to the low-frequency end, and punch in a beeping tune.

The range of wavelengths from 0.3 m to about 1 mm is called **microwaves.** These are used today for everything from carrying phone conversations and cooking hamburgers to catching speeders (with radar). Because they readily penetrate the atmosphere, microwaves are commonly the means for communicating with space vehicles. The parabolic dish-shaped antennas seen on tall buildings and atop towers are microwave receivers and transmitters. There's a lot of serious thought being given to a satellite system that could convert solar

The electromagnetic-photon spectrum.

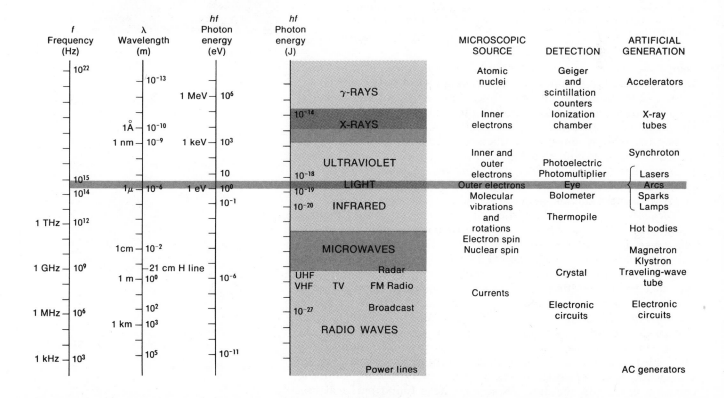

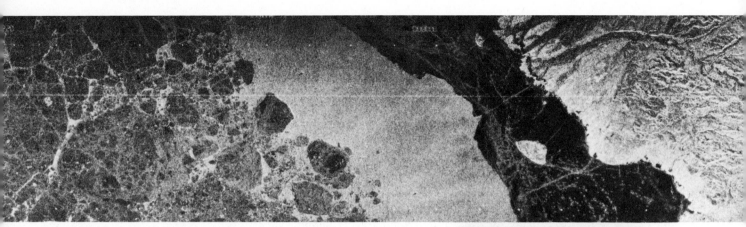

A photo of an 18 by 75 mile area northeast of Alaska. It was taken by the Seasat satellite 800 kilometers (500 miles) above the Earth. The overall appearance is somewhat strange because this is actually a radar or microwave picture. The wrinkled gray region on the right is Canada. The small, bright shell shape is Banks Island, embedded in a black band of shore-fast, first-year sea ice. Adjacent to that is open water, which appears smooth and gray. The dark gray blotchy area at the far left is the main polar ice pack. There are no clouds because the radar "sees" right through them.

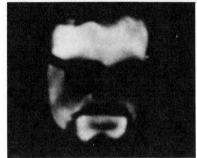

Infrared thermograph of the author. The dark, cool regions are the beard, moustache, and sunglasses. Warm areas appear white. Keep in mind that this is an infrared image—it could have been taken just as well in a pitch-black room.

energy into microwaves, which would then be beamed down to Earth, where that energy would be used to generate electricity.

In 1800 the renowned astronomer (and musician and runaway from the Hanoverian Foot Guards) Sir William Herschel made a shocking discovery. Using a prism, he had been studying the amount of thermal energy conveyed by the different colors of sunlight, and he was amazed to find that his thermometer registered its greatest increase just next to (or "beneath") the region illuminated by red light. Herschel rightly concluded that he was observing the effects of an invisible radiation, now called **infrared** (beneath the red).

Because it is so readily absorbed (i.e., converted into thermal energy) by molecules whose oscillatory frequencies as a rule are in this range, IR is often misleadingly spoken of as "heat waves." Infrared is the energy copiously emitted by warm objects, from glowing coals to radiators. Roughly half the radiant energy from the sun is IR. We ourselves are emitters of IR (a fact the military has exploited in the form of sniperscopes that "see" in the dark). There are photographic films that respond to IR; TV systems, known as thermographs, that produce continuous infrared pictures; IR spy satellites that look for rocket launchings; and "heat-seeking" antiaircraft missiles that are guided by infrared. Wherever subtle variations in temperature are of concern, as in detecting breast cancer or simply spotting a burglar sneaking around in the dark, IR systems have found practical use.

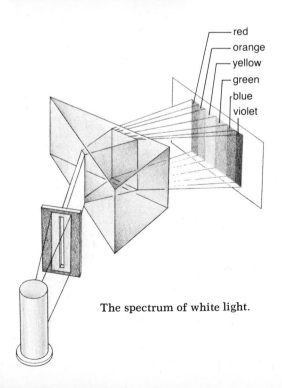

red
orange
yellow
green
blue
violet

The spectrum of white light.

The infrared region extends from wavelengths of about 1 mm down to 780 nanometers (1 nm $= 10^{-9}$m), which corresponds to frequencies of roughly 3×10^{11} Hz up to 4×10^{14} Hz.

The narrow band of the spectrum that we humans see is often referred to as **light.** That's really a rather inaccurate specification, because the retina also responds to X-rays—we can "see" X-ray shadow patterns directly cast on the retina although the eye cannot form images in the usual way. Moreover, many of us can see—if only poorly—into the infrared and ultraviolet as well. Thus let's fix the meaning of the word "light" as that tiny range of electromagnetic radiation from 780 nm to 390 nm.

Newton was the first to realize that white light is actually a mixture of all the colors of the visible spectrum; that the prism does not create color, as had been thought for centuries, but simply fans out the white light, separating it into its colors. What we perceive as the colors of light are the various frequency regions extending continuously from about 4×10^{14} Hz for red, through orange, yellow, green, and blue, to violet at about 8×10^{14} Hz. The human eye is most sensitive to yellow-green (one reason that yellow street lights are so common), and it gradually diminishes in its response at both higher (blue, violet, UV) and lower (orange, red, IR) frequencies.

Granted that the wavelength of even red light at 0.000000780 meters is rather small (780 nm is about 1/100 the thickness of this page), it's not inconceivably small. For that matter, on an atomic level the wavelengths of light are immense, several thousand times the size of an atom. If a uranium atom were scaled up to the size of a pea, a single wavelength of red light would then be about 54 feet long. This disparity is one of the crucial factors in determining the way light reflects off material objects. Remember that most of what we see, by far, is reflected light. Eyeballs evolved on a planet bathed in sunshine; if conditions had been otherwise, IR-eyes would probably have been more useful. On the other hand, if we "saw" radio waves instead of light, most things around us, including ourselves, would be more or less invisible. Roll up in a ball around a portable radio and prove that for yourself; the radio will keep right on playing as if you weren't even there, as if you were transparent.

Because light is usually neither emitted nor transmitted but reflected, it produces a richly colored, shadow-filled, high-contrast picture of the surrounding world. Still, you can be misled by your narrow range of vision—for example, into believing that objects are solid in a literal sense. You know now that your hand is mostly emptiness, despite the fact that it "looks" continuous enough at the operating wavelengths of eyeballs. Ah, for Superman's X-ray vision!

In a flood of sunlight, where more than 10^{15} photons arrive each second on a one-centimeter-square area, the quantum nature of the process is easily overlooked. Yet in the visible frequency range, photons are already energetic enough to produce effects on a distinctly individual basis. For instance, the human eye can detect as few as ten photons and possibly even one. Quanta of light can break up delicate chemical bonds, and therefore substances like aspirin and

An ultraviolet photo of Venus taken by Mariner 10.

A very early (1896–1897) X-ray setup used by Dr. R. Reynolds in his medical practice. This particular instrument was carefully crafted on a framework of polished wood. It was powered by seven wet cells in series, one of which is just visible on the floor in back.

wine are often stored in dark bottles. And of course, we see light in the first place because it can trigger chemical reactions in the retina.

A year after IR was discovered, Johann Ritter found yet another invisible radiation, just "beyond the violet," which has come to be known as **ultraviolet.** This is the so-called black light of luminous poster fame, the radiation that tans skin and activates the synthesis of vitamin D within it. In the frequency range from about 8×10^{14} Hz to 3×10^{17} Hz, these UV photons each carry enough energy to ionize atoms and rip apart chemical bonds. At wavelengths smaller than about 290 nm, UV is germicidal; i.e., it kills microorganisms. Nowadays there are ultraviolet photographic films and microscopes, even celestial telescopes designed especially to view UV.

All the current concern about the Earth's ozone (O_3) layer stems from the fact that this gaseous envelope absorbs what would otherwise be a lethal stream of solar UV. Anyone who has ever been blistered by a sunburn has firsthand knowledge of the potency of ultraviolet.

Wilhelm Conrad Röntgen discovered **X-rays** quite by chance on November 5, 1895, while working with cathode rays generated by a Crookes tube (Section 10.1). In the darkened room he noticed that one of his luminescent screens (a piece of paper coated with a barium salt) lying some distance away was glowing brightly. The conclusion was inescapable: Some invisible penetrating radiation was being emitted by the tube. Röntgen became the instant hero of the age and winner of the first Nobel Prize in Physics. Within months his marvelous rays were at work everywhere in the hands of quacks and the "qualified" alike. Unfortunately, however, no one was really qualified. Without the slightest inkling of their profoundly dangerous nature, people used X-rays for everything conceivable, from removing facial hair to examining luggage. Too often the results of that cavalier attitude were horribly tragic. And of course, the Victorian ladies were often warned about the only danger anyone seemed to grasp (however ridiculous), namely, that of lurking X-ray peeping Toms. Incredibly, X-rays were being used to treat acne even into the 1950s, but it's only now that many of the victims of such treatment are developing cancer, particularly of the thyroid.

Extending in frequency from about 3×10^{17} Hz to 5×10^{19} Hz, X-rays have exceedingly short wavelengths, roughly equal to the size and spacing of atoms in a solid. They are highly penetrating and potentially very hazardous. The individual photon energies are so large that X-rays can interact with matter in a clearly granular fashion, almost like bullets of energy.

Of course, this radiation will expose ordinary film; anyone who has sat biting an X-ray plate in a dentist's chair knows that from experience. Most often, X-ray photos are simply shadow castings, but modern techniques for focusing the rays are producing images of distant sources like the sun. Orbiting X-ray telescopes have given us a new eye on the universe.

Gamma rays constitute the highest-energy, lowest-wavelength electromagnetic radiation. They begin where X-rays leave off and go indefinitely up in frequency and down in wavelength (smaller than the size of an atom). Emitted during the nuclear transitions (or readjustments) that take place when a radioactive atom decays, each gamma-ray photon carries so much energy that it can be detected individually with little difficulty. It is here that the particlelike behavior dominates and the wavelike response recedes into obscurity.

These potent blasts of energy are the ultimate hazard to all forms of life. Ironically, although high-energy photons can cause cancer, cancer cells are more easily destroyed by gamma rays than are normal cells. This selectivity has made controlled exposure to gamma rays a powerful cancer treatment. What kills also heals.

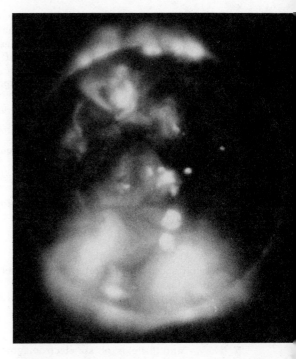

X-ray photo of the sun taken March 1970. The limb of the moon is visible in the southeast corner as a dark arc.

EXPERIMENTS

1. Use an electronic calculator as a source of radio waves (p. 473) to be picked up by an AM receiver. Position the two so that the radio makes a loud noise when set between stations. Now see what effect you get by inserting different materials between transmitter and receiver. Try a book, your hand, or any other nonconductor, and then slip in a sheet of aluminum foil. Make some holes in the foil. Does that change things appreciably?

Momentarily touch the terminals of the coil from Experiment 5 in Chapter 19 to a 1.5-volt battery, and the radio will pick up the pulse of electromagnetic energy, just as it picks up the emissions from nearby car ignitions as static. Rub a comb or plastic rod to charge it, and the radio will clatter its response to the electromagnetic waves sent out by the sparks.

2. Fill a sink basin or tub with some water and play with waves for a while. Let a drop fall on the still surface to create circular wavefronts. Observe how the waves reflect off the sides and how they overlap and pass each other.

REVIEW

Section 20.1

1. State which four of the following best belong together: light, positrons, death rays, X-rays, gentian violet, infrared, delta rays, gamma rays.

2. A time-varying magnetic field is accompanied by _____.

3. A time-varying electric field is accompanied by _____.

4. Why is the name of Heinrich Hertz always mentioned in any discussion of Maxwell's electromagnetic wave theory?

Incidentally, David Hughes actually performed the feat first (1879), but his work, which was not nearly as thorough as Hertz's, went unpublished for years.

5. Is the space surrounding you at this moment being crisscrossed by electromagnetic waves?

6. Define what is meant by a *wave*.

7. Can a single pulse be a wave?

8. Distinguish between *longitudinal* and *transverse* waves. Incidentally, it is possible for a wave to have both these characteristics at once. Ocean waves are primarily transverse, but they have a longitudinal element as well.

9. Sound is classified as a _____ wave.

10. Light is a(n) _____ electromagnetic wave.

11. Sound, which can be thought of as a series of _____ fluctuations, travels in air at a speed of about _____.

12. What is a *periodic wave*?

13. Describe what is meant by the *wavelength* of a periodic wave.

14. Discuss the meaning of the term *frequency*, and state its units.

15. A high-pitched sound has a(n) _____ frequency.

16. State the relationship between the frequency, wavelength, and speed of a periodic wave.

17. As a rule, the higher the frequency, the _____ the wavelength.

18. What is the ordinary frequency range for sound (as heard by the human ear)?

Section 20.2

1. At what speed does infrared radiant energy from the sun propagate through space?

2. In an electromagnetic wave in space it is the _____ and _____ fields that are undulating.

3. A(n) _____ charge radiates.

4. What is an *oscillating dipole*?

5. The $\mathcal{E}$- and $\mathcal{B}$-fields of an electromagnetic wave are _____ to each other and to the propagation direction.

6. What is meant by the term *electromagnetic spectrum*?

7. About how long are the longest radio waves thus far detected?

8. The short wavelength limit of radio waves is about _____ cm.

9. What range of frequencies does your FM radio respond to? Incidentally, TV pictures are transmitted as an AM signal, but the sound portion is FM.

10. What are *microwaves*?

11. The spectral region *beneath the red* is known as the _____.

12. What kind of electromagnetic radiation do you yourself emit?

13. List a few contemporary uses of IR-detecting equipment.

14. The wavelength range of infrared is from _____ mm to _____ nm.

15. An electromagnetic disturbance with 2.5-cm wavelength falls in the region of the spectrum known as _____.

16. The human eye is most sensitive to _____ light.

17. Most of what we see is in _____ as opposed to directly emitted light.

18. What keeps the surface of the planet from being bombarded by lethal solar UV?

19. What is the name of the most energetic form of electromagnetic radiation?

QUESTIONS

Answers to starred questions are given at the end of the book.

Section 20.1

1. Describe Maxwell's contribution to the theory of light. What role did Faraday play in the saga?

2. Why was the speed of light so crucial to Maxwell in the interpretation of his theoretical results?

* 3. I once saw a motion picture of a large, dense crowd of people waiting to catch a glimpse of Lindbergh. As his car approached, a group in the back surged forward to see a little better, and a compression wave of people spread out across the gathering. What kind of wave was it, longitudinal or transverse? Explain.

4. Although the search goes on even at this moment, no one has yet *conclusively* detected *gravity waves*. Reasoning from electromagnetic theory, describe what such a phenomenon might be.

5. Draw a sequence of sketches of a single water wave moving past a ship. Show the motion of the ship.

* 6. Ordinary 60-Hz AC power lines radiate electromagnetic waves. Given that their speed is about 186,000 miles per second, how long are such waves?

7. Explain why an earthly observer sees sunlight consisting of plane waves.

8. Describe the interrelationship between electric and magnetic fields in an electromagnetic wave.

* 9. Considering the discussion of currents, in what sense can you say that Galvani's frog was the first receiver of radio waves? Explain what caused the frog to jump every time lightning flashed.

10. The figure on page 479 depicts optical levitation. What can you say about the ability of light to exert force, apply pressure, and transfer momentum?

Section 20.2

* 1. Will a charge initially moving in a circle at a constant speed radiate electromagnetic energy? Explain. By the way, Brookhaven National Laboratory has built a "synchrotron light source" that generates high-intensity UV and X-ray illumination. A large storage ring 150 feet in diameter provides a circular "racetrack" for high-energy (2500 MeV) electrons.

2. The Stanford Linear Accelerator imparts energies of up to 20,000 MeV to electrons accelerating along its straight

The tiny starlike speck is a minute (one-thousandth of an inch diameter) transparent glass sphere suspended in midair on an upward laser beam.

two-mile course. Will these particles lose some of that energy along the way by emitting electromagnetic radiation?

3. Make a rough sketch of the magnetic field lines accompanying an oscillating dipole as it moves through a complete cycle. *Hint:* Remember the right-hand rule (Section 19.3).

4. List the major subdivisions of the electromagnetic spectrum. State briefly something about each that you found of interest.

* 5. Discuss the relative sizes of light waves and X-rays, as compared with the typical dimensions of an atom.

6. How do we know that photons of light have enough energy to trigger chemical reactions?

7. What is the range of wavelengths corresponding to X-rays?

8. Microwave technology has been actively developing over the last fifty years. Discuss its primary applications.

9. Why is the definition of light as "the region of the spectrum that humans see" not a particularly good one? Someone who has had a lens removed from the eye because of a cataract is better able to see into the UV because the lens is responsible for much of the absorption of long-wavelength UV.

* 10. In what sense can we say that light physically has different frequencies but not different colors?

MATHEMATICAL PROBLEMS

Answers to starred problems are given at the end of the book.

Section 20.1

1. Calculate in meters the wavelength range of sound (i.e., just the audible region) in air. At 0°C the speed is 331 m/s.

2. The speed of sound in water at 15°C is about 1.5 km/s. Suppose that we were to design a sonar system to detect objects about 1 m or so in diameter. That would mean using waves with wavelengths of 1 m or less. What frequency range would we need?

* 3. The high range of ultrasonic frequencies (which is of course well beyond anything we can hear) extends into the hundreds of MHz. Calculate the wavelength of such a disturbance at 600 MHz in air. Compare that to the wavelengths of light.

4. The frequency range of vibrations below the audible is called infrasound. A rock band with a good electronic amplification system and plenty of power can actually resonate your internal organs (5–10 Hz), a phenomenon the reader may well have already experienced. What wavelength range in air does that correspond to?

Section 20.2

1. How long does it take a beam of gamma rays to travel 10 feet in vacuum?

* 2. What frequency gamma rays will have wavelengths about the size of an atomic nucleus?

3. What is the energy of a UV photon with a frequency of 3×10^{17} Hz? *Hint:* $E = hf$.

* 4. Light begins at about 780 nm. How much is that in inches?

5. Verify: If an atom is scaled up to the size of a pea, the wavelength of red light will be approximately 50 ft.

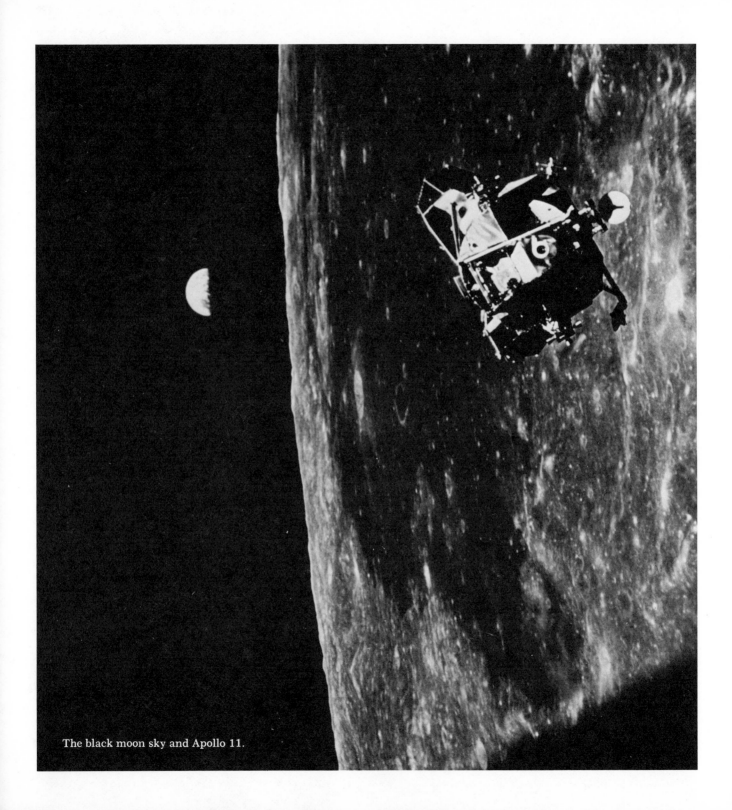

The black moon sky and Apollo 11.

This final chapter deals with the behavior of light as it interacts with and propagates through matter. The first section, entitled "Reflection and Refraction," describes those two primary notions from an atomic perspective. It defines the concept of **index of refraction** and continues with an explanation of **scattering,** particularly as it relates to blue sky and red sunset.

We usually think of light traveling in air and reflecting off, say, the surface of a piece of glass, but the reverse is also possible. Light certainly can originate within the glass and still be reflected back at the interface. Accordingly, both the operation of a simple mirror and the idea of **total internal reflection** (the crucial concept behind the booming technology of **fiber optics**) are examined. The next subject is **color** and the appearance of things in general. For example, why is ice clear and snow white? The last segment is an introduction to **lenses**—how they form images and how they function in such devices as eyeglasses and telescopes.

The second section is an introduction to three important classes of optical phenomena, the study of which established the wave theory of light in the nineteenth century. **Interference** produces those familiar swirls of color that grace soap bubbles, oil slicks, and even peacock feathers. **Diffraction** can be simplistically thought of as the bending of light waves around obstructions as is commonly manifest in the complex structure of shadows. Just look very carefully at the shadow of your hand in direct sunlight, or hold a pencil between your eye and a distant lamp, and "watch" the light bend around its edge. **Polarization** is perhaps most widely known via Polaroid sunglasses (and there are still faint bespectacled memories of those dreadful 3-D movies of the fifties). Still the phenomenon occurs everywhere. Even skylight is partially polarized, and so is the glare on a bald head.

The degree to which a substance is opaque or transparent to any kind of electromagnetic radiation is determined by the structure of its atoms and the manner in which those atoms group themselves. Light that traverses a piece of clear glass does not simply sail through the interatomic spaces. The process is far more subtle than that.

CHAPTER 21 Light

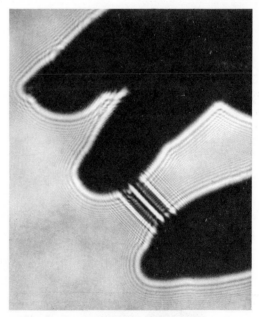

The shadow of a hand-held paper clip under plane wave illumination.

21.1 REFLECTION AND REFRACTION

481

In the beginning God created the heavens and the earth. The earth was without form and void, and darkness was upon the face of the deep; and the Spirit of God was moving over the face of the waters. And God said, "Let there be light," and there was light.

GENESIS

AUGUSTIN JEAN FRESNEL
(Physicien),
Membre de l'Académie des Sciences.
Né à Broglie (Dépt de l'Eure) le 10 Mai 1788.

Augustin Jean Fresnel

Every atom can be imagined as a kind of oscillator, an electron cloud bound to its nucleus by a springlike Coulomb force so that the whole thing can vibrate. And as with any mechanical system, atoms have **natural frequencies** at which they will oscillate when struck a sharp blow of some sort. Tap a wine glass with something rigid, and it will ring at its natural frequency. That is a rather special quantity because it is also the rate of oscillation at which the system (the goblet) most effectively responds to a pulsating external excitation. In other words, if we try to pump energy into the wine glass, it will vibrate most effectively, taking up the most energy, when driven by an oscillation that matches its own natural (or **resonant**) frequency.

That's how Enrico Caruso (and more recently Ella Fitzgerald supposedly) shattered such glasses—by driving them with powerfully sung notes at or very near their resonant frequencies. And that is also why the doors on my china closet vibrate when the bass is turned up on the hi-fi and why the ashtray in my car clatters annoyingly to the beat of the radio. As a matter of fact, troops marching over bridges are usually ordered to break step in order to avoid driving those structures into resonance with the pounding low-frequency cadence.*

The pulsating electromagnetic field of an incoming light wave can exert forces on an atom's electron cloud, setting it into oscillation. The result is the conversion of some of that radiant energy into vibrational kinetic energy. But this can happen in general only when the frequency of the light matches a resonant frequency of the atom. In effect, an atom has certain energy levels or natural frequencies at which it will ring like a bell—or a wine glass (Section 11.3). That ringing atom is said to be in an **excited state,** a condition that typically lasts about 10^{-8} or 10^{-9} second before a photon is emitted and the atom drops back to its usual ground state. However, if the illuminated substance is dense enough—as it often is, especially when it is a solid or a liquid—collisions between adjacent atoms may occur during that relatively long time. Consequently, the excitation energy may be transferred into thermal energy before it has a chance to be reradiated. In fact, that is just what the word **absorb** means in this context—*the conversion of energy from radiant to thermal.* The excited atom returns to the ground state in a nonradiative transition, and the photon that initially came in is transformed into random atomic motion of the substance.

Every material has such natural frequencies at which energy can be absorbed. Microwaves are strongly absorbed by water, which makes them ideal for cooking moist things (meat, fish, and the like), and yet water is quite transparent to light. Germanium is a metallic-gray, opaque element, and yet it's almost perfectly transparent to

* At least two bridges collapsed in this way during the mid-nineteenth century—one near Manchester in England, the other in Anjou in western France.

IR. We cannot see through a chunk of graphite, but X-rays sail right through it. And that also tells us something very important about our human light-picture of the world around us. What we happen to see with light is not the only vision of the universe.

Clear colorless glass usually has resonances in both the IR and UV, strongly absorbing certain frequency bands in each. For example, most glass of any appreciable thickness is quite opaque to UV wavelengths of around 100 nanometers. The atoms resonate and absorb that radiant energy, and little or none passes through the glass. You can easily prove to yourself that glass absorbs radiant energy by simply feeling a windowpane that has been warmed in the sunshine. It's a waste of time to try to get a sunburn behind a window. By contrast, the frequencies of visible light are not near a resonance in glass, and so light traverses that substance with hardly any loss at all—but of course that's why it's clear and colorless in the first place. Moreover, if you've ever listened to a portable radio in a windowed room, you already know that glass is transparent to radio waves as well.

When an atom is struck by a photon whose frequency does not match one of its own natural frequencies, it will generally not be raised to an excited state. Instead, the atom will take up that blast of radiant energy (resulting in a situation that can be likened to a ground-state oscillation, formally called a virtual transition). It will then immediately fire out a photon of the same frequency and will thereafter calm down again. This *nonresonant* response is the prominent mechanism that gives rise to much of what we ordinarily see.

That swallowing up and firing out over and over again, from one layer of atoms to the next, is the way light progresses through a clear piece of glass or plastic or ice or water or anything else. When a beam of light impinges perpendicularly on a piece of glass, most of the energy reradiated by the atoms continues inward in that same direction, and that light is described as **refracted** or **transmitted.** However, a small fraction (about 4% or so) of the light that is reemitted by the atoms comes back out in the opposite direction and is known as the **reflected** beam. It's this light that allows us to "see" the glass itself or at least to determine where its surface is. The photons that emerge, whether reflected or refracted, are not the ones that entered. With each atomic collision a photon vanishes, and a brand new duplicate copy emerges.

The atoms within glass do not respond to all the different-frequency nonresonant photons in exactly the same way during this ritual of swallowing up and firing out. What happens is rather complicated—it involves advancing the wiggling fields in one case and retarding them in another—but the effect is simple enough. Because the frequencies of light are below the UV resonance, each time light is reemitted in the glass, it will be "held back" by the atoms. As a result, the light appears to have traveled through the glass at some slower speed, v, even though no single photon ever moved through the interatomic void at any rate other than c—indeed, they exist only at c.

AMAZING PREDICTIONS

Greater things than these can be performed by refracted vision. For we can give such figures to transparent bodies and dispose them in such a manner with respect to the eye and the objects that the rays will be refracted . . . toward any place we please, so that we shall see the object near at hand or at any distance. . . . Thus from an incredible distance, we may read the smallest letters; and may number the smallest particles of dust and sand . . . and many things of the like sort, which persons unacquainted with these principles would refuse to believe.

ROGER BACON

(1220–1292)

Scattering of some of the energy out from a plane wave in the form of a spherical wave.

The ratio of c to v—that is, c/v—is an important practical measure of the relative slowing down of light in any medium as compared with its rate in vacuum, a quantity called the **index of refraction,** n. Gases at atmospheric pressure are quite tenuous and have so little effect that v is very nearly equal to c, and n usually differs from 1 only out in the fourth decimal place. Water has an index of 1.33, and most transparent solids (glasses, salts, and plastics) have indices of around 1.5. Diamond has an extraordinarily high refractive index of 2.4—light traverses it at less than half its vacuum speed.*

The lovely blue color of our sky and our fireball orange sunsets arise from nonresonant **scattering,** the swallowing up and firing out of light by atmospheric molecules, particularly N_2 and O_2. These molecules have UV resonances, and as sunlight comes blazing through the atmosphere, that random jumble of oxygen and nitrogen will scatter it in all directions. The nearer in frequency some portion of the light is to the resonance, the more it will be scattered. Therefore, these molecular oscillators most effectively scatter the violet and blue components of white light (which are close in frequency to their own UV resonances). A fair percentage of violet and blue are flung every which way out of the beam of sunlight, whereas green, yellow, orange, and red are successively less affected. Since sunlight starts out weak in violet, and our eyes don't respond very sensitively to it either, the light scattered out into space and down into your window has a definite blue tinge. Of course, invisible UV is scattered as well. That's why you can get a bit of a sunburn even while sitting in the shade.

---

* X-rays, whose frequencies are higher than the resonances, are effectively sped up by glass and will emerge *as if* having traveled faster than c! Experimentally measured values of the index of refraction for X-rays are indeed less than 1.

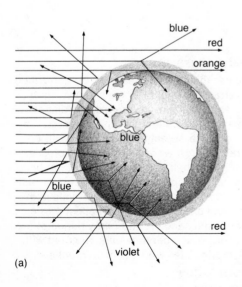

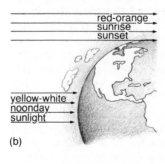

(b)

The scattering of blue light by the Earth's atmosphere.

Looking directly at the sun low in the sky as it rises or sets, we see it in light that has passed through a great thickness of air, in light that has lost much of its violet and blue. We see it in the unscattered reds and oranges that still remain. If there were no atmosphere, the bright daytime sky that bathes the planet with light in all directions would vanish. On the moon there is blazing sunlight, dark shadow, and a black star-studded sky (p. 480).

Imagine a light wave impinging at some angle on a smooth piece of glass. A fraction of the energy will be reflected; the rest will be transmitted. By the way, all the angles we will be dealing with are measured from a perpendicular to the surface. The ancient Greeks already knew that *the reflected beam comes off at an angle equal to the angle made by the incoming or* **incident beam.** That much is easy to observe with a flashlight or even better with a laser beam. The imagery of antiquity was in terms of straight-line streams of particles, a metaphor that later got translated into the Latin as radii and reached English as **rays.** Nowadays we define a ray not as an actual beam of light but *as a mathematical direction, a line following along the flow of energy.* In a medium that is uniform (homogeneous) the rays are straight, and if the medium is also the same in every direction (isotropic), *the rays are simply perpendicular to the wavefronts.* For a point source, rays flow outward radially, whereas with plane waves the rays are all parallel.

From One Medium to the Next

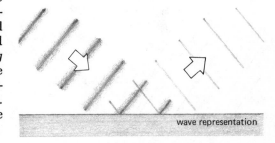

wave representation

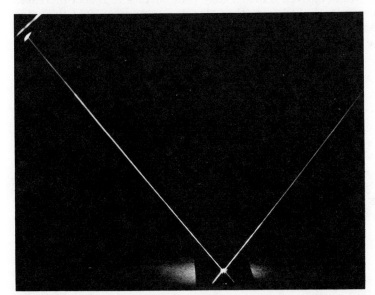

Specular reflection. The angle of incidence equals the angle of reflection.

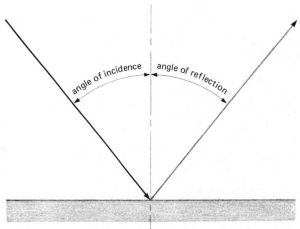

angle of incidence   angle of reflection

When the reflecting surface is smooth, the light reemitted by millions upon millions of atoms combines to form a single well-directed wave, and the process is called **specular reflection.** However, if the surface is rough (i.e., its irregularities are large in comparison with λ), although *the angle of incidence equals the angle of reflection* for each individual ray, they all come off in every which way, constituting what is known as **diffuse reflection.** Because this page is a diffuse reflector, it can be seen from a wide range of angles, and it still does not allow the source of room light to produce a glaring mirror-like reflection.

When you stand in front of a flat mirror (any polished surface), every illuminated point on your body serves as a source of spherical wavelets of reradiated light or, equivalently, as a source of a cone of rays. Some of these rays will reflect off the mirror in just the right direction to enter your eye. But the eye-brain receiver, accustomed to straight-line propagation, will "see" the source as if it were behind the mirror. Every point on your hand will have its corresponding image point behind the mirror, recreating the appearance of that hand. Rays reflecting off the mirror but not reaching your eye may be received by someone else looking into the mirror, who will then see your image as well. The fact that the mirror is flat and the fact that the angle of incidence equals the angle of reflection combine to produce an undistorted, rightside up, life-size image standing just as far behind the surface as you are in front of it.

The ability of even a moderately poor surface to reflect increases as the incoming light approaches glancing incidence. Hold this page horizontally at eye level so that light from a lamp grazes off it into your eye, and you'll see a bright image of the bulb reflected in the paper.

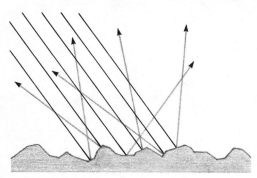

Diffuse reflection.

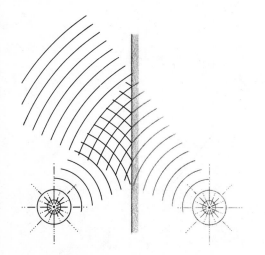

A cone of spherical waves from a point source reflecting off a mirror.

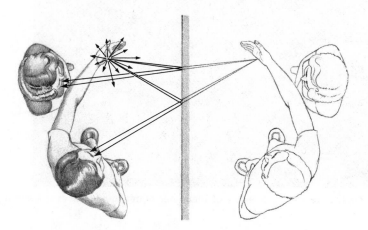

Images in a plane mirror drawn using rays.

Having considered the waves reflected off a piece of glass, let's now follow the light as it crosses the interface and enters the more dense medium. Two very interesting things happen as a result of the difference in the indices of refraction of the two substances (i.e., the difference in the speeds of the wave). First, the portion of the wavefront still in the air travels faster than the portion moving into the glass. Consequently, the region in air advances farther, and the wavefront bends. In effect, as the lower portion gets bogged down in the glass, the segment in the air whips around it, with the result that the transmitted beam propagates at an angle different from the angle of the incident beam.

When light enters a medium of higher index, it bends toward the perpendicular line, and when it enters a less optically dense medium, it bends away from the line. For example, the light coming from a coin at the bottom of a cup of water will be deflected away from the perpendicular as it emerges into the air. An observer will "see" the coin as resting higher in the liquid then it actually is— a phenomenon that makes spearing fish rather tricky. Similarly, a pencil half-submerged in water seems to bend upward as if it were rubber. All of this bending of beams obeys well-known laws and can be calculated with great precision.

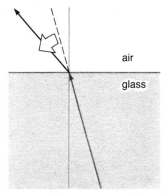

Without the water in the cup to bend the light, the coin would not be visible at that angle. Try it! This experiment was first suggested by Ctesibius of Alexandria (50 B.C.).

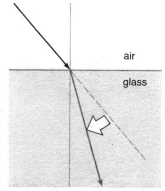

The refraction of a light ray toward the perpendicular.

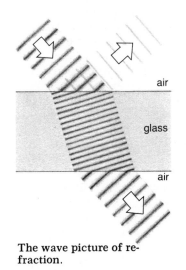

The wave picture of refraction.

The reflection and refraction of light incident on a block of glass.

The refraction of a light ray away from the perpendicular.

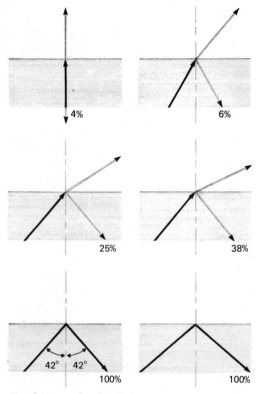

4%

6%

25%

38%

42° 42°

100%

100%

Total internal reflection.

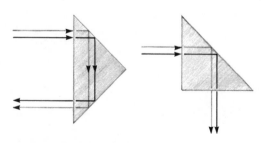

These prisms do their thing via total internal reflection. They are constructed so that the incident angle on the reflecting face exceeds the critical angle.

The other interesting thing that happens (see figure) is that the wavelength decreases in the more dense substance. Since $v = \lambda f$ and the number of waves per second, f, does not change as the disturbance moves across the interface,* a decrease in v necessitates a decrease in λ. Note that this doesn't mean the color changes—that's fixed by f, i.e., by the energy. If a color is referred to by its wavelength (for example, the bright red 632.8-nm emission of a helium-neon laser), either the transmitting medium is assumed to be vacuum or it must be specified as well. If not, the reference to color is meaningless.

It's a bit like the relationship between mass and weight. Telling someone you weigh 500 lbs is not much good unless you specify in which astronomical gravity field the measurement was made. Experience certainly tells us that color does not appreciably change as light goes from one transparent medium to the next. Under water a red (650 nm) bathing suit still looks red, even though the wavelength is decreased by about 25 percent (to 489 nm, which in a vacuum would be a pretty blue). The bathing suit is red in the water and red in a vacuum. Even though the wavelengths of the light it reflects are different, the frequencies are the same.

Although it is more familiar to imagine light in air arriving on glass, it certainly is possible for the beam to originate in the optically more dense material. When that occurs, a curious and rather useful thing happens. As the incident angle becomes larger, the transmitted beam bends more and more away from the perpendicular and toward the interface. As that takes place, the transmitted beam weakens and the amount of light reflected back into the dense medium increases. When a particular incident angle is reached (known as the **critical angle**), all the light striking the interface will be reflected back, none of it being transmitted. This **total internal reflection** persists for all incident angles greater than the critical angle, which for air and glass is about 42°. Reflecting prisms cut with 45° angles take advantage of this phenomenon to redirect light (since the incident angle will exceed the critical angle).

The most exciting application of this scheme is in *fiber optics.* Light channeled into a thin glass or plastic fiber will generally hit the inside surfaces at orientations exceeding the critical angle and so be reflected back and forth thousands of times as it advances each foot along the fiber. Light so captured can be piped great distances over winding, even knotted, paths. Flexible bundles of fibers can carry images, or each strand can be used to transmit information,

---

* If the light is exceedingly bright it can drive the oscillating electron clouds into a complex motion that generates other frequencies as well, but that is a rather special situation. Clear colorless crystals have been developed, however, into which we can shine a red light beam, and out will come a green beam. But again, this is not what happens ordinarily.

as a phone cable does. Much of the industrialized world is now beginning to shift toward the era of optical communications, of light flashing along fibers, replacing electricity moving in wires—for transmitting not power but information. The much higher frequencies of light allow for incredible increases in capability. For example, at present just a handful of TV stations in each area can be accommodated at the frequencies used in broadcasting. If light was used instead, perhaps 75,000,000 stations could be going at it all at once in each locale.

Light entering a thin fiber of glass is literally captured within it.

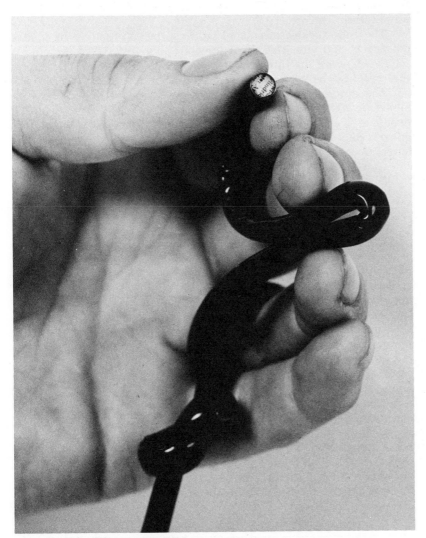

A bundle of thousands of carefully arranged thin glass fibers transmitting an image, even though the fibers are knotted and sharply bent.

Looking down into the edges of a stack of very thin, flat cover-glass slides held together by a rubber band. Note how it works as a light guide.

Red, White, and Blue . . . and Green

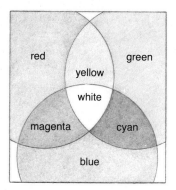

Three overlapping beams of colored light.

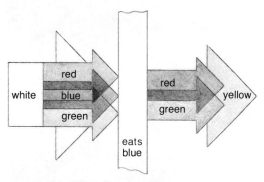

Yellow stained glass absorbs blue light, passing yellow.

The manner in which colored beams of light blend to produce new hues is quite different from the mixing of paint pigments, inks, and dyes. The former is known as *additive* because each beam adds color; the latter is *subtractive* because each pigment absorbs the colors it doesn't reflect.* In fact, because most of us have little experience with colored light, we almost have to see it to believe it—to believe, for example, that a beam of red overlapping a beam of green will produce a region of bright yellow. Of course, the corresponding pigments would have resulted in a dark brown. The eye responds to red and green together just as it does to the single frequency about midway between, yellow.

The widest range of colored light can be conjured up by mixing red, green, and blue, and these three are therefore called the **primary** colors. They are the three components (emitted by three phosphors) that produce the whole gamut of hues seen on a color TV set.

The accompanying figure shows many of the basic results obtained by mixing light: Red plus green is yellow; red plus blue is magenta (a sort of reddish purple); blue plus green is cyan (a bluish green); and all three primaries together make white. Using the first letters of each, we can write all this as

$$R + B + G = W,$$
$$M + G = W, \text{ since } R + B = M,$$
$$C + R = W, \text{ since } B + G = C,$$
$$Y + B = W, \text{ since } R + G = Y.$$

Any two colors that together produce white are said to be *complementary*, and the last three symbolic statements exemplify that relationship. The greatest range of hues can be produced with inks and dyes by mixing magenta, cyan, and yellow. These are the "primary colors" of the paint box although they are very often mistakenly spoken of as red, blue, and yellow.

Suppose then that we obtain a piece of stained glass with a resonance arranged to exist in the range of frequencies corresponding to blue. Looking through it at a source, we can consider the incoming white light to be composed of red, blue, and green. The medium strongly absorbs blue, passing everything else—red and green —hardly diminished. But red and green combined is yellow, and so out comes yellow light. The stained glass looks yellow. Yellow cloth, paper, paint, dye, plastic, and ink all absorb blue. Therefore, if you peer through a yellow filter (e.g., a piece of clear yellow plastic) at something blue, the only light coming from the object will be blue, and that will be absorbed—the object will appear black. *All the colors together produce white; the absence of all of them means no light at all, black.*

---

* Mix all the colors of paint together and you get no color—black. Mix all the colors of light together and you get all color—white.

Much of the color that surrounds us arises from exactly this sort of selective absorption, but let's consider the phenomenon of whiteness first. *The presence in a light beam of a wide range of frequencies will be perceived by the eye as* **whiteness.** It may be the blue-white of a star like Rigel or the yellow-white of a faded page. We easily adjust to color variations, hardly noticing the myriad different whites. Since a photographic film cannot change its perception, and the color of a white sheet of paper is different when lit by a bulb or by the sky, there must be both indoor and outdoor color films. Nonetheless, the central aspect of whiteness is the presence of a range of different visible frequencies.

This page is white; so are salt and snow; clouds are white; powders and pills and cloth are white; even the foam on a glass of beer is white—and yet all these diverse perceptions share a common origin. They are all composed of grains or fibers or particles or bubbles that are large in diameter compared with the wavelengths of light but otherwise quite small. Moreover, despite the common opinion that things white are inherently opaque, each grain or fiber or bubble is actually transparent. Sprinkle a single layer of salt on this page, and the print will still be legible. Crush a piece of clear glass, and the grains, like grains of sand, will appear white. There is no such thing as a single particle of white pigment that by itself is opaque white. That may seem paradoxical, but it's true.

Whiteness arises when the incident white light is reflected back at the viewer without any preferential absorption. All colors come in and all scatter off equally. When white light enters a structure composed of many layers of transparent objects whose surfaces are oriented in every direction, it is *diffusely scattered.* What emerges is white light, and the material looks white. Whether reflected or transmitted, it is light composed of a broad range of frequencies, and it is diffuse. Indeed, the difference in appearance between a white surface and a mirrored surface is one of diffuse versus specular reflection—they both reflect the full color range of the incident light.

To make white paint we need only to mix a powdered clear material, the pigment, into a clear liquid vehicle like oil or the newer acrylics. If the indices of refraction of the two are nearly equal, the pigment will simply vanish into the liquid, and the "paint" will be transparent and of no use. If the indices are markedly different, there will be plenty of reflection at each of the countless interfaces, and the paint will be a bright white.

To color paint or cloth or anything else, we need only to dye the surfaces of the transparent grains or fibers. For example, if the dye absorbs cyan, every time the initially white light passes through a film of the dye, it will lose more of its blue and green. Instead of all the colors reflecting back at the viewer, only red will remain, and the cloth or paper will appear red. **Selective absorption** is the primary mechanism that colors everything from balloons and book covers to lips and roses.

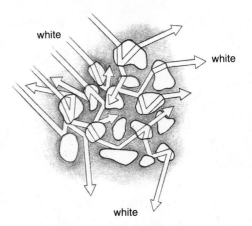

Scattering of white light off layers of small transparent objects.

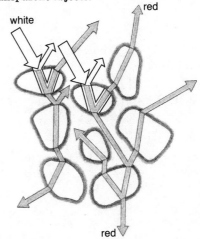

The appearance of red-dyed fibers under white light illumination.

Lenses

A lens is a refracting device whose primary function is to reshape wavefronts—often, though not always, in order to form images. Suppose as a starter that we wished to create a *burning-glass* (a lens commonly used to start fires in ancient times) like the one alluded to by Aristophanes in his comic play *The Clouds* (424 B.C.). In essence, such a gadget simply creates a tiny image of the sun where the energy is then so concentrated that it ignites absorbing materials. Almost everyone has played that game with what we call a magnifying glass, although most of us take a while to realize that white paper reflects all too well.

The wavefronts of sunlight, having expanded out 93 million miles, are fairly flat, and so our problem is to convert plane waves into converging spherical waves that will collapse to a single point. We need to bend forward the outer extremes of the flat wavefront relative to its central region. That can readily be done by interposing a material such as glass. Because glass has a higher index than air, the light traveling within it is slowed down. If the piece of glass is thicker in the middle than at its edges, the wavefront will be slowed most at its center and least at its periphery, bending just as desired. Alternatively we can think of a bundle of parallel rays impinging on the lens, refracting, emerging, and arriving at a point. In either conception the light converges to what is called a **focal point.** There are two such points, one on either side, each a distance away from the center of the lens, known as the **focal length.** This sort of device is said to be a **positive** or **converging** lens, and it will work perfectly well as a burning-glass.

If the lens is thinner at its middle than at its edge, the wavefront will be advanced in the center, bellying out into an expanding spherical wave. This sort of **negative** or **diverging** lens constitutes the second main classification. Incidentally, virtually all lenses are produced with surfaces that are segments of spheres, not because that configuration is ideal for the job—it's not—but because any other shape is exceedingly difficult to grind.

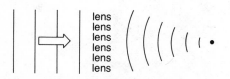

We would like a lens to transform incoming plane waves into converging spherical waves.

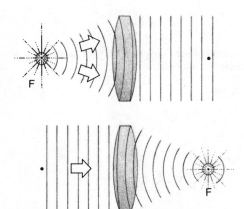

The operation of a converging lens in the wave picture.

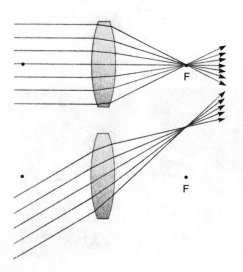

A parallel bundle of rays is brought to a focus by a converging lens.

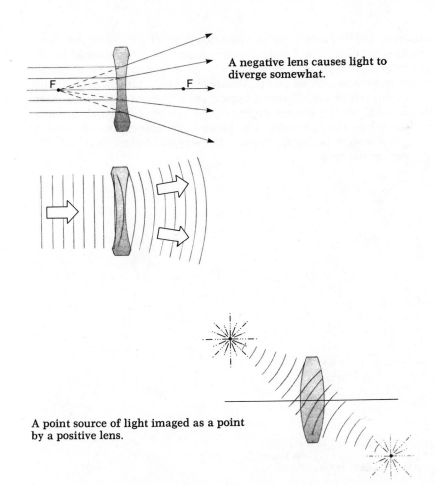

A negative lens causes light to diverge somewhat.

A point source of light imaged as a point by a positive lens.

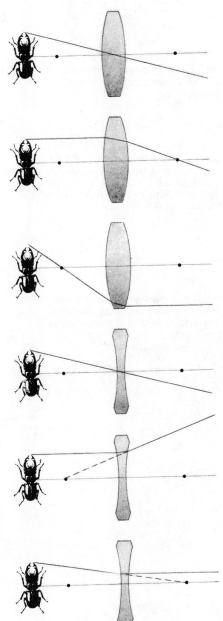

A single positive lens is also capable of taking the light diverging from a source point and causing it to converge to an image point. That is the crucial ability that allows each and every point on the object to be transformed into a corresponding image point, thereby creating a light picture of the object. For a thin lens there are only two rules to keep in mind, and with them we can determine the size and shape of any image. (1) A ray heading toward the center of a lens goes straight through. (2) Any ray that enters or leaves a lens parallel to the central axis will pass through a focal point (or can be extended to pass through one).

Now suppose that a bug is sitting beyond the focal point of a lens, as in the figure. We can locate the image of the top of its head by tracing any two emanating rays through the lens, realizing that all the other rays emitted from that point and collected by the lens will converge to the same spot, forming a point on the image. Among the easiest rays to construct are one straight through the center of

Tracing a few key rays through a positive and negative lens.

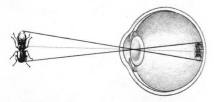

A camera lens projects an image on a viewing screen very much as the eye does on the retina.

When an object is between one and two focal lengths from a positive lens, its image will be inverted, real, and larger than life. That's how a slide projector works and why the slides go in upside down.

As an object approaches a lens, rays received from it are increasingly divergent. In other words, rays that diverge a great deal when coming from a distant object simply don't get into the lens.

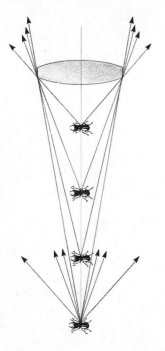

the lens and the other through the focal point (F) emerging from the lens parallel to the axis. Where these two intersect will be the image of the bug's head, and the same process can be performed for every point down to its rump. What results is a larger-than-life, upside-down image, which would appear on a piece of paper held there, much as a moving picture appears on a screen. Images that can be projected like this are called **real images.** The cornea and the lens in your eye work together in exactly this way to produce real, inverted images on the retina.

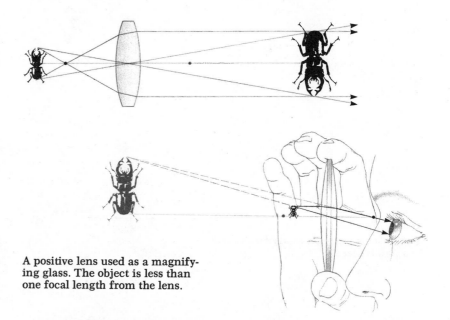

A positive lens used as a magnifying glass. The object is less than one focal length from the lens.

Actually the type of image formed by a given lens is determined by the distance of the object in comparison with the focal length. The closer an object comes, the more its rays entering the lens diverge, the more bending is required in order for the lens to bring the rays into convergence, and the larger the resulting image is.

The focal length of an eyeball is about the size of that eyeball. Anything farther away than two focal lengths from a converging lens will form a real inverted, *minified* image—convenient if you want to squeeze a picture of the New York skyline onto the relatively tiny retinal screen. And of course, most everything you look at is farther than two eyeball lengths away. The same configuration is used in the photographic camera, where the retina is replaced by a roll of film.

A bug located exactly two focal lengths from a converging lens has a real, *life-sized* image two focal lengths beyond the lens. This is the setup used in copy machines.

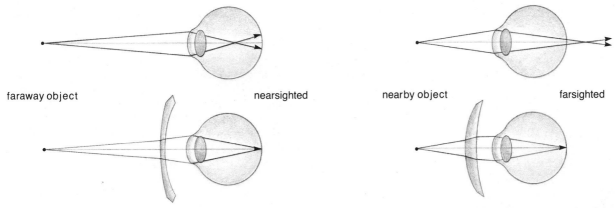

faraway object   nearsighted

nearby object   farsighted

With the bug sitting between one and two focal lengths, its image is still real and inverted, but now it's *magnified*. That's the arrangement used in slide projectors, and that's why the slides go in upside down.

As the bug moves toward the lens, coming closer to one focal length, its image moves ever farther away. At the moment our bug gets exactly to the focal point, its image moves off to infinity and simply vanishes—there is no image.

When the bug comes nearer than one focal length, the lens can no longer bend the rays enough to make them converge. At that point someone peering into the lens would see the beastie in diverging rays, very much like the light coming from a mirror. The image produced by the eye on the retina is real, but the magnified, rightside-up image formed by the lens is not. That image (which appears to be on the same side of the lens as the bug itself) can be viewed and studied just as if the bug had grown several times larger, but it cannot be projected onto a screen, and so it is said to be a **virtual image.** This is the familiar application of the converging lens as a *magnifying glass.*

The *nearsighted* eye, the eye that can clearly view close objects, is one that generally provides too much convergence; that is, the images of distant objects, where the incoming rays have little divergence, are formed in front of and not on the retina. Consequently the picture formed on the retina is blurred. That problem can be corrected simply if the rays are made to diverge a little by passing through a negative lens before entering the eye.

The *farsighted* eye, the eye that can clearly view distant objects (whose rays therefore diverge only slightly) provides too little convergence. Its images of nearby objects (from which it receives highly divergent light) "form" behind the retina, and so what appears on the retina is a blur. To correct that condition we need only to increase the convergence with an additional positive lens. If you can project the image of, say, a lamp onto a piece of paper with an eyeglass lens, then the lens is positive and the wearer is farsighted. Negative lenses do not form real images.

The correction of near- and farsightedness using eyeglass lenses. The nearsighted eye has trouble seeing things far away. The farsighted eye blurs the images of nearby objects.

A real image projected on the viewing screen much as the eye projects its image on the retina.

The minified, rightside-up, virtual image formed by a negative lens.

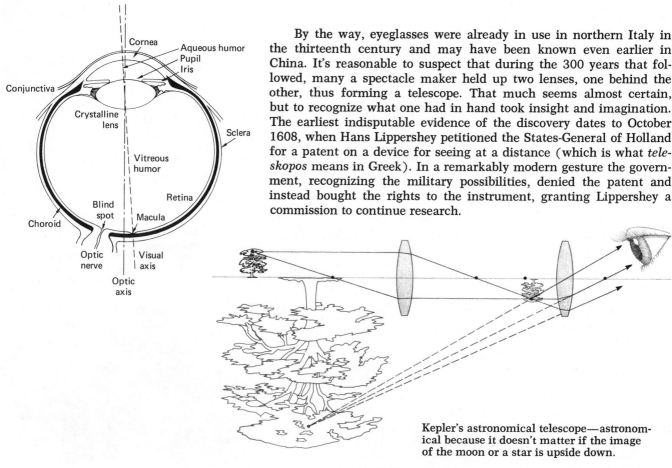

By the way, eyeglasses were already in use in northern Italy in the thirteenth century and may have been known even earlier in China. It's reasonable to suspect that during the 300 years that followed, many a spectacle maker held up two lenses, one behind the other, thus forming a telescope. That much seems almost certain, but to recognize what one had in hand took insight and imagination. The earliest indisputable evidence of the discovery dates to October 1608, when Hans Lippershey petitioned the States-General of Holland for a patent on a device for seeing at a distance (which is what *teleskopos* means in Greek). In a remarkably modern gesture the government, recognizing the military possibilities, denied the patent and instead bought the rights to the instrument, granting Lippershey a commission to continue research.

Kepler's astronomical telescope—astronomical because it doesn't matter if the image of the moon or a star is upside down.

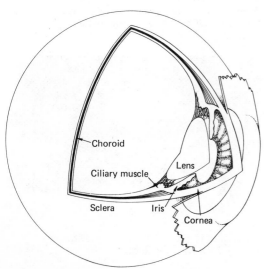

Galileo constructed his first telescope in 1609, but it was Kepler who brought the lens from the domain of the shopkeeper into the sphere of science. It was in his *Dioptrics*, published in 1611, that the operation of the lens was finally framed in the light of a rigorous mathematical theory. He explained the workings of the telescope and even went on to design an improved version of his own, an instrument consisting of two positive lenses. Its operation is quite easy to understand. The object being viewed is located far away from the front converging lens, called the *objective*, which itself has a relatively long focal length. That first lens creates a real, inverted, minified image, which is then viewed by a second lens, called the *eyepiece*. In other words, for the eyepiece the object is tiny, upside down, and nearby. As long as this intermediate image falls at less than one focal length from the eyepiece, that lens will function as a magnifying glass. It will create the final image—inverted, virtual, and perhaps considerably magnified. In effect, the telescope puts the distant object under a magnifying glass.

After a long obscurity, the wave theory of light was born again in the agile hands of Dr. Thomas Young, one of the most brilliant minds of the late eighteenth century. At Cambridge he was called "Phenomenon Young." He studied under Joseph Black at Edinburgh and lectured at Rumford's Royal Institution, becoming its first professor of physics. As a physician, his early interests in the eye gradually gave way to an exploration of the nature of light itself.

By the end of the eighteenth century, Young was already lecturing on the wave theory and had introduced the very important **principle of interference:**

> When two undulations, from different origins, coincide either perfectly or very nearly in direction, their joint effect is a combination of the motions belonging to each.

Two waves of the same frequency overlapping so that peak aligns with peak will combine in what is called **constructive interference** to produce a single larger disturbance. By contrast, waves aligned so that the peaks of one more or less overlie the valleys of the other undergo **destructive interference.** They will tend to counteract one another; the resultant wave will be the difference of the two and may even vanish altogether.

21.2 LIGHT AS WAVE

Interference

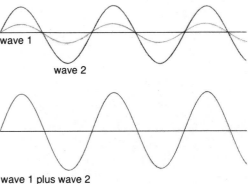

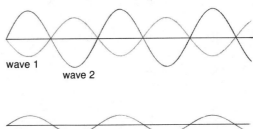

Constructive interference of two overlapping waves.

Destructive interference of two overlapping waves.

Overlapping water waves pass each other and sail away unruffled.

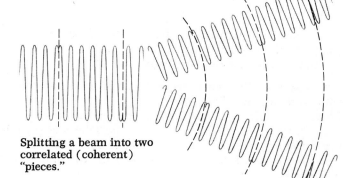

A simplified representation of ordinary
light as a series of unrelated wave trains.

$\sim 10^{-8}s$ $\sim 10^{-8}s$

ordinary light

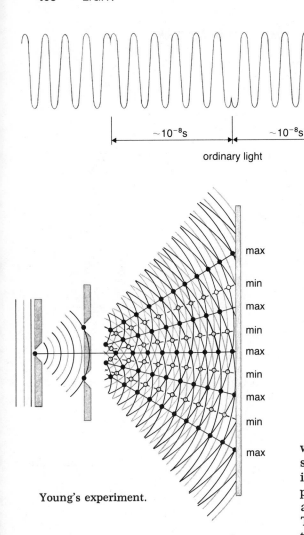

max

min

max

min

max

min

max

min

max

Young's experiment.

Splitting a beam into two
correlated (coherent)
"pieces."

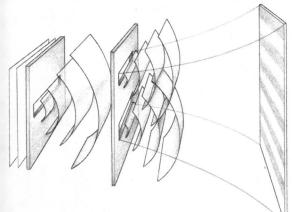

The problem in testing these conclusions as they applied to light
was that Young needed two waves that maintained a fixed relationship between their respective risings and fallings. An ordinary source
is composed of the emissions of countless atoms, each firing out tiny
pulses of light lasting less than 10^{-8} second. The resulting beam as
a whole is a jumble of wiggles that changes randomly and rapidly.
The smooth succession of hills and valleys is there, but it's not sustained for any more than an instant before it changes and begins
again. Two such ordinary sources, two flashlights, would be correlated
only over intervals of time shorter than 10^{-8} second. If their beams
overlapped, they would interfere, but the pattern would change, and
after 10^{-8} second or so, it would change again and then again so
rapidly that it would be totally unobservable. To get around this difficulty, Young cleverly split off two "pieces" from a single beam and
used them to interfere with each other. No matter how they varied in
time, since they had the same source, they remained coordinated.

Young's **double-slit** experiment is one of the classics of physics. A
source followed by a filter provided light composed of a narrow band
of wavelengths (i.e., light that could be loosely thought of as having a
single frequency). By shining this on a narrow slit, he produced a cylindrical single-colored wave. Another opaque sheet with two slits in

it led to the generation of two coordinated cylindrical waves, which propagated out, overlapping and interfering throughout the region beyond. A viewing screen placed anywhere in that space revealed a pattern of bright and dark bands corresponding to places where peak overlay peak to bring light, and where peak met valley to cancel it.

This lovely experiment is quite inexplicable if we imagine light as a stream of particles in the classical sense. Particles don't cancel one another and don't veer off from their straight-line paths. The believer in luminous corpuscles would expect to see two well-defined bands of light, one behind each slit, and that's all!

Interference from the thin air film existing between two clean microscope slides.

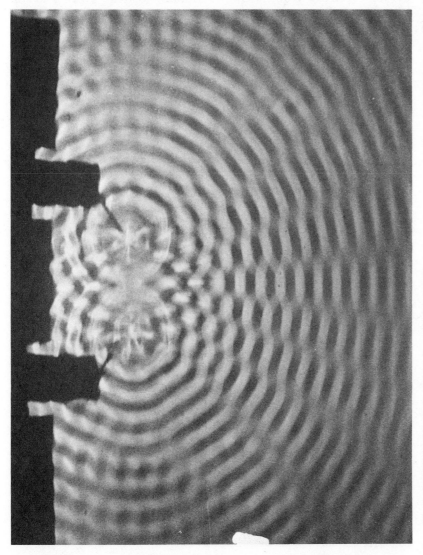

◄ Water waves from two point sources create the classic *Young's experiment* interference pattern in two dimensions.

A wedge-shaped film made of liquid dishwashing soap, showing interference fringes. The top part is drained thin.

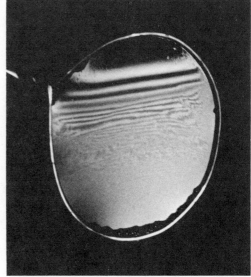

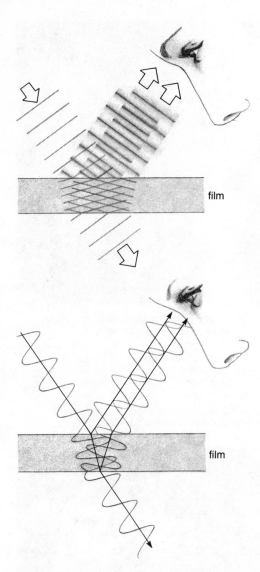

The wave and ray representations of thin-film interference. Light reflected from the top and bottom of the film interferes to create a fringe pattern.

film

film

Besides this triumph, Young was able to explain the colors arising from thin films; the colors of soap bubbles and oil slicks on wet pavement; the iridescence of peacock feathers and mother-of-pearl. Long before, Hooke and Newton had both independently studied thin-film phenomena, and both had recognized the underlying repetitive nature so beautifully displayed in the bands of concentric color—a nature that suggested even then that something was vibrating at the very heart of the process.

Stated rather simply, Young argued that a thin film has the effect of splitting a wave into two correlated segments, one reflected from the top surface, the other reflected from the bottom. These come off in the same direction and are made to overlap at some point on the viewer's retina. Because they travel different routes, depending on the film's thickness, they may ultimately interfere constructively or destructively. In light composed of a single wavelength, the surface of the film will appear to be covered with bright and dark regions. In white light those regions become colored. Wherever an oil film is exactly the right thickness for the two waves of emerging red light to undergo constructive interference, the film will appear to reflect a blotch of red light—and so on across the spectrum and across the film.

Young's superb accomplishment brought him instant enmity and almost universal rejection. An uninformed writer in the *Edinburgh Review* described the masterwork as "destitute of every species of merit." Under the pall of Newtonian infallibility the pedants of England were not quite ready for the new wisdom of Young.

Diffraction

When an object is illuminated by a beam of light from a point source, it casts a shadow of some sort on a distant screen. But marvelously, that shadow is an intricate complex of bright and dark regions. The Jesuit Francesco Grimaldi had noticed as much in the mid-1600s, and he was convinced by what he saw that light was a wave. This *deviation from straight-line propagation*, something he called "diffractio," is *a general characteristic of all wave phenomena that occurs when a portion of a wavefront is obstructed.*

Waves bend around obstacles, and the larger the wavelength is, compared with the obstacle, the more the waves bend. Ten-foot-long sound waves easily pass and quickly close around trees and telephone poles and the like. There are no perfectly sharp sound-shadows in which one can stand and not hear the tumult beyond.

Working by himself, a young man in France named Augustin Jean Fresnel (totally unaware of any of Young's earlier results) developed a mathematical wave theory of his own. In 1818 he entered it in a competition sponsored by the French Academy. The judging committee consisted of such luminaries of French science and mathematics as Pierre Laplace, Jean Biot, Siméon Poisson, Dominique Arago, and

Joseph Gay-Lussac. Poisson, an ardent antagonist of the wave description, managed to deduce a remarkable and seemingly untenable conclusion from Fresnel's theory. He predicted that a bright spot should then appear at the very center of the shadow of a circular opaque obstacle; such an absurdity must surely disprove the entire theory.

Arago, who was not one to accept anything on face value, went to the laboratory to learn the truth of Poisson's death blow. And there, wonder of wonders, at the center of the shadow he saw the "absurd," a bright spot of light. Fresnel was right; light behaved like a wave. In the end his paper won first prize.

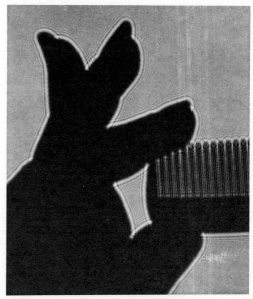

Shadow of a ⅛ inch diameter ball bearing. The bearing was glued to an ordinary microscope slide and illuminated with a He-Ne laser beam. There are some faint extraneous nonconcentric fringes arising from both the microscope slide and lens in the beam.

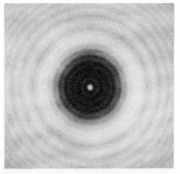

The shadow or diffraction patterns of various square apertures, decreasing in size from (a) through (f). All were illuminated by the plane waves from a He-Ne laser.

Diffraction pattern of a hand-held comb. This is simply the shadow cast on a wall by a distant point source.

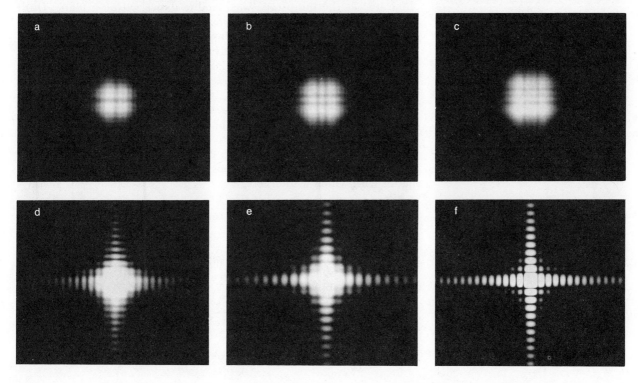

Polarization

A curious discovery in 1669 set the stage for yet another great insight concerning light. A crystal found then and known as Iceland spar (or calcite) remarkably produced a double image of everything seen through it. That strange separation, or *polarization*, was first studied in depth by Huygens. Recognizing that there was some basic asymmetry, Newton attempted to understand it as arising from the fact that "Every Ray of Light has therefore two opposite Sides." Then in 1808 Etienne Malus quite by accident discovered that this two-sidedness of light appeared when light was reflected from certain materials. Fresnel, joined by Arago, conducted a series of experiments to see if the state of polarization of light had some affect on interference. Their positive results were utterly bewildering. Because they ardently believed that light was a longitudinal wave, these asymmetries were totally inexplicable. It was a dark hour in the camp of the wavers.

For the next few years Fresnel and Arago, at last allied with Young, wrestled with that stubborn problem until finally Young had the crucial revelation: Light was a transverse wave.* The two-sidedness was a manifestation of the fact that transverse waves can oscillate in different directions while still traveling in the same direction—something longitudinal waves cannot do. We can wiggle a jump rope straight up and down in a vertical plane or back and forth in a horizontal plane. In general, a wave so restricted is said to be **plane-polarized.** A plane-polarized light wave has its electric field oscillating in a plane just as the jump rope was fixed to dance within a plane. Ordinary light is *unpolarized* in the sense that it can be imagined as composed of waves oscillating in planes at every possible orientation, which change rapidly and randomly, too rapidly to measure.

Calcite is one of many materials each of which has within it two special perpendicular directions determined by planes of atoms within the crystal. When unpolarized light interacts with this medium (which, unlike glass or water, is not the same in every direction), all the radiant energy can distribute itself into two plane-polarized beams oscillating perpendicular to each other. As a result, the two separate images one sees through calcite are each polarized. In this instance asymmetries in the light arise from asymmetries in the matter with which it interacts.

---

* None other than R. Hooke also held that view.

Double image formed by a calcite crystal.

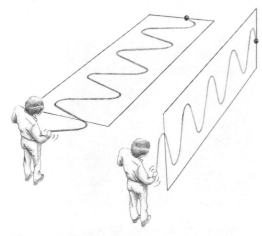

Plane-polarized waves.

Images in sodium chloride (table salt) and calcite single crystals.

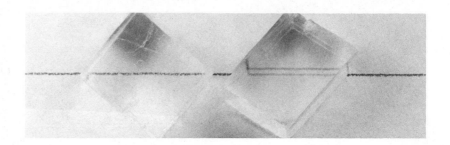

When light reflects at most angles off a nonconducting surface, such as glass or water or the top of a bald head, the wave is interacting with an asymmetrical distribution of matter and becomes polarized to some extent in the process. Indeed, a beam coming off a glass surface at about 56° will be almost totally plane-polarized, oscillating parallel to the interface. That was what Malus stumbled onto and recognized as he peered through a calcite crystal at the sunset glinting off the distant windows of the Luxembourg Palace.

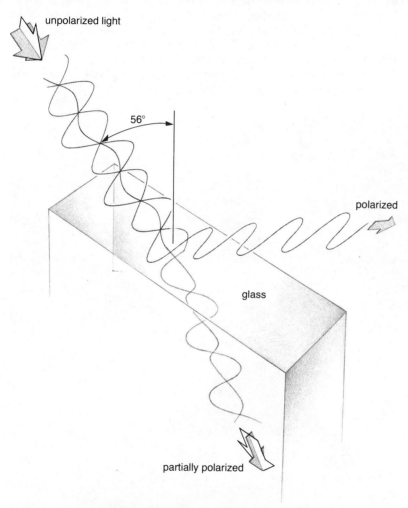

The polarization of light that occurs on reflection from a dielectric, such as glass, water, or plastic.

Light reflecting off the puddle is partially polarized. When viewed through a Polaroid filter whose transmission axis is parallel to the ground, the glare is passed and visible (a). When the Polaroid's transmission axis is perpendicular to the water's surface, most of the glare vanishes (b).

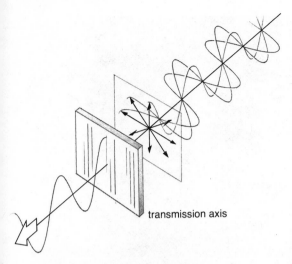

Unpolarized light passing through a plane-polarizer such as a Polaroid filter.

A pair of crossed Polaroids.

Old-fashioned cellophane has its long-chain molecules aligned in one direction and will therefore play lovely tricks with polarized light. Place a crumpled wad of it between crossed Polaroids. The photo shows a piece of ordinary glossy cellophane tape stuck to a barely visible glass microscope slide.

Today the most commonplace polarizer of light is the *Polaroid* filter devised in 1928 by Edwin Land (the founder of that big camera company with the same name). The modern versions of the filter are sheets of polyvinyl alcohol, stretched to align their long chain molecules and dyed with iodine to provide free electrons. The result is an endless array of lined-up molecules along which electrons can flow as if in microscopic wires. An incoming light wave vibrating parallel to these "wires" sets their electrons oscillating as they would in an antenna. Electromagnetic energy is soon lost, in part to Joule heating. Only light whose pulsating electric field is *perpendicular* to these "wires" can effectively traverse the filter, and that direction is called the *transmission axis*. These filters appear gray, simply because they absorb half the incident light. Of course, if two such filters are held with their transmission axes perpendicular, no light at all will emerge. What one passes, the other will absorb.

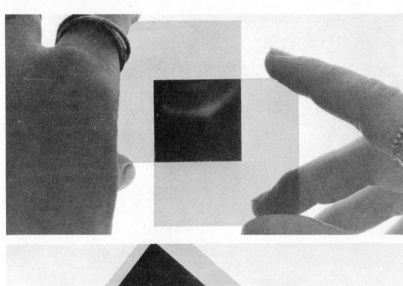

A pair of crossed polarizers. The upper one is noticeably darker than the lower one, indicating that scattered sky light is partially polarized.

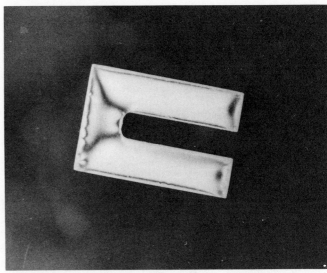

Stress within a material can cause shifts among its molecules, which then affect the transmission of polarized light. Here a piece of clear plastic between crossed Polaroids is being squeezed to increase that stress.

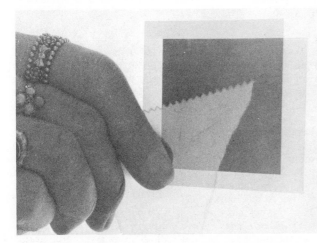

A piece of waxed paper between crossed polarizers. Light, having passed through the first filter, is plane-polarized, but the hodgepodge of fibers depolarizes the beam. Try it with a piece of wet ordinary paper.

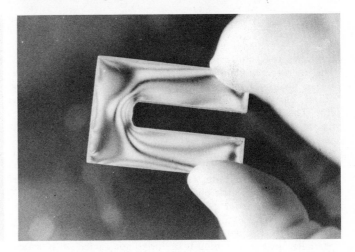

This explanation of polarization was the missing link in the puzzle. By 1825 the particle theory had only a few tenacious advocates; light became a transverse aethereal wave. But of course the story did not end there. Happily, it never ended at all! In time there was electromagnetic theory and then quantum theory and then—

EXPERIMENTS

1. *Natural Frequencies.* If you gently stroke the rim of a clean wine glass with an equally clean finger moistened with water, it will ring with a loud sustained tone. The process is very much like bowing a violin string; it has to be done by feel—not too hard and not too soft. Try it. The effect is marvelous and well worth the effort.

2. *Scattering.* Atmospheric scattering can be simulated by

see the seam run right down the middle of their faces.) You can verify this with another procedure. Look out a window with both eyes open. Hold a finger up at arm's length so it obscures something in the view behind it. Close one eye and then the other. When the image of your finger seems to jump across the background, you are seeing it with your weaker eye.

putting a teaspoon or so of milk in about a quart of water. The microscopic globules will scatter blue light rather nicely. Use a flashlight as a source, and put the liquid in a container that will provide a beam path of 8 or 10 inches if you can. The light coming out the sides and top will be white tinged with blue, and the direct beam will be orange-red. Just as the scattering process partially polarizes skylight, it partially polarizes the bluish waves emerging perpendicular to the beam. Of course, if the batteries are low, the filament temperature of the bulb will be low, and the beam will be reddish (i.e., have little blue) to start with.

The smoke from a lighted cigarette rising in front of a black background will appear bluish white because the particles are microscopic and they scatter preferentially. Smoke particles held in the mouth or lungs get coated with water vapor, and these large droplets scatter white light as a result of multiple refractions. When the smoke is exhaled, the blue tinge vanishes and the smoke appears white.

3. *Reflection.* Hold two plane mirrors together so that they are at right angles. Look directly into the corner and adjust the mirrors until your face looks normal. Now wave with your right hand and watch the image wave its right hand on the opposite side. Try combing your hair while looking into this thing. Whichever of your eyes is the stronger will have the seam running through its image in the mirror. Close that eye and the seam will jump to the other eye. (A few people

4. *Interference.* Young's experiment is easy to duplicate. Just punch two small pin holes (about the size of a period) separated by a distance equal to the diameter of one of the holes in a piece of thin cardboard. Hold it very close to your eye, and look through the holes at a distant street lamp at night. The fringes will appear perpendicular to a line connecting the holes and will rotate if the card is rotated.

Thin films can be made easily with a drop of liquid dish detergent in a little water. Put a straw in the solution, blow some bubbles, and make a froth. After a few seconds of settling, the bubbles will become thin enough to show lovely colors in ordinary light.

Take a very close look at a smooth but worn (not polished) surface under direct sunlight. A fingernail, an old coin, or the roof of an old car will all show a very fine granular pattern of tiny colored dots. This so-called *speckle effect* arises from the interference of light reflected off the ridges of the surface. The same effect can be seen much more clearly with laser light reflected off a wall.

5. *Diffraction.* Put a droplet of water on a piece of glass (on your eyeglasses if you wear them), hold it very close to your eye, and look through it at a distant street light at night or at the sun as it sets or at any point source. You will see the glob surrounded by dark and bright fringes that are fixed to it and rotate as you rotate the glass. People who have driven at night with rain on their spectacles have seen these

diffraction patterns before. You may also see some very small concentric ring systems and threadlike groupings that move around, particularly when you blink. Look at a broad source and squint until you can see the shadows of your eye lashes and everything is a blur. Within that field will be those moving bacterialike images known as *floaters*. What you are seeing is the diffraction patterns, the shadows, cast on your retina by cells floating in and on your eyeball.

Look at a distant point source at night through a piece of nylon cloth, and each little repeated box will diffract the light into a cross. Fog up a piece of glass with your breath, and look through it at a point source. The hemispherical droplets will produce a colored system of concentric rings.

REVIEW

Section 21.1

1. What is meant by the term *natural frequency* as applied to an object capable of vibrating?

2. An atom is usually in its _____ state.

3. Germanium is transparent to _____.

4. For ordinary white light impinging perpendicularly on a piece of glass, about _____ percent is reflected back. There will always be some reflected light, no matter what the angle.

5. Describe the important aspects of the process of *nonresonant* scattering.

6. Define the term *index of refraction*.

7. The higher the refractive index, the _____ the speed of light therein.

8. Light is reflected such that the angle of _____ equals the angle of _____.

9. Define what is meant by a *ray*. How is it different from a beam of light?

10. When light goes from a medium with a high index to one with a low index, it bends _____ the perpendicular.

11. How should you aim if you wish to fire a bullet at a target that appears near the bottom of a lake several feet below and in front of you?

12. To what does the term *critical angle* refer?

13. Which three are the *primary colors* when we talk about mixing light?

14. Two colors are *complementary* if _____.

15. Distinguish between something that appears black as opposed to white.

16. What is the mechanism that gives rise to much of the color appearing around us?

17. From its appearance, what can you say about the structure of a piece of chalk? Of a cloud?

18. The whiteness of an object is the human eye's response to _____ reflected light of a wide range of _____.

19. What is the primary function of a lens?

20. Distinguish between a *converging* and a *diverging* lens.

21. A parallel bundle of rays entering a positive lens along its axis will converge at its _____ _____. With a negative lens, the rays will appear to diverge from its _____ _____.

22. Does the eye form a real image on the retina?

23. How do we correct farsightedness?

24. What is a *virtual* image?

Section 21.2

1. Distinguish between *destructive* and *constructive* interference.

2. Can two light waves totally cancel each other at some point in space?

3. Why is Young's experiment a strong argument in favor of the wave theory?

4. What is *diffraction*?

5. Do sound waves experience diffraction?

6. The ability of a wave to bend around a particular obstacle is determined for the most part by its _____.

7. The light entering a lens (whether in a camera or a telescope) is only a portion of the wavefront emanating from the source, so we can anticipate the presence of a _____ pattern.

8. The diffraction or shadow pattern of a steel ball illuminated by a point source will have a _____ at its very center.

9. Light can be polarized because it is _____.

10. What is meant by *plane-polarized* light?

11. Is the light from an incandescent bulb polarized?

12. The amazing thing about calcite is that it is capable of producing _____ images of something seen through it.

13. Can light become polarized either totally or partially by reflecting off a dielectric?

14. What is a Polaroid filter; that is, what does it do?

15. Discuss what is meant by the *transmission axis* of a Polaroid filter.

16. A common analogy for the Polaroid filter is a picket fence. When a jump rope oscillates vertically, the disturbance passes through the slats, whereas a horizontal oscillation is blocked. Why is this a terrible representation?

QUESTIONS

Answers to starred questions are given at the end of the book.

Section 21.1

1. When a jet plane flies overhead, why do things in a home near the airport vibrate? Can you think of anything relating to resonance that happens to a car, usually going at about 55 mi/hr, when the tires aren't balanced? Explain.

* 2. A particular piece of quartz is clear and colorless. What does that tell you about the resonances of its atoms?

3. What is the speed of light in glass of index 1.5?

4. Why is the sky blue? What makes it whitish gray on a hazy (or polluted) day?

5. When we view the sun at high noon, why doesn't it look bright red, as it does at sunrise?

6. Distinguish between *specular* and *diffuse* reflection. Light is _____ when reflected off your skin and _____ when reflected from a window pane.

* 7. Look at both sides of a polished spoon, and describe the image as best you can.

8. Confirm by calculation that a beam of light with a vacuum wavelength of 650 nm will have a wavelength of only 489 nm when traveling in water.

9. Given incident white light, what color passes through a filter that absorbs green? If a filter appears bluish green, what color does it absorb?

* 10. Given incident white light, what color will be transmitted by a cyan-colored filter? In other words, what color must be removed from white to leave cyan? What will be passed by a magenta filter? What will be passed through a cyan filter followed by a magenta filter?

11. What will be passed by a stack of three successive filters, magenta, cyan, and yellow? Draw a diagram showing the progress of the original primaries through these filters.

12. If a piece of glass has a resonance at the red end of the spectrum, what color will it pass? Gold has an IR resonance and so strongly both absorbs and reflects red. The astronauts' helmets have visors that are coated with a thin film of gold in order to reflect IR and yet remain fairly transparent to visible light. The idea is not to overload the air-conditioned suits. While on the moon, one of our astronauts, having forgotten that the visor behaved like sunglasses, excitedly announced he had found a colored rock. What color did he mistakenly think it was?

* 13. Consider the difference between a rough (or matte) finish and a smooth one. Discuss why you can see your face reflected in a black patent leather shoe, but not in a piece of black velvet. By the way, the reflection involves a layer of surface atoms only about $\lambda/2$ thick.

14. A bug is sitting on the axis of a converging lens at a distance of two focal lengths from its center. Draw a ray diagram locating the image, and describe that image in detail—size, orientation, location.

15. Describe how a telescope works.

* 16. Explain how the rippled surface of the Ganges River on a lovely January morning spread out the reflected image of the sunrise in the photo.

17. Large paraboloidal radio antennas are often constructed as open mesh work. Explain how they manage nonetheless to produce specular reflection.

18. Is Venus in Velasquez's painting looking at herself in the mirror? Explain.

Section 21.2

1. Suppose that two waves of the same frequency traveling in almost the same direction arrive at a point such that

one is ahead of the other by a half wavelength. What kind of interference can you expect to occur at that point?

* 2. Waves that maintain a fixed relationship with each other in their respective risings and fallings are said to be *coherent.* Are two candle flames coherent sources, and will their light observably interfere? Incidentally, two different laser beams have produced measurable interference patterns.

3. Young's experiment produces a series of bright and dark bands of light called fringes. The central one, the one on the axis midway between the slits, is always bright. Explain.

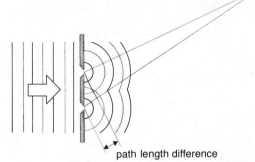

path length difference

4. The accompanying diagram shows Young's double-slit experiment illuminated by single-colored plane waves (this arrangement doesn't change anything from the previous setups). What will the resultant light reaching some point, P, be like if the path-length difference is one half wavelength? How would the result change if the wavelength was halved? *Hint:* The waves start out rising and falling together, so it is the path-length difference that determines how they will interfere at P.

* 5. Suppose that a thin film is just the right thickness at one spot for green light beams reflected from the front and back surfaces to come out with one a half wavelength ahead of the other so that they cancel each other. What color will that spot appear when the film is illuminated with white light?

6. Camera lenses are often coated with a transparent dielectric whose thickness is one-fourth the wavelength of yellow-green light. The wave reflecting back at the film-glass interface will ultimately traverse that coating twice and so be shifted a half wavelength with respect to the wave reflected at the air-film interface. Thus the two waves arriving peak-to-valley in the air cancel each other, and no yellow-green is reflected. What color will the lens appear in reflected light? What color will it transmit into the camera most efficiently? Why do you think light in the yellow-green is usually selected to be passed by these *antireflection coatings*?

7. Whenever it rains, one can see marvelous multicolored patterns in the gasoline and oil splashes that appear on blacktopped roads and parking lots. What causes the phenomenon? Describe the role played by the rain water and the blacktop.

8. Waves very much larger than an object essentially behave as if the object were not there, hardly reflecting off it and easily closing around it, leaving little or no shadow region. Would you expect to be able to see a pin with radio waves or microwaves? Why do you think bats use ultrahigh-frequency sound to reflect off targets? Estimate the typical frequency of this rodent sonar, presuming that bats eat moths.

9. Describe the state of polarization of a light beam reflecting off glass at around 56°. How could you use this knowledge as a handy means for locating the transmission axis of a Polaroid filter?

10. Why are Polaroid filters of very limited value in polarizing high-power laser beams? *Hint:* The devices used instead are prisms made of clear calcite or a similar crystalline material, which separate the beams into two plane-polarized portions.

* 11. Glare is more or less specularly reflected light, often but certainly not always coming from a horizontal surface. How would you design Polaroid sunglasses to eliminate this inconvenience? Would they work with the glare on a store front window?

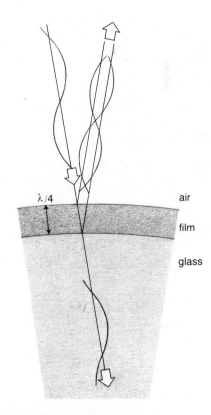

MATHEMATICAL PROBLEMS

Answers to starred problems are given at the end of the book.

Section 21.1

1. Assuming a reflection of about 4% at each interface, determine what fraction of the light incident perpendicularly on a sheet of glass in air will emerge, having passed through both surfaces? People who wear eyeglasses (particularly nearsighted people) often learn to raise them in situations where they need as much light as possible and must overcome the loss due to reflection.

* 2. If the speed of light in a substance is effectively 1.88×10^8 m/s, what is the index of refraction of the medium? Incidentally, that index matches the range of common flint glasses.

3. How long would it appear to take light to traverse a glass block that is 10 cm thick and has an index of 1.5?

4. What is the length of the smallest flat mirror in which you can see your entire body? Answer this with a ray diagram.

5. Suppose that you walk directly toward a flat mirror at one foot per second, beginning at a point 10 feet away. What distance separates you and your image at the start? What is your relative speed with respect to the image? How long will it take to collide with the image? Where will the collision take place? How much ground will be covered by the two moving entities?

* 6. Suppose that you are standing between two roughly parallel floor-length mirrors in a clothing store. If the mirrors are separated from each other by five feet and you are standing three feet from the mirror you are facing, how far away from you will the image of your back appear, as reflected in the image of the rear mirror?

Answers to Questions

Section 1.1

2. An obvious problem is the lack of rigorous reproducibility. In that respect ESP and related phenomena border on the magical, and one might wonder if it's even reasonable to study them scientifically.

4. We can affect the course of natural occurrences by utilizing our understanding of Nature's laws and in that sense by "obeying her." We don't force events; we simply arrange to let them occur. We don't force Uranium-235 to explode; we merely clear away the obstacles that ordinarily keep it from exploding naturally.

Section 1.2

6. In deriving his second law, which is quite correct, Kepler made several errors that effectively canceled each other. The search for Pluto, like that for Neptune, was begun for good theoretical reasons, and yet when the ninth planet was finally found, it was nowhere near where it "should have been." None of this is quite serendipity in the usual sense. Perhaps a new word is called for that means looking for something in the wrong way and finding it anyhow.

Section 1.3

8. Yes, because we really cannot *prove* it absolutely false in the first place. For example, the kinetic theory of heat, the idea that heat is related to the motion of atoms, was put aside about 200 years ago, only to be resurrected in the second half of the 1800s. Facts change as our understanding changes.

Section 2.1

2. First, weighty objects fall toward the Earth's center, and second, the "planets" and stars obviously revolve around it as well.

4. According to Aristotle, air rushing in to fill the dreaded vacuum propels the spear. The flat-ended one should therefore be more effectively pushed, but of course, air friction produces just the opposite result. The pointed, low-drag spear travels farther.

Section 2.2

6. Any object in free flight is influenced by gravity and will change speed. Then again, the Earth is curved, and objects in powered motion will generally follow that curvature and so, changing direction, will change velocity.

11. Average speed is distance over time. The speeds would be equal if half the distance was covered in half the time by the second person. In this case, twice the time is taken so the speed is one-quarter as great.

14. A ball just at the very instant it's dropped is presumably accelerating downward but has not yet begun to move.

20. The second car always remains one second in time behind the first and will pass any point on the track one second later. Since their speeds are both increasing, the corresponding distance separating them must also continue to increase if it's to remain equivalent to one second's worth.

Section 2.3

7. The Earth is actually closer to the Sun in January than in June and so, on average, it moves faster in winter than in summer.

13. The planet sweeps out equal areas in equal times.

Section 3.1

2. The followers of Aristotle had always been able to rationalize away the mottled appearance of the moon as being due to local effects such as the Earth's atmosphere—after all, the moon was divine and perfect. To imagine the moon to be Earthlike was to challenge the whole system.

Section 3.2

3. Doubling the time quadruples the distance.

6. By the law of inertia, the passengers, tending to continue in their straight-line motion, will be thrown forward.

8. The speed is 0, and the acceleration, which is presumed constant, is 32 ft/s^2.

11. $90° - 20°$, or $70°$.

Section 4.2

7. A certain amount of effort goes into overcoming friction, but it also takes a force in order to accelerate.

14. The bowling ball has more mass and will require a greater force to achieve some acceleration.

16. The rope is motionless, and so the net force is zero. No. Then the internal tension within the rope increases beyond 10,000 lbs, but unless the rope moves (e.g., the hook falls out), the net force is still zero.

22. The box would collapse because you would have to exert a downward force greater than your weight.

Section 4.3

4. The cart, pushed backward by your foot as you *accelerate,* will roll back. By contrast, while you are moving at a constant speed, the force exerted on the cart is negligible, and it will remain motionless.

8. During the impact there will be a transfer of momentum, $\Delta(mv)$, from the massive combination of racket and body to the ball. Because of the ball's low mass, it will attain a relatively high speed.

Section 5.1

1. It was reasonable from the inverse-square behavior of emanations from a point source, such as the light from a candle flame.

4. Doubling the distance quarters the density of the spray.

Section 5.2

1. Doubling the speed requires that the force be multiplied by four. With half the mass, only half the force is needed. Doubling the radius halves the force.

5. Doubling the mass doubles your weight. Doubling the distance quarters the force of gravity, i.e., your weight, so you would weigh half your weight on Earth.

8. Centripetal force varies inversely with R whereas the gravity force varies inversely with R^2. Setting the two equal $(GmM/R^2 = mv^2/R)$ yields a speed that varies inversely with $\sqrt{R}$. As distance increases, planetary speed decreases.

12. The released apple will follow the ship in the same orbit. Fired at a somewhat increased speed, it will sail off into a large elliptical orbit about the planet. It will always come back to the point of injection (where it was initially thrown) with that maximum speed. When thrown backward (or equivalently slowed down), it will drop into a smaller elliptical orbit. It will return to the injection point, where its speed will be a minimum value or, if slowed enough, it will impact the ground.

Section 6.1

4. Given that the classical picture is incorrect before we start, the answer is 12:00. The light wave that reaches the astronomer carrying the image of the clock face sweeps past you in the ship at the same instant. Of course, lots of motionless people along the route also saw it read 12:00 as the wave progressed toward Mongo.

Section 6.2

3. The aether still had two main functions by the turn of the century. It was the medium for light waves, and it was the backdrop of the universe at absolute rest. When absolute rest was put aside, the aether went with it. Light waves were then envisioned as traveling across the electromagnetic field itself.

6. The second postulate demands that all observers measure the speed of light to be c, regardless of their motion—c is the only absolute in Wonderland.

Section 6.3

3. Imagine a straight line connecting the locations of two events in space, and set our two observers in uniform relative motion along a second line perpendicular to the first. If the events occur simultaneously to either observer, they will be simultaneous for the other as well.

Section 6.4

3. Since the pions were traveling at nearly the speed of light with respect to the Earth, a ground observer would have seen their little clocks running appreciably slower. Thus what the pion "thought" was its normal 0.026×10^{-6} s lifetime was observed to correspond to 160×10^{-6} s on the Earth-based clocks. The pions' life cycle was a moving clock, and it was seen to run slow.

6. The top of the atmosphere and the Earth's surface represent the end points of a ruler along which the pion is moving. Accordingly, there is a length contraction, and the pion "sees" the atmosphere as only 7.8 m thick.

10. There is no absolute answer. Length is a quantity that is relative, depending on the motion of the observer. The geography book distances are only for observers who are stationary with respect to the planet.

13. Presumably everything would be quite normal for Sam within the space ship. As for you on the Earth, you would see the beam advancing toward the front wall at the speed c (second postulate) and the front wall, in turn, receding from the light. The beam could never catch up and arrive. Indeed, if the ship moved faster than c, the light would eventually illuminate the back wall. Both observers would see quite different occurrences, and the laws of physics would apparently be different, thereby violating the first postulate. We must conclude that such speeds equal to or in excess of c are unattainable—or the postulates are wrong.

Section 7.1

3. Yes, he increases the books' gravitational PE by doing work on them, by exerting a force over a vertical displace-

ment. When he takes the elevator, it's the elevator that does the work, not our young salesman, since the elevator provides the displacement.

6. During the accelerating phase, a force is necessary and work is done. It takes no force, however, to maintain the constant speed, and so no work is done during that segment of the run.

11. Ideally—with no friction losses—yes. Practically, however, since there are friction losses ordinarily, the longer the route, the more energy must be supplied, and therefore, the more work needs to be done.

Section 7.2

2. Just imagine an object being pushed, accelerated down the aisle of a speeding train. Two observers, one inside and one at rest outside, will each see a different change in the object's kinetic energy, $\Delta(KE) = KE_f - KE_i$, since KE varies as v^2, not as v. Moreover, each will see a different displacement of the object and therefore a different amount of work done. But in each case, $W = \Delta(KE)$.

5. The work done, 100 lb $\times$ 100 ft, equals the change in KE since the road is level and friction negligible. Thus, if you knew the car's mass, you could predict its final speed.

Section 7.3

3. At a height of 20 ft its gravitational PE with respect to the ground is 3000 lb $\times$ 20 ft, and that is the amount of work it can deliver as well. The PE with respect to the well bottom is simply 3000 lb $\times$ 30 ft.

7. Setting the PE equal to $F_w \Delta h$ assumes that F_w, the weight, is a constant independent of height. Still, both g and weight (F_w) are strictly not constant as the height varies. Generally, though, the deviation is negligible for terrestrial problems.

Section 7.4

1. (a) At the top of the first hill.
(b) This really depends on the particular construction—generally at the bottom of the lowest valley, provided it's not near the end of the ride, where a lot of energy will already have been lost to friction.
(c) No.
(d) Yes. In a system losing energy, there's no reason the maximum KE has to occur when PE = 0. Just make each valley lower than the one before by enough to regain the energy lost to friction, and the speeds in each can be made equal and the highest can be reached.
(e) No.
(f) They all have to be lower than the first.
(g) No—because of friction losses.

Section 8.1

2. One could pound a nail with a hammer and so make it hot by doing work on it. This is quite different from putting the nail in contact with something at a higher temperature, thereby supplying heat and again raising the nail's temperature.

5. Anything that heats a volume of air increases the average kinetic energy of the molecules as they move around randomly. If these are made to stream out of the dryer, we get a hot gust of air.

Section 8.2

3. The high-speed particles are absorbed, losing their kinetic energy via collisions with the material of the core. The core atoms are left with that energy in the form of vibrational KE. This is thermal energy, and the core heats up.

6. At −40 degrees the two scales match.

9. It imparts KE to the air molecules and therefore actually heats the air. The apparent cooling effect comes only from the more efficient evaporation of perspiration by the breeze.

Section 8.3

1. Yes. Just imagine a teaspoonful of water versus a lake.
3. It will drop $2\frac{1}{2}$ degrees C.
6. Yes, if it's undergoing a phase change, for example, while melting ice.
9. They melt a little on their surfaces. When two cubes then come in contact, that liquid film cools and freezes again.

Section 8.4

2. Initially the total momentum of the system is zero since the same amount is directed in opposite directions. If the collision were elastic (as with rubber cars), they would spring apart with equal speed and zero momentum. But it is inelastic, so they stick together, and the momentum, as ever, remains zero. The KE of the cars goes into the random KE of their molecules, i.e., thermal energy.

7. As the water descends, it increases in speed and KE. On impact that KE is converted to thermal energy, and the temperature will rise.

Section 8.5

3. Heat flows from a higher to a lower temperature. If we could come up with something that would remain appreciably colder than the sea, the scheme would work. The atmosphere is cooler than steam, so we can keep running steam engines that dump heat into the air. But there is no readily available large pool of low-temperature stuff for a ship to use—practically at least.

9. Yes. Every heat engine must exhaust a certain amount of waste heat somewhere—to the air or a nearby river. The challenge is to do it in the least detrimental way.

13. When the system is open and we exchange energy, the entropy of the environment increases—all those little circular breakfast crunchies are turned into a disordered blur. When the body is isolated, its own entropy must either increase or at best remain the same. The result (from the first law) is death and (from the second) generally dust.

Section 9.1

1. The alchemist was a serious, pious, ritualistic, dogmatic, devoted practitioner. The puffer was an experimenter, a no-holds-barred putterer, who would try anything to make gold, with little or no real concern about the spiritual.

4. Aristotle's fifth element was the divine substance, the *quinta essentia*.

Section 9.2

1. Helium, oxygen, nitrogen, neon, and argon are among the gases in the air we breathe.

4. Fire was the essence of change, and since all earthly things are in a state of constant change, it makes a rather reasonable element.

Section 9.3

3. The periodic table was a statement of a pattern of order among the elements. As a graphic statement of a law without any real theoretical understanding, it was akin to Kepler's achievement.

6. CO, CO_2, O_2, N_2, N_2O, NO, NH_3.

8. It implies spaces within the water into which molecules of sugar can nestle, and that suggests that the water is not continuous.

10. Besides being quite arbitrary, grams and kilograms are inconveniently large for dealing with individual atoms. The mass of an atom is certainly a more basic quantity than the mass of a lump of metal in a vault near Paris.

Section 10.1

1. Coulomb provided a description of the interaction between charged objects. Like the gravitational interaction between masses, this is a fundamental concept, a basic property of matter.

4. Since the electron was a negative charge far smaller in mass than an atom, it meant that these specks were probably atomic components. That in turn implied the presence of some kind of positive charge since atoms on the whole are neutral.

Section 10.2

2. It was already clear at the beginning of this century that tremendous amounts of energy were locked up in the atom. The discovery of radioactivity made the point quite certain.

Section 10.3

2. High-speed massive alpha particles could not possibly be scattered backward by a "raisin pudding" atom. Neither the electrons nor the tenuous positive charge in which they were embedded represented an appropriately dense, massive collision center.

Section 10.4

7. Thorium
9. Bismuth
10. ^3He
11. It was simply an isotope of thorium.

Section 11.1

2. Technically the insight was that the oscillators absorbing and emitting energy within a blackbody only did so in specific blasts (or quanta). Much more would be made of this strange notion later on.

Section 11.2

3. The energy of a photon varies with its frequency. UV photons have a higher frequency, and so each one individually has a higher energy than a quantum of light.

7. Fluctuations in the intensity of the light produce corresponding fluctuations in the number of electrons emitted (i.e., in the output current of the device).

Section 11.3

2. When an atom takes up a certain amount of energy (e.g., during a collision with another atom), an orbital electron in the ground state may be raised to some higher excited state. The process of returning to the ground state can take place via the emission of the excess energy in the form of a photon.

6. The sodium passes an electron to the fluorine, completing the outer shells of both. The oppositely charged ions then attract. By contrast, the hydrogen and chlorine each share an electron to form HCl.

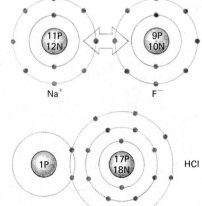

5. Glass is a nonconductor, and the charges placed on it tend to stay put. If you rub the middle, the ends remain neutral.

9. If you touch the elephant with a neutral proof plane, it will take on some of the excess charge, which can then be carried to the pith ball or anywhere else.

Section 17.2

2. Q/3

4. It will remain unchanged.

8. The positively charged sphere induces an equal negative charge on the inner surface; i.e., electrons are drawn to that surface, leaving the outside equally charged and positive.

13. The belt becomes charged as it rubs against each of the two pulleys. The central idea is that as the belt moves up, the charge it carries can be wiped off by a conductor and channeled to the inside of the dome. From there it naturally flows outward via mutual repulsion. Because the excess charge is distributed over the *outside* of the sphere, more charge can continue to be introduced into the inside, where it will experience no repulsion from the previously deposited electricity.

Section 17.3

3. Yes. The charge will pick up speed and move in a straight path along a force field line. Things are different when the field lines are not straight. Once set in motion, the charge will tend to continue in a straight line via the law of inertia, and at the same time, it will be deviated by the curved lines of force. The result is a combination of both.

4. If the force field was not perpendicular, some portion of it would act along the surface, pushing on the charge. The latter would then move, redistributing itself until all motion ceased and the field finally became perpendicular.

7. $\mathcal{E} = F_E/q$; that is, force over charge, and so its units are newton/coulomb; to get an equivalent form for a newton, we make use of the fact that work equals force times distance; hence joule = newton · meter, and therefore newton = joule/meter; accordingly the units of $\mathcal{E}$ are (joules/meter)/coulomb, but 1 joule/coulomb = 1 volt, which yields volts/meter.

11. At any point in space there is a single *net* electric force acting on a test-charge. That is, when held at that point, the test-charge experiences a resultant force in some direction, and it is this that determines the field direction. Even when there are many source charges and many overlapping $\mathcal{E}$-field contributions, there will always be only a single resultant field at each point in space.

Section 18.1

1. 100 C/5s = 20 A.

5. In order to conserve space, most parallel-plate capacitors consist of a rolled-up layered structure, such as foil-paper-foil.

9. The capacitance is a measure of the ability to store charge. The larger the plates, the more charge a capacitor can hold at a given voltage. Remember, it's the mutual repulsion of the charges on each plate that gives rise to the voltage.

14. If you place four 1.5-volt dry cells in series, the output voltage adds up to 6 volts. Four D cells in parallel will deliver 12 amp-hours at 1.5 volts.

Section 18.2

1. The flow of heat along a metal rod is proportional to the temperature difference across its ends, just as the flow of charge is proportional to the voltage difference. It is the motion of free electrons that transports charge and heat.

6. Since $\mathcal{V} = I\mathcal{R}$, doubling the voltage doubles the current if $\mathcal{R}$ is constant. Holding $\mathcal{V}$ constant and halving $\mathcal{R}$ doubles I.

10. The total resistance increases as the number of bulbs in series increases. Since $\mathcal{V} = I\mathcal{R}$ and $\mathcal{V}$ across the bulbs is a constant 1.5 V, the current decreases as $\mathcal{R}$ increases.

14. 10 volts; $10\,V = I(2\,\Omega)$, $I = 5$ A; $10\,V = I(10\,\Omega)$, $I = 1$ A; $10\,V = I(5\,\Omega)$, $I = 2$ A; $P = I\mathcal{V}$. Therefore maximum power is dissipated by the 2 Ω resistor and minimum by the 10 Ω one.

Section 18.3

2. It becomes zero twice in each cycle and therefore 120 times per second.

7. (a) No. (b) Yes. (c) Yes, now the voltage across terminals A and B would be zero. (d) Yes, if the bulb is in and intact you would be shocked. (e) With the switch closed, A and B are at the same voltage, and you can't be shocked by touching them.

10. The bird would become charged. There's a negligible potential difference between its two feet, while it is standing on the wire. Yes, as long as it stepped along with both feet falling on the rail and didn't touch anything grounded. With one foot on the rail and one on the ground it would have serious problems.

13. Very large currents can pass through the wrench, which is essentially a short circuit. The resistance is highest at the contact points, and it's there that the power developed is highest and the temperature the greatest.

Section 19.1

3. Place them in the configuration of a T. The magnet has its poles at its ends and will not attract a piece of iron at its middle, where the draw to each pole cancels the other. Thus, if they stick, the bar that makes the stem of the T is the magnet; if they don't, the crossbar is.

7. The magnet would be horizontal (i.e., zero dip) around the equator and that, coupled with the zebra, suggests Africa.

13. No. Over the length of a typical magnet, the Earth's field is essentially constant, acting nearly equally and oppositely on the two poles. In the nonuniform field of a bar magnet, it will indeed move.

Section 19.2

2. A monopole would experience a force in a uniform magnetic field. Thus a north monopole would accelerate along the field direction, just as a positive charge would accelerate along an electric field.

4. The two magnets are slowly separated, being drawn toward each end of the steel bar. After this is done several times, the bar's end reached by the north pole of the magnet itself becomes a south pole, and vice versa.

9. Suppose that you wished to create a field-free region. You would need only to make a closed shell of a highly permeable material. The field would concentrate within the shell, and the space enclosed would then be well shielded. By contrast, a diamagnetic material, in effect, pushes some of the field lines out of the region it occupies. There is less field inside than outside such a material. Under certain circumstances, superconductors can be perfectly diamagnetic, not allowing any $\mathcal{B}$-field to penetrate within them.

Section 19.3

4. The field direction is, by convention, from north to south—here left to right. The force is clearly vertical. Applying the righthand rule, we see that a positive charge would move downward, so this one must be negative.

7. Putting your right thumb in the direction of the current, your fingers curl in the direction of the $\mathcal{B}$-field, which is downward outside the coil and upward inside it. That matches the field of a bar magnet, whose north pole is on top.

11. Current flowing in the direction shown in the figure passes through the mercury baths and out of the device. Looking downward, the left side has a counterclockwise $\mathcal{B}$-field, and the single north pole rotates, moving along it (see Question 2, Section 19.2). On the right side there is a downward current in the $\mathcal{B}$-field of the magnet, and therefore the hanging rod experiences a perpendicular force, which causes it to rotate clockwise.

Section 19.4

2. Into heat, mostly via friction although there are transient currents and Joule heating from these (overlooking a negligible amount of radiation).

7. No, it will lose energy setting up a current as it drops through the coil.

10.

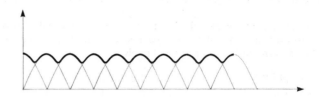

Section 20.1

3. Longitudinal. The medium (the crowd) moved in the direction of the wave.

6. $v = f\lambda$; 186,000 mi/s $= (60$ Hz$)\lambda$; $\lambda = 186,000/60 = 3100$ miles.

9. A wire tied to the spinal cord served as an antenna with the frog's nervous system as the receiver. The electromagnetic wave created every time there was a blast of lightning induced a current in the antenna, which caused the frog to twitch. Nowadays we simply hear these stray signals as static on AM radios or see it as interference on our (AM) TV pictures.

Section 20.2

1. Yes, indeed. It's changing direction, therefore changing velocity, therefore accelerating, so it radiates.

5. Light waves are several thousand times the size of a typical atom; X-rays are about the size of an atom.

10. Color is the human eye-brain response to optical frequencies. Yellow light has a frequency range of roughly 500×10^{12} Hz to 520×10^{12} Hz, and yet if we overlap a beam of red and a beam of green we "see" a yellow, even though no light in that specific frequency range is present. Seeing color is actually a complicated physiological phenomenon, which is not at all well understood.

Section 21.1

2. It is not absorbing in the visible region of the spectrum and therefore has no resonances in the visible.

7. When you look into the concave bowl from a few inches away (beyond the focal point, which is probably about $\frac{1}{2}$ inch from the surface), your image is minified and upside down. Very close in (less than the $\frac{1}{2}$-inch focal length) the image is rightside up and magnified. At the focal point, the image is a great blurred smear (which is necessary if, in passing through that point, it is somehow to flip over). The image on the convex side is always rightside up and minified (unless the object touches the spoon, whereupon it becomes life-size, as it obviously must for a mirror of any shape).

10. A cyan-colored filter passes cyan (i.e., blue + green) and therefore it "eats" red, since B + G + R = W. A magenta filter passes red + blue and "eats" green. If white goes in, blue and green will emerge from the cyan filter, but only blue will make it through both.

13. Contrary to popular belief, a real-life black surface does not absorb all the light that impinges on it. Indeed, if the very outer layer is smooth, a few percent of the incoming light will be specularly reflected. Thus a piece of shiny black glass, leather, plastic—whatever—will serve as a poor but functional mirror. The effect can be destroyed simply by making the surface slightly rough.

16. The day this 6:00 A.M. sunrise-on-the-Ganges photo was taken, there was a light breeze and the river's surface was slightly rippled with small waves. The rough surface diffused the image, allowing sunlight reflected along a long band to reach the observer's eye. If the river were perfectly smooth, only light reflected from a small region would reach the eye, producing the usual mirror image.

Section 21.2

2. No. The light wave coming from a candle flame varies rapidly and randomly, so no two candles can put out light waves that stay locked in rhythm long enough to create an observable interference pattern.

5. Since G + R + B = W, R + B = W − G = magenta.

11. Glare reflected from a horizontal nonconducting surface is polarized in that horizontal plane. Polaroids with vertical transmission axes will "eat" the glare. If the surface is vertical, the glare is vertically polarized and will pass through the sunglasses and be seen.

Answers to Mathematical Problems

Section 2.2

1. 11.2 m/s
3. 249.4 ft/s
6. 3195.7 hr
12. 1.64 s
14. 32 ft/s^2
18. $v_{av} = (200 + 0)/2 = 100$ mi/hr; $\Delta d/\Delta t = (100 \times 1.467)$ ft/s; since $\Delta d = 5$ ft, $\Delta t = 0.034$s.

Section 3.2

1. 72 ft
5. 144 ft, 176 ft
8. 4 s total flight time, 256 ft down range
10. 10 m = $\frac{1}{2}$ 9.8 t^2; $t = 1.43$ s; $v = 15/t = 10.5$ m/s.
11. It will land 2 s later, right on top of Big John.

Sections 4.2 and 4.3

3. 960 − 900 = 60 lbs.
5. $F \Delta t = m \Delta v$; $m = 0.5 \times 10^{-3}$ kg; $\Delta v = 62$ m/s.
10. 20 kg × 9.8 m/s^2 = 196 N.
14. .046 kg × 70 m/s = $F \times 0.5 \times 10^{-3}$ s; $F = 6440$ N. $a = 70/0.5 \times 10^{-3} = 140,000$ m/s^2.

Section 5.2

3. $F_c = mv^2/R = 10789.7$ N.
8. $mv^2/R = GmM/R^2 = ma_G$; $v^2/R = a_G$.

Section 2.3

1. 1/52
4. $T = 2\sqrt{2} = 2.8$ Earth years.
5. $D^3/T^2 = 1.85 \times 10^{13}$; 9.66 days.
7. 8

10. $v_{esc} \simeq 2384$ m/s.
14. $mv^2/R = GmM/R^2$; $v^2 = GM/R$. Circumference = $2\pi R = vT$; $v = 2\pi R/T = 19924$ m/s, where $1y = 31,536,000$s; $M = 6 \times 10^{30}$ kg.

Section 6.4

1. $\Delta t_M = 2 \times 10^{-6}/\sqrt{1 - (0.998)^2} = 2 \times 10^{-6}/6.32 \times 10^{-2} = 31.6 \times 10^{-6}$ seconds.
4. 1800 mi/hr = $\frac{1}{2}$ mi/s; $v^2/c^2 = (0.5/186,000)^2 = 7.2 \times 10^{-12}$; $\frac{1}{2} v^2/c^2 = 3.6 \times 10^{-12}$; $\Delta t_M \simeq \Delta t_S(1.0000000000036)$.
6. $L_M/L_S = 44\% = \sqrt{1 - v^2/c^2}$; $1 - 0.1936 = v^2/c^2$; $0.898c = v$.
11. $m_M/m_S = 1/\sqrt{1 - (9 \times 10^7/3 \times 10^8)^2} = 1.048$.

Section 7.1

2. $W = 10N \times 10m = 100$ N-m. The amount of work is the same, and since there is no friction, it must go into overcoming inertia, i.e., accelerating the body. In each case there is a force in the direction of the displacement and therefore work done.

5. Power is work over time or force × distance/time or force × speed; power = 10,000 J/s = (1000N) v; $v = 10$ m/s or 22.4 mi/hr.

8. 220 × 10^{9}W/100W = 220 ×10^7: 1 J/s = 1W; 220 × 10^9 J/s. That was easy. Now to get the per capita share, just divide by the total population.

Section 7.2

3. $m = 4000/32 = 125$; 60 mi/hr $\times 1.467 = 88.02$ ft/s; $\Delta(KE) = \frac{1}{2} 125(88.02)^2 = 484,220$ ft-lb; power $= \Delta(KE)/$ time $= 484,220/10s = 48,422$ ft-lb/s; or $48,422 \times 1.36 \simeq 65,850$ W.

6. $74,000 \times 10^{12}$ J $= \frac{1}{2} 3.5 \times 10^6 v^2$; $v = 205,635$ m/s.

Section 7.3

4. For every foot of rope the monkey pulls past itself, it rises $\frac{1}{2}$ foot, as does the weight. It exerts 10 lb through 20 ft of tugging and therefore does 200 ft-lb of work. The weight rises 10 ft (as does the monkey); a 10-ft length of rope ends up on the floor. The total increase in gravitational PE is 200 ft-lb.

Section 7.4

2. The initial $KE_1 = \frac{1}{2} mv_1^2$ must equal the potential energy at the maximum height, namely $mg\Delta h$; therefore $\frac{1}{2} mv_1^2 = mg\Delta h$ and $v_1 = \sqrt{2g\Delta h}$.

5. $\Delta(PE) = mg\Delta h = 120$ lb (10 ft) $= 1200$ ft-lb. This is the energy stored. $\Delta(PE) = KE_1 = \frac{1}{2} mv_1^2$, and so the speed she recoils at is $v_1 = \sqrt{1200 \times 2/m}$, where $m = 120/32$. Therefore $v_1 = \sqrt{1200 \times 2 \times 32/120} = \sqrt{640} = 25.3$ ft/s. She will rise 10 ft or 14 ft above ground—again assuming no losses.

Section 8.2

1. 5/9 (Fahrenheit $- 32$) $=$ Celsius; 5/9 (98.6 $- 32$) $=$ 37°C; $273 + 37 = 310$ K.

3. -183°C, -297.4°F

Section 8.3

2. $75 \times 4200 = 315,000$ J.

5. (a) $(90° - 10°)/2 + 10° = 50$°C.

(b) Δheat $=$ specific heat $\times m \times \Delta T = 1 \times .25 \times 80$°C $= 20$ Cal.

$$20 \text{ Cal} = 1 \times .75 \times \Delta T$$
$$\Delta T = 26.6°$$
$$\text{Final temperature: } 36.6°\text{C.}$$

6. (a) 800 Cal
(b) 800 cal
(c) 1600 cal drawn from water will melt the ice, leaving it at 0°C. Removing 1600 cal drops the water temperature by $-1600 = 200(1) \Delta T$ or $\Delta T = -8$°C down to 12°C. Mixing 20 g at 0°C with 200 g at 12°C yields

$$20(T - 0°) = 200 (12° - T)$$
$$T = 10.9°\text{C.}$$

Section 8.4

1. 10 kg $\times$ 540 Cal $\times$ 4200 J/Cal $= 22.68 \times 10^6$ J
$mgh = \frac{1}{2} mv^2 = 22.68 \times 10^6$.
$(10)(9.8) h = \frac{1}{2}(10)v^2 = 22.68 \times 10^6$
$h = 2.3 \times 10^5$ m, $v = 2130$ m/s.

4. $\frac{1}{2} mv^2 = (4200 \text{ J/Cal}) [m(330 - 30) 0.03 + m6]$
$v = 355$ m/s.

Section 8.5

1. efficiency $=$ work-out/heat-in $= 1000/10 \times 4200 = 2.4\%$.

3. $40\% = 300 \times 10^6/$heat-in.
heat-in $= 750 \times 10^6$W, lost power $= 450$ MW.

5. $mgh = 700,000 \times 9.8 \times 200 = 1.37 \times 10^9$ J
$85\% \times 1.37 \times 10^9 = 1.17 \times 10^9$ W.

Section 9.2

2. If you weigh 150 lb, that's 105 lb (47.6 kg) of water. One mole of H_2O has a mass of 18 g, so you contain 2646 moles, each with 6×10^{23} molecules, for a total of 1.6×10^{27} molecules.

5. 6×10^{22} atoms per gram $\times 10^{33} \times 10^{11} \times 10^{11} = 6 \times 10^{77}$.

Section 9.3

4. H combines with O to make H_2O in the weight ratio of $2:16$ or $1:8$. Therefore 5 g of H will unite with 40 g of O to form 45 g of water, with nothing left over.

8. $17(C) + 21(H) + 1(N) + 4(O)$
$17(12) + 21(1) + 1(14) + 4(16) = 303$.

9. Francium has a half-life of 21 minutes, so we are dealing with a span of three half-lives. After 21 min there would remain $10^{-6}/2$ grams; after 42 min, $10^{-6}/2 \times 2$ grams; and after 63 min, $10^{-6}/2 \times 2 \times 2$ grams, that is, $\frac{1}{8}$th of whatever we started with.

Section 10.1

1. $m_e = 9.1 \times 10^{-31}$ kg $= 9.1 \times 10^{-28}$ g; $1/9.1 \times 10^{-28} = 1.1 \times 10^{27}$.

Section 10.2

3. $KE = \frac{1}{2} m_e v^2 = \frac{1}{2} 9.1 \times 10^{-31} (2 \times 10^5)^2 = 1.82 \times 10^{-20}$ m/s.

Section 10.3

4. The mass of a proton is 1.67×10^{-27} kg and from the previous problem there are roughly 2.39×10^{38} of them per packed cubic centimeter. Thus 1.67×10^{-27} kg $\times 2.39 \times 10^{38} = 3.99 \times 10^{11}$ kg; 1 kg weighs 2.205 lb; 8.77×10^{11} lb or 439×10^6 tons.

Section 11.2

2. 3×10^{-19} J $= hf = 6.63 \times 10^{-34} f$; $f = 4.5 \times 10^{14}$ Hz.
3. $(6.63 \times 10^{-34})\, 1.5 \times 10^{15} = KE + 7.3 \times 10^{-19}$; $KE = 2.6 \times 10^{-19}$ J.

Section 11.3

2. At least 1.6×10^{-18} J.
4. This is the energy needed to lift an electron out of the ground state up beyond all the excited states to infinity. No.

Section 11.4

1. $mv = h/\lambda$; 2000 kg $\times 20$ m/s $= 6.63 \times 10^{-34}/\lambda$; $\lambda = 1.66 \times 10^{-38}$ m.

Section 11.5

1. $\frac{1}{2} m_e v^2 = 2 \times 10^{-15}$ J; $m_e = 9.11 \times 10^{-31}$ kg; $v = 66.3 \times 10^{6}$ m/s; $\Delta x\, \Delta v = h/m$; $\Delta v = 6.63 \times 10^{-34}/9.11 \times 10^{-31} \times 10^{-10}$; $\Delta v = 7.3 \times 10^{6}$ m/s; $\Delta v/v = 11\%$.

Section 12.2

2. Density of water is 1.0×10^{3} kg/m³.

Section 13.3

3. $p = d \times \rho$; $p = 35{,}000$ ft $\times 64.4$ lb/ft³ $= 2{,}254{,}000$ lb/ft² or 15,653 psi.
5. $p = 100$ m $\times 9.8$ m/s² $\times 1.03 \times 10^{3}$ kg/m³ $= 1 \times 10^{6}$ N/m².
7. 20 ft $\times$ 10 ft $\times \frac{1}{12}$ ft $= 16.66$ ft³; $16.66 \times 62.4 = 1040$ lb per inch of settle. Volume $= 20 \times 10 \times 1 = 200$ ft³; raft weight $= 200$ ft³ $\times 25$ lb/ft³ $= 5000$ lb; equivalent volume of H₂O $= 5000/62.4 = 80.1$ ft³; depth $= 80.1/200 = 0.4$ ft.
11. The large end; $F/10 = 2000/500$; $F = 40$ lb.

Section 14.2

2. The net outward force is the difference arising from the pressure inside and out, that is, 14.7 lb/in² $\times (5-1) \times 1$ in² $= 58.8$ lb.
4. Since we are given the mass density we'll have to multiply by g to get the weight density. $p = \rho g h = 1.36 \times 10^{4} \times 9.8 \times 0.76$ m, where 0.76 m $= 29.92$ in; $p = 1013 \times 10^{2}$ N/m².

Section 14.3

1. Density of mercury from the table on p. 283 is about 850 lb/ft³; $850 \times \frac{3}{12} = 212.5$ lb/ft²; dividing by 144 gives 1.48 lb/in² as the drop in pressure.

4. Weight of air displaced $=$ weight of gas $+ 1000$ lb. For helium $V(0.08 \text{ lb/ft}^3) = V(0.0111 \text{ lb/ft}^3) + 1000$ lb; $V = 14{,}514$ ft³; for a weight of $14{,}514 \times 0.0111 = 161.1$ lb. For hydrogen $V = 13{,}441$ ft³, and its weight would be 75.3 lb.

Section 14.4

1. $p_i V_i = p_t V_t$; $p_i\ (1) = (1 \text{ atm})\ 10$; $p_i = 10$ atm. The depth h at which the pressure is 10 atm is found as follows: The pressure in lb/ft² is 10×14.7 lb/in² $\times 144$ in²/ft² $= 62.4$ lb/ft³ $\times h$; $h = 339.2$ ft.

Section 15.1

2. Change in rest-mass is zero; $E/c^2 = 100$ J$/9 \times 10^{16}$ m²/s² $= 1.1 \times 10^{-15}$ kg.
3. 100 kcal $= 418{,}600$ J $= mc^2$; $\Delta m_S = \Delta m_M = E/c^2 = 418{,}600/9 \times 10^{16} = 4.65 \times 10^{-12}$ kg; $\Delta KE = 0$.

Section 15.2

1. (a) $^{14}_{7}\text{N} + ^{4}_{2}\text{He} \rightarrow ^{1}_{1}\text{H} + ^{17}_{8}\text{O}$.
   (b) into mass of particles.
   (c) 2.325×10^{-26} kg $+ 6.645 \times 10^{-27}$ kg $- 1.673 \times 10^{-27}$ kg $- 2.822 \times 10^{-26}$ kg $= 2 \times 10^{-30}$ kg. $E = 2 \times 10^{-30} \times 9 \times 10^{16} = 1.8 \times 10^{-13}$ J.

Section 16.1

2. $\qquad ^{235}_{92}\text{U} + ^{1}_{0}\text{n} \rightarrow ^{90}_{37}\text{Rb} + ^{143}_{55}\text{Cs} + 3\,^{1}_{0}\text{n}$.

protons: 92 0 37 55 0
neutrons: 143 1 53 88 3

5. $\dfrac{4\pi r^2}{\frac{4}{3}\pi r^3} = \dfrac{3}{r}$; 3, 1.5, 3/4; 5000g/18.7 g/cm³ $= 267.4$ cm³ $= \frac{4}{3}\pi r^3$; $r = 3.997 \simeq 4$ cm $\simeq 1.57$ in.

Section 16.2

2. $4 \times 1.673 \times 10^{-27}$ kg $- 6.644 \times 10^{-27}$ kg $= 4.8 \times 10^{-29}$ kg; $\Delta mc^2 = 4.3 \times 10^{-12}$ J.

Section 17.2

4. $F_E = 9 \times 10^9 (1.6 \times 10^{-19})(-1.6 \times 10^{-19})/(0.5 \times 10^{-10})^2$
   $F_E = -9.2 \times 10^{-8}$ N.

Section 17.3

2. $F_E = q\mathcal{E}$; 40×10^{-6} N$/10^{-6}$ C $= 40$ N/C.
5. $W = F_E \times \Delta d = \Delta$PE; $\mathcal{V} = \Delta$PE$/q = F_E \times \Delta d/q$; but $F_E/q = \mathcal{E}$, so $\mathcal{V} = \mathcal{E} \times \Delta d$.
6. Using the results of the previous problem, $\mathcal{V} = 4 \times 10^5$ N/C $\times 10^{-3}$ m $= 400$ N-m/C $= 400$ volts.

Section 18.1

1. 6000 C/1 min $\times$ 60 s/min = 100 A.
3. Capacitance (C) is given by $C = q/\mathcal{V}$; $1 \times 10^{-6} (100) = 10^{-4}$ coulomb.

Section 18.2

1. Since $\mathcal{R} = \mathcal{V}/I$, 1 ohm = 1 volt/amp; but
   1 V = 1 N-m/C = 1(kg-m/s^2)m/C
   1 A = 1 C/s,   so that
   1 Ω = 1(kg-m^2/s^{2}C)/(C/s) = 1 kg-m^2/C^{2}s.
3. $\mathcal{V} = I\mathcal{R}$; 100,000 = 2 $\times$ 10^{-6} $\mathcal{R}$; $\mathcal{R} = 10^5/2 \times 10^{-6} = 5 \times 10^{10}$ ohms.
8. $P = I\mathcal{V}$; therefore
   30 = I 6   and
   I = 5 amps.
   Since $\mathcal{V} = I\mathcal{R}$ 6 = 5 $\mathcal{R}$ and $\mathcal{R}$ = 1.2 ohms.

Section 18.3

1. $P = I\mathcal{V}$, 1400 + 1300 = $I \cdot$ 115, I = 23.5 A, and the fuse pops.
4. $P = I\mathcal{V}$, 100 W = I(120 V), I = 0.83 A; the total power drawn is 200 W, and the total current is 2(0.83)A; $P = I^2\mathcal{R}$, or 200 = (1.66)$^2\mathcal{R}$. The combined resistance is 72.6 Ω, whereas the individual resistances were each twice that.

Section 19.3

2. The force on *each* charge is $q_e v \mathcal{B}$, and there are nL such charges. Therefore, the sum force is $nLq_e v\mathcal{B}$.
5. $F_M = qv\mathcal{B} = mv^2/R$, $q\mathcal{B} = mv/R$, and so $R = mv/q\mathcal{B}$.

Section 20.1

3. $v = f\lambda$, v = 331 m/s, 331 = 600 $\times$ 10$^6\lambda$, λ = 331/600 $\times$ 10^6 = 5.52 $\times$ 10^{-7} m or 552 nm. That's the same wavelength as green light.

Section 20.2

2. Take the nucleus to be roughly 10^{-15} m across; $v = f\lambda$; 3 $\times$ 10^8 = f(10^{-15}); f = 3 $\times$ 10^8/10^{-15} = 3 $\times$ 10^{23} Hz.
4. λ = 780 nm = 780 $\times$ 10^{-9} m; 1 m = 39.37 in; thus 780 $\times$ 10^{-9} $\times$ 39.37 = 3.07 $\times$ 10^{-5} in.

Section 21.1

2. $n = c/v$, $c \simeq$ 3 $\times$ 10^8 m/s; therefore n = 3 $\times$ 10^8/1.88 $\times$ 10^8 $\simeq$ 1.6.
6. The image (M-2) of the back mirror (M-2) will appear 5 ft behind the front mirror (M-1). The image of your back reflected in M-2 will be 2 ft behind it or 10 ft from you.

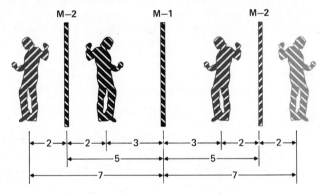

PHOTOGRAPH ACKNOWLEDGMENTS

Chapter 1

Opposite p. 1: Cluny Museum, Paris. 3: Ealing Corporation and Dr. F. Miller, Jr. 4: Ann Landers, *Newsday,* © Publishers-Hall Syndicate. Reprinted by permission. 6: Burndy Library, Norwalk, Conn. 10: O.K. Harris Gallery, N.Y., N.Y./Eric Pollitzer, Photographer. 12–13: Courtesy of the Trustees of the British Museum. 16: Yerkes Observatory Photograph, University of Chicago, Williams Bay, Wisconsin. 17: Burndy Library, Norwalk, Conn. 20: The Philip H. and A. S. W. Rosenbach Foundation, Philadelphia. 21 top: Marshall Henrichs. 21 bottom: National Bureau of Standards. 24: Crown Copyright, Science Museum, London. 27: Copyright, Newsday, Inc. Reprinted by permission.

Chapter 2

29: Marshall Henrichs. 30: Cliché des Musées Nationaux, Paris. 33: Burndy Library, Norwalk, Conn. 35: Bibliothèque Nationale, Paris. 36: The Chapin Library, Williams College, Williamstown, Mass. 38–39: The Metropolitan Museum of Art. Gift of Charles Bregler, 1941. 38 bottom: Michael R. Glaser. 41: Dr. Harold E. Edgerton, MIT, Cambridge, Mass. 48: The Chapin Library, Williams College, Williamstown, Mass. 51: Royal Danish Ministry for Foreign Affairs. 52: Burndy Library, Norwalk, Conn. 59: NASA.

Chapter 3

60: Biblioteca Marucelliana, Firenze. 62: Yerkes Observatory Photograph, University of Chicago, Williams Bay, Wisconsin. 63: Western History Collections, University of Oklahoma Library. 65: Burndy Library, Norwalk, Conn. 67: Educational Development Center, Newton, Mass. 68, 69: U.S. Navy photographs. 72: Cliché des Musées Nationaux, Paris. 74 top: Biblioteca Ambrosiana, Milano. 74–75 bottom: Dr. Harold E. Edgerton, MIT, Cambridge, Mass. 78, 79: Burndy Library, Norwalk, Conn. 82: U.S. Air Force photograph.

Chapter 4

88: Burndy Library, Norwalk, Conn. 89: Cambridge University Library. 91: Systems Engineering Laboratories, Florida. 97: Dr. Harold E. Edgerton, MIT, Cambridge, Mass. 102: NASA.

Chapter 5

115: Burndy Library, Norwalk, Conn. 125: NASA. 127: U.S. Air Force photograph. 135: Martin Seymour.

Chapter 6

136 right: California Institute of Technology, courtesy AIP Niels Bohr Library. 140: AIP Niels Bohr Library. 148, 149: Marshall Henrichs. 152: Stanford Linear Accelerator Center. 154: Fred Stein.

Chapter 7

160: Yankee Atomic Electric Co., Rowe, Mass. 163: Crown Copyright. Science Museum, London.

Chapter 8

178: The Royal Institution, London. 188: Marshall Henrichs. 193: Dr. Harold E. Edgerton, MIT, Cambridge, Mass.

Chapter 9

206: Marshall Henrichs. 208: National Gallery of Art, Washington, D.C., Rosenwald Collection. 216: The Metropolitan Museum of Art, Purchase, Mr. and Mrs. Charles Wrightsman Gift, 1977. 218: Photo. Science Museum, London. 219: Crown Copyright, Science Museum, London. 221: Burndy Library, Norwalk, Conn. 223: Information Dept., Embassy of the USSR, Novosti Photo. 227: Gregg Snigur. 232: Brookhaven National Laboratory.

Chapter 10

234–235: Albert Crewe, University of Chicago. 236: Gregg Snigur. 238 right: Photo. Science Museum, London. 239: Cavendish Laboratory, University of Cambridge. 241: AIP Niels Bohr Library. 244: Cavendish Laboratory, University of Cambridge. 246–247: General Electric. 248: Burndy Library, Norwalk, Conn.

Chapter 11

252: Royal Danish Ministry for Foreign Affairs, Press Dept. 256: The Metropolitan Museum of Art, The Cloisters Collection, Gift of John D. Rockefeller, Jr., 1937. 261: Institut International de Physique Solvay. Courtesy AIP Niels Bohr Library. 263: Royal Danish Ministry for Foreign Affairs, Press Dept. 269: From the PSSC film *Matter Waves,* Educational Development Center. 270: French Embassy Press & Information Division, New York. 271: AIP Niels Bohr Library. 274: AIP Niels Bohr Library. Archives for History of Quantum Physics. 277: Linda J. LaRosa.

Chapter 12

278: Photo. Science Museum, London. 289 left: NOAA.

Chapter 13

292: Marshall Henrichs. 296: Photograph by Lorus and Margery Milne. 297: Suzanne Gorgen. 307 top: U.S. Navy photograph. 309: Marshall Henrichs.

Chapter 14

312: U.S. Naval Observatory photograph. 317: Photo Deutsches Museum München. 318: The British Library. 319 bottom, 320: Pat Ingrassia. 322: The Royal Institution, London. 324: Gustav Lamprecht. 325: Kitt Peak National Observatory. 329: Marshall Henrichs.

Chapter 15

330: U.S. Naval Observatory photograph. 332–333: U.S. Navy photograph. 336: Cavendish Laboratory, University of Cambridge. 338–339: CERN European Organization for Nuclear Research. 341: Argonne National Laboratory. 344: AIP Niels Bohr Library. Fankuchen Collection. 346–347: Brookhaven National Laboratory. 350–351: Argonne National Laboratory.

Chapter 16

352: Argonne National Laboratory. 357: Pittsburgh Post-Gazette. Courtesy of AIP Niels Bohr Library, Lande Collection. 358–359, 360, 361, 362–363: Argonne National Laboratory. 363 bottom: Battelle-Northwest. 364: United Press International photograph. 366, 367, 368 top, bottom left: U.S. Air Force photographs. 368 bottom right: U.S. Army photograph. 369: NASA/Ames Research Center. 370–371: Reproduced by courtesy of the Trustees, The National Gallery, London. 372: U.S. Air Force photograph. 374: © 1933 N.Y. Herald Tribune Co. 375: Novosti Photo. 376: Argonne National Laboratory. 377: Lawrence Livermore Laboratory.

Chapter 17

380, 382: Burndy Library, Norwalk, Conn. 383: Jim Lennon. 390 top: NOAA. 390 bottom: Burndy Library, Norwalk, Conn. 393: The Royal Institution, London. 395, 396, 397: Harold M. Waage, care of Physics Dept. Jadwin Hall, Princeton University. 403: Burndy Library, Norwalk, Conn. 404: John G. Trump, High Voltage Research Laboratory, MIT.

Chapter 18

406: Physics International, a Rockcor subsidiary. 407, 408: Burndy Library, Norwalk, Conn. 413: Photographie Bulloz, Paris. 420: Circus World Museum, Baraboo, Wisconsin. 421: Courtesy of the Wellcome Trustees. 424: General Electric.

Chapter 19

432, 434: Burndy Library, Norwalk, Conn. 441 bottom, 442: General Electric Research and Development Center. 444: Burndy Library, Norwalk, Conn.

445: From *PSSC Physics*, D. C. Heath and Co., Lexington, Mass., 1965. 447: Air Force Geophysics Laboratory. 448: From *PSSC Physics*, D. C. Heath and Co., Lexington, Mass., 1965. 449: Michael R. Glaser. 451: The Royal Institution, London. 458: Burndy Library, Norwalk, Conn.

Chapter 20

462: Marshall Henrichs. 467: Bell Laboratories. 472: Michael R. Glaser. 474 top: The Jet Propulsion Laboratory and NASA. 475: NASA. 476: Crown Copyright. Science Museum, London. 477: Dr. G. Vaiana and NASA. 479: Bell Laboratories.

Chapter 21

480: NASA. 482: French Cultural Services. 485, 486: Donald Dunitz. 489 left: American Cystoscope Makers, Inc. 497: Joseph V. Romano. 499 left: From *PSSC Physics*, D. C. Heath and Co., Lexington, Mass., 1965. 501 top left: Marshall Henrichs. 503: Martin Seymour. 508: Reproduced courtesy of the Trustees, The National Gallery, London.